†

COSMOLOGICAL BALANCE UNIVERSE:
TRIUNE DYNAMIC EQUILIBRIUM

A Unified Theory
Copyright © 2020

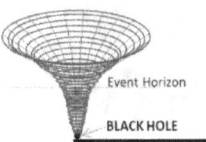
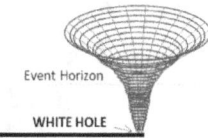

Authored by: Mark D. Calvo

Forewords by Jeff Yee and Steven Stoddard, Ph.D.

COSMOLOGICAL BALANCE UNIVERSE:
Triune Dynamic Equilibrium
A Unified Theory

Authored by Mark D. Calvo

Edited by Malcolm Garland

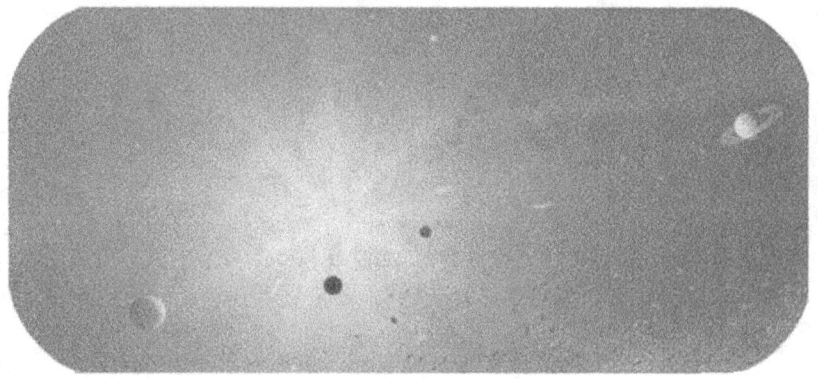

Published by Amazon Publications

Copyright© 2020 by: Mark D. Calvo

This book, licensed for your personal enjoyment and education, presents a simple unified theory of the universe for scientific advancement. No part of this book may be reproduced, or stored in a retrieval system, or transmitted, or referenced in any form of any kind, or by any means, mechanical, electronic, photocopying, recording, or otherwise, without the express written permission of the publisher or author. If you are reading this book, and did not purchase it, then please properly purchase your own copy. Thank you for respecting the hard work of the author, and for rating and commenting on this book.

All rights reserved.

Cover design by Mark D. Calvo

Dedication

To my father,

September 8, 1924 – August 1, 2015

For mechanical insights and basic physics.

JMJ

Glory be to the Father (the Creator), and to the Son, and to the Holy Spirit, as it was in the beginning, is now, and ever shall be, world without end. Amen!

Table of Contents

Copyright ii
Dedication iii
Table of Contents iv
Start Reading 1
Forewords 2
Preface 4
Introduction 8
PART 1: FUNDAMENTAL INSIGHTS 12
Chapter 1: Matter and Its Natural Properties 12
   1.1 Molecules, Atoms, and Particles 14
   1.2 Micro Elements and Quantum Theory 19
   1.3 Macro Elements of the Universe 24
   1.4 Space and Time 26
   1.5 Matter and Energy Interaction 31
   1.6 Red Shift and Light Refraction 34
   1.7 What is the Force of Gravity? 39
   1.8 Why is Sun's Corona So Hot? 41
Chapter 2: Physical Law in Cosmology 42
   2.1 Kepler's Laws and Newton's Gravity 44
   2.2 Stars and Their Light 48
   2.3 Black Hole Attributes and Radiation 51
   2.4 General and Special Relativity 54
   2.5 String Theory and Others 58
   2.6 Inexplicable: Dark Matter, Dark Energy 61
   2.7 White Hole Attributes 64
Chapter 3: Gravity and the Milky Way Galaxy 65
   3.1 Galaxy Development Timeline 67
   3.2 First Generation Stars 70
   3.3 Second Generation Stars 72
   3.4 Timeline to Third or Fourth Generation Stars 74
   3.5 The Solar System to its Current Age 76
   3.6 Rotational Speed of the Galaxy 80
   3.7 Solar-Planetary Rotational Torque Theory 84
   3.8 Stellar-Planetary Gravity Influence on Weather 97
PART 2: GRAVITY OF THE GRAVITON 107
Chapter 4: The Natural State of Balance 107
   4.1 Balance in the Natural World 108
   4.2 Balance in the Micro Elements 113

4.3 Balance in the Macro Elements 115
4.4 Balance in Formation of Galaxies and Universe 117
4.5 Balance in Galilean and Einstein Relativity 119
4.6 Balance in Gravity and Antigravity 124
4.7 Balance in Quantum Mechanics is Gravity 127

Chapter 5: Theory of Cosmological Balance 138
5.1 The Basic Theory 140
5.2 The Cosmological Balance Defined 143
5.3 Application of Cosmological Balance 145
5.4 Observational Tests 148
5.5 The Physics of Creation 152
5.6 Energy, Mass, and Momentum 155
5.7 Planetary and Galactic Orbital Periods 159
5.8 Quantum Graviton Unified Theory 161
5.9 The Function of a Field Theory 181

Chapter 6: Theory and Proof of Balance 183
6.1 Two Sides of Balance Explained 185
6.2 Visible Side of The Universe 187
6.3 How Black Holes Excrete vs Evaporate 192
6.4 Invisible Side of the Universe 200
6.5 How White Holes Excrete vs Evaporate 202
6.6 Depiction of the Cosmological Balance 205
6.7 Mathematics and Proof of Cosmological Balance 207
6.8 Newton Gravity Effects in Black Holes 209
6.9 Antigravity Interaction Theory 214
6.10 Mercury's Orbital Anomaly Solved 216
6.11 Representation of Gravity Waves 219
6.12 Gravitational Force Analogy 221

PART 3: PROOF AND PRINCIPLES 227
Chapter 7: Inspired Theories Analyzed 227
7.1 Expansion of the Universe? 229
7.2 Lighter Elements and Background Radiation 231
7.3 Microwave Radio Static Source 233
7.4 Possibility of Big Crunch? 236
7.5 String Theory Loophole 237
7.6 Special and General Relativity Quandary 240
7.7 Bending of Light by the Sun 256
7.8 Red Shift of Distant Galaxies 266
7.9 Age of the Universe 271

7.10 Age of the Milky Way Galaxy ... 274
7.11 Galactic Orbital Equation ... 275

Chapter 8: Space Balances Gravity Antigravity ... 279
8.1 What Makes Outer Space Appear Void? ... 283
8.2 Structure of Space ... 286
8.3 Pressure and Energy of Space ... 292
8.4 Vacuum Force of Space ... 294
8.5 Importance of Space ... 301
8.6 Cosmological Balance Creation ... 303
8.7 Cosmological Balance Equation ... 305
8.8 Space-Matter Affects Time & Curvature of Space ... 312
8.9 Space-Matter Affects Acceleration and Time ... 316
8.10 Antigravity and Antimatter: It's About Time ... 323

Chapter 9: Cosmological Balance Principles ... 328
9.1 Principal Types of Matter ... 332
9.2 Principle of Black and White Holes ... 338
9.3 Principle of Light Bending & Energy Absorption ... 342
9.4 Principle of Gravity and Antigravity ... 345
9.5 Principle of Tidal Forces and Mercury's Anomaly ... 348
9.6 Principle of Balance in Types of Matter ... 351
9.7 Principle of Planetary Motion and Gravity ... 355
9.8 Principle of Planetary & Galactic Orbital Period ... 358
9.9 Principle of Cosmological Balance Equation ... 362
9.10 Principle of Basic Acceleration and Gravity ... 365

Conclusion and Way Ahead ... 369
C.1 The Way Ahead ... 375
C.2 Adjunct Thoughts ... 372
C.3 Consciousness and Intelligence ... 374

Appendices ... 375
Appendix A: Units, Equations, and Constants ... 376
Appendix B: Commentaries and Future Challenges ... 379

Acknowledgements ... 380
About the Author ... 381
Glossary ... 383
Bibliography ... 389

Carl Sagan 1934-1996 said,

"Somewhere, something incredible is waiting to be known."

Stephen Hawking said:

"There should be no boundaries to human endeavor. We are all different. However bad life may seem, there is always something you can do, and succeed at. While there's life, there is hope."

Neil deGrasse Tysons said:

"Any time scientists disagree, it's because we have insufficient data. Then we can agree on what kind of data to get; we get the data; and the data solves the problem. Either I'm right, or you're right, or we're both wrong. And we move on."

Forewords

Mark provides a unique framework to explain the mysteries of the universe. Although some aspects of it are counter to currently adopted theories and explanations within academia, his approach is grounded in extensive research of experiments and analysis found in mainstream physics documentation. It is welcoming to see a fresh perspective that attempts to solve the inadequacies of current models.

For those of us in applied sciences, an understanding of the mathematical language that governs the universe is essential to creating technology and products that are useful to humankind. One can make an argument that the last century - since the quantum revolution - has provided very little in understanding the basic components of the universe and its mathematical model. In mechanics, many of the equations used in engineering date back to Newton's era in the 1600s, with the exception of some special cases required for gravity. In electronics, many of the equations in engineering date back to Maxwell's era in the 1800s.

The world needs scientists and engineers that challenge current models and seek alternative explanations. While it is not possible for every alternative theory to be valid, it is refreshing to see new ideas to solve some of the holes in the quantum explanation.

> Jeff Yee
> Author, *The Particles of the Universe* series

FOREWORD

As Carl Sagan said, "The universe seems neither benign, nor hostile, merely indifferent." This book takes a challenging look at the universe. There are some ideas here that may sound outright impossible. That said, the cosmos offers seemingly as many questions as answers. Mark's work probes those questions in ways that challenge the established scientific tenets.

If you are a "questioner" of the myriad mysteries of deep space, then this is a worthy read. I think you will find it intriguing and the sort of thing that makes you ask even more questions even as you debate the author.

Science is a funny thing. The theories get established by the world's brightest minds. Until they are unproven by other brilliant scientists sometime later. The solutions of the past are not always pertinent to the problems of the present. Mark's premise in the Cosmological Balance Theory is that we need to carefully consider all the information at hand and recognize that the data are not necessarily complete. This enables us to craft opinions for debate, which drives future experimentation and analysis. For example, Mark points out that a replacement theory would have to deal with both General Relativity and Quantum Mechanics or integrate both of them. The Cosmological Balance Theory has to make the same predictions as would replace the existing theories and cover the gaps and discrepancies others have missed in order to withstand the scrutiny of tests. A good physicist or theorist should always build on what others have discovered, or effectively proven, and use that information to develop new theories. Mark's intent is to kick-start a discussion so that "we" can work together to get it right.

The concepts, ideas, hypothesis, and theory presented in this book are to the best of the author's own abilities. Mark wrote this book primarily for inquisitive readers who desires new insights into the universe as a whole. This is a book for those who question the established science of the cosmos. While basic high school mathematics and science skills are quite helpful in understanding this edition, they are not necessary.

As you read this book, you might offer additional questions, debate various points, or find that Mark did not present the math or ideas in it clearly enough. If that is the case, please write a comment or review of the book. Feel free to upload it publicly onto the amazon.com comment page. Mark will read and consider your comments in his future editions.

Steven A. Stoddard, PhD

Preface

"A book, too, can be a star, a living fire to lighten the darkness, leading out into the expanding universe."

— Madeleine L'Engle

This book challenges some of mainstream science long-standing foundation concepts of matter, space and time, and creation, in hopes to advance humankind on the road of scientific progress. Cosmological Balance Theory declares that there are three types of matter in the universe. There exists matter with gravity with which we are all quite familiar, those with antigravity (reverse gravity) with which we will learn about, and finally there existed beforehand those that make space what it is, this third type of matter separates the firmament of the heavens from the firmament of the earth, solar systems, stars, and the galaxies. Per Sir Isaac Newton, gravitational force governs all matter we know and see in the visible universe, drawing all objects together in space, and on earth. The second type of matter, opposite of ordinary matter, is the highly contested antigravity antimatter, which has a mirrored construct of antimatter particles. Anti-gravitational force, likewise, governs all antigravity antimatter within the invisible part of the universe, pulling all antigravity objects together in space. I say "invisible" only because we are not able to directly see, or detect, that part of the universe from our point of view, or with our instruments. The third type of matter, yields to the movement of both gravity matter and antigravity antimatter, consubstantiate all matter; I will call space-matter or **hygratium**. Space-matter exists in a vacuum gaseous form, it just occupies, and "floats," in deep space to fill the void, making space what it is, and existed beforehand. It is the source of, or the push vacuum force behind, the perceived gravity, and antigravity, force. Space-matter is a buffer between the other two types of matter, and thereby surrounds and curves around their massive objects. All gravity, and antigravity objects, sharing a common origin are inherently quantum entangled with other objects of their type, instantaneously, and infinitely connected, almost as if telekinetically one entity.

The constant interactions of these three types of matter have always existed and will continue to sustain each other indefinitely. There was never a Big Bang per George Lemaitre, nor singularities per Stephen Hawking; the universe is not expanding at an accelerated rate, per Edwin Hubble, nor collapsing on itself, and the cosmic microwave, and the abundance of lighter elements in deep space are from a different source. The curvature of space and time together, as defined by Albert Einstein, is questionable. Time is universal throughout space, steadily and consistently ticking away. It speeds neither up, nor slows down, with an increase, or decrease in gravity, or antigravity forces, or with increase, or decrease in velocity, and acceleration. Space itself or more specifically space-matter curves around massive gravity matter and antigravity antimatter; open space is spatially flat and uniformly dispersed everywhere else, an observation confirmed by NASA with 0.5% margin of error, and its visible matter isotropic. Finally, massive black holes and white holes excretion, not hawking radiation or evaporation, are essential to the Cosmological Balance Universe recycling and renewing itself throughout time. Our existence came to be because of this enduring balance, one universe, three types of matter.

To set the tone, the reader must be made aware that some of our greatest ideas, and solutions, come from the strangest of places. Our history is riddled with laymen presenting, what many considered outlandish ideas, only to find out that what was presented is in fact accurate and correct. These people, some of which risked all, gave humanity the boost it needed to move us to the next level of consciousness, and benefited the progression of science. Just because most scientists currently agree on a theory, does not necessarily make it the right concept we needed for scientific advancement. It could have been flawed since conception, but no one with sufficiently strong argument stood up to challenge or discredit it. For this reason, scientists, or theorists, with sound concepts, who were not strong or eloquent enough to defend their justified theory, were shot down by the educated academia, and swept away under the carpet, as failed ideas, never to be given another opportunity to redeem their creditability. Whoever speaks with the most articulate voice, and convincing arguments, earns the respect of colleagues, and gathers the most public supporters, and possibly wins the Nobel Prize. So, beware of the results announced by the scientific community, because accuracy is not always their foremost objective; they are more concern of recognition, and fame achieved, even at the expense of concealing truth. False publicity by academic trendsetters does more harm than good.

In reading the Cosmological Balance Theory, we need to look carefully at, and pay close attention to all the data before our eyes, and our instruments, and determine, either I am wrong, and they are right, or I am right, and they are wrong, or we are both wrong, or we are both right to some degree, or another. Whatever the case, we must solve each dilemma together, and move on to the next set of challenges, and transcend upward in our infinitesimal position in this vast universe. Carl Sagan elegantly said *it, "The universe seems neither benign, nor hostile, merely indifferent."* The universe itself does not care if our leading astrophysicists, and cosmologists are taking us down a "rabbit hole," led by the mathematical "carrots" of brilliantly written equations, and eloquently presented arguments, but our future generations, and humanity, as a whole, is counting on them to come up with the appropriate, and correct cosmology. The solutions of the past are not always pertinent to the problems of the present. One thing to keep in mind is that a replacement theory would have to replace both General Relativity and Quantum Mechanics or integrate both as each works perfectly separately, and also replace any other proven acceptable associated theories prior to those. The principle of consistency also known as the principle of non-contradiction demands that new theories be as good as old ones, while fixing their errors and discrepancies. Recent observations have confirmed General Relativity's prediction of gravity waves, or so scientists have said in the published New York Times article February 2016, but not exactly as predicted. A new theory must make the same predictions as what it replaces, and covers the gaps and discrepancies others missed, to withstand the scrutiny of tests. A good theorist should always build on what others have discovered, or effectively proven, and use that information to develop new theories. So, let us all work together to get it right.

In chapters 1 through 3, I will briefly present basic information, in the best possible light, necessary to establish a common knowledge base among readers, concerning the resources, pertinent data, and observations of what we know about matter, and its natural properties. I will start with basic facts of the visible micro and macro universe, to include its building blocks, from the smallest subatomic particles, quarks, and electrons to the largest galaxies, and the black holes within them, and their associated physics, and natural laws as background and mathematics for follow-on discussion. I will argue how gravity causes matter in a newly forming protostar to heat up, and begin fusing hydrogen into helium, and present the mathematics explaining the pressure, and the temperature buildup, necessary to ignite the star. Then conjecture the death of stars, and the creation of black holes, and the Hawking evaporation process. Finally, the last section of the chapter will discuss Einstein's Special and General Theory of Relativity, and mention other competing unified theories, such as String Theory, "Dark Matter," and "Dark Energy." Chapter 3 will review galaxies and their orbiting stars, investigate of the timeline necessary for enough matter, dispersed in deep open space, to consolidate to form a barred spiral galaxy the size of the current Milky Way Galaxy, and compare it to science 14-billion-year age of the universe. Then examine generations of stars' lifespan from birth process to expiration, and the timeline for their debris to reassemble into stellar system and argue that a massive galaxy's creation is beyond 14 billion years. Educated readers with a graduate degree in Physics, Astrophysics, Mathematics, or Cosmology, and are confident may scan through chapters 1 and 2, as supporting information; your start point is chapter 3, where the Cosmological Balance hypothesis and theory is introduced. This book will discuss, and attempt to solve numerous issues baffling our astrophysicists, and cosmologists, while presenting a new concept, and a possibly controversial theory challenging some of the current, and accepted science. This book provides an alternate explanation of why the universe is perceived as expanding at an accelerating rate, presents solutions to anomalies, and explains why the universe is homogenous and isotropic and why average temperature is consistent.

In chapter 4, I will discuss the natural state of balance in the universe, reviewing the balance in the microscopic world, the balance in the macro elements of the grand universe, the balance in the natural world, the balance of super symmetry, relativity, gravity, and antigravity, and conclude with the balance in the formation of the solar system, the galaxies, and the universe. Through extrapolation I define the balance to the force of gravity, antigravity, and the existence of the neutral graviton. In chapter 5, I will present the thesis and outline the basic theory, the Cosmological Balance Theory, and its application and supporting observational tests, the physics of creation, the energy, mass, and momentum relationship, and finally present the function of a field theory. The rest of the book will set forth aggressive detailed visualization for the theory's defense. In chapter 6, I will explain the two sides to the universe by talking about the visible side first and explain how black holes evaporate or eject material in another way, besides Hawking Radiation, I will call excretion, and discuss the invisible side of the universe. In chapter 7, I will relook the Big Bang tenants and discuss the universe expansion, the source of the microwave background noise in deep space beyond the galaxy; re-evaluate the abundance of light elements; and remnant background radiation, the age of the universe, and finally the formation of the Milky Way galaxy and its age. Chapter 8 will discuss the composition of space-matter, and its importance toward creation. Chapter 9 will present a condense synopsis of the Principles. In the conclusion and way ahead, summarize the book and list the tenants supporting the theory, and close with comments.

In reading this book, pay close attention to examples showing balance in the universe. For every action, there is a reaction; for every positive there is a negative; for every magnetic north there is a south pole; for every right side there is a left; for good there is evil; and so, on and so forth. Everything seems to be counterbalanced in nature with their opposite. All four natural forces should therefore have counterbalancing elements. The balance found within nature is the reason for naming this book, The Cosmological Balance. It presents and attempts to prove the hypothesis with enough reckoning to call it a theory. It is a pursuit to comprehend the invisible and visible universe and how each side's destiny affects the other. Research used within this book relied heavily on source data, and information, from numerous credible sources listed in the Bibliography, the National Aeronautics and Space Administration (NASA) Official website www.nasa.gov and augmented lastly from the user-edited Wikipedia (Wikipedia e. , 2001) at https://en.wikipedia.org as an online encyclopedia.

My goal and objective of this book is to pass the idea discussed here within, and the flaming torch it represents, to whomever to confirm the clues in the galaxy, and the universe, supporting the validity of my hypothesis, and theory, and help me work out all the presentations, necessary to make it a scholarly document. That action requires someone schooled in the physics field to transform it sufficiently from pseudoscience language to that of a professional physicist in his or her own technical language. With that said, I hope that the evidence, and the convincing arguments, presented in this book can win you over. If I can spark that flame to but one astrophysicist, or cosmologist, who can carry the new theory, the torch, to the next level, I would be ever grateful. I also understand that my presentation of this book may be ridiculed, or scorn upon, by mainstream science, but it is a worthwhile story which must ultimately be told. This book carries with it a revolutionary, and all encompassing, theory of the universe, as revealed to me, through my unique interpretation of observations, analysis of readings, and personal quest for knowledge in this field. Self-educating is a challenge, however, freedom from the obligation to adhere to the accepted mainstream science views, and academia standards taught by university physics, astrophysics, and cosmology professors, allows for out-of-the-box thinking, and open interpretations, without grading repercussions as a graduate, or post-graduate student. Most of and above all, I hope you find this book highly stimulating, and thought provoking, as well as informative, and entertaining.

Like many other previously controversial theories in the past that opposed accepted mainstream science thoughts and concepts, I expect academia to declare this one as a pseudoscience by the educated "doctors" of physics, astrophysics, and cosmology, especially since it did not come from among the renowned within their academia itself. They are entitled to their comments. Nonetheless, science is all about explaining data as well as developing theories the data confirms for maximum insight into the laws and nature of reality. It is not about the credentials of who discovered the new concept, and the supporting math. Only the passage of time, repeated tests, and experiments, confirmed observations, and gradual acceptance by students, and teachers of science, can cause the naysayers to change their minds. In any case, do not let them stop you from seeking the truth. I urge you to read on, see my full theory, conduct your own research, develop your own analysis, and draw your own conclusions. Enjoy reading the Cosmological Balance presentation of the universe.

- Mark Calvo, 2019

Introduction

"You are a function of what the whole universe is doing in the same way that a wave is a function of what the whole ocean is doing."

— Alan W. Watts

Humanity has made tremendous progress over the last few centuries discovering, and in some instances, re-discovering science, and the necessary math to support our developing, and newly found knowledge. Despite all our advancements, we still have a long way to go to become a successful galactic, or intergalactic, space-going race. Such a venture requires intense diligence, accomplished in reasonable steps. The content and knowledge presented in this book revealed to me after many years of studying, and reviewing mathematics, physics, science, and cosmology breakthroughs, is my unification theory. At some point, science crashes headlong into questions of origin and intent that have no answers - practical or theoretical. The universe as I believe it is almost without end by either distance or time, somewhat infinite in both respects. We should nevertheless explain the universe based on what we already know, and modify that definition with each idea revealing observation, and newfound knowledge, a unified theory of eternal and renewable universe in Cosmological Balance.

During my initial writing of this book, from January to March 2015, I attempted to get comment, or a letter from Neil deGrasse Tyson. My correspondences stated that I had a possible solution to the dark matter, and dark energy dilemma, with which scientists have been wrestling; no response received. I then reached out to Professor Fred Adams, Physics Department at the University of Michigan, USA, then to Professor Stephen Hawking, Department of Applied Mathematics and Theoretical Physics at Cambridge University, UK, whom since passed away, and then to Mr. Bill Nye, Chief Executive Officer, of the Planetary Society, and several other prominent scientists and their staff to request review, comment, or feedback pertaining to the content of this book. Mark had waited over and beyond a reasonable amount of time for their reply, and copyright adherence, before offering a portion, or the entire book, to them for review, or comment before proceeding with the publication of this book. Since distributing prepublication of the manuscript, I have not received any reply from all the academic physicists, mathematicians, astrophysicists, and cosmologists, or their colleagues that I emailed or attempted to contact in person. On completion of my manuscript, I submitted an extract paper on planetary torque to a physics university professor, who reviewed it, and commented that it was imitation science not worthy publishing in a scientific or physics journal, obviously due to lack of references to modern physics theories, equations, and technical terms. This solar planetary torque paper is included within. Nevertheless, I applied the solar planetary torque theory to the path and intensity of cyclones, hurricanes, and typhoons and found a direct correlation, shared those results with a NOAA hurricane hunter, who suggested for me to publish my findings in the Atmospheric Scientific of America. However, the chief editor of that journal declined my article as it was in variant to wave buoyancy theory. With that said, I have moved forward with the comments received thus far from university certified physicists, and cosmologists, from within my network of friends, and colleagues, to self-publish my findings here within.

INTRODUCTION

I applaud the scientific advancements the truth seekers gave us thus far, and the doors they opened. I say truth seekers, and honest scientists, for their claims, and theories were not for fame, and fortune, but for the benefit of humanity. For today, we know that the planet is not flat, that the planet is not the center of the solar system, that the solar system is not the center of the galaxy, and that the galaxy is not one of two galaxies. We now know that the galaxy is among billions, if not trillions of other galaxies scattered throughout the visible universe. These are just a few of the many scientific, and technological breakthroughs, we needed. Many of these brilliant ideas, concepts, and theories, we now accept today, as fact, were once declared improbable, and even listed among the false precepts, or even pseudoscience, until they were proven valid. Starting from generally simple ideas, sparked by sheer inspiration, science developed theories to unify previously existing, and formerly thought to be stand-alone concepts. Examples of these unifications: Faraday's experiments in electricity, and magnetism leading to electromagnetism; Newton's observations of celestial, and terrestrial motion, solving the force of gravity; Maxwell's light, and electromagnetism, combining in the electromagnetic field; Glashow-Salam-Weinberg's weak nuclear force, and electromagnetism, interacting as the electroweak force; and relativity, gravity force, and electromagnetic field, merging together in Einstein's special, and general relativity. These are but a few examples of how our science evolves with every item of newfound knowledge. Examples of pseudoscience not originally proposed by certified scientists, that became accepted mainstream science are the concept of the Atom, the Big Bang Theory, and the Plate Tectonics to name a few. This book, the Cosmological Balance Theory, offers a simple unified theory, combining all, and when accepted could be considered, as one of the potential contenders for the base trunk of the unification tree, if trimmed appropriately.

We should explain the universe based on what we already know, and modify that definition with each idea revealing observation, and newfound knowledge. This book builds upon reality and presents the foundation for a new unified theory of an eternal and renewable universe, the Cosmological Balance. Before we dive into the details of the actual Cosmological Balance hypothesis and theory, let us remember, and reflect on the discoveries made thus far in our "modern" times. It is "modern" only in respect to our own generation's current awareness and knowledge. As you may be aware of, the visible universe, and the world, are teeming with wonderful discoveries, and many more unanswered questions. With each new discovery, new questions arise, and another generation of scientists, and theorists, take on the new challenge to find much needed answers. It is through this building of knowledge base that we find inspiring ideas, pursue the quest to resolve issues and dilemmas, and intelligently and accurately learn of the universe, and the world around us. Some of us remember that the construction of the Large Hadron Collider brought fear to many. Some thought its first usage would bring the end of the world. They believed that micro black hole created would grow exponentially as they consumed matter around it. It turns out that these fears were false. Some say we were lucky the LHC opened the doors to new discoveries, not disaster. With the LHC, we have broken apart atoms into their basic component particles, documenting as we find and define them. In addition to identifying sub particles, we have identified the many atomic elements on earth and placed them onto a periodic table. We have chemically created molecules and new materials. We have also extracted the essential elements, and smaller molecules, from within other larger more complex molecules.

We have identified the composition of the planet earth, and explored most of the world we live in. Yet there are still places unknown. Nature and climate have always intrigued us. We can predict to a certain degree of probability, and accuracy, the weather patterns of the planet, and track climate changes. With this information, we have warned the public of pending severe weather, and saved lives. We have also documented numerous species, and the migration patterns of thousands of animals. We have studied the oceans and its tides. We know with a great degree of accuracy how gravity affects the planet and how the planet affects the moon, the sun, and adjacent planets, Mars, and Venus. Yet there are still undiscovered things about the planet, and its unexplained freak weather anomalies, not to mention what extraterrestrial challenges lay before us.

We have studied the planets in the solar system, identifying and documenting facts about them. We count eight planets, and three dwarf planets, in the solar system, and search for one other possible outer planet with a 1500-year orbit to no avail. We analyzed their orbits around the sun, their rotational speed, their size, surface gravity, temperature variations, and their many moons. We have sent probes to these planets to take close up and detailed images and sent probes into deep space with a message from earth, spoken in many languages and sounds to include symbols in the universal language of mathematics. To ensure survivability of our species, we are doing our best to find, and track, as many asteroids, and comets that are possible near-earth collision candidates. Our scientists are working on methods to redirect these objects away from the earth, or to destroy them in their tracks. Despite our great advancements, there are things about the solar system that are yet still unknown. We should all applaud the great work done in the name of, and for the benefit of humanity and its future.

We have examined the sun and learned about its lifespan. By studying other stars of similar size, we learned about the sun's age, and determine how much time remains before it runs out of fuel. Scientists conjecture that it is about 4.5 billion years old. The sun has enough fuel to continue burning for about another five billion years. We know that the sun is a nuclear fusion furnace held together by gravitational forces, where hydrogen fused in several stages into helium atoms, and molecules. This fusion process releases photons of varying degrees of energy, which eventually make it to the surface of the sun, and ejects outward as "light," and other wavelengths of the electromagnetic spectrum in all directions. The sun also radiates tremendous amounts of energized particles, and heat, during solar mass ejections.

We have gathered significant amounts of data and learned much about the galaxy. We have discovered the Oort cloud around the solar system, named after Jan Oort. We have calculated the distance to the nearby planets, the Oort cloud, and to neighboring stars, and beyond to the best of our ability, and technological accuracy. Using gravity equations, we have calculated the path of each star as it orbits around the galaxy, to include its weaving up and down motion above and below the galactic plane. We know that the galaxy consists of billions and billions of stars of various sizes, and in various stages of their lifespan. We have estimated the distance in light years from one edge of the galaxy to the opposite edge. We have conjectured that at the center of the galaxy exists a super massive black hole. In addition, there exist numerous other smaller black holes throughout the galaxy. We know that there are billions and billions of other galaxies visible to us. We have strived to measure the distances between galaxies in parsecs, and light years, with ever-increasing advance technology, and precision instruments. A parsec is equal to about 3.26 light years (3.086 × 10^{13} kilometers). The oldest galaxies discovered so far is estimated by cosmologists to be at least 13 billion years old, the time it takes its light to reach us on earth. We know that all galaxies fall into three general categories: elliptical (spheroid), barred, and spiral galaxies. There are variations of these spiral, barred, and elliptical galaxies. We conjecture that the Milky Way Galaxy is a barred spiral galaxy. Modern cosmologists have conjectured that galaxies in the visible universe are moving away from each other. Many hypothesize that the universe is expanding, and most likely, will not come together to die in a big crunch. They have identified that there is an invisible mass within the universe with invisible energy. They call this mass, dark matter, and the energy, dark energy, stating that because of these unknown elements the universe is expanding.

Although we have come a long way, we have yet to rediscover many things our ancestors knew. We do not know how the pyramids, the stone hedge, and numerous massive structures of stone around the world were built, and for what purpose. Our scientists have only recently rediscovered and announced that they have finally demystified the over 2,000-year-old Greek geocentric astronomical calculator, retrieved from a shipwreck near the island of Antikythera in 1901, over a century ago. They declared that the reconstructed calculator, consisting of several brass gears with prime number teeth, deduced from x-rays images, predicted future occurrences of astronomical events, like lunar eclipses. Even in modern times, we do not know what Edward Leedskalnin discovered to enable him to build the Coral Castle all by himself in Homestead, Florida, as he took his secrets to the grave with him in 1951. Did he rediscover and use the Egyptian pyramid secrets to build the Coral Castle? As we had mentioned earlier, we have yet to discover everything there is to discover here on earth. We have yet to build a spaceship capable to transport humans across the great expanse of space, between stars, within one individual's lifespan.

Maybe our future advancements will help us grow as a race, and take us to the next step, hopefully to improve humanity. Maybe one day, we will obtain a deep, and profound, sense of knowing or awakening. Hopefully on that day, we will be proud of reaching a level, and have a strong sense of accomplishment, in achieving our newfound place in the universe. This powerful knowing comes from sharing as well as accumulating information for the benefit of all. It is through this oneness of humanity can we achieve scientific advancement, and hopefully peace. After all, we are all made from the same elements created within stars. To bring readers to equal par, we are compelled to review the basic knowledge, then move on to macro-elements and physical laws of the universe itself, as understood according to mainstream, studying the equations and proofs of several renowned scientists, and their influence on the next generation of geniuses. Then we will take the bold step together to introduce this new hypothesis and theory, and present supporting mathematics, break new ground with a unified equation of the universe, provide a way ahead, and finally close with additional comments. Enjoy the read!

PART 1: FUNDAMENTAL INSIGHTS

Chapter 1: Matter and Its Natural Properties

"The beauty of a living thing is not the atoms that go into it, but the way those atoms are put together."

— Carl Sagan, *Cosmos*

Let us start with ordinary matter, which we can touch, feel, and see; something we all know very well. Our interaction with matter has become second nature to us. Many of us do not even think about its composition as we move about. However, scientists needed to know more about the world around us to include why certain things function the way they do, and why some things were ailing us, and why some things made us feel stronger, and more efficient. Scientific investigation, observation, documentation, and proof are their livelihood. There was a time in our modern history that we could not ever visualize smaller objects other than through a microscope. Even our strongest microscopes could not show us objects smaller than a living cell, or a piece of sand, or dust. It was not until 400 B.C. did a philosopher named Democritus propose the Greek word atomos (wikipedia, Democritus, 2015), meaning indivisible, to describe the atom. For hundreds of years, his prediction held true. Of course, today we know that the atom can be divided, although for but a split second, into its sub-atomic particles[1].

Several of the discoveries we are learning today, some scientists and theorists say, appears to have been known by our ancient ancestors. It was only a few decades ago that we created the atomic bomb. Yet, there is evidence on earth of nuclear explosions, thousands of years ago, in India (wikidot, 6500 BC), and Middle East (s8int, 1992). There are numerous other examples that our ancient ancestors may have known more about the universe than we do today. For example, the Dogon People displayed information about the star system Sirius in their dance rituals (wikipedia, Dogon Knowledge of Stars, 2015). Our recent discoveries are only confirming the data they had shown us (unknown, 2010). These and other rediscoveries are surprising us all. For example, ancient Egyptians knew how to add, subtract, multiply, and divide numbers using binary system our computers employ today (West, 1990). Egyptians also exhibited extensive knowledge of the internal functions of the body, constructing temples reflecting that knowledge, down to the smallest organs and blood vessels, as if they understood their purpose, and the matter and elements within them. Egyptian scholars documented, retained, and studied this knowledge at the Ancient Library of Alexandria[2]. What do science know about matter? All matter is

---

[1] The world's most powerful atom smasher, the Large Hadron Collider (LHC) provides a window into the composition of subatomic particles within micro universe.

[2] The Ancient Library of Alexandria or the Royal Library of Alexandria in Alexandria, Egypt, was one of the largest and most significant libraries of the ancient world, dedicated to the Muses, the nine goddesses of the arts. The library was part of a larger research institution called the Museum of Alexandria, where many famous thinkers of the ancient world studied. It flourished and functioned as a major center of Egyptian scholarship from its construction in the third century BC until the Roman conquest of Egypt in 30 BC, when it was burnt to the ground.

composed of arrangements of smaller molecules and atoms, which is further made up of even smaller particles, protons, neutrons, and electrons. Some of the matter that exists on the earth are invisible to our eyes, such as air or clear glass at a certain angle, but most matter are visible in either its liquid or solid form. We know air is there by its effects, such as wind, sound carrier, temperature medium, and the sustainment of life[3].

Ordinary matter surrounds us as we go about daily business. Its existence enables us to move about and use our five senses to carry out our day-to-day functions. Everything we pick up, feel, breathe, smell, see, hear, and taste is due to the interaction of these molecules and atoms with our own. We interact with ordinary matter from the first moment we wake up, brush our teeth, eat breakfast, drive, or walk to work, contribute to society, and when we go back to our home, and prepare for bed, or wherever we sleep. Whether we are flying by airplane to Heathrow Airport London, or to Incheon Airport Seoul, or to any airport in the United States, or traveling by car, bus, train, bicycle, or foot, or riding an animal sightseeing in Brussels, Belgium or Amsterdam, Netherlands, we see innumerable examples of contact with normal matter influenced by gravity. Without gravity, the sun would not be shining, and we would not be here reading this book, or be able to plan future trips to Cambridge, England; Paris, France; Madrid, Spain; Rome, Italy; Cairo, Egypt; or any place in the world. Gravity affects us all. To top it off, the materials and matter we see all around us on earth are in fact remnants of the death of a first, or later generation star. We owe our substance to these dead stars, and the debris they scattered on their last moments of existence. To quote Neil deGrasse Tyson, "We are part of this universe; we are in this universe, but perhaps more important than both of those facts, is that the universe is in us." No matter where we go on the planet or in space, we are bound to interact with the universe, both within and around us[4].

Ordinary matter on earth appears to us in three primary forms: gaseous, liquid, or solid. Some matter can assume all three forms, like water or H_2O. Some matter only appears in two of the three forms, like carbon dioxide C_2O, going straight from dry ice, a solid form, directly to vapor or gaseous state with just the slightest amount of heat. The conversion from solid to liquid and liquid to gas are known as phase transitions caused by changes in temperature over a wide enough range. Symmetry plays a crucial role during these phase transition. With enough heat energy added, the metal element lead, or Pb, only goes from solid to liquid form. There are other alternate forms of matter, which include plasma, super fluids, etc. Scientists studied the structures of atoms, separated the elements, and then documented the details, and characteristics, they observed. But science tells us there is significantly more about the world around us, explaining why certain things function in certain ways and sometimes act differently than expected. Scientific investigation, observation, documentation, and proof are their legacy, and the purpose of science. We must view results from every angle for signs of symmetry and balance.

---

[3] It took a long time for our scientists to discover the makeup of air. It will take even longer for them to figure out what exactly is in space that makes it what it is. Just because we cannot see it, touch it, or sense it, does not mean that it consists of nothing.

[4] The material to create our solar system came from the destructive death of previous generation stars, mixed with new materials like an abundance of hydrogen, helium, and other lighter elements.

1.1 Molecules, Atoms, and Particles

Richard Feynman, American Physicist 1918-1988, once said, "The world is made of atoms," a profound statement in and of itself, together with what most scientists agree upon, symmetry underlies all the physical and natural laws of the universe. Let us then take a closer look at the structure within the atom. Atoms consist of electrons, protons, and neutrons. Electrons are in constant motion around the nucleus, protons and neutrons vibrate within the nucleus, and quarks jiggle within the protons and neutrons, while gluons hold them together. The picture below is a snapshot of the distortion. If we drew the atom to scale and made protons and neutrons about a centimeter in diameter, then the electrons and quarks would be fraction of a hairline, and the entire atom's diameter would be greater than the length of over 3 kilometers or 3000 meters! We get 99.999999999999% of an atom's volume is just empty space[5]!

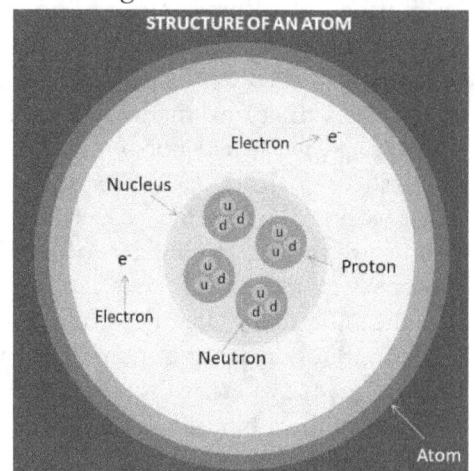

Since there is so much empty space within an atom, how does each particle tell the other particles what to do? Controls believe it or not are simple performed through electrostatic differences, electrical charge maintain attraction and repulsion, while the strong nuclear force provides structural integrity, and the weak nuclear force provides atomic bounding. These controls are instantaneous and of short range. Without them, atoms would not exist, and the material in and around us would not accumulate, and we would not be here.

Take, for instance, the periodic table of elements. It arranges the elements in a symmetrical pattern, linking or grouping those with common characteristics, properties, and similarities together. The science of chemistry tries to answer many of our questions, and in doing so, uncovers more questions the deeper we understand things. The more we investigate and learn about the micro universe, the more we understand that the internal structure of the atoms themselves affect chemistry and its rules. An atom's structure and characteristics are based on the composition of the electrons, photons, neutrons, and protons within them, which in turn are hinged on the movement of the smaller subatomic particles, and so on a so forth. As one opens the outer large Russian doll shell, they only find a smaller complete doll within it, revealing layer after layer the deeper one goes inward. So far, our scientists have identified the quark and the electron as the smallest of subatomic particles. Are there smaller point particles or something that creates what look like strings within them? Alternatively, do they consist of something still unknown? Quantum mechanics observations suspect point particles.

The periodic table lists the known elements found or created on earth. An element is a basic structure that is inseparable and consists of more than one atom bonded together. For example, the element oxygen exists as two atoms of oxygen bonded together. In

---

[5] The scientist who discovered that atoms contained significant amounts of empty space immediately became concerned of atoms passing through other atoms. These fears soon passed when they start learning of the forces of nature atoms possess.

another example, the element nitrogen exists as two atoms of nitrogen bonded together. If you separate them, you no longer have the molecule of oxygen or the molecule of nitrogen respectfully. You get an ion of oxygen or nitrogen, which immediately bonds itself to another atom or molecule to become stable. Therefore, an atom is the smallest complete part of an element or the smallest particle of a chemical element that can exist. It consists of protons, neutrons, and electrons (Heritage, 2015). The chart of all the elements we have found or created on earth so far[6], the periodic table (Helmenstine, 2015) arranges the elements in an organized fashion based on their atomic number (number of protons in the nucleus), electron configurations, and any recurring chemical properties are displayed with the same color.

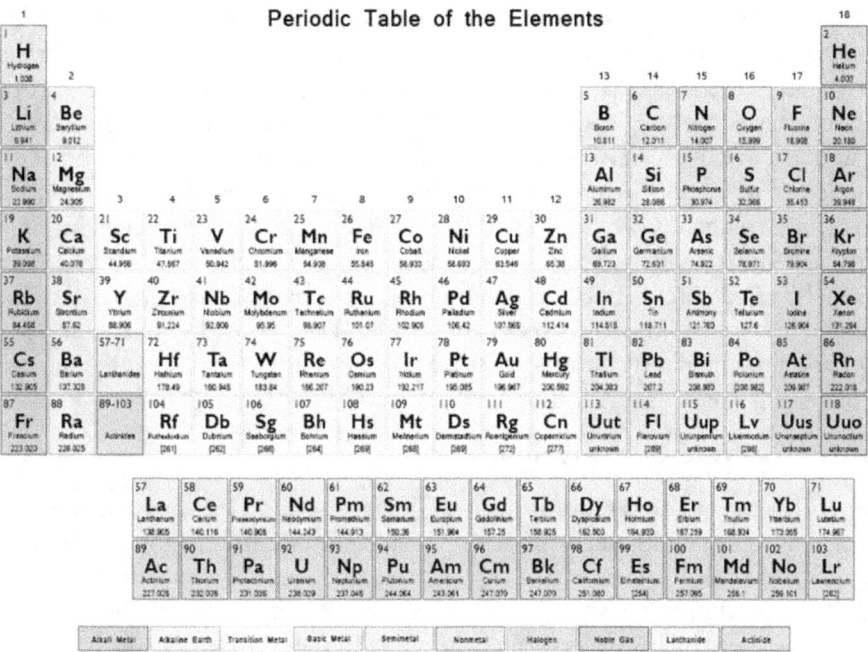

A molecule is an arrangement of two or more atoms that are chemically linked together, (H_2, C_2, CO_2, etc.). A compound is a molecule that contains more than one element not chemically bonded (H_2O and $C_6H_{12}O_6$, etc.). Compounds are not on the periodic table of elements. Vitamin C is another example of a compound; it consists of $C_6H_8O_6$. Chemistry is the field of science (Wikipedia, Chemistry, 2015) that study how matter bonds to other matter, specifically at the atomic level. These bonds are chemical in nature. Chemistry studies matter at this level, its properties, how and why these substances combine or separate to form other substances, the energy needed to make these bonds, and the interaction of the substances. In a matter of speaking, we are all chemists. Our bodies are chemical factories converting one or more chemicals into another to produce the material and chemicals for our bodies to function as intended by nature. The medical professions, such as doctors, nurses, physician assistants, veterinarians, are required to study chemistry to understand the basic concepts of our bodies, before administering and

---

[6] Many of the elements were discovered by breaking them away from within compounds. For instance, mercury oxide powder was heated up with magnified sunlight to release the oxygen and leave behind the liquid mercury.

practicing remedies. Chemistry is part of our daily lives whether we know it or not. Chemists call the assemblage of most chemical bonds a "right-handed" configuration, depicting them to lean to the right. What this means is that when looked at under the electronic microscope most of the atoms and elements lean to the right side. If you take the same chemical molecule combination, and make the atoms reassemble, or chemically bond in a different configuration, you will get an entirely new compound with entirely new characteristics, and properties. On earth, it is quite rare to have "left-handed" chemical bonds. Living organisms and the organic material they produced are left-handed molecules[7]; this makes us different from inanimate objects.

The five branches of chemistry are physical, organic, inorganic, analytical, and biochemistry. Physical chemistry combines physics with chemistry. Organic chemistry studies compounds using or incorporating the element carbon. Inorganic chemistry concentrates on materials made up of metals, and gases, without the carbon atom. Analytical chemistry uses qualitative and quantitative observation to find and measure physical and chemical properties of compounds. In addition, biochemistry studies the process that occurs within living organisms, plants, and beings. Then there is the totally unrelated field of alchemy (Wikipedia, Alchemy, 2015), the study of changing metals into gold. Although most modern scientists do not accept Alchemy as a science, Alchemists developed the structure of basic laboratory techniques, theory, terminology, and experimental method, some of which are still in use today by scientists. Alchemists predated modern chemistry, atomic theory, the chemical element and substance, the periodic table, and the conservation of mass.

Atomic Particle Composition

Atoms consist of combination of protons, neutrons, and electrons arranged to balance their charges. Photons hold electrons within electron field around the nucleus. Electrons are negatively charged. Protons are positively charged, and neutrons are neutral in charge (Wikipedia, Atom definition, 2015). Scientists sometimes refer to the photon as the "messenger particle" of the electromagnetic field; the photon is also the carrier of information in radio and television broadcast waves.

The atom's nucleus consists of a group of protons and neutrons held in balance by the strong nuclear force. Excessive number of protons, or excessive numbers of

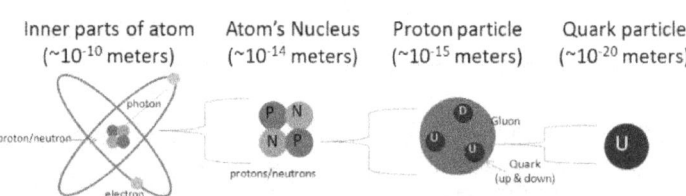

Sub-components of Atoms

Inner parts of atom (~10^{-10} meters) — Atom's Nucleus (~10^{-14} meters) — Proton particle (~10^{-15} meters) — Quark particle (~10^{-20} meters)

neutrons, will cause the nucleus to be imbalanced, and radioactively decay, (or fall apart piece by piece) until it reaches a stable balance of protons and neutrons. If an atom loses electrons as it decays, it will require the loss of additional neutrons and or protons to reach another stable level. If it has excessive neutrons in the nuclei then it might convert one neutron to a proton, and emit a negatively charged beta particle, or electron, neutrino, and

---

[7] As living intellectual beings, we have an inherent connection to the stars. In ancient times, our ancestors studied the skies and the stars. This connection is routed within the fact that we consist primarily of left-handed molecules, which are influenced by the universe around us.

CHAPTER 1: MATTER AND ITS NATURAL PROPERTIES

anti-neutrino in the process. If it has an excessive number of protons in the nuclei then it might convert one proton into a neutron, and emit a positively charged beta particle, or positron, neutrino, and anti-neutrino in the process[8]. The radioactive decay process also emits alpha, beta, and gamma radiation. An example of this is radon gas, radioactively decaying over time, until it becomes the element lead, a stable metal. The nucleus of the atom consists of protons and neutrons. A proton consists of three quarks: one down quark and two up quarks, held together by three gluons. A neutron consists of two quarks: one down quark and one up quark, held together by two gluons. A photon consists of two quarks: one up quark and one down quark. An electron has less mass than a quark and has yet to be seen. In addition, charge characteristic wave hold electrons at a defined spherical distance from the nucleus by photon energy levels. The combined charge of each grouping of electrons holds them at different spherical orb distances. First two electrons are at the closest level to nucleus, the next group at another level higher, and so forth. Electrons move to higher levels when they absorb photons, and likewise electrons drop levels by releasing photons. Quarks are the smallest known particle we found thus far, with mass between 4.1-5.8 MeV. The value MeV = $1.60217657 \times 10^{-13}$ joules. One MeV is one mega-electron volts or one million electron volts. Below is a chart depicting the sub-atomic particles and forces discovered so far (ordinary matter; one of three types of matter).

ATOMIC PARTICLES

Quarks	u	c	t	γ	Force Carriers
	up	charm	top	photon	
	d	s	b	g	
	down	strange	bottom	gluon	
Leptons	v_e	v_μ	v_τ	Z	
	electron neutrino	muon neutrino	tau neutrino	Z boson	
	e	μ	τ	W	
	electron	muon	tau	W boson	

Each column of the chart above depicts the three energy levels of each sub-atomic particle (Freudenrich, 2015). For example, low energy quarks are up or down quarks. Medium energy quarks are charmed or strange quarks. Moreover, high-energy quarks are top or bottom quarks. The same applies to Leptons. Low energy leptons are electron and neutrino. Medium energy leptons are muon and muon neutrino. In addition, high-energy leptons are tau and tau neutrinos. Quantum chromo-dynamics or QCD (Wilczek, 2000) goes on to define three colors and six flavors associated with quarks, they are red, blue, and green with fractional electrical charges, and the flavors are paired combinations of these three colors[9]. Similarly, force carriers have three different energy levels to interact with the three quarks, and three leptons levels.

A similar chart depicts the three energy levels of anti-matter sub-atomic particles. There are the low energy anti-quarks, up or down; the medium energy level anti-quarks, charmed or strange; and the high energy level anti-quarks, top or bottom. The same list exists for anti-leptons: low energy level, positron, and anti-neutrino; medium energy level, anti-muon, and anti-muon neutrino; and high-energy level, anti-tau and anti-tau neutrinos. Then there are the three energy levels of anti-force carriers to interact with the three anti-quarks, and the three anti-leptons levels. Scientists speculated that the same Quantum Chromo-Dynamics or QCD also would hold true with anti-matter defined by three colors

---

[8] As scientists rip apart atoms into the atomic components and subatomic particles, they learn that the smaller subatomic particles contain much stronger forces that sustain them than the larger atom or the components from which they came.

[9] The chart showing quarks, leptons, and force carriers is in a state of flux, updated as new discoveries are announced.

and six flavors, which would be paired combinations of these three colors. Anti-Matter and its subatomic particles do not occur naturally in the visible universe. When anti-matter and matter meet each other, both annihilate each other, as conjectured by scientists. Their existence, first proposed in science fiction books, TV shows, and movies, and then confirmed within the confines of the Hadron Collider experiments. In these experiments, the anti-matter shows up on the instrument's readout for but a fraction of a second and then eliminated when colliding with normal matter, their counterparts.

Gravitons have yet to be identified (wikipedia, Graviton definition, 2015). Gravitons, a sought-after hypothetical elementary particle responsible for gravitation forces in the macro universe, also mediate the force of gravity within quantum field theory. If it exists as an individual particle it is expected to be mass-less and have unlimited range. Just as the Higgs particle makes up the Higgs field, physicists believe that gravitons constitute the gravitational field. When scientists discover it, they believe it will unite quantum theory with gravity, as quantum gravity. Brian Greene relates the gravitational field synonymous with the shape of space, where the effects of gravity is less noticeable in flat space and stronger in curved space around massive objects (Greene, The Fabric of the Cosmos: Space, Time, and the Texture of Reality, 2004, p. 333). Gravity is less noticeable in flat space because the force is equal in all directions, a feature common in the perfect fluid of space. Then there is the "god" particle, the Higgs boson, a particle that gives all matter the quality of mass. The difficulty of subatomic particles moving through the Higgs field is direct reflection of the particles' mass; traveling mass-less particles experience no resistance. Physicists have gone so far to say that they attribute the 2.7-Kelvin degree average intergalactic temperature to the presence of the Higgs particle[10]. And of course, let us not forget the neutrinos that the sun and all the stars make, billions and billions of these ghostly particles passing through atoms without interacting with them, or at least most. Neutrinos and antineutrinos have the same mass, massless, and neither has an electric charge. Therefore, for a given neutrino type (there are three of different mass), gravity and electromagnetic forces are no different for a neutrino versus an antineutrino. Generally, the possibility of a neutrino or an anti-neutrino interaction with ordinary matter is incredibly small, approximately about 10^{-19} per centimeter. Then there is the opposite of gravitons, anti-gravitons with anti-gravitational force, which scientists have yet to identify. Since gravity exists, then there is a good possibility antigravity too might exist as a natural balance. The atomic particles chart depicts only one-third of all types of matter, the other types will be discussed in Part 2 and Part 3. To make sense of this, scientists created mathematics specifically for the uncertainties, quantum mechanics.

---

[10] The Higgs particle, field, and ocean permeate space. The Cosmological Balance Theory goes so far as to say it is part of space itself.

1.2 Micro Elements and Quantum Theory

Ordinary normal matter and its natural properties have shown that symmetry governs all, physical and natural laws. The sphere, for example, is a perfect symmetrical object in every aspect. No matter which way it is turned, it is always a sphere. Obviously, not everything is completely symmetrical at the surface, but if you look down deep enough, it reveals an internal symmetry. Symmetrical tendencies are the result of gravitational influence while gravity itself is the result of uniformed symmetrical interactions of particles within atoms. When scientists and physicists break molecules and atoms apart, in whatever means they can come up with, they learn that each atomic and subatomic piece consists of one or several subcomponents further consisting of an arrangement of still smaller particles presumed to be spherical. Observations have led us to believe that the smallest of all spherical objects is a point particle, not a one-dimensional string. Typically, people see strings as a fine line of point particles connected end-to-end over a given length, or the superfast movements of point particles creating what may look like a "string" and form the patterns calculated by string theorists. And then there are those who insist point particles are standing waves.

Modern physics uses counterintuitive Quantum theory (Rouse, 2015) to explain the nature and behavior of matter and energy at the atomic and subatomic level. Scientist sometimes refer to this theoretical basis as quantum mechanics and quantum physics, a measurement of systems in which it is essential to view the objects as interacting and connected, rather than isolated systems. The field of quantum mechanics deals with the particle phenomenon at such small levels where normal physics principles scientists say breaks down. It is the difference between the visible, seen by naked eye, and the microscopic world, viewed through the aid of devices. In the quantum world or quantum reality where nature has no access to absolutes, one cannot directly observe interactions, so scientists examine these particles interactions using high tech specialized instruments. In this reality, quantum mechanical behavior is fundamental to the microscopic world as well as possessing a mystifying multi-valued relationship with other particles.

The fact that all the other objects in the universe determine the properties of any particle or object is known as the quantum reality principle (Thomas, Hidden In Plain Sight: The simple link between relativity and quantum mechanics, 2012). We rely heavily on relationships to realize or perceive what in the world is real. The properties we assign to each object describe this relationship. In other words, when we see with our eyes and pick up an object with our hand, we determine from that contact that the object is real. This feeling of reality, based on the assumption that our hand is also real, enables us to relate to the world around us. Properties without relationships with other objects are meaningless and lack reality according to this definition. Quantum physics discovered that mass too results from interaction of particles with the Higgs field. Similarly, the interaction of particles with other particles produces mass, a reality based on assumed reality of the other particles. For example, the property of a negative charge requires the defined existence on the opposite, a positive charge, and vice versa, a mutual definition. The interaction of a new multi-valued charged particle with existing charged particles in the universe around it determines which charge that new particle permanently assumes.

These new quantum discoveries sometimes lead to new equations and solutions, to resolve issues. In this case, scientists developed new mathematics to bring order to the uncertainty of this microscopic quantum world. Quantum mechanics particles used wave functions to explain the interactions of particles as a role of chance, versus the unrestricted chaos of a lawless universe. This new mathematics deals with handling various probability outcomes, which are not exact predictions and are fundamentally random. Brian Greene explains, "Only quantum laws (quantum mechanics) were capable of resolving a host of puzzles and explaining a variety of data newly acquired from the atomic and subatomic realm… where the universe participates in a game of chance (Greene, The Fabric of the Cosmos: Space, Time, and the Texture of Reality, 2004)." Since nature is based on symmetries, quantum physics results reflect that same characteristics.

Quantum physics is unique, bringing with it, "weirdness" without comparison, so hard to imagine because it shatters our concept of reality, the micro universe. The macro universe, however, moves with the same laws of physics, predictability, and regularity of every particle, and object as defined by the Newtonian frame of reference, with which Einstein agreed. Quantum mechanics, on the other hand, breaks this mold, telling us that we cannot know exact location and exact velocity of every particle. In physics, scientists define locality as the physical space of influence from one object to another, and that Quantum mechanics declares that the universe interconnections are not local. Said another way, what happens over there can influence what happens over here, intertwined even if nothing travels between the objects[11]. These connections can persist even if they are at opposite sides of the universe (Greene, The Fabric of the Cosmos: Space, Time, and the Texture of Reality, 2004, p. 80). Such characteristics agree with Newton's definition of gravity as being infinite, and instantaneous. Quantum physics therefore declares that the universe is not local. "Mathematically, we describe the uncertainty principle as the following, where `x' is position, and `p' is momentum (Heisenberg, 2015):

$$\Delta x \Delta p = \frac{h}{2\pi}$$

The uncertainty principle, roughly speaking, tells us that if we can gain knowledge of the first feature then it compromises our ability to have knowledge of the second feature of that particle. Similarly, the more precisely we can find out how fast a particle is moving, the less we know about where it is located. In other words, the energy expended to locate an electron, or subatomic particle, or used to determine its velocity, will most likely affect, and move that particle faster, or in a different spin direction about more than one axis. Interacting with one of the object's feature contaminates, or changes, the outcome of the others; you cannot measure the particle's definite position, and definite speed, or definite spin simultaneously and accurately.

In 1925, Erwin Schrodinger developed unique differential equations, which describes the evolution of those wave functions. He studied the paths of electrons passing through two slits on a card in front of a wall, plotted their impacts on the wall, and discovered they

---

[11] In quantum mechanics, particles that share a common origin are entangled and affect each other regardless of distance and time. By the same reasoning, if we go back far enough in time we can find a point where the particles share a common origin. This entangled theory supports Newton's gravity equations with infinite distance and instantaneous force.

acted like waves. Even if the electrons were released one at a time over a long period, the result consistently remained the same, a wave pattern like that by light with alternating lighter and darker bands. Places with small probability wave are locations where the electron is not likely to hit. By using Schrodinger equation, scientists can find the wave function, which solves a particular problem in quantum mechanics. Sadly, it is usually impossible to find an exact solution to the equation, so scientists use certain assumptions to obtain an approximate answer for the problem. The best quantum mechanics can do is to predict this or that outcome. Schrodinger's Equation." (OregonUniv, 2015)

$$i\hbar\, \partial/\partial t\, \Psi(r,t) = -\hbar^2/2m\, \nabla^2\, \Psi(r,t) + V(r,t)\Psi(r,t)$$

Where "i" is the imaginary number square root of negative 1, "$\hbar$" is the Planck's constant divided by two π, "$\psi(r,t)$" is the wave function, defined over space and time, m is the mass of the particle, ∇^2 is the Laplacian operator, and $V(r,t)$ is the potential energy influencing the particle.

One example showing the dual properties of subatomic particle like electrons and protons is the experiment using parallel vertical thin slits on a panel placed in between the particle emitter and the wall, where the particles passing through the slits create a wave interference pattern of lighter and darker vertical stripes. It does not matter if the particle emitter sends one proton or electron in increments of time over a long period, the resulting pattern on the wall remains the same, wave interference. Brian Green explains that electrons or protons, or light, all seem to know where the previous particle contacted the wall and deviate from the point to create the pattern. Brian goes on for several pages explaining the situation and calls it the magic of weirdness possessed by the subatomic particle. Nowhere in his discussion did he mention about the material used in which held the cut slits, and how an electron or proton passing near the edge statically electrified that material, thereby causing the next electron to deviate to another part or opposite the charged area hence causing the wave interference pattern[12]. Schrodinger told us he cut two slits on a card, but we are still unsure if the card was plastic, paper, or cardboard, and if its presence and emanating waves affected the results. It is amazing how our great scientists can overlook the simplest natural reasons of intersecting waves on moving particles just to develop their own agenda, math, and theories. Maybe such reasons have already been considered, tested, ruled out, and declared irrelevant. In any case, Neils Bohr, Danish Physicist, October 7, 1885 - November 18, 1962, summarized it well in his Principle of Complementarities that travelling subatomic particle, in fact, has both wavelike and particle like aspects.

Quantum mechanics also tell us that particles are entangled, where the feature of one influence the feature of the other. "The intuitive reason is that the two photons (or any other particles) are spatially separated, their common origin establishes a fundamental link between them (Greene, The Fabric of the Cosmos: Space, Time, and the Texture of Reality, 2004, p. 116)." This interconnection applies to all particles sharing the same origin, making them act and function as if they are one entity. We can also extrapolate this quantum concept to the function of gravitons, attraction, where there exists no direct

---

[12] In the study of quantum physics or quantum mechanics, the slightest things within the experiment can change the outcome of the results even the act of trying to observe the particle can change the location, speed, and or direction of the particle's path.

physical or transmitted signal between particles and yet linked and drawn instantaneously to each other as if they are one. Hence, since all ordinary matter are composed of a variety of particles sharing the same common origin, they are all attracted to all other particles nearby, inversely proportionate to the distance between the objects of which they are part. An object in space consisting in billions of particles all sharing similar origins in essences acts as one entity just as though it were one particle. Therefore, an external push force exerted upon one of two equally massive objects in close proximity in space toward the other acts as an attractive force, gravity, where the second object also receives a similar push toward the first. Likewise, the motion of one object toward an equally massive object triggers the second object to begin movement toward the first; Newton called this attraction the force of gravity. The same relationship applies to the spin of the entire object in space, where the rotation of the sun partially influences the rotational direction of the planet, although mainstream science declares the sun applies no physical torque to the planet. Quantum entanglement affects the motion of particles as well as contributes to the pull movement of massive objects[13]. Quantum mechanics has some compatibility with special relativity but not with general relativity. However, the blending of quantum mechanics with general relativity yields nonsense results and artificially divides the universe into two completely independent realms (Greene, The Fabric of the Cosmos: Space, Time, and the Texture of Reality, 2004, p. 336). Fortunate for us, not all quantum theorists think the same way. This difference leave room for more theories to be developed and circulated, and eventually one that will capture the interests of the majority, one that unifies most if not all theories. The fact that all other objects or particles in the universe determine the properties and characteristics of any particle or object around it is clearly an application of symmetry in action.

Quantum theorists call it the reality principle, we see it as a micro symmetrical system of the universe. Why do particles in the quantum world "jump" into and out of existence as they move probabilistically from place to place? Why do they possess this "weirdness" characteristic? Why do these particles not act or move like normal objects or larger particles of matter in the macro world? Quantum theorists likely attribute these strange movements to what they call the "uncertainty principle," and what others call the connection quantum entanglement. Lastly, there is the less popular quantum theory, the Wave Structure of Matter (WSM). In a book by (Wolff, 2008) *Schrodinger's Universe: Einstein, Waves, and the Origin of the Natural Laws*, Milo tells us that wave energy interactions fully explain the formation of all matter and its natural properties. He views all matter particles as stable "standing waves" constructed from spherical intersecting in and out wave energy, where all waves originate from the energy density of space has infinite amplitude. Milo declares, "The structure and behavior of all the matter follow two basic Principles:

WSM Principle I says that there exists a quantum wave medium throughout the universe. Waves in the medium form the basic particles ant their structure determines the rules of behavior of the particles (as predicted by Erwin Schrodinger, *Life and Thought, 1989,* where he stated that particles are just 'schaumkommen' or appearances). That is

---

[13] Since the majority of the object's mass are contained within the nucleus of atoms, the direction the nucleus moves affects the quantum entangled direction of the second object, causing attraction at 1 meter per kilogram.

quantum wave structures are real: discrete particles are not. Instead, all matter is a wave structure in a quantum wave medium and material bodies are only appearance to us."

WSM Principle II reiterates the Mach's Principle, a discovery by Bishop Berkeley and Ernst Mach a century ago. The math form of their discovery provides the mechanism that determines the density of the wave medium. The simple rule is that *Medium density is proportional to the sum of the wave intensities of all the matter in the universe*.

WSM Principle III can be deduced from principle II. It is so useful and helpful in thinking about particle behavior that is stated separately: '*The waves everywhere in space tend to adjust themselves to a minimum amplitude*', or the Minimal Amplitude Principle (MAP)."[14]

Many in the scientific community oppose this idea do not understand the wave structure of matter concept or are not accustomed to seeing it. Wolff states that the wave structure of an electron fits all the laboratory-measured properties observed better than commonly accepted classical point particle configuration, particles formed totally of waves, each with infinite amplitude that spans the entire fabric of space. In a somewhat contradiction, he claims the four natural laws originate from quantum waves of the basic particles (electrons, protons, etc...) affected by other matter in the universe; particles and natural laws result from in-wave and out-wave interactions. Doppler Effect result of relative motion ($\lambda = h/mv$) alters the interactions of in and out waves $[1-(v/c)^2]^{-\frac{1}{2}}$ of particles (resonance) and redefines the momentum and the mass of these particles. (Wolff, 2008, p. 38), Milo states "Schrodinger's equation and the (natural) laws are derived from the quantum wave structure of the universe." Hence, by simply replacing the notion of point particles with a *spherical wave structures* Milo reveals a universe of quantum wave structures. He declares that the wave structure of the basic element of matter (the electron) replace the old notion of non-existent discrete "particles." I, however, beckon the existence of a unified theory, which overcomes this rift.

---

[14] Milo Wolff claims that every subatomic particle and every atom of matter in the universe is structured and joined together by these rules of quantum waves, thereby redefining what we perceive as solid particles as interlaced 'structured waves' traveling within the space medium.

1.3 Macro Elements of the Universe

"I bet you could sometimes find all the mysteries of the universe in someone's hand."

— Benjamin Alire Sáenz, *Aristotle and Dante Discover the Secrets of the Universe*

Gravity affects all matter in the visible universe. Ordinary particles and sub-atomic particles of matter floating within a region of space is pulled together, per Newton, by mutual gravity to form ordinary matter consisting of the smallest element such as hydrogen and helium, to small chunks of gaseous and solid matter to form the basic building blocks to stone like debris. The mutual gravitational attraction of sufficient amounts of material collapsing inward can cause the entire mass to start rotating around a common barycenter. Enough building blocks then gravitationally assemble to form asteroids and comets, and eventually planets orbiting around a much larger central orb mass. That central orb of the rotating disc is primarily composed of hydrogen and helium, and several other remnants debris from previous stars. When that central sphere gets large enough to ignite into a nuclear fireball, it begins to fuse its hydrogen into helium elements and shines with starlight going out in all directions[15]. A star is born, and the newly forming planets get bathe with the newly emitted energized light and heat. Lighter gases that are not bound to planets escape outward into deep space by the new solar winds. The high-energy photons created by the nuclear fusion take hundreds if not thousands of years to make it to the star's surface before finally heading outward at light-speed as light and radiation (heat) to the planets. Depicted below are the building blocks of the Solar System:

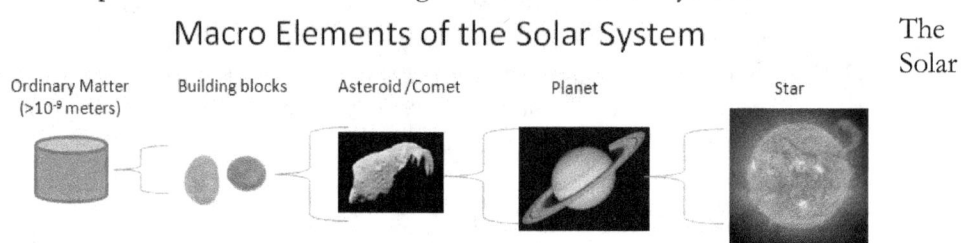

System contains a star at its center and is orbited by one or several planets, asteroids, comets, and other debris. Most solar systems contain a second star and sometimes a third star and are known respectfully as a binary or three-star system. If the star system is composed of two stars of similar sizes then both stars orbit around a point called the barycenter (wikipedia, Barycentric Coordinates (Astronomy), 2015). Three-star systems would also have more complicated orbital pattern. Theoretically, a planet-like orb of just about any size could even exist sitting right at the barycenter between two orbiting stars[16]. Orbital patterns or motions like this can sometimes be misinterpreted or inferred that a black hole is at the barycenter of these stars, especially if we are unable to see the planet or planet-like orb due to the glaring light from the nearby stars despite the planet-like orb has a molten surface. An orb consisting of primarily heavy elements about the mass of the sun

---

[15] Gravity is the dominant force in the macro universe. Gravity affects light. The physics of light and the three other natural forces only influence the micro world; they do not rule over gravity and therefore do not significantly affect the macro universe.

[16] In solar systems where there is only one massive star and one or two huge planets (less than 5% the mass of the star), its barycenter is usually contained within the star itself. More specifically, the star wobbles around that barycenter.

but denser would most likely be categorized as a dwarf star due to its extremely hot molten surface. In some instances, this center orb might have sufficient mass and gravitational pull to crush itself into a black hole of stellar masses.

Stars close enough to each other are locked mutually together by gravity. Normally star systems are gravitationally bound to other nearby star systems in what is known as the local star group. These local star groups travel around the galaxy together, sometimes going up above and down below the galactic plane or flatten disk of a galaxy. Sometimes stars within the local group race past the other stars and then gravitationally pulled back causing other stars to move forward in a tug-of-war dance. Some scientists also believe that "density waves" are responsible for compressing gases together and provide the energy to birth stars. Density waves are high concentrations of stars and large debris with strong gravitational influences. Below is a depiction of macro elements from stars and solar systems to the visible galaxies in the universe:

A galaxy is composed of a great number of stars and their planets, gaseous materials, and debris orbiting about a common center. Galaxies come in three basic shapes: elliptical, barred, and spiral. Cosmologists conjecture that the Milky Way Galaxy is a barred spiral galaxy. Galaxies in the local group, mostly bound together by a very weak gravitational pull, are homogeneous and isotropic throughout the universe (Baumann, 2014). Moreover, the Milky Way Galaxy is moving on a collision course towards the Andromeda Galaxy. The two galaxies will eventually merge into one, as their black holes tug each other into a mutually spiraling dance inward, and finally form one massive black hole at its center. Distant galaxies however are not as fortunate, many of them are moving away from each other. Stephen Hawking said, "It was quite a surprise, therefore, to find that the galaxies all appeared red-shifted." Cosmologists declared that Doppler Effect was the reason for this red shift. "Every single one was moving away from us (Hawking, The Theory of Everything: The Origin and Fate of the Universe, 2002, p. 23)." Per mainstream scientists, the universe is expanding[17]. NASA images of stars and their classes.

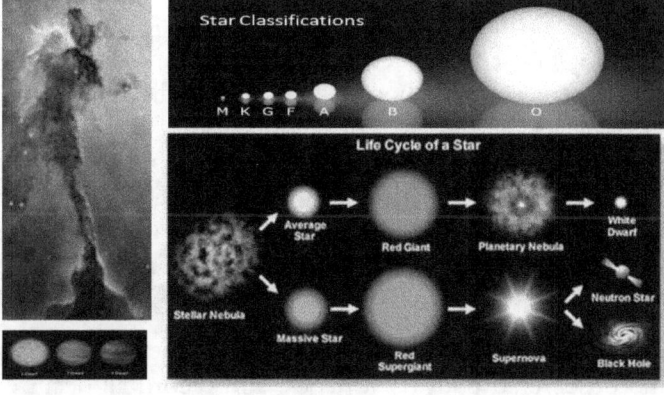

---

[17] According to mainstream science, space is empty and void and as such, there can only be one explanation causing the red shift detected from distant galaxies, the Doppler Effect. Edwin Hubble, supported by mainstream scientists, therefore concluded that the only reason for the red shift light is that these galaxies are speeding away. However, recently evidence is pointing to another reason; matter in space is absorbing photon energy and red shifting the light from distant galaxies.

1.4 Space and Time

Space is regarded as containing three dimensions (3D): length, width, height. Time is considered something like a fourth dimension, a dimension that we cannot physically see, but nonetheless measure with the ticking of clocks. Time is illusive and slips past us unknowingly. What exactly is time and how fast does it fly? Paul Davies in *That Mysterious Flow,* (Davies, 2002) presented another view of time, as block time, in which both past and future are as fixed as the landscape, like that in Kansas before the tornado took Dorothy away to the land of Oz. Time, he imagined, is something Davies called *timescape*, where the entire existence of time in the universe is laid out from beginning to end, past, present, and future together depict a temporal direction. In Paul's view, there is no passage or flowing of time, with every event fixed and predestined to occur, just as in the frames of a film at the movie theater, no set duration speed, and no lasting current moments. Quantum mechanics, on the other hand, tells us the "now" universe is not completely deterministic because all events have certain probabilities of occurrence, some call free will. Nevertheless, the illusion of time remains a mystery to both physicists and philosophers alike[18]. Explaining time with time as in second per second is as meaningless as explaining gravity with gravity-laden effects in rubber sheet 3D image.

Although the perception of time is illusive, the observer's view does not change the sequence of events or alter the passing of time itself. In Brian Greene's book, The Fabric of the Cosmos (Greene, The Fabric of the Cosmos: Space, Time, and the Texture of Reality, 2004, pp. 128-139), he describes time in terms supporting Einstein's view as laid out in Special Relativity, that time for one observer is different from time in terms of another observer particularly evident when traveling at great speeds close to the speed of light. Brian starts by placing two observers at a distance from each other say 500 meters then describes that time from observer A point of view is unique and different from observer B point of view. Observer A sees an accident occur right in front of him in 0.5 billionth of a second of the event, while observer B sees this same accident occur 2 billionth of a second ago, where the minor difference in distance introduces a minor difference in time perspective. He then goes on to say that Chewy on another planet 10,000 light years distance visualizes and experiences the same event on earth at a greater time difference even upon walking a short distance of one kilometer spanned 150 years or so either way depending on the direction of his gait, saying because of the sheer distance the minor angle of the slice taken of the "loaf" of time becomes significant and makes it easier for the normal person, who has difficulty visualizing how excessive speed affects time, provides his readers another approach to time difference or time dilation in SR. Brian said that in Newton's absolute space and absolute time everyone's "now-slice" coincides and reminds readers, in a relativistic universe, there is a big difference.

In the details Brian and Einstein used they disregard that all the events that observer A and observer B both try to clock, and record all occurred in ordinary ticking of absolute time, and not one of them occurred in slower or faster increments of time than other

---

[18] Time will always remain a mystery to humanity, until we are somehow able to break its one law, the smooth ticking forward direction, by traveling either jumping forward or skipping backward. Much like the speed of light, no one can know exactly what happens until we are able to move close to or travel at that speed. Einstein uses these two facts to advance his theories to try to shape the unknown.

events. Just because observer A and observer B see the events at different durations and perspectives after they occurred does not change when the event occurred. Specifically, if the moon received an impact from a small asteroid at exactly 12:30:45 am on July 20, 2014, we would see it on earth at slightly different times depending on where we are on the planet. The observer closes to the moon sees it at about 1 and ½ seconds recording it at 12:30:46.5 Eastern Standard Time (EST), and observers just coming over the horizon sees the event occurring closer to two seconds after impact and clocked it at 6:30:47 pm Hawaii Time (12:30:47 am EST). Similarly, we see the sun as it was eight minutes ago when we detect a solar mass ejection. The speed of light itself changes the perception of the sequence of events for each observer. The point is that it does not matter how each observer on earth clocks or times the event and records it, their actions does not change when the event occurred. Events all around us, whether these events are as close as objects within our reach or on the moon or as far as 13 billion light years distance away, all occurs in absolute universal time as defined by Newton where every observer experiences the normal steady passage of time from their unique view. Thus, excessive observer distances or extreme speed does not change when the viewed events occurred around the observer; time flows normally for all neither speeding up nor slowing down with difference in gravity or speed. The mirage they presented in the illusion of Special Relativity is not everyday reality or universal view, and of course is completely opposed to absolute time[19]. Accepting difference in "now-views" of different observers is not contrary to Newton's definition of absolute time; it defends it.

Brian Greene supports Einstein's perception of time as defined in the Special Relativity Theory. Rather than explaining the time difference is in terms of speed, Brian explained it in terms of distances. Events occurring in the observers' immediate surroundings are what that observer first detects as some events that occurred within nanoseconds in their immediate past. Observations occurring at increasing distances translate to increasing time into the past. Brian Greene depicted a sliced loaf of "bread" to represent events in time where he slices it at different angles to show different "now" views and then stretches the loaf widthwise to show how extreme distance from one side to the other becomes greater differences in time. Brian goes on the say that the view can be seen from either side of the slice, peering into the past as well as into the future. Viewing into the future is in error. The "now" view along the angle of the slice can only go from the present observer viewing into past events only. The loaf concept has its flaws. In addition, any one observer cannot see every event that occurred within their slice as Brian Greene eluded. An observer in a straight perpendicular slice of his current time cannot see everything that has occurred within his slice, at best he is only able to see events next to him that occurred within a couple of nanoseconds. Observers can only see an expanding wedge from his current time into past and the wider the view of events.

---

[19] In Einstein's Special Relativity, he uses the qualities of light to define observational differences between two observers in constant motion traveling near the speed light. Each observer sees the other's clock running slower than their and therefore Einstein concludes that their excessive speed slows both of their clocks, called time dilation. Einstein explains that gravity affect light leaving massive objects, thereby causing gravity time dilation. Both are mirages. Detailed mental experiments using the Cosmological Balance definition will reveal there is no time dilation in either excessive speeds or massive gravitational fields.

Cosmological Balance differs from Brian Greene and Einstein view of time and time dilation and agrees with Newton's perception of an absolute time. All events including the motion of all particles occur throughout the entire universe from moment to moment in a duration that is set forth by absolute time itself, and then moves onto the next moment in time, unceasing and relentlessly flowing forward with the arrow of time. These events transmit their momentary information outward in all directions in concentric spheres at the speed of light to all observers wherever they are. Light emitted from these occurring events is what observers detect and record within their "now" view. Light traversing throughout space carries with it a flowing historical sequence of past events, much like the frames of a movie film. If we watch one section of the sky over a long period, we will absorb a time segment of the information transmitted from long ago, a historical view of events as they unfolded back then[20].

Time, and what observers perceive as the "now" view, flows in the direction of the "arrow" of time from one grid to the next grid at the speeds many times faster than that of light, clocked by durations set forth in absolute time. Each event within the same time grid is separated by some given distance, where the dots depict space or distance between events. Each observer is only capable of seeing past events that have already occurred from the closest ones just nanoseconds old to the furthest historical event at the greatest distance they are capable of viewing or detecting, some billions of years old. Each enlarging circle with spherical surface resides on an increasing concentric layer of equal time into the past.

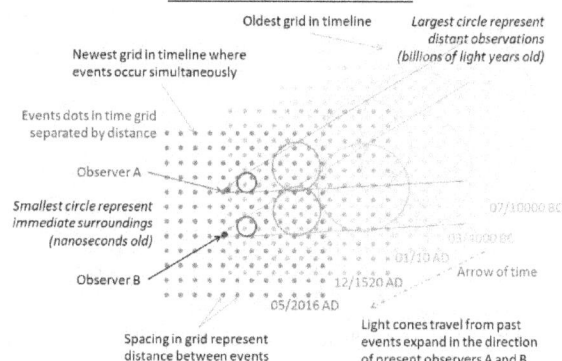

In the above picture of the spherical cones of "now" views of time, we see that observer A and observer B experience two versions of "now" based on their locations and direction of view. Each grid of dots represents a group of events that all occur simultaneously at 05/01/2016 at exactly 13:42:40 hours, where each event dot separated by a given distance. Observer A and Observer B both see events nearest to them first within nanoseconds of their occurrence represented by the smallest circle (with spherical surface) in the diagram. The further away they look from their locations the further back in time the event they are observing occurred. The next larger circle (with spherical surface) represents events that occurred at a greater distance from their location and sometime in the near past, for example a small asteroid hitting the moon or pieces of new comet hitting Jupiter. The middle larger circle with spherical surface represents events further into the past, like the light coming from Alpha Centauri. The next to the largest circle with spherical surface represents light from events more distant than that, possibly from a super nova in the Andromeda Galaxy. Finally, the largest circle with spherical surface represents light from events that occurred millions if not billions of years ago. Observer A can see all events that occur within his or her spherical cone of "now" view, while observer B sees all

---

[20] Time as we know it only flows in one direction forward. As such, no matter which portion of the sky we observe, we are only able to see the light from the past, never from the future.

events that occur within their own "now" cone of view. In this depiction of time, each observer can only look backward in time at events that have already occurred within his or her spherical cone of view. They can never look forward in time at pending future events that have not happened. Similarly, light emitted from past events travel toward the present observer in a light-cone (also a spherical cone) of its own, where its intensity weakens as it traverses space. The further light from the past travels to the present observers, the less time differences each observer detects and the closer they agree on simultaneity of the distant event. By the same "now" cone view rules, observer C in a distant galaxy some 19 million light years from earth cannot see us as we presently are, but only see the light the galaxy stars emitted as they were 19 million years ago. So, Greene's view of Chewy in the future can never happen in a normal universe.

A more perfect four-dimensional depiction of the "now" view can be drawn with increasing concentric spheres where each layer outward equates to a more distant past sequence of events. Here is a rough depiction of how four observers view a collection of time information to build their "now" perspective (two observers on planet earth, one on planet X, and one in a spaceship {not shown}). In the image, observers A

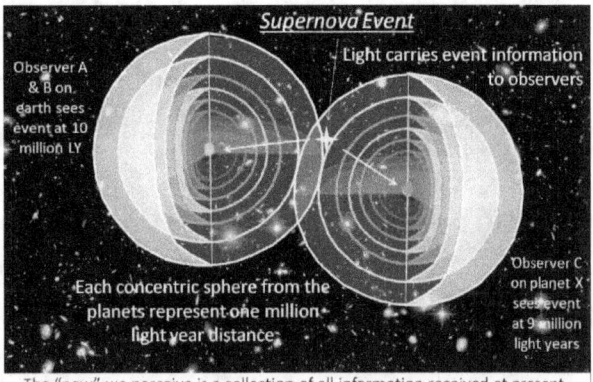

and B on planet earth detect a super nova event which they figure occurred some 10 million light years distance from the planet, and of course 10 million years in the past. Observer A sees the flash just nanoseconds slightly before observer B at the edge of horizon coming into view. Observe C in another galaxy sees the same super nova event occurring some 9 million light years distance from planet X. Observer C detects the event in their "now" moment about one million years before Observers A and B on planet earth who both record it in their "now" view. All witnesses to the super nova explosion received the information of the event via light waves traversing open space. Observer D, an astronaut in a spaceship about 3 million light years from planet X heading to earth sees the same super nova and records it in the starship log, as an event some six million light years distance from the ship's present location in their "now." All observers eventually get to see and record the event as they receive the information arriving via EM waves[21].

In Thomas' third book (Thomas, Hidden in Plain Sight 3: The secrets of time, 2014), he reveals to us four principles of time. Time flows at the speed of light in one direction and governs the second law of thermodynamics. Time exists in the universe; we measure it and use it to identify "the when" of any event. Momentum in the time dimension is energy or time gives us energy as it flows while the total energy of the universe remains at zero. No observer is special when it comes to time. Thomas does not explain how time travels that fast nor how time is energy. How can time travel at the speed of light? My

---

[21] The observed times collected from information received from distant objects does not and cannot change the date and time the events occurred. It happened when it happened, at the duration it occurred and the order it occurred, regardless of what we think as observers.

interpretation leads me to believe that time or the internal clock that provides it is embedded within each one of us, all living creatures and plants, and within every inanimate object as well, and every sub-atomic particle. We all somehow sense time passing by. How can that be? Time flows at the speed of light simply through the vibrations of quarks within protons, neutrons, gluons alike and governs the motion of electrons around their orbits like clockwork, all of which travel at the speed of light. This is nature's internal clock within all matter, the quantum mechanics vibration and universal synchronized ticking. It allows everything to exist for and if alive eat, breathe, grow, live, and move at the now moment. There is a saying, if you live in the past or future, you are never present. Responses in the now moment improves chances of survival, eat or be eaten, fight or flight, act accordingly to current situations, the crucial point leading to success, and the responses to past moments spells failure. Time is hereby universal. Without the vibrations of time and energy the universe would not exist.

Any mathematical model that combines or uses space and time elements into a single interconnected continuum is known as spacetime or space-time continuum; thereby astrophysicists have significantly simplified a great number of theories and described the workings of the universe in a more uniformed way, at galactic and super-galactic levels. Per Einstein, space-time continuum equations help us to understand the gravitational interactions within the galaxy and between galaxies. If we treat time as constant, we could use Euclidean space equations instead of space-time equations to define classical mechanics and astrophysics that are independent of the observer's motion. Time can never be separated from the observer. So, to be more accurate the passage of time and the gravitational attraction should be considered to function together. Although gravity is considered an instantaneous force according to Newton, where time and the duration thereof plays a significant part in the motions of objects caused by gravity per Einstein. When applied to outer space, the concept of space-time is relevant. It combines space and time into a single abstract universe, three dimensions of space and one of time. The four dimensions are independent components of a coordinate grid, to pinpoint a certain spot in space at a certain point in time, when the coordinate is relevant. An experiment revealed that time slows at higher speeds of the reference frame relative to another reference point. Per Einstein, objects appear shorter in length as they approach speeds closer to the speed of light to a distant observer viewing the speeding object from the outside as it moves past. Time if viewed from the outside also slows down for individuals traveling within the fast-moving object but remains the same if viewed from within the object itself. For the person traveling close to the speed of light, time outside of their ship passes more quickly, according to Einstein. The term dilation is used to describe this slowing of time known as Einstein's SR theory. GR (Einstein, 1920) equation is:

$$G_{\mu\nu} = \frac{8\pi G}{c^4} T_{\mu\nu}$$

where $G = 6.673 \times 10^{-11}$ N $(m/kg)^2$, is the Newtonian gravitational constant. Einstein's work tells us that all frames of reference are equivalent, provided we are all willing to include possible gravitational effects. In accelerated or in non-inertial frames, the forces will appear the same and are indistinguishable from gravity forces. Einstein declared that space-time is a dynamic entity, where matter curves it, and the way in which it is curved determines the motion of the matter within it. All bodies are affected in the same way within it by that curvature, while the gravity effects are independent of the nature of the body[22]. Changes in matter distribution deforms space-time.

1.5 Matter and Energy Interaction

The universe as we know it is made up of normal gravity matter and energy. Matter is anything that has mass and takes up space (and time). Because we deal with matter routinely in our everyday lives, matter is pretty much self-explanatory and taken for granted, but energy on the other hand is a bit more complex to explain. In physics, energy is the ability to move something, or do work, or make a change to the matter[23]. The amount of energy something has refers to its capacity to cause change or something to happen. Energy has a few properties. It is always conserved; it cannot be created or destroyed. It can be transferred, however, between objects or group of objects by interaction of forces. Energy comes in many different forms such as heat, light, chemical energy, and electrical energy. It can be kinetic or potential energy, the ability to bring about change or to do work. Kinetic energy is that of motion. An object moving has kinetic energy as it rolls on the ground or moves through space. An object in motion tends to stay in motion. Potential energy is the energy contained in an object not in motion. A beer bottle on a table standing upright on its top end has potential energy. If the table is shaken hard enough, the bottle will tip over, and fall onto the table with the pull of gravity and make a noise. A spring also has potential energy, especially if it is stretched. Other types of energy include, chemical, electrical, thermal, electromagnetic, and of course nuclear. Gunpowder has more weight than the sum of its elements alone since it has stored energy. This is an example of chemical energy.

Thermodynamics, the study of energy, has two **Laws of Thermodynamics:** First Law of Thermodynamics: Energy can be changed from one form or type into another. Energy cannot be created or destroyed. The total energy and matter in the universe remain constant, merely changing from one form into another. The First Law of Thermodynamics is also known as the Law of Conservation. The Second Law of Thermodynamics states that "in all energy exchanges, if no energy enters or leaves the system, the potential energy of the state will always be less than that of the initial state." This is commonly called entropy. Entropy is a measure of disorder, the natural inanimate environment state of balance: living cells are structured and so have low entropy, structure not found in inanimate environment. The flow and accumulation of energy maintains order and life. Entropy wins when organisms stop taking in energy in the form of food and sunshine, and die, and its atoms return to the natural balanced state, inanimate soil, and earth from which it came. In my view, entropy is a measure of return to original state of balance found in inanimate elements, soil, and earth. What scientist label as "disorder" is the original balance state; life and manmade objects are not naturally found in inanimate landscapes (like that of Mars).

Albert Einstein, interlinked mass, and energy into his one famous equation: $E=mc^2$, in which "E" stands for "energy," "m" stands for "mass" and "c" stands for the speed of light. Matter and energy are interchangeable, where energy is the product of mass and this

---

[22] In Einstein curvature of space-time, time is "curved" by excessive speed and massive gravity, and space is "curved" by massive objects and deformed as observers travel excessive speeds. We debunk both.

[23] This simple definition of energy tells us that energy cannot stand alone; hence matter cannot be converted to total energy as Einstein eluded in his equation $E = mc^2$.

arbitrary number c^2. This equation, the possible existence of energy without matter or mass, was one of the elements that paved the way for the Big Bang Theory. However, we all know that matter and energy are inseparable, energy affects the state of matter and motion of matter affects energy. Matter and energy normally obey physical law in the visible universe. The addition or absorption of energy by matter makes the matter weigh more and gives it greater potential. It changes the electron orbit levels of the atoms and molecules within the matter. In doing so, it absorbs more photons to hold the electrons at the new higher levels. Matter accelerated by energy does not absorb that energy, its speed increases not the object's mass. Matter, energy, momentum, and time are all interconnected. Remember time is universal and absolute, as predicted by Newton. Without the steady ticking vibrations of time embedded within all quarks shielded by a super strong sub-atomic force and the energy they give us; the universe would not exist. In other words, if all quarks or the smallest sub-atomic particles in the universe suddenly stop vibrating or jiggling at the beckoning of God, they would cease to exist and with their disappearance all atoms and matter would cease to exist, time halts or fails to flow at the speed of light, no energy produced, and the universe as we know it would completely vanish, or maybe not. Vanish and return to its original source, the nothingness of space.

What happens if we find out that Einstein's famous energy matter equation was slightly off? This equation is only applicable to the visible universe, and not necessarily applicable to ordinary matters' effect on space. His equation also does not consider the invisible part of the universe, dark matter, and dark energy, and as such, it needs to be modified to accommodate the missing pieces. Einstein thought he toppled Newton with his theory of General Relativity. Newton's equation on gravity, still stands the test of time, it remained brilliantly stated and on target. Newton declared that gravity between anybody of two masses was inversely proportional to the square of their distances, operated instantaneously and was of infinite range, and required absolute space and absolute time to work. What he did not know was the mechanism that made it work that way. Einstein's work skims the wave top of the concept that pulls it all together, the unified theory of everything, the Cosmological Balance. How can the physics of light and these light particles or waves affect the gravity of massive objects? Is gravity alone sufficient to bend light traveling near a star or galaxy's gravitational field? Are matter and energy truly interchangeable? How does the magnetron within a microwave oven work? Why does the sun appear redder during sunset and sunrise? What does light refraction do to light? How can lenses bend light without breaking it into its hue components or colors? What happens to light as it moves from one clear or transparent medium into another? Can gravity alone bend light like a lens, or is there another factor causing the bend? The answers to these questions, resolved by observations and insights of symmetrical balanced systems, are explained in the next few sections are the basis for this new unified theory.

Magnetron and Microwave Signals

An excellent example of the transfer between matter and energy is the magnetron device within a microwave oven. Microwave ovens have become a standard appliance within the modern home of today. Our younger generation cannot imagine how they can get through their daily routine without one. Meals and snacks are easily prepared within them in a matter of minutes or seconds. Most people, however, have no idea how they work. Microwaves in the typical household microwave oven are no shorter than 1-

millimeter wavelengths or about 0.04 inches. Let me try to explain how the magnetron works. It functions with some characteristics of an old tube television emitting electrons to a phosphorous screen where impact dots light up to form part of the picture, and some characteristics like a flute player blowing over the mouth of the flute at just the right angle to cause the tube to vibrate at a particular frequency to produce a sound.

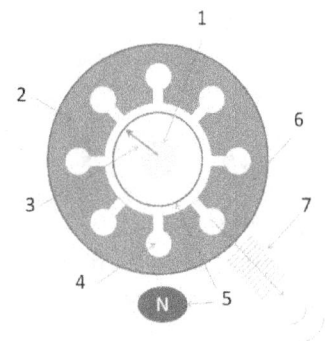

Basic Magnetron

Here is a simplified diagram of the magnetron within a microwave oven. Each of the numbers in the diagram is explained below: Item 1 – The cathode center rod (yellow color) is heated up and stimulated to emit electrons, which take with it a photon to jump the gap. Item 2 – The surrounding anode compartment (orange color) is charged positively to attract the electrons from the cathode. Item 3 – Normally, the electrons would boil off the cathode and fly directly across to the anode in straight lines as shown by the thick line with arrowhead. However, we could introduce a couple of things that change its path. Item 4 – First, we added slots and drilled holes into the surrounding anode compartment as shown above. These resonant cavities become the frequency chamber somewhat like the flute's mouth and entry into the tube. Item 5 – Then we added an electromagnet at the base of and the length of the magnetron, with north on one end of the magnetron and south on the other end. This magnetic field causes the electrons emitted from the cathode to follow a curved path (dark thick blue circle) around the cathode. Item 6 – As the electrons and accompanying photons pass over the cavities, they enter into the chambers or the resonant cavities at an angle, like the flute player blowing into the mouth of the flute tube to produce sound. However, instead of sound, we obtain the desired microwave frequency to excite and heat up water molecules. Item 7 – The microwave radiation that the cavities produce is then funneled and channeled by a wave guide and concentrated into the cooking compartment of the microwave oven. For a microwave radio, it is beamed outward into the air by an antenna or dish to another receiver. The actual operations of the magnetron in a microwave oven are a bit more complicated than this[24]. It was accidentally discovered during the magnetron's radio signals testing phase that the nearby water molecules would heat up. Thanks to American engineer Percy Spencer, he incorporated this side effect of radar technology into what he invented, the Radarange, later renamed the microwave range. Remember how this magnetron device works. It holds one of the keys to the universe. Besides Doppler Effect, what else causes red shifted light? Atmosphere density. If all the stars coming over the horizon in the Earth's night sky appear to be red shifted, does that mean that they are racing away at tremendous accelerating speeds? No. For if we wait a few more minutes, they all appear to have normal light, and are no longer red-shifted. Did they stop racing away? No, their distance was constant. The further Hubble looks the more red-shifted these galaxies appear, not because they are racing away faster, but because there is more distance, and therefore more material in space between us and that galaxy that absorbs photon energy.

---

[24] The function of a magnetron within a microwave oven was discovered by accident when developers were trying to improve the operation of microwave radios. The Cosmological Balance Theory uses this simple design to explain why we "hear" and "see" microwave radiation in our sky, day after day.

1.6 Red Shift and Light Refraction

Let us look at what reddish light effect naturally occurs here on earth. Imagine sitting on your front porch at your house or at a window seat at a restaurant eating dinner. There you observe the sunset; the sun appears redder and redder, as it gets closer to the horizon. Technically, the sun is way below the horizon, but because of the magnifying effect of the thicker atmosphere, we get to see it go down for a bit longer. The increased amount of molecules in the atmosphere absorbs more photon energy at the horizon and causes the light to scatter more of the bluish component of the white sunlight, leaving behind the reddish light to penetrate the additional layers of atmosphere before reaching our eyes thereby coloring the sky with a reddish tint and making the sun appear redder. This scattering effect and photon energy absorption is most noticeable during sunset and sunrise but occurs throughout the day and is the reason the sky is blue. Therefore, the thicker the atmosphere the more photon energy it absorbs and the redder the sun appears, even to the point of turning the sky orange-red or red[25]. Diagram below depicts the effect:

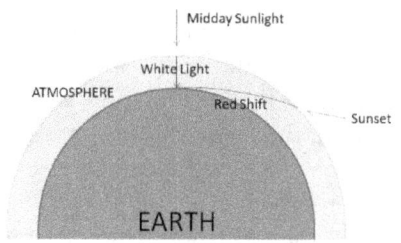

On earth, the density of the atmosphere at sea level is 1.2 kg/m³, and the thickness of the atmosphere is about 480 km at vertical with most dense part of the atmosphere below 16 km. This gives a 20 km column of air straight up in the atmosphere a density of about 12000 kg/m² * atmosphere cylinder 20 km long. At sunset we are looking at about three times that density 30000 kg/m² * 2.5-atmospheres, which of course gives us a significant red shift effect we know so well on the sun. The same effect also occurs during sunrise, but in reverse. The sun is redder when it first rises and then becomes yellower and then almost white as it moves upward toward vertical position in the sky. In the above analysis, we learned that as light passes through matter, its energy is absorbed, refracted, or reflected to some degree or another. And in some cases, the energy of the medium adds to or takes away from the energy of the light re-emitted, either speeding it up or slowing it down. This respectively results visually in a bluish or reddish tinting of the light reaching us. It does not matter how dense the medium is, the effect depends on the number of particles the light passes through and the energy of the material. That absorption determines our perception of the universe. Remember it.

Similarly, the presence of particles of matter provides some level of pressure and energy. A man weighing 100 kilograms at the beach in Perth, Australia could very well have a weight of 97 kilograms at a resort lodge on some mountaintop at near Everest. The same man may not even notice it, because he is too concerned about breathing due to the level of oxygen in the air at that altitude. Likewise, the air pressure pushing on the man at sea level may be one kilogram per square centimeters, while the air pressure pushing on the same man on the mountaintop reads about 0.97 kilograms per square centimeters. Assuming that the atmospheric pressure acts on the surface area of the man equally at the beach, we derive his surface area by multiplying the air pressure there at one kilogram per square centimeter, times 10,000, to convert to square meters, divided by his known weight

---

[25] The physics showing the red shift of sunlight during sunrise and sunset and during sandstorms applies with starlight from distant stars and galaxies, assuming space has scattered matter within it.

100 kilograms, which gives us 100 square meters. We then take his surface area of 100 square meters and divide it by the atmospheric pressure at the mountaintop in the Alps of 0.97 kilograms per square centimeters times 10,000 to convert to square meters; we get the adjusted weight of 97 kilograms there.

We just explained that atmospheric pressure accelerates mass toward the earth equals that of the force of gravity. The pressure of space-matter acting on the outer layers of earth's atmosphere should theoretically do the same thing, creating a spherical "bubble-like" layer or boundary around the earth. Space-matter is the source that causes the atmospheric pressure we feel at sea level and provides what we see as gravity and its effects throughout the universe. Before we leave this subject, let us take another look at the density of the gaseous matter in deep space. This known density contributes to the dissipation at extreme distant light and its diffusion, combined with the loss of photons energy, results in red shift readings, which cosmologists attributed to the Doppler Effect and the expansion of the universe, and not the low-density gaseous material found dispersed in deep space. From the planet surface, the sun appears to us during the hours of dawn and dusk as more orange-red color than at midday. This effect is the result of three times more air molecules absorbing more photon energy at those two times of the day. We can expect the same type of effect to occur as we view further and further out into deep space especially at extremely distant galaxies. Large amounts of hydrogen and other gaseous matter although dispersed uniformly over great distances in deep space would absorb enough light photon energy from distant stars and galaxies to cause a red shift, and beyond a certain point, say 13.8 billion light years away or more block out all light. Strongly associated with this reddening effect is the alluded cause, the refraction of light and all other EM waves. Refraction of light also plays a key role in determining bending around large, massive objects with differing layer density of matter and space around them. The next section will review what we know about this phenomenon.

Refraction of Light

Another example of the interaction of matter and energy is found when observing light passing from one clear medium into another clear medium with a different density. Light changes speeds as it propagates through the new medium. The speed of light (encompasses all electromagnetic radiation) in science is defined as the speed light travels in a vacuum (space), which measures at 299,792,458 meters per second (usually rounded up to 3.00×10^8 m/s). If light travels through another medium such as air, glass, plastic, diamond, or anything clear, it changes to a different speed and direction depending on the density of that new medium or its index of refraction. In other words, refraction is defined therefore as the bending of the light wave's path as it transitions from one clear medium or material to another clear or translucent medium. The refraction occurs at the boundary where density changes and causes a change in the speed of the light wave as it crosses that boundary. Multiple layers of density as in the atmosphere cause multiple bending. Again, the tendency of a ray or beam of light to bend or angle toward one direction or another is dependent upon whether the light wave speeds up or slows down upon crossing the new medium's boundary. This index of refraction is defined as

$$n = c/v, \qquad (1.4.1)$$

where c is the speed of light in a vacuum and v is the speed of light in the medium.

We all see the effect of refraction in our everyday lives, whether we wear eyeglasses or not. Yes, eyeglass lens are refractors. In the science classroom, the most common demonstrated refraction example that teachers use is a prism, which produces a rainbow of colors. For that matter, the rainbow we see in the sky after it rains is another example.

What also causes refraction? Refraction is cause by transition between mediums with different densities. Changes in density within that medium will also decrease or increase the speed v and therefore more or less refract the light respectively, as well as its wavelength. Mirages are an example of refraction caused due to the different density of heated air at the surface of a scorching hot desert sand or road when compared the temperature of the rest of the atmosphere above it; the heated air is less dense than normal atmosphere above. People who have seen mirages usually think they are looking at a pool of water or some other objects not actually there; these images are caused by a refraction of the blue sky or some distant objects' light being bent toward their eyes by the difference between the heated air density and the rest of the atmosphere's density. Reverse mirages occur over the ocean or lake waters during summer months, where cold currents, carrying icebergs or glacier ice pieces, can suddenly cool the air immediately above the ocean or lake making it more dense while the air above that stays relatively warmer and less dense. This effect can hide objects in the water from view until it is too late to react. Titanic ship disaster is an example of this; by the time, their crew saw the iceberg it was too late.

The equation showing changes in wavelengths are

$\eta = \lambda/\lambda_m,$ (1.4.2)

where λ is the wavelength of the electromagnetic (EM) radiation in a vacuum and λ_m is the wavelength same EM radiation in the medium.

Despite light's changes in speed and wavelength, the frequency of the light or EM radiation remains the same for the most part. However, a large distance of medium shifts the frequency toward the red. As the light encounters every molecule along the way it loses a tiny bit of photon energy, so prolong interactions with molecules cause the red shift to occur and become more noticeable.

The relationship between speed, frequency and wavelength is

$v = f * \lambda,$ (1.4.3)

where v is the speed, f is the frequency, and λ is the wavelength.

Below is a depiction of the changes in speed that occur as light passes between one medium into another with different densities (note: angles in diagram below are not drawn to scale):

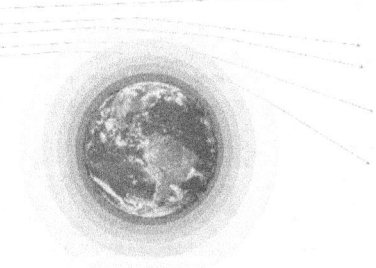

LIGHT REFRACTION ANGLE CHANGES WITH AIR DENSITY

Snell's Law defines the relationship between refraction angles and density as:

$n_1 \sin\theta_1 = n_2 \sin\theta_2$ or $n_2 / n_1 = \sin\theta_1 / \sin\theta_2$ or $v_1 / v_2 = \sin\theta_1 / \sin\theta_2$

Here are some measurements taken by mainstream science:

Density (water=1): $1 \text{ g/cm}^3 = 1000 \text{ kg/m}^3$

Mean density of entire Sun 1.41 g/cm^3 = 1.41 x 10^3 kg/m^3

    Interior (center of the Sun) 160 g/cm^3 = 1.60 x 10^5 kg/m^3
    Surface (photosphere) 10^{-9} g/cm^3 = 1.0 x 10^{-6} kg/m^3
    Chromosphere 10^{-12} g/cm^3 = 1.0 x 10^{-9} kg/m^3
    Low corona 10^{-16} g/cm^3 = 1.0 x 10^{-13} kg/m^3
    Sea level atmosphere of Earth 10^{-3} g/cm^3 = 1.0 kg/m^3

LIGHT PROPAGATION THROUGH A MEDIUM

The mechanism, which transports light through a medium, is like the way that any other wave, such as sound or the wave in water, is transported; it moves by going from particle-to-particle as it travels through the medium. Unlike sound in air or buoyancy waves in water, light is self-perpetuating particle motions and freely traverses empty space. An electromagnetic wave (or light wave to include the visible frequencies) is produced by a vibrating electric charge with a photon magnetic wave. As the light wave self-propels (perpetual motion) itself through the vacuum of empty space, it travels at a speed of c (3 x 10^8 m/s), defined above. When the wave encounters a particle of matter, part of the energy is absorbed and causes the electrons within that atom to increase its vibration motion. If the light wave's frequency does not match the electron's resonant vibration frequency, then the light wave is reemitted at the other side of the atom in the same direction and the same frequency as original wave. The light wave continues to move from atom to inter-atomic empty space to another atom again as it passes through the entire medium. The cycle of absorption and reemission continues as the energy moves from particle to particle through the medium. light travels between the inter-atomic void at a speed of c; with time delay during absorption and reemission process by the atoms of the material consequently lowering the net speed of travel through the medium, $v < c$, where v is the speed of light traveling in the medium. If the atoms in the gaseous medium are more energized or excited, then more collisions between light wave and atoms are expected, causing additional acceleration of the light's speed and a slight increase in bending angle as the energy is transferred. Density also affects the Index of refraction and greater optical results. The optical density of that material affects the speed of the light passing through it. The optical density of a medium is not the same as its physical density[26]. The physical density of a material refers to the mass/volume ratio. The optical density of a material relates to the tendency of the atoms within the material to absorb energy of an EM wave in the form of vibrating electrons before reemitting it as a new wave.

Amount of refraction depends upon:
1. Density of the material.
2. Angle at which the light enters the material.
3. Wavelength - causes colors in white light to separate as it goes through a prism.
4. Light will bend away from the normal when passing into a less dense medium.
5. Light will bend toward the normal when passing into a denser medium.
6. Refraction occurs at Sunrise and Sunset.

The more optically dense the slower that wave will travel through the medium, refract.

---

[26] This important characteristic allows for the curvature of space (densely packed space-matter) to have superb optical density without being extremely packed like the molecules within a glass lens.

Vacuum 1.0000
Air 1.0003
Water 1.333
Ethyl Alcohol 1.36

Plexiglas 1.51
Light Flint Glass 1.58
Dense Flint Glass 1.66
Diamond 2.417

DUE TO REFRACTION:

During sunrise, the sun will appear to rise about two minutes sooner than it does. Moreover, on sunset, the sun will appear to set about two minutes later than it does. As the sun moves upward from the horizon to the vertical position, the angle of refraction gradually goes from 0.5 degrees to zero. Even stars seen in the atmosphere during the night appear slightly off from their exact position in the sky. 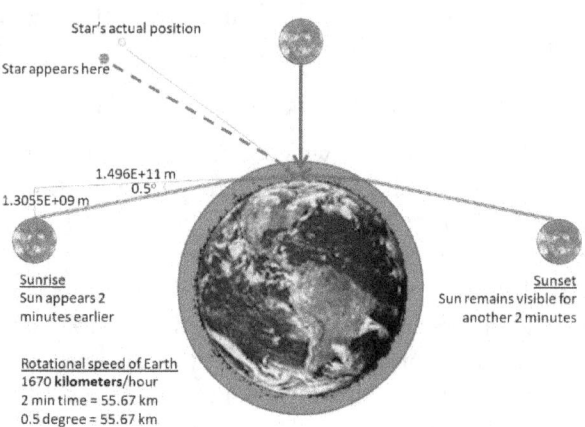 Depicted are examples of refraction as seen from the planet surface. Let us take a look at this phenomenon off planet. Hence, from the perspective of an astronaut in the international space station or one standing on the moon's surface during a partial lunar eclipse, the astronaut sees the sun or a distant star respectively appearing to come up from behind the earth, say about 10 km above the earth's surface, but due to refraction is actually looking at an image of the sun or the distant star, which is still behind the planet[27].

The slope difference between the image and the actual sun or distant star is about 1 degree angle, and about four minutes of time to actual sunrise or star-rise over planet earth. The diagram above show this. The further up the sun rises the less angle of refraction occurs until the sun clears the upper layer of the earth's atmosphere. Once clear of the earth's upper atmosphere and the curvature of space around the earth, the refraction index becomes zero, that of empty space. On earth, this image displacement effect is also known as a mirage.

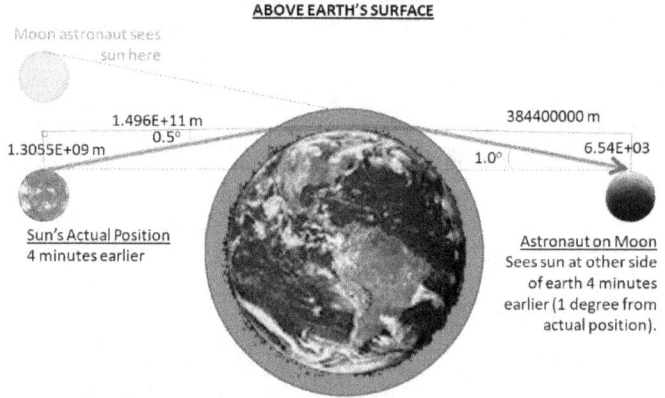

---

[27] The light refraction caused by the earth's atmosphere and the curvature of space around the earth both affect starlight passing through it. Atmospheric refraction changes the lights color more than bending by the curvature of space (space-matter around the earth).

1.7 What is the Force of Gravity?

Even after defining gravity and its properties in Principia, Sir Isaac Newton hated his own theories about gravity being an "action at a distance" force. He strongly believed there must be some unknown material in empty space that connects objects and provides the gravitational force between bodies of matter. He was one of the first scientists to suggest there was some mysterious substance in space some called the aether (sometimes spelled ether) that connected all objects in the universe." How does gravity work? Why is it instantaneous? Is there an outside mechanical source of gravity as alluded to by Sir Isaac Newton? Is gravity instantaneous and infinite in range? Why is the gravity constant set at 6.6734E-11 Newton times meter squared per kilogram squared? The answers to these questions are literally invisible to our instruments. Although scientists and academia have said that Einstein's GR theory supersedes Newton's, it is not widely used. Due to its simplicity, physicists use Newton's equations to map out objects in space.

Gravity (also called gravitation) is a natural phenomenon that affects all physical bodies, mutually attracting the two bodies to each other. On the planet Earth, gravity pulls the physical object downward toward the earth and thereby gives it weight, this force of gravity keeps it grounded. Gravity enables us to do everything we do in our everyday lives. Without gravity there would be no sun, no planet earth, no solar system, no galaxy, and for that matter no visible universe. If gravity were just a tiny bit stronger, the sun would burn through its fuel and burn out too quickly for life to take hold on earth. If gravity was just a bit weaker, the sun might not have formed and ignited for that matter it might not have the strength to hold the planets in place (Greene, The Elegant Universe: Superstrings, Hidden Dimensions, and the Quest for the Ultimate Theory, 2003, p. 13). Even the slightest increase to the mass of the electron would cause it to plunge into the proton within the nuclei and hydrogen, the primary fuel of the sun, could not exist and fuse; we owe our existence to this delicate gravitational balance. According to physicists and current astrophysicists, the general theory of relativity (introduced by Einstein) portrays a "more accurate" equation by which gravity is the result of the curvature of space-time. For most applications disregarding the curvature of time, Newton's law of universal gravitation well approximates gravity. Hence, I side with scientists that believe gravity effects are instantaneous. If the Sun were to suddenly disappear completely without a trace, its planets will immediately begin to travel in a straight tangent line perpendicular to the lost gravitational force. This concept cannot be proven or disproven scientifically due to our inability to travel anywhere near the speed of light, Einstein took advantage of it and this dilemma and used both to further his childhood dreams and theories. Did his imaginary experiments do justice to science or set us back one century? He also adamantly opposed quantum theory because its proof of entanglement supports Newton's instantaneous claim and therefore threatened to topple his curvature of space-time theory[28]. In pursuit of a

---

[28] Einstein spent several years working on General Relativity Theory, literally lost in energy-momentum tensors, simply because he would not let go of basic assumptions, he originally made that eventually turned out to be wrong. Once he realized these assumption errors, everything fell into place and he was able to publish GR. Likewise, his assumption of S/GR time dilation and physics of light along with speed of light constant prevented him from developing a unified theory. Letting go of these errors could have opened the doors to solving the mysteries of the universe more perfectly. Cosmological Balance takes this path.

theory of everything, the merging of general relativity and quantum mechanics (or quantum field theory) into a more general theory of quantum gravity has become an area of active research. Scientists hypothesized the gravitational force mediated by a mass-less spin-2 particle, called it the graviton in which gravity would have separated from the electronuclear force during the grand unification epoch, per Hawking. Scientists have not found the graviton particle.

Although, scientists consider gravity as the weakest of the four natural forces in the universe, it is the most influential over great distances. Gravity is a force not a "field" in the traditional sense of the word. It is unlike electrical or magnetic fields, which both operates at the maximum speed of light; gravity is an instantaneous force supported by the innate micro attraction of entangled quantum particles trying to merge and fueled by infinite space itself. By comparison, gravity is the weakest of the four fundamental forces of nature. The gravitational force is approximately 10^{-38} times the strength of the *strong nuclear force* (i.e. gravity is 38 orders of magnitude weaker), 10^{-36} times the strength of the *electromagnetic force*, and 10^{-29} times the strength of the *weak nuclear force*. In other words, if the strong force is 1, then the relative strength of the electrostatic or electromagnetic force is about 10^{-3} and the gravitational force is approximately 10^{-45} compared to the strong nuclear force. Gravity believed to be of negligible influence on the behavior of sub-atomic particles and plays no role in determining the internal properties of everyday matter in the visible universe, other than in black holes. Is it and why is this so? I will disprove this later. For now, it suffices to say that gravity is the dominant force at the macro-scale level of the universe. It causes the formation, shape, and trajectory (orbit) of astronomical bodies, including those of asteroids, comets, planets, stars, and galaxies. Gravity is the only force acting on all particles of matter in the visible universe. When Newton presented in his gravity equation that one body instantaneously attracted another body over infinite distance, called action at a distance, he did not realize that that initial attraction tendency was caused by quantum entanglement properties inherent within like matter of common origin. He did, however suspect that there was some other element responsible for the mechanical aspect or force of the motion of gravity. Together these two elements constitute the force captured by Newton's gravity equation. Gravity is instantaneous and infinite in range because space's vacuum force and energy functions uniformly[29]. Gravity is always attractive and never repulsive, and to the best of our knowledge and scientific experiments it cannot be absorbed, transformed, or shielded against its effects. Einstein undermined the "never repulsive" aspect of Newton gravity law with the cosmological constant he amended to the GR equation in which he declared that under the right circumstances of negative pressure gravity can be repulsive, thereby trying to add some stability to the universe. Einstein, upon learning of Edwin Hubble's announcement that the universe was expanding, then publicly withdrew the cosmological constant as his biggest blunder. Even though electromagnetism is far stronger than the force of gravity, according to academia, EM is not relevant to astronomical objects (net charge of zero).

---

[29] Newton wrote his gravity equations based on what he and others before him observed and documented. He did not develop his theory based on imaginary experiments, only solid data, which he defined as laws. He induced that there was some mechanical force that provided gravity, but due to limited data and observations simply left it out of his equations stating that such force was instantaneous and of infinite range without explanation of how. Einstein did the same when he deduced that massive objects curve space-time, without explaining how or what about space "curved," other than light.

1.8 Why is Sun's Corona So Hot?

The atmosphere of the sun consists of four primary layers. The photosphere or upper surface layer is relatively cooler compared to the plasma within the sun. The chromospheres layer is relatively cooler than the photosphere. As energy move upward into the transition region the temperature increases from the chromosphere high temperature to the corona higher temperature. In the corona, the expansive outer layer of the solar atmosphere that extends millions of kilometers from the sun's surface, temperatures reach millions of Kelvin. The surface, by contrast, is a tepid 6,000 Kelvin (around 5,700 degrees Celsius). At this point, the corona traps the heat there where it begins to build up, until a solar mass ejection occurs. The solar mass ejection takes with its large amounts of plasma and heat away from the sun into space. Although astronomers have developed a few possible explanations in recent years, no one can say precisely how or why the corona gets so hot. 'NASA'S two leading hypotheses that seek to explain the corona's extreme heat. One idea holds that magnetic field lines twist and braid as they skitter around the sun, building up tension and unleashing massive amounts of energy when they finally break like rubber bands. Alternatively, magnetic waves rolling from deep within the sun could transfer energy and heat into the chromospheres and corona. "We don't know whether any of these are right or whether we need new concepts," Priest says. "It's the observations from IRIS that are going to be able to determine that (Boyle, 2013).'" Another hypothesis, the heat accumulation is simply due to a gravity bubble layer consisting of some unused gas byproduct expelled from within the sun, or simply density layers of the curvature of space itself. If gravity from a black hole can stop light and radiating heat from escaping, then the sun's gravity is sufficient to stop its convection heat from radiating into space but not all its photons of light[30].

On earth, heat radiates slowly upward from molecule to molecule into the atmosphere and eventually into space. The earth's gravity holds and traps some of that heat within the lower atmosphere layers, at least until sunrise. The gravity of the sun is significantly greater than that of all the planets. As heated matter and particles try to escape the sun's upper edge of the corona, the sun's immense gravity pulls them back downward, essentially creating convection heat loops there. The corona "surface" tension film bubble, far above the loops, surrounds the sun and stops the radiating heat from escaping, thereby steadily increasing the corona's temperatures. This gravity boundary, far above the loops seen over the corona and is invisible to us, allows light photons to pass through. It is a clear colorless additional layer of solar atmosphere not yet detected or named. Again, only solar mass ejections have sufficient electromagnetic force to carry excess heat and plasma to escape speed beyond the corona layer's upper film or gravitational influence. Without these solar mass ejections, heat energy would not be able to escape the sun's gravity; the corona layer would continue to heat up and expand into the upper area of the chromospheres, thereby shrinking its band height. Yes, the sun's gravity and the density of space-matter around the sun causes the corona's high temperature.

---

[30] One of the qualities of large density of space-matter surrounding a massive object is that it reflects and deflects light; black hole is an example. Space-matter surrounding the sun, also known as curvature of space, reflects sunlight back into the sun's corona and thereby reflecting heat back and causing it to increase in temperature.

Chapter 2: Physical Law in Cosmology

"Through our eyes, the universe is perceiving itself. Through our ears, the universe is listening to its harmonies. We are the witnesses through which the universe becomes conscious of its glory, of its magnificence."

— Alan W. Watts

We know that scientific law or physical law is a theory produced from given facts, pertinent to a specific group or class of unusual events, and defined by a statement where those unusual events always happen and only when specific conditions are at hand. Physical laws are the resultant conclusion based on repeated scientific experiments and observations over several years, in which the scientific community accepted universally the conclusion within. Describing our environment with these physical laws is the goal of science. Essentially, natural law and physical law are one, and the same concept. In applying natural and physical law, we must rely on a bit of faith, assuming that these laws of physics will remain valid now and for all of eternity. For the present moment, we have no evidence or reason to doubt that it will ever change over time. Physical laws and the structure of the universe must therefore be intertwined and dependent on each other. Although physics equations work in either the forward or backward directions without distinction, the "arrow of time" only takes us in one possible direction, always forward in time, and never backward or reverse chronology (Greene, The Fabric of the Cosmos: Space, Time, and the Texture of Reality, 2004, p. 13). Hence, all physically laws obey the time element, and its path forward[31].

It is quite amazing that the Fibonacci number patterns, for example, occur so frequently in nature (they are found in flowers, shells, plants, and leaves, to name a few). This phenomenon appears to be one of the principals, "laws of nature." Fibonacci sequences appear in biological settings, in two consecutive Fibonacci numbers, such as branching in trees, arrangement of leaves on a stem, the flowering of artichoke, the fruitlets of a pineapple, an uncurling fern and the arrangement of a pinecone. Pascal's Triangle is described by the following formula:

$$a_{nr} \equiv \frac{n!}{r!(n-r)!} \equiv \binom{n}{r}$$ where $\binom{n}{r}$ is a binomial coefficient.

Fibonacci numbers 1, 1, 2, 3, 5, 8, 13…. Sequence pattern: The next number is the sum of the previous two (shown here).

The four most influential natural laws govern all matter, as we know it. Three of four examples of natural laws found to influence the microelements of the universe are the force that holds the protons and neutron together in the nucleus, the strong nuclear force; the electric and magnetic force or the

---

[31] Time moves in only one direction, forward. Although physical law allows for playing events in either direction, however, the arrow of time shows us the occurrence in the forward motion.

electromagnetic force; and the force that holds atoms and molecules together, the weak nuclear force. At sufficiently hot temperature, the weak nuclear force and the electromagnetic force merge into one force called electroweak (Thomas, Hidden In Plain Sight: The simple link between relativity and quantum mechanics, 2012). Another example of natural law or physical law is the law most pertinent to the macro universe is the force of gravity. Newton's law of universal gravitation states that any two bodies (m_1, m_2) in the visible universe attracts each other with a force (F_1, F_2) that is directly proportional to the product of their masses and inversely proportional to the square of distance (r) between centers.

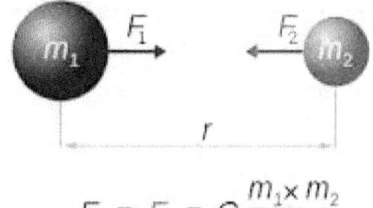

$$F_1 = F_2 = G \frac{m_1 \times m_2}{r^2}$$

The gravity force of a sufficiently large mass is even capable of attracting and bending a beam of light. Earth's gravity force bends light ever so slightly, approximately 4.9 Pico meter over a 300-meter distance. Although gravity is weak, it has an infinite range and considered instantaneous per Sir Isaac Newton. Time moves at the very least the speed of light at the atomic level within all matter. Energy is the motion or potential motion of matter, and matter in motion has momentum. The operation of a magnetron in a microwave oven is an example of an electromagnetic energy emission, demonstrating the relationship and interchange between matter and energy. Is time and space really intertwined as Einstein predicted, or is there another hidden attribute? Is time universal and consistent as predicted by Newton? Is time absolute? Does time flow in only one direction, forward? Can time flow or run backwards? How is time relative? How can an observer's perspective of time on a distant object alter the actual flow of time on the surface of the observed object? How could one observer's "now" view differ from another? Is space in fact absolute? Is space flat, or curved, or both? What exactly is curvature of space? How can nothingness have curvature? What is the best representation of these concepts?

According to the equation depicted above, gravity waves do not recede from the objects; they go towards the objects! This action tugs the two objects toward each other. See Appendix B for additional details on Newton's Principia and how he developed this equation. Gravity is both a push force and a pull force. I will explain exactly how it operates in the Cosmological Balance.

2.1 Kepler's Laws and Newton's Gravity

Johannes Kepler applied the concept of torque and polar coordinate system to the motion of planets orbiting around the sun and discovered three laws. They became known as Kepler's laws of planetary motion.

1. The orbit of a planet is an ellipse with the Sun at one of the two foci. In other words, the planets follow the path of ellipses, and that a circular orbit is a specific type of ellipse where both foci are one in the same location.

2. A line segment joining a planet and the Sun sweeps out equal areas during equal intervals of time. Similarly, a ray from the sun to a planet sweeps out equal areas in equal times.

3. The square of the orbital period T^2 of a planet is proportional to the cube of the semi-major axis a^3 of its orbit. This constant of proportionality is independent of the individual planets, or stated in another way, each planet has the same proportional constant. Kepler's formula now reads $T^2 = (4\pi^2/(G*M)) * a^3$, where M is the mass of the sun and G is the universal gravitational constant Newton discovered.

Most of the planetary orbits observed are almost circular. Observations and calculations are required to establish if they are ellipses. Since Johannes Kepler's calculations on Mars proved that it was an ellipse, he inferred that heavenly bodies with wandering movement (planets) including those farther away from the Sun also have elliptical orbits[32]. The image depicts Kepler's three laws with planets orbiting the sun.

1. Both orbits are ellipses with focal points f_1 and f_2 for the first planet m_1, and f_1 and f_3 for the second planet m_2. The Sun s_1 is located in focal point f_1.

2. The two shaded slices A_1 and A_2 have the same surface areas. The time it takes planet p_1 to sweep arc of A_1 is equal to the time it takes to sweep arc of A_2.

3. The total orbital times for planets p_1 and p_2 have a ratio $a_1^{2/1} : a_2^{2/1}$ or 1:4.

Kepler's work published sometime between 1609 and 1619 improved the heliocentric theory of Nicolaus Copernicus, explaining how the planets' speeds varied, and using elliptical orbits rather than circular orbits with epicycles. Among his laws is the conservation of angular momentum, which is the quantity that plays the same part in rotational mechanics as linear momentum does in linear mechanics. In other words, the angular momentum vector of a planet is a constant vector, because the sun does not apply torque to the planet per physics teachers.

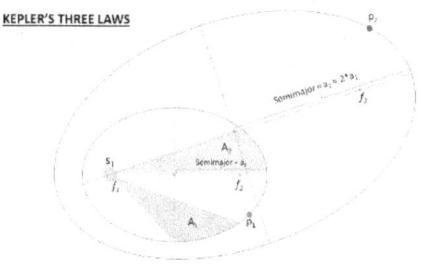

KEPLER'S THREE LAWS

In 1687, Isaac Newton used Kepler's planetary relationships in the solar system to a good approximation, to develop his own laws of motion and law of universal gravitation in Principia. Note that in the above chart Kepler's law estimates are a good approximation

---

[32] Kepler's planetary motion work was one of the most influential contributors to Newton's gravity equations. Although today, we show Kepler's equations with Newton's gravity constant.

but not exactly accurate as we will learn in Chapter 3 of this book. The orbital period was calculated with this formula $T = \sqrt{[(4\pi^2/(G*M)) * a^3]}$. Orbital period of Mercury is $\sqrt{(4\pi^2/(6.6734 * 1.9889E+30) * (5.791E+10)^3)} = 7,600,262.18$ seconds, which is equal to (7,600,262.18 seconds) / (86,400 seconds/day) = 87.9659974 earth days or 7,600,262.18 seconds / (31,556,952 seconds/year) = 0.2408 earth years. The chart above shows all the calculations completed for each of the planets, noting this formula disregards the gravitational influences of other planets. Kepler's estimate of Mercury's orbital period is 87.966 earth days; Venus is 224.66 earth days; Mars is 686.75 earth days (1.88 years); Jupiter is 4335.83 earth days (11.87 years); Saturn is 10,828.14 earth days (29.65 years); Uranus is 30,803.09 earth days (84.34 years); Neptune is 60,216.24 (164.87 years); and Pluto is 90,605.65 earth days (248.07 years). Nonetheless, Kepler's laws remain as part of the foundations of modern astronomy and physics. Johannes Kepler formulated these laws based on polar and torque principles. Using these polar coordinates, we review how torque changes an object's rotational state, then apply the polar coordinates to objects in motion, and then simplify formulas to get the same motion obtained rectilinearly. Kepler equation defines the orbital movement period of the planets around the sun. Below is a spreadsheet showing the orbital periods as strictly calculated with Kepler's third law:

kg	orb	orbital per = s	Radius = m	orbit (days)	orbit(yrs)
1.9889E+30	Mass of sun	10020.57907	6.9630E+08	0.115978924	0.00031754
3.2972E+23	Mass of Mercury	7600262.175	5.7910E+10	87.9659974	0.240842721
4.8673E+24	Mass of Venus	19410607.04	1.0820E+11	224.6598037	0.615097651
0.0000E+00	Mass of Earth and Moon	31556993.23	1.4960E+11	365.2429772	1.000001307
6.3900E+23	Mass of Mars	59335514.72	2.2790E+11	686.7536426	1.88026761
1.8981E+27	Mass of Jupiter	374615852.8	7.7850E+11	4335.83163	11.8711038
5.6830E+26	Mass of Saturn	935551096.1	1.4330E+12	10828.13769	29.64643405
8.6810E+25	Mass of Uranus	2661386927	2.8770E+12	30803.08943	84.33599438
1.0240E+26	Mass of Neptune	5202683454	4.4980E+12	60216.24368	164.8664755
1.3090E+22	Mass of Pluto	7828328315	5.9063E+12	90605.65179	248.0698489

Newton's Gravitational Properties

Newton's theory of gravitation, main article titled: Newton's law of universal gravitation: By Sir Isaac Newton, English physicist who lived from 1642 to 1727. In 1687, English mathematician Sir Isaac Newton published Principia Mathematica or simply known as *Principia*, which hypothesizes the inverse-square law of universal gravitation. To present his hypothesis and the solution properly he invented calculus. All Newtonian mechanics and equations in Principia were built on the following three laws of motions[33]:

LAW I: *"Everybody perseveres in its state of rest, or of uniform motion in a right line, unless it is compelled to change that state by forces impressed thereon."* This law states that an object or body remains at rest or in uniformed motions in a straight line unless another force external to it

---

[33] Newton's three laws of motion define the rules used to explain gravity. All three laws and everything else he presented in Principia all pointed to gravity as being instantaneous and infinite in range, as well as based on absolute space and absolute time. Without these qualities, Newton would not have been able to discover his gravity equations (the pull force). Similarly, Einstein's General Relativity equation in linear form simplifies into Newton's gravity equation. This fact should persuade our astrophysicists that gravity cannot and does not travel at the speed of light as Einstein claimed. Good mathematicians and physicists should never violate the Principle of Consistency.

acts upon it. This resistance property is known as **law of inertia** and the mass of that object is a measure of its inertia, which is also referred to as inertia mass.

LAW II: *"The alteration of motion is ever proportional to the motive force impressed; and is made in the direction of the right line in which that force is impressed."* This law states that when a net force acts on an object, it will accelerate that object or body in the direction of that force, called **force law**. The relationship between the mass m, acceleration a, and the force F is written $F = m*a$. Similarly, the relationship between mass m, the momentum p, and the velocity is written $p = m*v$.

LAW III: *"To every action there is always opposed an equal reaction; or the mutual actions of the two bodies upon each other are always equal and directed to contrary parts."* This law states that when one object or body exerts a force on another object or body, the second object exerts an equal force in the opposite direction against the first object. This law is simply the **action-reaction law**, and applies to objects in all situations, whether the objects are in motion in a uniform velocity or accelerating, or stationary.

In Newton's own words, "I deduced that the forces which keep the planets in their orbs must [be] reciprocally as the squares of their distances from the centers about which they revolve: and thereby compared the force requisite to keep the Moon in her Orb with the force of gravity at the surface of the Earth; and found them answer pretty nearly." (Newton, 1687) The equation is the following:

$$F = G \frac{m_1 m_2}{r^2} \quad \text{or} \quad F = G (\{m_1 * m_2\} / \{r^2\})$$

where F is the force, m_1 and m_2 are the masses of the objects interacting, r is the distance between the centers of the masses and G is the gravitational constant (6.673×10^{-11} N (m/kg)^2). In Newton's equations, the objects m_1 and m_2 are considered **point masses**, as if all their gravitational masses were concentrated at the center of the objects themselves[34]. Newton's theory enjoyed its greatest success when astrophysics used it to predict the existence of Neptune based on motions of Uranus that could not be accounted for by the actions of the other planets. From Newton's equation, we derive the **escape speed** V_e that an object must achieve in order escape the gravitational pull of the larger body M, such as a planet, where it will not fall back onto M or go into orbit around it. It is written as V_e equals the square root of $2 \times G \times M / R$ or $V_e = \sqrt{(2GM/R)}$ where G is the gravitational constant, M is the mass of the larger object, and R is the radius of the larger object. Escape speed refers to an object overcoming the larger object's gravitational pull in any direction without self-propulsion or engines, like a cannon ball fired from a mountaintop into space. Escape velocity refers to an object with a certain trajectory leaving and escaping the planet's gravitational pull.

A discrepancy in Mercury's orbit pointed out flaws in Newton's theory. By the end of the 19th century, it was known that its orbit showed slight perturbations that could not be accounted for entirely under Newton's theory, but all searches for another perturbing body

---

[34] The term "point mass" refers to how Newton treats objects in his gravitational equation $F=Gm_1m_2/r^2$, or more specifically as if all the mass of each object was concentrated at its center point. We will correct this later to mean that "point mass" refers not to a point but to the Schwarzschild radius of each object or a group of objects where it the mass of the object or objects seems to be concentrated and the center of this radius is the average of the center of "gravity." For example, the Sun, Mercury, Venus, Earth combined together affect the orbit of Mars.

(such as a planet orbiting the Sun even closer than Mercury) had been fruitless. We know that Mercury's orbit is elliptical, and that this ellipse long axis slowly rotates around the sun, a precession of perihelion totaling 5600 arc-seconds per century. Physicists calculate that about 5030 of this is due to classical behavior or gravitational pull between the Sun and the planet Mercury. Another 530 arc seconds are due to gravitational interaction between Mercury and the other planets. This leaves about 40 to 43 arc seconds of orbital deviation per century unaccounted. The issue was resolved in 1915 by Albert Einstein's new theory of general relativity, which accounted for the small discrepancy in Mercury's orbit. Although Newton's theory, according to academia, has been superseded by the Einstein's general and special relativity, most modern non-relativistic gravitational calculations are still made using the Newton's theory because it is simpler to work with and it gives sufficiently accurate results for most applications involving sufficiently small masses, speeds and energies. Why is this? The Einstein equations are the mathematical embodiment of this idea of curvature of space around massive objects. Given the initial positions and velocities of all bodies, the solution to Einstein equation predicts their future relative positions and velocities of the same bodies. In the limit where the energies are not too large and when the velocities are significantly below the speed of light c, the predictions of Einstein's equations are indistinguishable from those obtained using Newton's theory[35]. However, according to academia, at excessive speeds and/or energy levels a significant deviation occurs, and Einstein's theory, not Newton's, describes the observations or predictions we expect to find. Despite the success of Einstein's equations, they do not solve the anomaly of excessive orbital speeds of stars around galaxies.

As depicted here, Newton's gravitational theory works well for objects outside the radius of m_1 and m_2. If m_1 enters m_2, then equations now show that m_1 is only pulled gravitationally by the inner mass m'_2 with radius r_1 and goes to zero as r_1 goes to zero. If we are to ignore the pressure exerted

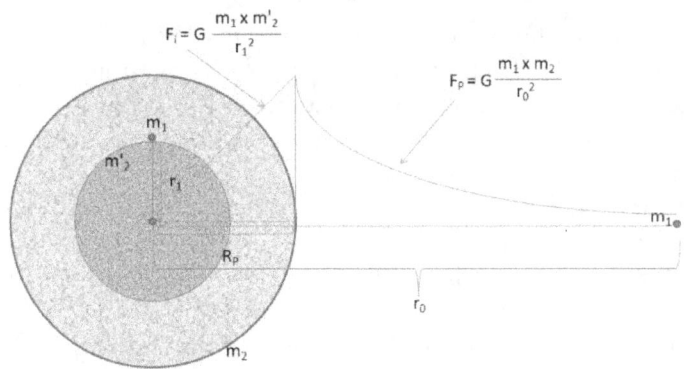

by the mass above m_1, then the force of gravity on m_1 decreases as the length of r_1 decreases. The chart above depicts the strength of gravity inside and outside mass m_2. As we have said, the above equations on the left side shows only the effect of gravity on m_1 and ignores the pressures of the mass above m_1 as it moves closer to the center of m_2. A scuba diver can surely attest that the pressure of water above him/her increases as they descend into the depths of the ocean. Most scuba divers measure the increased pressure in terms of atmospheric pressure or mercury pressure. Using this technique, scientists predict similar pressures within the Sun. These extremely high pressures and temperatures are responsible for fusion of hydrogen into helium. In the next section, we will look at how to quantify the pressure exerted by the mass above an object inside m_2.

---

[35] The Cosmological Balance Theory finds that sweet balance between Newton's Principia and that of Einstein's General Relativity work within a simple unified theory.

2.2 Stars and Their Light

Humankind has always looked up to the stars. Some have even consulted them for predictions of future events hence we have astrology readings. We all have that inner longing to find out if we are or are not alone in the universe. Today, you would have to leave the city area and go into the dark countryside to get a good look at the brilliance of the stars. Look into the night sky and you will see stars in all directions shining and twinkling in the dark. What is the light that we are seeing, and how does it get all the way from the distant stars to here? Why do some stars appear to twinkle or fade in and out even in a clear moonless night, appear to be fuzzy and some appear to be clear and crisp?

Stars are vast nuclear balls of hot plasma held together by gravity. They start their lives primarily made up of mostly hydrogen and helium, with trace amounts of other elements[36]. Mutual gravity holds the star together and compresses it inward toward the center. Without some push back force, the stars would just compress themselves down to the size of Venus, the Earth, or even smaller. However, as gravity causes the star to compress smaller, the friction between molecules causes it to heat up in its core. When the core of the star reaches about 10 to 15 million Kelvin degrees under sufficient pressures, hydrogen fusion begins. In this process, stellar pressures, and temperatures fuses hydrogen atoms together through a multi-stage process to form helium and heavier elements. This reaction in turn produces more heat and energy, which pushes outward and expands the star, and thereby releases high-energy photons. This balance supports and holds the star's spherical shape. A star like the Sun releases 3.86×10^{26} joules of gamma radiation every second.

These photons of energy trapped inside the star eventually find their way out. Over a journey that can take more than 100,000 years, atoms within the star continuously emit and absorb these photons until they break the surface. Each of these jumps can cause the photon to lose some energy. When they finally reach the surface of the star, they have lost a tremendous amount of energy to the numerous atoms they encountered and have fallen from high-energy gamma rays down to the visible wavelengths. The star releases these photoemissions in all wavelengths of the electromagnetic spectrum, affected by the number of atoms the photons encountered on their way to the surface[37]. Typically, this is how science books portray an electromagnetic wave:

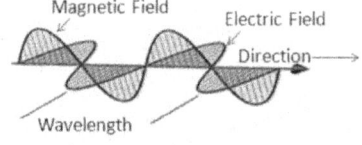

Electromagnetic Wave

Once the photons reach the surface of the star, they are henceforth released, and free to traverse the vacuum of space. Photons continue to lose energy with every molecule it encounters along the way to include those in the solar atmosphere, particles in space, and those within the planet's atmosphere. Photons travel through space with a companion

---

[36] Stars are primarily made of the most abundance ordinary matter elements in the universe, hydrogen and helium and trace amounts of other elements, some of which are remnants of previous stars. Mutual gravity alone takes millions if not billions of years to pull enough mass together to ignite into a star.

[37] Photon energy loss does not stop at the surface of the star. Photons escaping from the star's surface continue to lose additional energy as they pass through molecules in the solar atmosphere, climb out of the star's gravitational field, and pass through the curvature of space on its way outward toward the planets and open space.

electron in a continuous electromagnetic pulse and move at the speed of light. Unless they encounter something, they will keep traveling in a straight line for millions and even billions of years. To the photon itself, time does not pass as it traverses the great distance, as envisioned by Einstein. Large gravitational bodies will bend the photons' path as they travel past them. When you step outside in a dark area and look up at a distant star that could be a few hundred light-years away, your eyes are the first things these photons have bumped into and ended their journey since they left the surface of that star! This is an amazing experience for the first timer or avid stargazer.

One might ask. How does a star really "burn" fuel and give off light? To answer this question, we must look at the natural forces within the star, which is a delicate structural balance of gravity and nuclear explosions. Now, let us apply Newton's gravitational equations to see how pressure and temperatures build up within a star, namely the sun, as an example. Image depicts the two methods[38]:

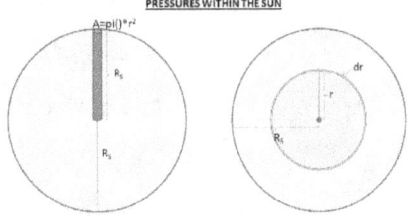

In the left image of the sun in the figure above, we see a cylinder drawn with an area $A = \pi r^2$ with length R_S from surface of the sun to center of the sun. The Volume of that cylinder is $V = \pi r^2 \times R_S$. Assuming the sun has a constant homogenous density ϱ and Newton's law of gravity $F = G \times m \times M_S / R_S^2$. Where gravity constant $G = 6.673 \times 10^{-11}$ N (m/kg)², the mass of the sun $M_S = 1.9891 \times 10^{30}$ kg, and the mass of hydrogen $m_a = 1.7 \times 10^{-27}$ kg, density of sun $\varrho = 1410$ kg/m³, the radius of the sun $R\odot = 695,800$ km, and set r=1.1 then the area $A = \pi r^2 = 3.801327111$ m.

The equation below roughly estimates the pressure p(r) at the center of the sun:

$p(r) = \varrho \times G \times M\odot / R\odot$

$= 6.673 \times 10^{-11}$ N (m/kg)² $\times 1410$ kg/m³ $\times 1.9891 \times 10^{30}$ kg/ 695,800,000m

$= 2.68975E+14$ N/m²

However, the equation below is more accurate.

Using Newton's equation, the pressure within the cylinder near the center of the sun is approximately,

$F = G \times (\varrho \times A \times R_S) \times M_S / R_S^2 = G \times (\varrho \times A) \times M_S / R_S$

$= 6.673 \times 10^{-11}$ N(m/kg)² $\times 1410$ kg/m³ $\times 3.801327111$m $\times 1.9891 \times 10^{30}$ kg/695,800km

$= 1.02246E+15$ N/m²

We can also look at the right side of the image in the figure above at the force of gravity on a layer of the sun of thickness "dr" at radius r with mass m caused by the inner sphere of mass M we get:

---

[38] Note that scientists do not use Einstein's complex equations to calculate pressures and temperatures within a star such as our sun. Newton's equations work sufficiently accurate enough even to predict what happens within a neutron star, and possibly within a black hole.

$m = \varrho \times V = \varrho \times 4\pi r^2 \times dr$

$M = \varrho \times (4/3) \pi r^3$

We get:

$P(r) = (4/3) \pi \varrho^2 * G *$ integral from r to R_S of r^2 dr $= G *(4/3) \pi p^2 (R_S^2 - r^2)$

$$P(r) = G *(4/3) \pi p^2 \int_r^{R_S} r^2 \, dr = G *(4/3) \pi p^2 (R_S^2 - r^2)$$

So, the pressure at $p(0) = G *(4/3) \pi p^2 (R_S^2 - 0^2)$,

And with $M = \varrho \times V = \varrho (4/3) \pi R^3$

Applying force $F = G \times M \times m / R^2$, we get the same rough pressure as before:

$P = \varrho \times G \times M\odot / R\odot = 2.68975\text{E}+14 \text{ N/m}^2$

Now let us look at the temperature within the sun. Using Boltzmann constant $k = 1.4 \times 10^{-23}$ J/K, the temperature within the sun can be approximated with the formula:

$T = G * m_a * M\odot / (k * R\odot)$

where m_a is the mass of a hydrogen atom (1.7×10^{-27} kg) we get,

$= 6.673 \times 10^{-11}$ N(m/kg)^2 x(1.7×10^{-27} kg)x1.9891×10^{30} kg/(1.4×10^{-23} J/K x 695,800 km)

$= 23164034.52$ K $= 2.3 \times 10^7$ Kelvin Degrees

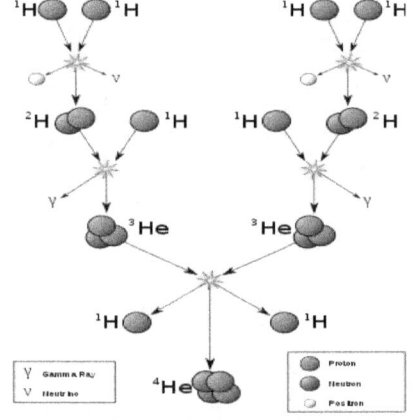

Under such extremely high pressures and temperatures within the sun, hydrogen fuses into helium through the following steps: First, two hydrogen atoms ^{1}H fuses together into one ^{2}H atom. Second, another ^{1}H fuses to the ^{2}H atom to form one helium ^{3}He. Third, two helium ^{3}He are fused together to form one helium ^{4}He and in doing so releases two hydrogen ^{1}H atoms. For each step of the fusion process, a photon is either emitted or absorbed to convert atoms. Each time the photon is absorbed it loses some of its energy and reemitted with lower energy levels. Scientists sometimes refer to this process as hydrogen burning.

Large stars have greater mass, greater pressures, and higher temperatures than the sun and therefore fuse hydrogen to helium and heavier elements like oxygen, carbon, nitrogen, etc. These stars go through a carbon-nitrogen-oxygen fusion cycle. Some of these massive stars can even fuse these heavier elements like silicon into much heavier ones like iron. At that point, iron starts to absorb the high energy levels given off by fusion of lighter elements and chokes the star on the way to its end, reaches a critical stage and go super nova, and collapse into a neutron star, or black hole[39].

---

[39] Many new elements are created during that last few seconds of fusion as the star releases tremendous amounts of energy and the star explodes in a massive supernova. Elements such as gold,

2.3 Black Hole Attributes and Radiation

"John Michell, *wrote a paper in 1783 in the Philosophical Transactions of the Royal Society of London*, pointing out that a star that was sufficiently massive and compact would have such a strong gravitational force that light could not escape." Hawking goes on to say that, "such objects are what we call black holes, because that is what they are—black voids in space (Hawking, The Theory of Everything: The Origin and Fate of the Universe, 2002, p. 46)." Today, we know that a black hole is far from being void. A black hole is the remnant of a super massive star, above the Chandrasekhar limit, that went supernova. In mathematical terms, scientists define it as a region of space-time with such a strong gravitational pull that no particle or electromagnetic radiation that passes the event horizon can escape from it, not even light. The Einstein theory of general relativity predicts that a mass sufficiently compressed, can deform space-time. That mass is the black hole. According to Einstein's general relativity, black holes are singularities where all its matter is crushed into one point, enormously massive, and yet tiny (Greene, The Fabric of the Cosmos: Space, Time, and the Texture of Reality, 2004, p. 337).

Depicted here is the before and after picture of a super massive star before going supernova, and an image of what it looks like after a black hole is formed per the Einstein fabric of space-time.

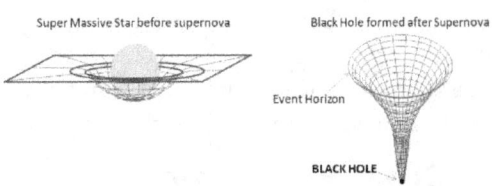

The event horizon of the black hole is the edge of the region where no escape is possible due to the gravitational tidal forces there. Crossing the event horizon will tear apart the object entering it. Scientists nicknamed this stretching action "Spaghettified." Since no light escapes or is reflected from the black hole, it appears completely black. Per Einstein gravitational time dilation, the deeper one looks into the black hole the further back in time one sees. In other words, a newly created black hole should remain fixed in space-time, while its event horizon moves with time through space, creating a trail like a wormhole tunnel. Quantum field theory in curved space-time predicts that area emits the "Hawking radiation" at a temperature inversely proportional to its mass, in the order of billionths of a Kelvin degree for black holes of stellar mass. It is so small that it is practically impossible to detect and observe.

Scientists expect the collapse of very massive stars at the end of their life cycle.to form black holes of stellar mass. After a black hole has formed, it can continue to grow by absorbing mass from its surroundings. By absorbing other stars and merging with other black holes, super-massive black holes of millions of solar masses (M☉) may form. Cosmologists generally agree that super-massive black holes exist in the centers of most galaxies. An unknown artist's view of a black hole:

Despite its black center, which appears invisible when compared to the blackness of space, scientists infer the black hole's presence through its interaction with other

---

lead, and alike that are stable stay intact. While unstable radioactive elements decay into more stable elements.

surrounding matter and with the electromagnetic radiation such as light, gamma, or the Hawking radiation. Scientists expect the matter falling onto a black hole to tear apart as it nears the event horizon and can form an accretion disk heated by friction[40]. This disk is visible as light, infrared, and other electromagnetic radiation. If there are other stars orbiting a black hole, their orbit can be used to determine its mass and location. In this way, astronomers have identified numerous stellar black hole candidates in binary systems and established that the core of the Milky Way contains a super-massive black hole of about 4.3 million M$\odot$ or about 4.3 million times the solar mass of the sun.

Physical insight into the Hawking radiation process may be gained by imagining that particle-antiparticle radiation is emitted from just beyond the event horizon. This radiation does not come directly from the black hole itself, but rather is a result of virtual particles being "boosted" by the black hole's gravitation as one is pulled in and the other ejected outward. By this process, the black hole loses mass, and, to an outside observer, it appears that the black hole has just emitted a particle. According to Hawking theory and predictions, a black hole of one solar mass (M$\odot$) has a temperature of only 60 Nano-Kelvin (60 billionths of a Kelvin); in fact, such a black hole would absorb far more cosmic microwave background radiation than it emits. A black hole of 4.5×10^{22} kg (about the mass of the Moon) would be in equilibrium at 2.7 Kelvin, absorbing as much radiation as it emits. Smaller primordial black holes would emit more than they absorb and lose mass.

Black Hole Radiation

According to Stephen Hawking, "Nothing lasts forever, not even black holes. Black holes will evaporate over vast periods of time. But exactly, how does this happen?" Hawking radiation is a radiation that is predicted to be released by black holes, due to quantum effects near the event horizon. What this means is that when two gravitationally bound orbiting particles exists at the event horizon, one can fall inward while the other flings outward into space. Over vast periods of time, the theory states that this trickle of ejected particles makes the black hole evaporate. Moreover, since its emitting energy and gamma radiation, the black hole needs to give up a little bit of its mass to provide it; otherwise, it would be violating the law of thermodynamics. Scientists named it after the physicist Stephen Hawking, who developed the theoretical argument of its existence in 1974. Hawking goes on to say that the rate of evaporation of a black hole depends on the mass of the black hole. For stellar mass black holes, it might take 10^{67} years to evaporate completely. For super massive black holes, typically at the center of galaxies, it could take 10^{100} years to evaporate. These are exorbitant long periods of time, but it is less than infinity. Black hole will continue to evaporate even when the universe is cold and desolate. Per Hawking, these black holes will fade into energy too weak and cold to spawn matter.

Hawking's work followed his visit to Moscow in 1973 where the Soviet scientists Yakov Zeldovich and Alexei Starobinsky showed him that according to the quantum mechanical uncertainty principle, rotating black holes should create and emit particles. Hawking radiation predicts the loss of the mass and the energy of the black hole, known as evaporation. Hawking stated that, "As the black hole loses mass, the area of its event

---

[40] Objects gravitationally attracted and caught in a black hole event horizon are predicted to collide with each other and heat up to the point of melting and "rip" apart into a stream of atoms, molecules, and even particles, which are either pulled into the black hole or flung outward as Hawking "radiation."

horizon gets smaller, but this decrease in the entropy of the black hole is more than compensated for by the entropy of the emitted radiation, so the second law (of thermodynamics) is never violated (Hawking, The Theory of Everything: The Origin and Fate of the Universe, 2002, p. 84)." Because of this, scientists expect smaller black holes that lose more mass than they gain through other means to shrink and ultimately vanish. Science predicts micro black holes to shrink and dissipate faster than ones that are more massive. Hawking also explains that rotating black holes can create jets of particles along the axis of rotation. "Matter falling into such a super massive black hole would provide the only source of power great enough to explain the enormous amounts of energy that these objects are emitting. As the matter spirals into the black hole, it would make the black hole rotate in the same direction, causing it to develop a magnetic field rather like that of the earth. In-falling matter near the black hole generates very high-energy particles. The magnetic field would be so strong that it could focus these particles into jets ejected outward along the axis of rotation of the black hole, that is, in the direction of its north and south poles (Hawking, A Brief History of Time, The Illustrated, 1996, pp. 125-6)."

This explanation needs a slight clarification. Because the black hole deep below the event horizon is rapidly spinning, it causes the particles of heavier matter (such as molten iron) to follow the same rapid spin direction around the black hole event horizon and thereby create a super intensive magnetic field. The black hole itself does not create the magnetic field for if it did the field force, which travels at the speed of light, would not be able to escape the gravitational attraction force of the black hole instead, form just above the event horizon. It is the matter, consisting of heavier elements like iron, trapped at the edge of the event horizon circling the black hole, which generates that magnetic field. Moreover, that magnetic field is what ejects the particles of matter near those magnetic poles outward in jets[41]. Let us consider the extreme potential of Hawking radiation. The maximum amount of Hawking radiation occurs when an entire star falls at a steep angle and wanders too close into the black hole's event horizon. In this instance, the black hole either pulls each particle or atom from the star into itself or flings it outward. We can easily imagine half of the mass of the star that was wandering too close to the event horizon of a black hole dragged in quickly into the black hole, while sending almost half of its mass flung outward in a massive ejection from the event horizon, but not from the black hole itself. Therefore, we can also expect large amounts of x-ray radiation to accompany such event-horizon solar mass ejections. As these ejected particles move outward and gain distance from the black hole the reconsolidate to form a smaller but massive object themselves. This ejection would appear as if numerous feeder streams of matter consolidated into a massive object -above or below the black hole- was being ejected and escaping from the black hole itself, but actually came from the event horizon only. NASA observed and recorded this event on March 15, 2016, from a black hole 500 billion of light years away.

---

[41] Hawking's radiation only touches the sheer power of the outer edges of the event horizon. The black hole itself has much greater forces stirring within it. The Cosmological Balance Theory explores those forces.

2.4 General and Special Relativity

In his General Relativity Theory, Einstein attributes the effects of gravity to space-time curvature instead of a force. The starting point for general relativity is the equivalence principle, which equates free fall with inertial motion and describes free-falling inertial objects as accelerating, relative to non-moving observers on the ground. In Newtonian physics, no such acceleration occurs unless a force is influencing at least one of the objects.

According to Brian Greene, "The two theories, Special and General Relativity, are among humankind's most precious achievements, and with them Einstein toppled Newton's concept of reality (Greene, The Fabric of the Cosmos: Space, Time, and the Texture of Reality, 2004, p. 10)." Einstein proposed that matter curves space-time, and that free-falling objects move along locally straight paths in curved space-time. Einstein called these straight paths, geodesics, analogous to the routes jet airplanes follow every day over planet earth. Like Newton's first law of motion, Einstein's theory states that if a force were applied on an object, it would deviate from a geodesic. For instance, we are no longer following geodesics while standing because the mechanical resistance of the Earth exerts an upward force on us, and we are non-inertial on the ground as a result. This explains why moving along the geodesics in space-time is considered inertial. Below is Einstein's equation, where space $G_{\mu\nu}$ tells ordinary matter $T_{\mu\nu}$ what to do and vice versa:

$$G_{\mu\nu} = \frac{8\pi G}{c^4} T_{\mu\nu} \quad \text{which was derived from} \quad R_{\mu\nu} - \frac{1}{2} R g_{\mu\nu} = \frac{8\pi G}{c^4} T_{\mu\nu}.$$

Einstein discovered the field equations of general relativity, which relate the presence of matter and the curvature of space-time and science named it after him. The Einstein field equations are a set of 10 simultaneous, non-linear, differential equations[42]. The solutions of the field equations are the components of the metric tensor of space-time, described as the geometry of space-time from which geodesic paths are calculated.

Einstein's 10 Field Equations

$T_1 = -a^2(\delta\rho + 2\rho A)$,

$T_2 = a^2(\delta\rho + 2\rho H_T)$,

$T_3 = \frac{a^2}{\sqrt{2}}[(\rho + p)v^{(-1)} - \rho B^{(-1)}]$,

$T_4 = \frac{a^2}{\sqrt{2}}[(\rho + p)v^{(0)} - \rho B^{(0)}]$,

$T_5 = \frac{a^2}{2\sqrt{2}}[(\rho + p)v^{(+1)} - \rho B^{(+1)}]$,

$T_6 = \sqrt{\frac{3}{8}} a^2 p(\pi^{(-2)} + 2H_T^{(-2)})$,

$T_7 = \frac{a^2 p}{2\sqrt{2}} (\pi^{(-1)} + 2H_T^{(-1)})$,

$T_8 = \frac{a^2 p}{3} (\pi^{(0)} + 2H_T^{(0)})$,

$T_9 = \frac{a^2 p}{2\sqrt{2}} (\pi^{(+1)} + 2H_T^{(-1)})$,

$T_{10} = \sqrt{\frac{3}{8}} a^2 p(\pi^{(-2)} + 2H_T^{(-2)})$.

ρ = The energy density of the flow of t-momentum in the t-direction.
P_x = The 'pressure in the x direction' of the flow of x-momentum in the x-direction.
P_y = The 'pressure in the y direction' of the flow of y-momentum in the y-direction.
P_z = The 'pressure in the z direction' of the flow of z-momentum in the z-direction.

Summarized as: $\left.\frac{\ddot{V}}{V}\right|_{t=0} = -\frac{1}{2}(\rho + P_x + P_y + P_z)$

Einstein's Field Equation is:
$G_{\mu\nu} = 8\pi G/c^4 \; T_{\mu\nu} = R_{\mu\nu} - \frac{1}{2} g_{\mu\nu} R$

where G, Newton's gravitational constant is 6.673 x 10^{-11} N (m/kg)² and $T_{\mu\nu}$ is the stress-energy tensor of a manifold, and the speed of light in a vacuum is about 299,792,458 meters per second (m/s). When Edmond Hubble declared that the universe was expanding, Einstein equation immediately explained the increased distances between galaxies resulted from the expansion of space itself, identifying that the further out they looked the more space existed between galaxies, like a balloon expanding. He did this by showing that general relativity accounts for all the "swelling" of symmetrical space from any location. Space is flexible and rubbery. According to Einstein, the

---

[42] It took Einstein several years to solve these tensor equations. He was simply "lost in the tensors" because he would not let go of a few basic assumptions he made when he started. Once he refuted those baseline assumptions, he was able to complete General Relativity Theory.

expansion of distant galaxy systems at the outer edge our telescope's view is due to the stretching of space itself. As such, general relativity says that time in one part of the universe with less expansion runs at a different rate that time in the part of the universe experiencing more expansion. General Relativity also states that for time on all equally expanding surface of balloon all clocks run at the same rate. In addition, Einstein did not initially address what such space swelling does to the star systems within a galaxy, namely what happens to the solar system, the sun, the earth, and all the atoms and molecules within each of us. This would be an unimaginable demise where gravity fails, so he corrects the situation by clarifying that space only expands where there are large regions of emptiness, otherwise gravity wins. The fact that Einstein believed space can expand and curve implies that space has some form or another type of matter and mass, since nothingness cannot expand nor curve.

Relativity has pondered many scientists for generations. Newton posed the question of relativity in the form of a bucket of water hanging from a twisted rope, when allowed to spin, moves with the water flat at first and then becomes concaved as it catches up with the spinning bucket. The question is why does the water concave on the walls of the bucket? Newton refers to *space itself*, where "he proposed that the transparent, empty arena in which we are all immersed and within which all motions take place exists as a real, physical entity, which he calls *absolute space*." Per Newton, an object is truly accelerating when it is accelerating with respect to *absolute space* (Greene, The Fabric of the Cosmos: Space, Time, and the Texture of Reality, 2004, p. 27). Space itself provides the true reference for defining motion. In Newton's words, "Absolute space, in its own nature, without reference to anything external, remains always similar and unmovable." We will find out later in this book how accurate Newton was. However, Einstein thought not.

Gravitational red shift was the first test of GR. Einstein used the Doppler Effect equation is $f_o = f_e (1 - v/c)$, where f_o is the frequency the observer away from the source records, f_e is the emitted frequency, c is the speed of light, and v is the speed of the light reaching the observer. For observers moving away from objects the $(1 - v/c)$ is less than 1 generates red shift, however, for $(1 + v/c)$ greater than 1 equals blue shifted light for objects coming toward the observer. With the Doppler Effect equation, the gravity field shortens the frequency of light moving upward through it. Photons lose energy as they climb out of gravitational field, and their frequency shortens, or reduces thereby, according to Einstein, we have gravity time dilation at the surface of the massive object. Atmospheric molecules and particles surrounding the massive object also absorb the light's photon energy and slow down its escape speed v. Even the curvature of space affects the escaping light's speed. To an observer at a distance away from the massive object emitting the light, time appears to run slower on the surface of the massive object due to the reduced frequency than at the observer location. Gravity time dilation prediction, where time runs slower in the greater gravity fields, is different from the one provided in Special Relativity where two observers traveling at a constant speed both see each other's time running slower than their own. However, another observer at an obtuse angle sees a redder shifted light in a shorter frequency coming from the surface of the massive object and deduces a higher degree of time dilation, an even slower clock. How can two time-dilations show different results? Further investigation questions its validity.

As an example, Einstein convinced academia first and then the scientific community to accept the constancy of the speed of light. Brian Greene highlights that mounting evidence from a variety of experiments dating back to 1880's supposedly confirms Einstein's prediction. Light speed remains the same regardless of the source of the photons or the specific motion of the observer. This claim is not based on realistic experiments conducted with actual observers retreating extremely fast away from the light source, or observers going towards the light at anywhere near the speed of light. Why because today's technology is insufficient to effectively conduct it. Einstein's simultaneity mental experiment contradicts this constancy light speed prediction[43]. Yet, the claim still stands. Mainstream science states the constancy of the speed of light to be true based on years and years of earth-based observations of distant stars in fast motion. Case closed according to Green (Greene, The Elegant Universe: Superstrings, Hidden Dimensions, and the Quest for the Ultimate Theory, 2003, p. 33). It is a by far the most significant paradox for Einstein uses it to stand on to topple Newton's longstanding gravity equations and Newtonian physics. For most readers, this property as presented is too hard to swallow. It goes against simple and straightforward Galilean relativity. Note that Einstein's simultaneity explanation is a direct application of Galilean relativity reinforced with the physics of light. However, if we look at Einstein's explanation of the constancy of the speed of light as seen from any observer, we find that it is in direct opposition to Galilean relativity. Can both be right? We will reevaluate light constancy speed premise later.

Although Einstein had excellent grades in mathematics and physics, he graduated far from the top of his class and was therefore unable to find work as a German physicist. After about a year of unemployment, the patent office in Bern, Switzerland finally hired Einstein as an assistant examiner patent clerk where he researched and kept up with the latest and greatest of developments within the physics field. From time to time, he wrote notes and stored them in his patent office desk drawer, presumably, ideas he may have pulled from or dreamed of as he reviewed patent documents, and then compiled them into his theory papers. There in Bern, he and his first wife Mileva Marić, who also was a fellow student and physics degree graduate at Zurich Polytechnic, together developed the Special and General Theories of Relativity (SR&GR); she ensured the papers were as error free as possible. The SR paper of 1905, originally titled "On the Electrodynamics of Moving Bodies", used Galileo Theory of Relativity, describing motion of items in a hull of a large ship sailing the oceans as the same whether stationary or moving at a constant speed. Einstein expanded on this and combined it with electromagnetism and light properties for a SR, operating in the absence of gravity, which insists that the laws of physics are independent of bodies traveling at constant speeds and motions of any observer. Einstein viewed time dilation and saw the results of the Lorentz contractions.

---

[43] Einstein's simultaneity mental experiment on an imaginary train confirms Galilean relativity when both equal distant observers on the train see the light flash simultaneously. Einstein then describes an outside observer on the platform. This observer disregards that the fact that the source of light is also moving along with the occupants of the train and makes the claim that the occupant in the back of the train sees the light a fraction of a second just before the occupant sitting toward the front of the moving train. So, if the platform observer considers and measures the movement of source of light and the movement of train occupants, then he or she might deduce simultaneity of the light flash reaching both occupants.

Special Relativity addresses how observers moving relative to each other measure events in space-time[44]. With the help of his fellow classmate Marcel Grossmann from Zurich Polytechnic and the Italian mathematician Tullio Levi-Civita, Einstein was able to publish The General Theory of Relativity paper of 1916, titled "The Foundations of the General Theory of Relativity," which considers other situations not addressed by Special Relativity. The acceptance and publication of these papers and others, enabled him to go back to attend the university as a graduate student, propelled Einstein into newfound status as a well renowned physicist, earn his PhD, and eventual employment in the field as a University Professor. After nearly a century of experimentation by leading physicists, they have concluded that any and all observers will agree that light travels at approximately 300 million meters per second or about 670 million miles per hour, regardless of the use of comparison benchmarks (Greene, The Elegant Universe: Superstrings, Hidden Dimensions, and the Quest for the Ultimate Theory, 2003, p. 31).

Many people have a mistaken impression of Einstein. Some believed he was just some physicists that only worked out of a patent office, with no formal academic training, who happen to develop two accepted papers, which revolutionized physics. This view of Einstein gives people false hope to do the same. But that myth of Einstein is incomplete, those who study his work and life are aware of his successful studies at graduate level physics during his time. His counter-intuitive ideas were actual extensions of other theories before him on which academia recognized its significance and published them worldwide. Albert Einstein himself said, "Two things are infinite: the universe and human stupidity; and I'm not sure about the universe." He pulled the wool over the eyes of humanity, blinded us with his eloquent equations, which hardly anyone outside academia uses, as they are just too complex, and deemed unnecessary for most situations. It seems that only physics graduates today may be the only ones trying to use them, by simplifying these equations, some set "c" to one, some use others shortcut work instead. Those who understand how his equations work will wholeheartedly attest to the brilliancy of the math. However, most just return to using Newtonian equations to solve most gravitational situations. See simulating extreme space-time website for additional insights (SXS, 2016) on simulations of black holes and Einstein's equations. Shan Gao too questioned and validated how Einstein came up with the postulate of the constancy of light in the Special Theory of Relativity in his book, Understanding Relativity. Gao investigated Relativity without light and eventually came up with conclusion that any and all observations of electromagnetic transmissions physically travelled at the fixed speed equivalent to the speed of light, c, as a universal constant of nature or an invariant speed as Einstein put it (Gao, Understanding Relativity: An Advanced Guide for the Perplexed, 2014). He used the minimal observable interval of space and time (MOIST) postulate to extrapolate the maximum signal speed c as invariant in every inertial frame. Gao's insights inadvertently exposed some implications where the constancy of light adjusts as it transitions between systems. Bottom line, Newtonian gravity is sufficient to explain all.

---

[44] Einstein's Special Relativity Theory paper relies heavily on the physics of light, primarily the Doppler Effect, where he extrapolates in mental experiments what observers would be seeing if they travelled near or close to the speed of light. In these experiments, he imagined objects lengthening or stretching, being distorted, and develops the math to prove his theory. Cosmological Balance investigates errors in SR & GR.

2.5 String Theory and Others

In string theory, string theorists predict microscopic strings that work together in either closed or open configurations form particles. The movement of these strings and their vibrations create cylindrical shapes, drawn to represent the interaction between particles. This theory eliminates the problems associated with point particles, describes how these strings propagate through space, and inherently connects freely with the interaction of gravity, on all particles. Brian Greene defines the strings in string theory as ultramicroscopic loops of energy (Greene, The Elegant Universe: Superstrings, Hidden Dimensions, and the Quest for the Ultimate Theory, 2003). The issue with String Theory that causes the most problem is its prediction that space should have at least ten or eleven dimensions, and exotic structures, a multi-verse concept. The updated version of string theory is superstring theory. In superstring theory, no subatomic particle is represented by a dot; particles are visualized as a tiny filament of energy or microscopic string. Like string theory, superstring theory also requires nine spatial dimensions and a time dimension, with the unique vibrations of each string determining the particle's properties (Greene, The Fabric of the Cosmos: Space, Time, and the Texture of Reality, 2004, p. 17). It is not a simple theory. Nonetheless, it is considered among mainstream physicists and cosmologists as the best contender for *Quantum Gravity* theory.

Another theory developed, similar to string theory but predicting loops is called *Loop Quantum Gravity* or Quantum Space-time, which presents dispositions to combine quantum mechanics with general relativity through the "quantization" of Einstein's field equations, with predictions on elementary particle properties. This theory visualizes space as an extremely fine fabric woven by a network of loops. This theory has morphed into something like string theory, which theorists present as the "theory of everything." Greene states that superstring theory, a refinement to the original string theory, might successfully merge the two, general relativity and quantum mechanics. Schwarz believed that he had uncovered the first plausible method to describe gravity in terms of quantum mechanical language using super thin strings, predicting the graviton via string (Greene, The Fabric of the Cosmos: Space, Time, and the Texture of Reality, 2004, p. 339).

In string theory, a particular vibration of a string or super thin strand of matter yields a particular particle. One pattern of vibration is detected as an electron, and another a quark, and another a gluon, and so forth for every type of particle in existence. These unique vibration patterns determine charge, mass, spins, and other characteristics of particles. The faster the vibrating energy the more massive the particle becomes. Excessive vibrating energy at some point causes the string itself to lengthen. Strings vibrate matter, as we know it, into existence, like an orchestra-playing symphony. With that said, string theory also limits how violent each jitter of the gravitational field becomes, this limit is just big enough to avoid catastrophic clash between quantum mechanics and general relativity. It is through the other dimensions within string theory that this gravitational force influences the universe. String lengths are no longer than Planck length and cannot be divided in half, while point matter particles have no specified or known size. With that said, no one has seen a string, and if string theory is right, no one will ever see one. This type of theory is more in the realm of philosophy and theology than science and physics. However, indirect evidence provides it some promise.

Superstring theory requires that the universe consist of nine dimensions, not three as we know it, plus the dimension of time makes ten. Brian Greene explains that the hidden dimensions were predicted by Kaluza-Klein theory, where each axis of the three original dimensions had an extremely tight clockwise and a counterclockwise direction avail to it making nine dimensions total. This number is a calculation derived within string theory; it is not an assumption, hypothesis, or educated guess. Five separate theories have emerged during the development of basic string theory, which became a challenge when presenting it as a unified theory until Edward Witten tied all five together into one, under M-theory. Witten created a common baseline showing through mathematical analysis that each of the five where interpretations of the m-theory, like one book printed in five separate languages (Greene, The Fabric of the Cosmos: Space, Time, and the Texture of Reality, 2004, p. 379). In addition, string theory contains other ingredients such as membranes, which are considered a bit more massive than the strings themselves.

In fact, both theories are based on the internal structure of sub-atomic particles and consequently the structure of everything in the visible universe, and then claim to be fulfill the sought after "theory of everything." Why describe particles in strings or loops? For many, it is too hard to see the connection between what goes on within and around a particle influencing the entire universe. Such motions and energy only provide the source of energy for everything else, without directly affecting how things necessarily function in the macro sense of the world and the universe. However, neither of these two theories is gaining the momentum and success they hoped to achieve among mainstream scientists, since neither properly addressed their supposed foundation of quantum mechanics[45]. For this book, we will reinvestigate this topic in Part 3 Chapter 7, after which no further discussion of them mentioned.

Again, closely related to structural patterns in string theory is the less known quantum Wave Structure of Matter theory. (Wolff, 2008, p. 29) Milo explains that standing spherical waves from space density generate all particles within the universe, and as such, he set forth the following eight conclusions:

1. Particles, natural laws, and the universe are an inter-dependent trilogy.

2. Measurement is a property of an ensemble (of at least five separate particles) of matter.

3. Particle properties require (a two-way perceptive or an instantaneous) communication between particles. Einstein labeled the means as the 'photon,' but no one knows exactly what it is.

4. There exists a means of continual (an electric force) communication between particles; he called this mechanism the scalar quantum-wave.

5. For each particle, a universe (assuming it is not infinite) is defined as the space and other particles within that space, which are able to communicate with the particle.

6. The measurement of time requires a cosmological clock (as proposed by DeBroglie – a cosmic clock or universal oscillator).

---

[45] Quantum mechanics provides valuable insights to solving the mysteries of the universe. String theory does not do it justice. Particularly, quantum mechanics exposes errors in general relativity.

7. The role of space (the medium of communication between particles provide time, length, and mass. Einstein's General Theory of Relativity (GTR), space affects matter, which in turn influences space. The properties and laws of quantum particles in space apply Einstein's GTR to the micro universe as well as to of the macro universe.

8. Definition of the finite universe: The radius R of the universe is equal to light speed divided by the lifetime T of the matter and space inside the communication range of its waves, this gives us $R = c/T$. This definition assumes the Big Bang theory is correct and that T is universal. It also assumes that no new matter is created continually, otherwise for different ages of T, we get different radius R, and different total number of particles in the universe. Hubble tells us the universe is expanding and that R is increasing exponentially with time.

Doppler Effect result of relative motion ($\lambda = h/mv$) alters the interactions of in and out waves $[1-(v/c)^2]^{-1/2}$ of particles (resonance) and redefines the momentum and the mass of these particles. (Wolff, 2008, p. 38) Milo states "Schrodinger's equation and the (natural) laws are derived from the quantum wave structure of the universe." Therefore, by simply replacing the notion of point particles with a *spherical wave structures* his book reveals a universe of quantum wave structures. Milo declares that the wave structure of the basic element of matter (electron) replace the old notion of non-existent discrete "particles." Despite all Wolff's reasoning and claims, he still openly admits there may be errors in this new theory and asks the help of his readers to identify those to him for correction.

(Wolff, 2008, p. 48) "If two waves have identical amplitudes, the reinforce each other. If two waves have opposite amplitudes, they cancel, and the waves annihilate each other. This explains why charges attract or repel each other." In a word, this is resonance, and the concept of anti-matter is just a form of matter that has opposite phases of the spherical waves of the particle resonances.

(Wolff, 2008, p. 59) "If two electrons are near one another, their identical waves add together producing maximum amplitude causing them to move apart seeking a minimum. If one is a positron, their waves will cancel each other producing a minimum amplitude they will be decreased as they move together." This WSM definition says that electrons always repel each other to seek minimum amplitude.

2.6 Inexplicable: Dark Matter, Dark Energy

"Why should things be easy to understand?"

— Thomas Pynchon

Neil deGrasse Tyson is an astrophysicist and one of television spokespersons for the scientific community. He explains the field of cosmology and physics in nonprofessional terms. Tyson has a show that talks about some of the biggest unanswered questions in science right now, as well as reviews a bit of history of previous mysteries that have since been resolved (Tyson, The Inexplicable Universe: Unsolved Mysteries, 2014). Vera Rubin claimed in the Scientific American "as much as 90 percent of the matter in the universe is invisible. She came to this conclusion when she discovered that the stars on the edge of rotating galaxy orbit at speeds similar or equal to stars nearer to the center of that galaxy, namely the Andromeda Galaxy. Vera then speculated that the galaxy resides inside a sphere of dark matter causing the stars to move faster than anticipated by gravity alone[46]. Detecting this dark matter will help astronomers better comprehend the universe's destiny." To put it plainly, our ignorance about dark matter's properties has become inextricably tangled up with other outstanding issues in cosmology and astrophysics. Issues like, how much total mass does the universe contain? How do galaxies really form? Is the universe destined to expand forever? Clearly, our unraveling of this dark matter issue, can bring light to our understanding of the size, shape, and fate of the universe. Mainstream physicists believe that that search could dominate astronomy for decades.

In Wave Structure of Matter (WSM) theory introduced in section 1.2, Milo Wolff declares space density standing wave energy produces matter in the universe, and their resulting out-waves affect local matter. (Wolff, 2008, p. 43) Milo restates WSM Principle II above in these terms, "The matter of the universe combines to tell the space medium what it is and in turn the medium tells all matter how to behave." If you really think about this, this definition tells us that the space medium cannot exist without the presence of all matter in the universe (or space is absolutely flat), and the absence of the density of the space medium leads to the non-existence of all matter, a feedback loop. In this definition, all matter must exist first before space, a contradiction to the basis of WSM, where in- and out-waves create matter. To explain this predicament, Milo goes on to say, "All matter in the universe tells space how to curve and in turn space tells matter how it must move." This too is another feedback loop. Space has no curves when matter is absent (flat space), and flatness of space tells any accidental quantum particles of matter (within itself) to do nothing (not to move); hence never form into matter.

So, however you look at it, all matter in the universe generates energy out-waves which flow toward all other matter as in-waves, and these in- and out-wave interactions tell all the matter in the universe how to behave and move, and space is the medium in which all matter and its quantum subatomic particles behaves and moves. The movement of particles creates out waves, which in turn enter other particles as in-waves that influence their behavior and movement and provide feedback to originating moving particle. Space

---

[46] Vera Rubin's prediction of dark matter and dark energy surrounding the galaxy like a halo provides one possible reason for the excessive speed stars orbit, but it does not fully explain why they all orbit in unison. Outer stars move in the same line as inner stars around the galaxy.

is the medium for these waves to move. (Wolff, 2008, p. 50) Milo inadvertently agrees, "Without particles to populate a universe, the universe could not exist because our concept of universe is a collection of particles... and understanding of the particles." The presence of particles leads to natural laws dominating the universe in a trilogy relationship: particles, laws, and the universe.

Dark Matter: According to mainstream science, dark matter is a hypothetical kind of matter that cannot be seen with telescopes but accounts for most of the matter in the visible universe. "We know that this matter is dark, in the sense that it does not interact significantly with radiation, both because we don't see it, and because it has not lost its kinetic energy sufficiently to relax into the disks of galaxies as has baryonic (nuclei and electron) matter. These particles are electrically neutral. Detailed studies of the dynamics of galaxy clusters indicate that the dark matter particles must also be cold, in the sense that their velocities are highly non-relativistic (Weinberg, 2008, p. 186)." Dark matter neither emits nor absorbs light or any other electromagnetic radiation at any significant level[47]. According to the Planck mission team, and based on the standard model of cosmology, the total mass–energy of the known universe contains 4.9% ordinary matter, 26.8% dark matter and 68.3% dark energy. Thus, dark matter is estimated to constitute 84.5% of the total matter in the universe, while dark energy plus dark matter constitutes 95.1% of the total mass–energy of the universe. Extensive studies and simulations have been done to estimate the distribution of "Dark Matter" (Wang, Walker, & Pal, 2008).

Astrophysicists hypothesized dark matter because of noted discrepancies between the mass of large astronomical objects determined from their gravitational effects and the mass calculated from the observable matter (stars, gas, and dust) that they can be seen to contain. Dark matter postulated originally by Jan Oort in 1932, albeit based upon flawed or inadequate evidence, to account for the orbital velocities of stars in the Milky Way and by Fritz Zwicky in 1933 to account for evidence of "missing mass" in the orbital velocities of galaxies in clusters. The first to postulate dark matter based upon robust evidence was Vera Rubin in the 1960s–1970s, using galaxy rotation curves. Many other observations have shown the presence of dark matter in the universe, including gravitational lensing of background objects by galaxy clusters such as the Bullet Cluster, the temperature distribution of hot gas in galaxies and clusters of galaxies, and the pattern of anisotropies in the cosmic microwave background. Consensus among cosmologists, dark matter is composed primarily of a not yet characterized type of subatomic particle. Physicists are still searching for such particle.

Dark Energy: In physical cosmology and astronomy, dark energy is an unknown form of energy that permeates all of space and tends to exert on and accelerate the expansion of the visible universe. Dark energy is the most accepted hypothesis to explain numerous observations since the 1990s showing that the universe is expanding at an accelerating rate. The unexplained movement that "dark energy" causes is called "dark flow."[48] According to the Planck mission team, and based on the standard model of cosmology, on a mass–

---

[47] If "dark matter" does not absorb photon energy of light, then it must reflect, deflect, or let it pass through without changing its color or frequency.

[48] Dark energy is a phrase scientist use to describe an unknown force within the universe that somehow pushes objects and galaxies. Cosmological Balance refers to this unknown force as antigravity force.

energy equivalence basis, the observable universe contains approximately 26.8% dark matter, about 68.3% dark energy (for an estimated total of 95.1%) and 4.9% ordinary matter. Again, on a mass–energy equivalence basis, the density of dark energy (6.91 × 10^{-27} kg/m^3) is incredibly low, much less than the density of ordinary matter or estimated dark matter within galaxies. Yet, it dominates the mass–energy of the universe since it is uniformly spread across space.

Additionally, high-precision measurements of the expansion of the universe are still required to understand fully how the expansion rate changes over time and space. In the theory of general relativity, the evolution of the expansion rate is parameterized by the cosmological equation of state (matter, energy, and vacuum energy density for any region of space). 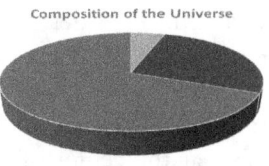 Measuring the equation of state for dark energy is one of the biggest efforts in observational cosmology today. "To take into account the possibility that the dark energy density is not constant, it has become conventional to analyze observations in terms of its pressure/density ratio $\varrho_{DE}/\varrho_{DE} = w$. Except in the case of a constant vacuum energy density, for which $w = -1$, there is no special reason why w should be time-independent (Weinberg, 2008, p. 55)." However, Shan Gao and a few other physicists have a different approach to dark energy. Understanding its origins is one of the most fundamental quests in cosmology and physics. Although many models proposed explain the cosmic acceleration, it remains a mystery. Gao highlights the holographic dark energy (HDE), which applies principles of quantum gravity and employs an event horizon for the entire universe, as the most promising alternative to the standard cosmological constant model. The HDE suggest, "The entire universe is viewed as a two-dimensional information structure 'painted' on the cosmological horizon (Gao, Dark Energy: From Einstein's Biggest Blunder to the Holographic Universe, 2014, p. Loc 285)," which was inspired by black hole thermodynamics. Gao and other physicists argued that space-time as a dynamical entity has quantum fluctuations, which contribute to vacuum energy, and hence may be the origin of dark energy. Dark energy comes from quantum fluctuations of (not in) space-time, therefore, produces a foamy consistency of space-time in small scales. The slowing of time predicted by Albert Einstein's general theory of relativity can also provide an alternate explanation of dark energy. Time dilation in response to gravity is directional in that matter in extreme gravity force will have slower clocks than an object in low gravity environments. The Einstein's theory of special relativity, on the other hand, describes reciprocal time dilation between two moving objects, where both moving objects' experience time slowing down relative to one another[49]. The effects of time predicted as faster in the past would thereby make the plotting of supernovas become rather linear. This implies that there is no rapid or increasing acceleration in the expansion of the universe. In this scenario, there is no reason to say that "dark energy" must exist. Hence, with Hubble peering further into the past, it may not be what exists now.

---

[49] Einstein predicted time dilation in both Special Relativity and General Relativity Theories. Both explanations are faulty in that they both describe outside observers detecting the viewed source clock slowing down. It does not matter what outside observer sees, the source clock speed remains the same on the surface of the massive object moving or not. Viewing a mirage does not make the image real.

2.7 White Hole Attributes

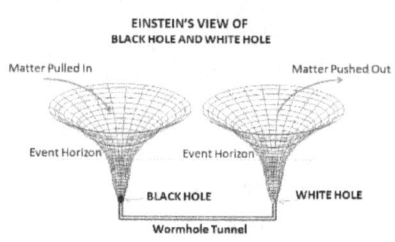

As previously discussed, a black hole is something so massive that pulls all matter from a place you do not want to get near, more specifically the event horizon. Nothing traveling past that point can escape not even light or electromagnetic radiation. Besides $E=mc^2$ and his Relativity equations, Einstein proved through complex mathematics, the existence of black holes. With today's advances in technology, we now have been able to find hundreds of these black holes and believe one massive black hole exists at the center of the Milky Way galaxy as well as in every other visible galaxy in the universe. What is astonishing; however, is what Einstein also proved through his mathematic equations: white holes also exist. Einstein declared them the exact opposite of black holes; he predicted white holes to "spit out" an incredible amount of matter from seemingly nothing[50]. White holes should be easy to find per that definition, yet none have been. If scientists found one, it may help us explain other unknown mysteries, such as where all the material come from that made the galaxies. Einstein introduced the idea of antigravity when he added the cosmological constant.

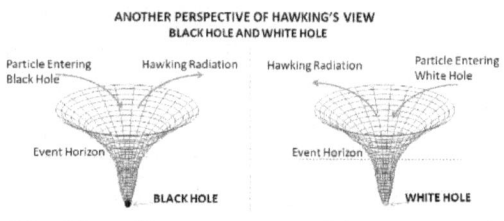

A white hole literally pushes matter away by excreting material outward. Einstein theory shows that a white hole is the reverse of a black hole; nothing can enter it, not even light. To him, a black and white hole connect via wormhole tunnel. The white hole's event horizon is a point where no object can travel or move past, despite how much fuel or energy you used. Steven Hawking's view is slightly different. Using quantum mechanics, he demonstrated that a black hole emits Hawking radiation and given sufficient time comes to thermal equilibrium and remains unchanged in the reversal of time. Hence, the reverse of a BH in thermal equilibrium is a BH in thermal equilibrium; meaning that a hole, black or white, is the same thing. Hawking's view of both.

Now let us draw Hawking's view of the black hole and the white hole without a wormhole connecting to the two massive holes together. The Hawking radiation expels about half of the particles that enter the black or white hole. We would get something like Einstein's view, except that the white hole per Hawking allows some particles to enter it and a type of Hawking radiation to leave[51]. Remember these two concepts, as we will elaborate and draw on a combination of both.

---

[50] Einstein's prediction of white holes spitting out matter from seemingly nothing is close to correct in that it describes antigravity ejection. To make it more accurate, white holes are exactly like black holes except it has a negative gravity (antigravity). Black hole ejects trillions upon trillions of micro white hole orbs and white holes eject trillions upon trillions of micro black hole orbs, both through wormhole tunnels which collapse 100,000 light years from the edge of their galaxies.

[51] Hawking's view of black and white holes is closer to Cosmological Balance Theory's perception. They both have Hawking radiations "evaporate", and "excretion", micro-orbs in wormhole tunnels.

Chapter 3: Gravity and the Milky Way Galaxy

"The universe is a pretty big place. If it's just us, seems like an awful waste of space."

— Carl Sagan, Contact

As we discussed earlier, a galaxy is a gravitationally bound system consisting of stars, stellar remnants, interstellar gas and dust, and matter not emitting or reflecting light. The word galaxy is derived from the Greek galaxias (γαλαξίας), literally meaning "milky", reference the Milky Way. Galaxies come in a range of different sizes from dwarfs with just a few thousand (10^3) stars to giants with one hundred trillion stars, each orbiting their galaxy's center of mass. Galaxies are usually categorized according to visual morphology: elliptical, spiral, irregular, and starburst. Many large galaxies are considered to have black holes at their center[52]. The Milky Way's central black hole, known as Sagittarius A*, has a mass four million times that of the Sun.

https://www.nasa.gov/mission_pages/galex/pia15416.html

There are approximately 170 billion (1.7×10^{11}) galaxies in the observable visible universe and counting. Most are 1,000 to 100,000 parsecs in diameter and usually separated by distances on the order of millions of parsecs (or megaparsecs). The space between galaxies is filled with a tenuous gas with an average density less than one atom per cubic meter. Most visible galaxies gravitationally bound into galaxy groups and clusters form in turn usually larger super-clusters. At the largest scale of the known visible universe, these associations arrange into sheets and filaments, which we perceive immense voids surround. Amazingly, these individual visible galaxies are also filled with about 99.9999% of void empty space. In other words, the galaxy's matter only occupies about one-millionth of the space that holds it. Yet, they have enough gravity to hold together. How is this possible?

Let us take into consideration the galaxy and its gravitational influences on its stars. Cosmologists still do not understand why the stars on the spiral arms of the Milky Way Galaxy move faster than predicted by influence of the gravitational force from the black hole and massive stars at the center of the galaxy. The spiral arms appear to move in unison like the rubber tire moves with the wheel it is on. Some of the stars pass others and then fall behind as they move up above and below the galactic plane. Some scientist state that there must be more dark matter beyond the edge of the galaxy rim to cause this gravitational anomaly.

---

[52] The universe filled with billions of galaxies is homogenous and uniform throughout. Galaxies twelve billion light years distance from us are just as developed as those in our local region are. Indicating that inhabitants on those galaxies are able to look outward at the rest of the universe away from our galaxy and see the same great homogenous spread in all directions. This characteristic alludes to a much older universe.

Some cosmologists believe that "density waves" push and hold together the spiral arms of galaxies. These density waves compress gases together and enable them to form first and second-generation stars and planets. Similarly, these scientists are unable to explain exactly what these density waves are and of what do they consist. Their explanations appear to be like stellar winds generated by supernovas or other stellar mass explosions. Since super massive stars much larger than the sun burn their fuels much quicker than the sun, they must have been recently born by these "density waves" in order to be shining today especially if they are in relatively close proximity to us. The star Betelgeuse is one such example. How did such a star recently get that much hydrogen in one place after 13 billion years of existence in the galaxy? Did it start as a star system consisting of various large, several smaller stars like the sun, and some even smaller stars orbiting a slightly larger star at the center, which eventually pulled in and consumed its orbiting stars as it grew in mass? On the other hand, did space or gravity just push together that much hydrogen all at once? In any case, the larger a star gets the more capable it becomes to gravitational influence its neighboring or local stars. Colliding and merging of two or more stars creates a more massive star with a much shorter lifespan than the originating feeder stars that dies with a more devastating end and leaves behind some type of stellar remnant.

Gravity binds all stars and other matter within these galaxies. Gravity plays an important part in holding the visible universe together or at least trying to slow the expansion of the universe. Sir Isaac Newton believed the force of gravity to be infinite and instantaneous. Newtonian gravitational equations work well when time is not a factor. Einstein in his Specific and General Theory of Relativity, further discussed in Part 3 Chapter 7, developed and solved the space-time continuum dilemma he imagined, primarily using Mercury's orbit discrepancy and the measured bending of starlight around the sun to justify his theory, a quandary worth reinvestigating later in this book[53]. It is time to affect a paradigm shift, starting with a discussion on the age of galaxies and from where they came.

---

[53] Principle of Consistency demands that any theory replacing another must be just as good and be able to fix errors of the theory it replaced. Einstein's Special Relativity Theory introduced speed time dilation, which did not exist in any previous theories. And Einstein's General Relativity Theory leaned on Special Relativity to solve energy-momentum tensors and introduced gravity time dilation. Time dilation, a feature of the physics of light or more specifically a mirage, is a micro universe or quantum misinterpretation error mistakenly carried over into the macro universe, the physics of the universe, gravity. Light does not control gravity.

CHAPTER 3: GRAVITY AND OUR MILKY WAY GALAXY

3.1 Galaxy Development Timeline

Since astrophysicists have been able to explain how stars and planets are born. We see that galaxies are not scattered randomly throughout space, but rather found in clusters, known as "super clusters", which are somewhat gravitationally bound. Scientists have two main theories to attempt to explain these galaxy formations. First, the gas left over from the big bang, if it occurred, grouped together to form galaxies with billions of stars. Second is that the gaseous material from the big bang created stars and planets, and they somehow migrated by mutual gravity into galaxies. Neither theory is universally accepted yet. More importantly than where do galaxies come from is the question of where the all the massive amounts of gaseous particles and atoms come from of which provided the building blocks for these galaxies. Did this material emerge from energy converting to matter? Did this gaseous material always existed since the dawn of time but could not pull together to form orbs, proto stars, and eventually stars, as discussed in the previous chapter? Did the universe start from a state of nothing, or did it always have energy and matter scattered uniformly like sheets and filaments? The latter is more feasible than the first. Why is this? Because energy cannot be created from absolute, empty cold dark nothing (outer space with a density of absolutely zero atoms, particles, or sub-atomic particles). We know, however, that the space is not empty. Matter exists in space. Scientists call it "dark matter" and the energy associated with it is "dark energy."

"Grand unified theories do not include the force of gravity. This does not concern us too much because gravity is such a weak force that we usually neglect its effects when we are dealing with elementary particles or atoms. However, the fact that it is both long-range and always attractive means that its effects all add up. Therefore, for a sufficiently large number of matter particles, gravitational forces can dominate over all other forces. Therefore, gravity determines the evolution of the universe. Even for objects the size of stars, the attractive force of gravity can win over all the other forces and cause the star to collapse (Hawking, A Brief History of Time, The Illustrated, 1996, p. 103)." Per mainstream cosmologists, high concentrations of "dark energy" can create particles, which can consolidate to produce protons, neutrons, and electrons, and eventually hydrogen atoms and molecules. With the mutual attraction of gravity, these molecules can assemble into stellar cloud nebulas[54]. Galaxies come from these irregular spreads of energy in space, and the production of matter with gravity. Their statement of creation is only possible if their definition of "dark energy" includes the rapid motion of dark matter subatomic particles, which it does not. Note: this claim of "energy" creating new matter out of nothingness is contrary to the definition of energy, which we know is the motion of matter. In cosmology, there is a cosmological model due to Bondi and Gold, and a version to Hoyle known as the Steady State theory (Weinberg, 2008, p. 45). It was thought to have been generally discredited by the expanding universe model alternative to the Big Bang theory of the universe and its origin. In steady state views, new matter is created continuously to keep its density ϱ constant as the universe expands, thus adhering to the

---

[54] Clearly, galaxies are created from an assembly of regional local stars in mutual gravitationally bound, which are each created from an accumulation of large amounts of gases, primarily of hydrogen and helium, along with other elements and molecules. Therefore, a galaxy comes from an extremely large area of space containing enough gases to span stars and simultaneously attract each other to form a system of stars.

perfect cosmological principle. It stands on the principle that the observable universe is the same in any time as well as any place. In this model, nothing-physical changes with time, so the Hubble constant is constant. While the steady state model enjoyed some popularity in the mid-20th century, the vast majority of cosmologists, astrophysicists, and astronomers now reject it, as the observational cosmic microwave background evidence found points to a hot Big Bang cosmology with a finite age of the universe, which the Steady State theory does not predict. The theory in this book is different from the old Steady State Theory.

Is big bang infallible? No. Its three supporting tenants are the cosmic background microwave, the abundance of light elements, and the expansion of the universe. What would happen to the steady state theory concept if we found that the observed cosmic microwave background evidence was in fact from another source? What would happen if we learned that the universe was not expanding? What would happen if we found out that the source of the abundance of lighter elements were from another source? What would happen to the big bang theory if we uncovered new evidence of an older universe? These discoveries would discredit the big bang and shed new light, on a variation of the old steady state concept[55]. Let us tackle the last question, the true age of the universe.

Assembly Timeline of the Galaxy

In Chapter 2: Physical Law in Cosmology, we talked about gravitational properties, we generalized how stars are born and solar systems formed, discussed what happens to a star at the end of its lifecycle, and then we briefly reviewed the remnants these stars leave behind, to include black holes. We then discussed the gravitational influences of the galaxy, its orbiting stars, and reviewed Einstein's General Relativity Theory. Now imagine an entire universe filled with just tenuous gaseous matter with gravitation properties only. Let us suppose that there are no forces other than gravity itself and the natural forces in particles and atoms, in other words no outside added push to compress gaseous clouds together. This chapter will look at the timeline necessary to form a barred spiral galaxy like the Milky Way Galaxy and compare it to mainstream cosmologist's timeline. In order to do this properly, we will have to re-evaluate the timeline necessary for enough matter to consolidate to form an entire spiral or barred spiral galaxy the size of the current Milky Way Galaxy and compare it to the mainstream science 14-billion-year age of the universe[56]. In doing so, we will be analyzing the timeline necessary to create the first-generation stars, followed by the lifespan of those stars from birth to death by explosion, and reviewing what remnants are left behind. We will then analyze the creation of second-generation stars and their lifespan, and the remnants they leave behind. Stars form in various sizes, and therefore have various lengths of life spans. We need to look specifically at the first and second-generations stars that are large enough to produce, when they exploded, all the materials and periodic elements naturally found on earth. Finally, we will look at the timeline to create the current sun and the solar system we live in, all the neighboring stars in the local group, and the rest of the stars in the galaxy shining today.

---

[55] Big Bang has been under attack over the last few years and is losing creditability. Its demise could affect off shoots like the inflationary theory. Scientists are reconsidering alternate theories, like steady state theory.

[56] The 14-billion-year timeline was chosen because it exceeds that from the acceptance of the Big Bang Theory, which of course is now being questioned by some scientists themselves.

We will look at several possible scenarios for such a galaxy the size of the Milky Way Galaxy to form. Did the galaxy form as one complete spiral or barred spiral galaxy, or did it form from the merging of several smaller galaxies in stages of growth? We know that the Milky Way and Andromeda Galaxies are on a collision course that will take about 4 billion years to come together[57]. This merger will produce a much larger galaxy; over a billion-year process with the exact shape of resulting galaxy unknown. Cosmologists speculate it will look like an elliptical galaxy. We observe the visible universe filled with examples of the merging of two or more galaxies at various stages of the process. The visible universe is also teeming with galaxies of various sizes and shapes, and combinations of shapes. Scientists speculate that all matter in the visible universe somewhat uniformly dispersed in sheets and filaments, implies that there are pockets of denser packed atoms in space and pockets of less dense atoms in space. Denser packed space would have consolidated first before less dense space. All this evidence implies that the Milky Way Galaxy did not start out as one huge galaxy but could have started out as an extremely large area or volume of fairly denser number of atoms and molecules floating in space. These molecules eventually pulled itself together under mutual gravity to form stars of various sizes, which then consolidated into large grouping of stars or globular cluster galaxy. These globular cluster galaxies when close enough to each other gravitationally attracted each other over time, collided, and merged to form a slightly larger galaxy taking on a more rotational characteristic. In due course, it continued to merge with other larger nearby galaxies and eventually took shape as the current barred spiral galaxy our cosmologists conjecture the Milky Way Galaxy is today. We propose that this merging process to build a spiral or barred spiral galaxy the size of the Milky Way surely took more than 14 billion years. If we strictly look at trying to build a barred spiral galaxy the size of the Milky Way Galaxy from gaseous material supposedly scattered uniformly throughout deep open space, it would be very difficult to imagine how such amounts of matter was able to pull itself together without a great force, push, or some type of density wave to help. However, it is much easier to imagine that amount of matter coming together in slow steady deliberate increments discussed in the previous paragraph above. Much like a tree incrementally grows trunk rings outward over time, some rings are thick, and some are thin, depending on circumstances in its life span[58]. The question is how much time would it take for that amount of matter in the galaxy to pull itself together on its own from an area of space containing that same amount of gaseous material spread uniformly at one atom per cubic centimeter? If this happened everywhere, then why do we still have deep open space between galaxies with the same material consistency? Why did the gaseous materials in those current open space areas not consolidate under its own gravity as well? Cosmologists conjecture that the density of deep space 13.8 billion years ago was slightly more than it is today, say one to two atoms per cubic centimeter.

---

[57] Based on astronomers' acceleration readings and agreed by NASA's analysts, we have about 4-plus billion-year remaining timeline to merger of Andromeda and Milky Way.

[58] The analogy of tree rings works well when the galaxy moving in the direction parallel to the galactic plane. Here, the edge would encounter and pick up material scattered in space. If a galaxy moved in the direction of its axis, then the entire span of the galaxy would encounter matter in space and consume it evenly.

3.2 First Generation Stars

Let us start by building a first-generation star close enough to other stars of various sizes so that they will consolidate into a globular cluster galaxy one-twelfth the size of the Milky Way Galaxy from gaseous material supposedly scattered uniformly throughout deep open space containing an average density of one to two atoms per cubic centimeter. Speculate how much time such amount of matter was able to pull itself together without any outside push or some type of density wave to help consolidate the matter. The first-generation star should be large enough that when it reached the end of its lifespan it was burning the last of its fuel making iron just before it died in an explosion. The final explosion of this star releases a great burst of energy giving the star a jolt, which further produces other heavier elements like lead and gold, and the like.

According to <https://en.wikipedia.org/wiki/Milky_Way>, the mass of the Milky Way Galaxy is estimated at Mass of 0.8 to 1.5×10^{12} M☉, with approximate number of stars at 200 to 400 billion ($3 \times 10^{11} \pm 1 \times 10^{11}$) or higher. Scientists estimated the deep open space to contain roughly one to two atoms per cubic centimeter with density as great as 1500 atoms/cm^3 and as small as 0.1 atom/cm^3. The atoms in deep open space are primarily hydrogen and some helium, created from high-energy subatomic particles stabilizing into larger particles and then atoms. Based on the high end of density, how much volume of area matches the mass of the one-twelfth of the Milky Way Galaxy? Based on the low end of density, how much volume of area gives us one-twelfth the mass of the Galaxy? Finally, what area provides for an average density of one to two atoms per cubic centimeter? Using formula volume = mass divided by density, or v=m/ϱ. On the high-end density of ϱ = 1500 atoms/cm^3, and galactic mass m = (1.5 x 10^{12} M☉)/12 or 2.48638E+41 kg, this gives us volume v = 9.98219 x 10^{64} cm^3 or 9.98219 x 10^{49} km^3 for 1/12th of the galactic mass. The low-end density is gaseous material we observe between galaxies, so spread out that it does not consolidate into a galaxy but becomes feeder resources for galaxies moving through it.

Starting the process from a large concentration of dense gaseous matter about 1500 atoms/cm^3 occupying an area of 1-meter cube of uniformly dispersed atoms in deep open space, attracted gravitationally to another large concentration of gaseous matter of the same density about five meters apart, begins to come together without an outside force pushing on it. We find that these two gaseous clouds take hundreds of years to close out the distance of just five meters with most of the molecules and atoms completely passing right by each other in a dance without coagulating to form an orb. Cosmologists also calculate that the two mass clouds must have a mass greater than "Jeans mass" and spherical radius less than "Jeans radius" or one-half the Jeans length in order for the merged clouds to continue to collapse on its own self and possibly form an orb and maybe a proto star. The formula for Jeans Length is:

$$\lambda_J = \sqrt{\frac{15 k_B T}{4\pi G \mu \rho}},$$ where k_B is the Boltzmann's constant ($1.3806488 \times 10^{-23}$ m^2 kg / s^2 K), T is the temperature of the cloud, ϱ is the cloud's mass density, μ is the mass per particle in the cloud, G is the gravitational constant (6.67384×10^{-11} m^3/ kg^1 s^2), and r is the radius of the cloud. In a spherical area, density ϱ represented as M / r^3, where M is the total mass of the cloud. This gives us λ_J = r, when $k_B T$ = GMμ / r. Therefore, Jeans' Length is the cloud's radius when thermal energy per particle equals

gravitational work per particle. At the critical length, the cloud neither expands nor contracts, it remains constant. If the radius is more than the critical length, the cloud expands and cools down. If the radius is below the critical length, the cloud eventually contracts and warms up. In both cases, this process continues for quite some time until equilibrium is achieved. Finally, if several feeder clouds are massive enough to pull themselves together into a combined dense cloud, it will contract more rapidly, warm up quicker and could possibly ignite into a proto star, a process that could take eons or as short as one or more billions of years[59]. Let us assume that two separate huge clouds of gas were successful in gravitationally collapsing into orbs. After about one billion years, we have the two large obs forming within the same region of space and moving basically in the same direction. These two spherical orbs begin to heat up when their internal pressure rises as both of them accumulate more and more of the surrounding gases into their structure. Within an additional 100,000 years they both ignite within close proximity to each other, just barely close enough to begin mutual attraction. Each of the two stars begins to vary slightly in a merging course toward each other.

Let us just suppose that these two stars the mass of the sun, one could be slightly smaller and the other slightly larger, ignite and become gravitational attracted to each other. Because both stars formed from clouds in the in the same general area of space they will be most likely moving in the same direction. The movement of the stars toward each other will therefore be a slow and long process. Once close enough, the smaller star will gradually pick up more and more speed in the direction of the larger star, until they gravitationally entangle in an ever-decreasing spiraling dance toward each other. This process could take hundreds of thousands of years if not a billion depending on the original proximity of the two and the angle of decline of the decaying orbit. The resulting merger creates a more massive star, with a shorter lifespan, possibly the size of the first generations star we are looking for[60]. So, at this point in the vast darkness of the area to become the galaxy we have but a few stars pulling together gravitationally to form a small globular cluster surrounded by massive average dense clouds trying to compress unsuccessfully on its own. As little as two billion years to as much as 6 billion years in worst-case scenario have passed and still no galaxy resembling the Milky Way Galaxy. Three to four more billion years pass and the largest of the first-generation stars begin to fuse heavy elements into iron, and start an inward collapse, its death is eminent. Pressures build up within the large star and it goes super nova, spreading new vital heavy elements in every direction. At this point, this corner of the universe has but a few stars scattered in small globular cluster galaxy, six to ten billion years have passed, and still no sign of a large galaxy resembling the Milky Way. "The bottom-up picture is supported by the observation that the commonest galaxies are dwarf spheroids, and that the galaxy and the Andromeda Nebula M31 each have about 20 smaller satellite galaxies (Weinberg, 2008, p. 414)." At six to ten billion years, we have but small groups of twinkling globular clusters somewhat gravitationally attracted to each other and above average dense gaseous clouds floating in the vicinity. These gaseous clouds receive the shock wave from the first super nova in this region of the universe.

---

[59] Eons refers to an extremely long time closer to infinite.

[60] The merger of two stars could also trigger a massive supernova when critical mass is achieved upon merging and iron is fused within the new core.

3.3 Second Generation Stars

To continue the way to create the solar system the way it is today, the set of second-generation stars must also be large enough that when it explodes it produces elements heavier than iron. Starting where we left off, the resulting group of supernovae sends material and shockwaves toward a large semi-dense cloud that was not able to pull itself together on its own. When the shock waves hit the edge of that cloud, it compresses inward from two or three different directions and begins an inward collapse. This collapse sends the cloud past the critical Jeans' density. Its matter swirls around and dances for about a hundred thousand years or so and finally settles into a significantly large proto star, which then ignites as it reaches the critical temperature to fuse hydrogen into helium. This star's center is composed of the leftover remnants from the first-generation stars and thereby accumulates its mass quicker around the dense center[61].

By this time anywhere between six billion one hundred years to eleven billion years have passed, and we are just beginning to see new large globular cluster galaxies numbering thousands of stars coming to life. These gravitationally bound new cluster galaxies begin an orbital dance, vying for dominance. The two central more massive galaxies of these clusters come together in one billion years into a spiral merger lasting three billion years. Their dominant massive stars collide and form a super-massive star, which becomes destine for a shorter lifespan of less than two to three billion years.

At this point, eleven billion to fifteen billion years have just passed, and still no sign of a galaxy that looks like the Milky Way. The super massive star created becomes the central hub of small elliptical galaxy, which is on a collision course with three globular cluster galaxies and another elliptical galaxy about 10,200 light years away. The collision and merger estimated to begin in three billion years will take another one to two billion years more to complete, about the same time that the super massive center star goes super nova. This time this super nova creates a large black hole of 150 solar masses, which is the beginning of the current super massive black hole at the center of the Milky Way Galaxy.

This sudden explosion, as the six small galaxies merge, sends shockwaves in every direction, causing stars to whiz past each other and in some cases collides to form larger stars. After another two to three billion years, the new proto–Milky Way Galaxy takes the shape of a medium-sized spiral galaxy two-thirds the current size.

---

[61] Denser or heavier elements and molecules from previous stars are able to consolidate more rapidly than lighter elements and provide a starter core for the surrounding gases to accumulate into a star.

To recount, at least fifteen billion to nineteen billion years have passed since the first star ignited, and the Milky Way Galaxy is finally starting to take shape. Its new size, mass, and gravity have taken hold of four other nearby globular cluster galaxies, now destine for a collision course, eminent to happen in another three billion years. All the while, the black hole at the center continues to grow by consuming nearby large stars prior to them going super novae. The black hole's mass at the center of the galaxy jumps to 2.5×10^4 solar masses. The black hole's voracious appetite enables it to consume more matter and grow rapidly. Therefore, any surviving orbiting stars assume faster and tighter orbits to overcome the black hole's ever-increasing gravity field. Time passes and the proto–Milky Way Galaxy then merges with the four other cluster galaxies, and now achieves the 9/10th's the full size it is today.

At this point, at least eighteen billion to twenty-two billion years have passed to create the Milky Way close to the size it is today. This massive new galaxy continues to absorb smaller globular cluster galaxies and grows closer and closer to its present size and mass[62]. On the horizon, the Milky Way Galaxy has now become gravitationally bound to and is on a collision course with another super massive spiral galaxy. Scientists expect this to happen in about nine billion years. Today, the distance between the two galaxies, where the second one, now named the Andromeda Galaxy, is estimated to be just fewer than five million light-years distance apart, slowly but steadily moves ever closer.

Other second-generation stars about two-thirds from the center of the new Milky Way Galaxy now go super novae and provide the material and debris to form a planetary nebula. This planetary nebula becomes the birthplace of the solar system. Shock waves then hit the dense cloud and after about one hundred thousand to two hundred thousand years, the solar system's proto star begins to take shape, five billion years ago, as conjectured by our cosmologists today.

If we take a step back and look at how the Milky Way Galaxy formed the oldest part of the galaxy are the center black hole and the material in its immediately surrounding vicinity. The youngest part of the galaxy is therefore form one-third to the outer edges. The stars that orbit the black hole at the center are sixth-, seventh-, or eighth-generation stars or later consisting of very heavy massive centers, expected to have short life spans. Their destructive deaths in the form of super novae will send super massive debris directly into the black hole at the center of the galaxy and continue to make it grow in mass.

---

[62] Roughly speaking our Milky Way galaxy could be much older than 18 to 27 billion years of age, which is surely a lot longer than the speculated 14-billion-year age estimated by the Big Bang Theory.

3.4 Timeline to Third or Fourth Generation Stars

At this point of our investigative analysis, we have calculated a timeline exceeding the one proposed by the big bang concept. In review, we have strictly looked at trying to build a barred spiral galaxy the size of the Milky Way Galaxy from gaseous material supposedly scattered uniformly throughout deep open space, without an outside force. It was difficult to imagine how such amounts of matter was able to pull itself together without a great force, push, or some type of density wave to help[63]. For all we know, eons passed long before anything even started to consolidate. The gaseous cloud could have been sitting idle for billions upon billions of years beforehand.

During our discussion, we have estimated that it took approximately twenty-three to twenty-seven billion years, in the best of circumstances, for that amount of matter in the galaxy to have pulled itself together on its own. This area of space only contained the same amount of gaseous material spread uniformly at one to two atoms per cubic centimeter, to as much as 1000 to 1500 atoms per cubic centimeter.

Again, according to https://en.wikipedia.org/wiki/Milky_Way, the mass of the Milky Way Galaxy is estimated at Mass of 0.8 to 1.5×10^{12} M$\odot$, with approximate number of stars at 200 to 400 billion (3×10^{11} ±1×10^{11}). Deep open space today is estimated to contain roughly 1 atom per cubic centimeter, but density as great as 1000 atoms/cm^3 and as small as 0.1 atom/cm^3. The atoms in deep open space, even today, are primarily hydrogen and some helium. We calculate with average density $\varrho = 1$ atoms/cm^3, and galactic mass $m = 1.5 \times 10^{12}$ M$\odot$, this gives us approximate volume v = 2.99466E+53, within dimensions like 8.385E+18 km x 8.385E+18 km x 4.257E+15 km. These dimensions in term of light years are approximately 886 thousand light years x 886 thousand light years x 450 light years. This area is over eight times the space the Milky Way Galaxy currently occupies. Clearly, the amount of time that amount of gaseous material surely took more than 23 billion years, if not more to form the Milky Way Galaxy.

MILKY WAY GALAXY

Not to mention that the structure of each galaxy is accumulated over time, it grows from the inner circle outward as it travels and moves through space, leaning heavily on the force of gravity to pull or push it from toward other existing galaxies. Mergers provide the material for small galaxies to grow into larger galaxies and then again into and become still larger ones. The more massive a galaxy becomes the more it attracts other smaller galaxies into it. The bottom-line up front is that the innermost part of a massive galaxy is usually where you would find the merger of large stars into huge super massive stars. Appearing to be young recently consolidated stars but consisting of material older than the rest of the

---

[63] The existence of density waves (or antigravity influence) helps speed up the consolidation of ordinary matter into spinning orbs capable of igniting into stars.

stars at the outer edge of the galaxy, like the age of rings on a redwood tree. To top this off, the center section of a black hole in the middle of a massive black hole is older than the layered components of the entire black hole. It formed at the very beginning of galaxy's formation.

Taking all of this into account, we can easily speculate that the black hole at the center of a super massive galaxy like Andromeda could conceivably be over 30 billion years old. This would make the fully formed galaxies at the 13 billion-light-year edge of the 29-year orbiting Hubble telescope view run anywhere from 45 to 100 billion years old, and more. Imagine that. Where does that leave us in respect to the Big Bang Theory? It takes us back to square one, no dice, and clearly not plausible. However, that is not going to stop the supporters of the big bang theory. Why stop there? The material in deep open space in the pre-matter pre-visible universe could have just been sitting idle for googol years or more before even starting to merge or clump together. Remember, we are talking about the smallest basic hydrogen atom, the atom containing the weakest amount of gravitational pull, and hence taking the longest to attract other hydrogen atoms, let alone achieve the Jeans critical mass to consolidate, clump up, and grow in sufficient quantities to form a protostar. Whom are we kidding when science says all this can happen within the timeline defined in the big bang theory?

INTERGALACTIC GASEOUS MATTER ABSORBS PHOTON ENERGY

Before we leave this subject, let us look at the density of the gaseous matter in deep space. This known density contributes to the dissipation at extreme distant light and its diffusion, combined with the loss of photons energy, results in red shift readings, which cosmologists attributed to the Doppler Effect and the expansion of the universe, and not the low-density gaseous material found dispersed in deep space. From the planet surface, the sun appears to us during the hours of dawn and dusk as more orange-red color than at midday. This effect is the result of three times more air molecules absorbing more photon energy at those two times of the day. We can expect the same type of effect to occur as we view further and further out into deep space especially at extremely distant galaxies. Large amounts of hydrogen and other gaseous matter although dispersed uniformly over great distances in deep space would absorb enough light photon energy from distant stars and galaxies to cause a red shift, and beyond a certain point, say 13.8 billion light years away block out all light. We will elaborate more on this red shift effect in Part 3 Chapter 7 when we discuss the perceived expansion of the universe.

3.5 The Solar System to its Current Age

In the previous two sections, we have argued that the timeline necessary for the composition of the earth and the solar system elements from the remnants of first, second, and/or third generation stars took much longer than eight billion years. We have re-examined each of those stars' lifespan from birth to expiration and the timeline for their debris to assemble into a solar system.

After a few more hundred thousand years, the proto star for the solar system, 4.8 billion years ago, now begins to enter its main-sequence phase, steadily burning hydrogen into helium. Among numerous forming small orbs and debris, all planets and planetoids begin to take shape, clear their orbital path, and push other neighboring planets toward the orbits they have today. The new sun burns cooler than today, allowing the smaller proto earth to cool more rapidly, and form oceans and an atmosphere at about 4.6 billion years ago. The proto earth at this point is wobbling in its orbit. Within a hundred thousand years more, the oceans begin to develop the first signs of early plant life[64]. Nevertheless, at 4.5 billion years ago scientists tell us, another planet the size of Mars collides with proto earth and creates a field of debris around the earth, which eventually consolidates into the moon. The collision erases all previous life on proto earth sending chunks of debris seeded with life elements into space. Due to the collision, the new earth spins faster at around six hours per rotation period and the new consolidated moon starts out its orbit close to the earth. In this collision, the proto earth was fortunate to retain its original prograde motion, west-to-east in relation to the stars, although at a more rapid speed starting at six hours per rotation period.

We however have a slight adjustment to the above scientific claim. The collision occurred much later, around 1 billion years ago. According to some scientists, the "Mars-size" planet Theia, created around 4.5 billion years ago, approached, and collided with proto-earth. Theia initially formed as a molten iron sphere at the location of the asteroid belt, was surrounded and bombarded by ice particles, which melted into oceans and rapidly cooled the surface of the iron orb causing it to develop a thick dense solid iron crust like an eggshell but only a few kilometers deep that sealed off its molten core. As Jupiter passed by, its gravity slingshot Ceres, a completely solid iron ball to its core but covered by ice and debris, into Theia and knocked it out of its orbital path toward Earth. Ceres' icy and debris crust and part of Theia's ice debris shell shattered into the now icy and rocky asteroids belt. Theia progressively degrading orbital path took it into an oblique angle collision course toward earth. Theia cushioned by the ice and rocky blanket around it collided with Earth and dug into proto earth's rocky surface. Theia's iron shell contacted proto earth's molten mantle. The friction between the two melted a hole in Theia's shell, spilled its core into Earth like the yolk of a cracked egg going into a bowl, while the remaining intact hard iron shell bounced into orbit where the left-over molten lava quickly

---

[64] The creation story follows both the scientific explanation of how our solar system evolved and confirms the explanation found within world religious stories, biblical writings, and Holy Scriptures in the book of Genius.

CHAPTER 3: GRAVITY AND OUR MILKY WAY GALAXY

patched the hole from within to restore the orb to its original spherical shape. The surrounding ice on Theia vaporized and the loose debris fell onto Earth's surface and onto the moon.

The colliding planetoid destroyed all early life (primarily plants) on proto-Earth, added one-half mass including its iron core into the new Earth, and kicked up enormous amount the debris; the loose debris, mostly dust and small rocks, eventually coalesced around the now hollow iron shell of the moon and became the relatively large moon. This hollow structured Moon explains why its craters are relatively shallow and why the Moon "rings" or vibrates like a bell when objects collide with it.

A planetoid collision sent the Earth into a rapid spin (scientists estimated 6 hours per day-night cycle). The presence of the newly acquired Moon incrementally slowed down the Earth's rotational spin over time while it gradually pushed the Moon upward into geosynchronous orbit around what became the Earth's equator where it remained for a few hundred million years. The Moon's 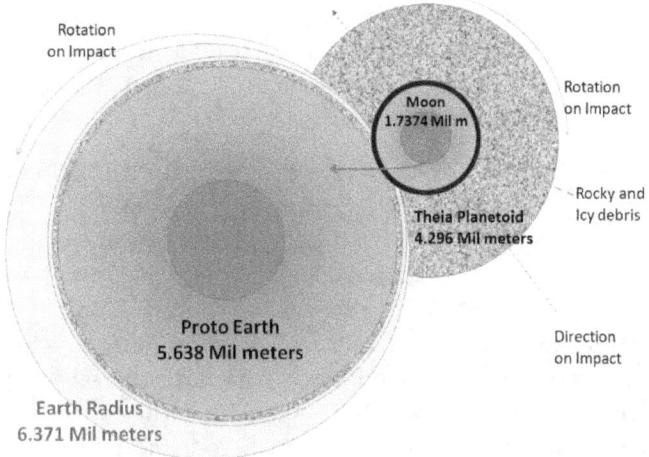 geosynchronous orbit steadily created one huge molten lava "massive mountain" wave on one side of the Earth, which locked onto the smaller lava wave in the shell of the moon and synchronized the moon's rotational spin to match its orbit; thus, the same side of the moon always faces the earth.

As the molten continent grew in height, the moon steadily moved upward in orbit and simultaneously slowed the Earth's rotational spin. Any water or gases within the hollow Moon would have fallen toward and settled onto that internal bulge closest to the side facing the Earth.

It would take the new larger earth another eight million years to slow down its rotation period, cool down enough to reform the oceans, develop an atmosphere, and for the second early life to re-emerge from it, this time creating oxygen as a byproduct. At around 800 million years ago, the Moon was in geosynchronous orbit with the Earth rotation estimated at 7.11-hour days. At 700 million years ago, new primitive life evolves to breathe this now abundant oxygen levels, and the earth's rotation period remained at 7.13 hours. At 200 million years ago, mammals and other complex creatures evolve, to include pre-dinosaur animals, the earth's rotation, due to low altitude tidal lock, slowed to 7.2 hours, and Pangaea began to split apart. At 65 million years ago, an asteroid hit the earth and plunges it into darkness and an ice age, killing off all the dinosaurs, and the earth's rotation period estimated at 7.58 hours and still locked with the Moon's orbit. At 12 million years ago, the Earth's rotation was 9.57 hours. At 6 million years ago, the Moon began to slip away from its geosynchronous state at 0.002-day orbit and the Earth's rotation rate slowed to 9.87 hours. At 4 million years ago, the Moon's orbit was 2.63 days, 3 million years ago at 11.33 days, and at 2 million years ago was 19.66 days. At 1 million years ago, Earth's day was 19.95 hours, and the Moon orbit was 25.35 days. At 700 thousand years ago, the Earth's rotation was 21.15 hours and the Moon's orbit was 26.36 days. At 600 thousand years ago, early Homo sapiens evolve, and the earth's rotation period gradually slows to 21.55 hours and the Moon was at 26.62 days. At 400 thousand years ago, the Earth's rotation was at 22.35 hours and the Moon's orbit at 27.02 days. At 200 thousand years ago, it was at 23.15 hours and Moon at 27.27 days.

At about 7 thousand years ago, the earth's rotation period begins to approach what it currently has today, 23.91 hours and the Moon at 27.35 days. As the moon moves more and more away from the earth, its gravitation influences have less and less effect in slowing down the earth's rotational speed. During this time frame a significant part of the earth's ocean was pushed over the land in a great flood, probably by a low angled colliding asteroid ricocheting off the earth's ocean surface waters back into space, like a pebble skimming across the pond. Such a grazed collision would leave behind no markers in the form of asteroid debris, but would be sufficient to kick up excessive amounts of water vapor into the atmosphere enough to rain for many days and nights, as well as send massive sunami waves up and over all the land masses around the world. Survivors documented this death by water event as the great floods. Note that the above creation scenario of the moon and proto earth, and the timeline, although speculative in nature, are the best guestimate.

About the time the mars sized planet collided with proto earth, the black hole at the center of the Milky Way Galaxy received a new string of feeder star materials from super novae and as a result sent debris and high quantities of mass into the black hole. The black hole's mass grows to 4.5×10^6 solar masses. This sudden consumption increase begins to change the structure of the Milky Way Galaxy into a barred-spiral galaxy the cosmologist conjecture it is today.

During earth's lifespan since its creation, it has undergone four life-changing events. First event, death of first early life by fire at one billion years ago as the moon formed. Second, death by air, more specifically suffocation by oxygen, when the new plant life produced and overwhelmed the ecosystem with tons of oxygen, which was poisonous to existing life then. Third, death by ice, an asteroid collision sent tons of debris into the atmosphere and triggered a long winter that caused the extinction of the dinosaurs. Fourth reset, death by water, when the global flood destroyed land life, an event documented by numerous cultures around the world.

The above solar system creation comparision analysis was based on scientific information available on the internet and simulated interaction between the moon and the earth during its sustained locked geosynchronous orbit. It is not factual, it is only speculative. Science used the current rate the moon is slipping away from the earth and

CHAPTER 3: GRAVITY AND OUR MILKY WAY GALAXY

rolled back in time to derive the impact that created the moon, approximately 4.5 billion years ago, assuming that this rate remained constant. Likewise, science used the clock decrement of 1.7 milliseconds per 100 years and deduced that the earth rotated at 6 hours per night-day cycle back then. Common sense and instincts tells us that this analysis is not totally accurate. Why is this? In rolling back the clock of time, one should consider that the closer the moon was to the earth the greater its tidal influences pushed the moon outward and likewise the moon's gravity slowed the earth's rotation more.

The closer the moon was to the earth the faster moon moved around the earth, or more specifically the shorter its orbital time, and the higher the ocean tides on the earth. So, as we reverse the clock from today into the past, we find that the collision, or partial collision that created the moon actually occurred sooner than 4.5 billion years, possibly as earlier as 6 million years ago to as late as 1 billion years ago. Such radical new insights demand an adjustment to the original theory of the creation of the moon. If we roll back 2000 years at 0.038 meter per year, we get the moon to be 2000 x 0.038 = 76 meters closer to the earth, but the moon is now orbiting 350 seconds faster. This additional increased acceleration translates to about 0.008098248 meter per year faster drift. Hence, 2000 years ago the moon was 76.04 meters closer to the earth. This acceleration moves up the timeline of collision generating the moon to about 2.1 million years ago. With this adjustment we now update the day-night rotation of the earth to 6 hours per day at the time of that collision.

Food for thought. Let us consider this same moon earth situation in another angle. In the lunar calendar before the development of the Gregorian Calendar, counting 13 orbits of the moon or moon-cycles was designated as one standard year, and one full day-night rotational cycle was declared as one earth day. Today a day is now seen as approximately 24 hours, and a year is approximately 365.2425 days. Since the 13-moon-cycles went faster or were of shorter-durations about 700 thousand years ago due to lower altitude lunar orbits, the first humans documented then to have lived several hundred years old is quite plausible and realistic per the timescale then to be perceived to outlive modern day man. A year today using the 13 moon-cycles of about 28 days is not the same as a "year" back then with 13 lunar cycles each at say 16-day orbits.

In other words, a year by today's standards, or one full orbit of the earth around the sun, was equivalent to about six or seven "lunar calendar years" back then when the first humans walked the earth; each "year" having 13 moon-cycles, where one lunar orbit around the earth took about 4 days to complete (using today's day) or 4.65 using the shorter day period back then, and a full earth orbit around the sun back then would have taken 423.47 days with each day around 20.7 hours long. So, a person living today to age 115 could have been declared to have lived to age 933, where 115 x 7 x 423.47/365.2425 = 933. As the moon moved to higher altitudes, the orbit period lengthen, and the documented human lifespan became shorter and shorter over time to what it is today.

At this current point in time in the creation of the solar system, today, at least twenty-three billion to twenty-seven billion years passed since the first gaseous clouds began to merge into the first star in the Milky Way Galaxy. The Milky Way is now almost at its full size today.

3.6 Rotational Speed of the Galaxy

We have just estimated the timeline necessary for the galaxy to assemble itself. However, missing in the discussion above is how that galaxy's stars can sustain excessive speeds as it orbits around the center of the galaxy, speeds greater than predicted by gravity force alone. Surely, the further out we go away from the mass of the black hole to the outer edge of the galaxy, the slower we expect the stars to orbit the center. However, Vera Rubin's observations of galaxies tell us otherwise[65].

Let us look at discussion in chapter 2.1 and use it in calculating the rotational period of each section of the galaxy by combining Kepler's third law ($T^2 = 4\pi^2/(GM) * a^3$) and the Schwarzschild radius ($R = 2GM/c^2$) black hole concept. We have already seen that the orbital period of planets within the solar system are determined by the length of its semi-major axis, a. We have also seen that gravity pulls matter, according to Newton, toward the center of mass of an object or more specifically at least toward the center of the Schwarzschild radius. We will also discuss later in Part 2 Chapter 5 of this trilogy that the matter within the Schwarzschild radius is homogenous, disperses more, and moves uniformly. This is a simplified depiction of the Solar System (not drawn to scale); the Schwarzschild radius increase is almost negligible with each planet's mass added.

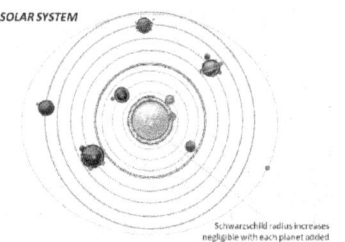

To analyze how the Schwarzschild radius affects rotational orbits let us re-examine the solar system in terms of Kepler's third law. The sun has a mass of 1.989E+30 kg and a radius of 6.963E+08 meters; it contains 99.87% of all the mass in the solar system and as such, we expect the outer planets to orbit more slowly than the inner planets, and observation confirms this expectation. We know that each planet's orbit around the sun affects the other planets adjacent to it. So, the motion of the Sun, Mercury, and Earth primarily affect Venus' orbit, and the Sun, Mercury, Venus, Mars and the Moon predominantly affect earth's orbit, and so forth. Schwarzschild radius formula of $R = 2 \times GM/c^2$ is used to calculate the radius of the same mass compressed into an orb with the same density as a black hole, where G is the gravitational constant, M is the mass of the object, in this case the sun and additional orb planets, and c is the speed of light.

kg	Mass of Orb	Accumulate	Schwarzschild	orbital per = s	Radius = m	orbit (days)	orbit(yrs)
1.9889E+30	Sun	1.9889146E+30	2.953601E+03	10020.57907	6.9630E+08	0.115978924	0.0003
3.2972E+23	Mercury	1.9889149E+30	2.953602E+03	7600262.175	5.7910E+10	87.9659974	0.2408
4.8673E+24	Venus	1.9889198E+30	2.953609E+03	19410605.43	1.0820E+11	224.6597851	0.6151
6.0471E+24	Earth and Moon	1.9889258E+30	2.953618E+03	31556952	1.4960E+11	365.2425	1.0000
6.3900E+23	Mars	1.9889264E+30	2.953619E+03	59335347	2.2790E+11	686.7517014	1.8803
1.8981E+27	Jupiter	1.9908246E+30	2.956438E+03	374614733.7	7.7850E+11	4335.818677	11.8711
5.6830E+26	Saturn	1.9913929E+30	2.957282E+03	935102200.8	1.4330E+12	10822.94214	29.6322
8.6810E+25	Uranus	1.9914797E+30	2.957411E+03	2659730347	2.8770E+12	30783.91605	84.2835
1.0240E+26	Neptune	1.9915821E+30	2.957563E+03	5199331719	4.4980E+12	60177.45045	164.7603
1.3090E+22	Pluto	1.9915821E+30	2.957563E+03	7823083930	5.9063E+12	90544.9529	247.9037

As an example, if we consider Mercury's orbit alone, the Schwarzschild radius of the sun itself calculates out to about 2.953601E+03 m with a mass of 1.9889146E+30 kg, and

---

[65] Jan Oort was one of the first astronomers to observe stars in distant galaxies orbiting much faster than predicted by gravity alone. She speculated that there must be some additional invisible or unknown mass outside the galaxy in some "halo" around it causing the stars' additional orbital speed.

CHAPTER 3: GRAVITY AND OUR MILKY WAY GALAXY

Mercury's orbital period, $\sqrt{(4\pi^2/(GM) \times r^3)}$ = 7600262.18 seconds = 87.966 earth days. To calculate Venus' orbit, we need to consider Mercury's mass, its orbit, and its gravitational impact on Venus. We then combine the mass of the Sun and the mass of Mercury to get the adjusted Schwarzschild radius of 2.953602E+03 meters with a mass of 1.9889149E+30 kg. With this combined mass, we derive Venus' orbital period as 19410605.4 seconds or 224.6598 earth days. As you can see from the above chart, the concentric slight increases of the Schwarzschild radius and the additional mass does influence the orbits of each of the planets per Kepler's third law[66]. This method is another rough way of showing the gravitational interactions between the planets. In the spreadsheet above: the orbital period of Mercury is 87.966 earth days; Venus has 224.66 earth days; Mars has 686.75 earth days (1.88 earth years); Jupiter has 4,335.82 earth days (11.87 earth years) and Saturn has 10,821.40 earth days (29.63 earth years). While Uranus has orbital period of 30,783.25 earth days (84.28 earth years); Neptune has 60,175.90 earth days (164.76 earth years) and Pluto has an estimated orbital period of 90,544.95 earth days (247.90 earth years). Planetary orbital periods calculated this way obtains more accurate numbers than that provided in chapter 2 section 1, which strictly used the mass of the sun alone for M in Kepler's third law equation.

Now let us consider the Milky Way galaxy as a whole with this method, starting with the black hole at the center of the galaxy, and move outward in concentric circles from .04 (outside event horizon of the Black Hole) at increments of .02. We also know that the stars nearest to the massive black hole at the center of the galaxy are orbiting the black hole at considerable speeds close to the speed of light. We also know that the inner sections of the Milky Way Galaxy contain more and larger stars than the other edges of the galaxy, and therefore more mass at the center of the galaxy.

Schwartzschild	kg	inner	section	has	mass fraction	Radius = m	orbital period (sec)	x 31556952=	years
1.38811E+10	9.34736E+36	Mass BH	2.45E-11	has	1.55789E-06	1.38811E+10	411.4327697		1.304E-05 years
7.26849E+14	4.89450E+41	Inner	0.040	has	0.081575	2.27053E+19	1.18944E+14		3.769E+06 years
7.19832E+14	4.84725E+41	Inner	0.060	has	0.0807875	3.40579E+19	2.19577E+14		6.958E+06 years
8.05426E+14	5.42363E+41	Inner	0.080	has	0.09039375	4.54106E+19	3.19593E+14		1.013E+07 years
9.37324E+14	6.31181E+41	Inner	0.100	has	0.105196875	5.67632E+19	4.14028E+14		1.312E+07 years
1.78349E+15	1.20097E+42	Inner	0.200	has	0.200162402	1.13526E+20	8.48955E+14		2.690E+07 years
2.67310E+15	1.80003E+42	Inner	0.300	has	0.300005075	1.70290E+20	1.27394E+15		4.037E+07 years
3.56408E+15	2.40000E+42	Inner	0.400	has	0.400000159	2.27053E+20	1.69860E+15		5.383E+07 years
3.74228E+15	2.52000E+42	Inner	0.420	has	0.420000079	2.38406E+20	1.78353E+15		5.652E+07 years
3.92048E+15	2.64000E+42	Inner	0.440	has	0.44000004	2.49758E+20	1.86846E+15		5.921E+07 years
4.09869E+15	2.76000E+42	Inner	0.460	has	0.46000002	2.61111E+20	1.95339E+15		6.190E+07 years
4.27689E+15	2.88000E+42	Inner	0.480	has	0.48000001	2.72463E+20	2.03832E+15		6.459E+07 years
4.45510E+15	3.00000E+42	Inner	0.500	has	0.500000005	2.83816E+20	2.12325E+15		6.728E+07 years
5.34611E+15	3.60000E+42	Inner	0.600	has	0.6	3.40579E+20	2.54790E+15		8.074E+07 years
6.23713E+15	4.20000E+42	Inner	0.700	has	0.7	3.97343E+20	2.97255E+15		9.420E+07 years
7.12815E+15	4.80000E+42	Inner	0.800	has	0.8	4.54106E+20	3.39720E+15		1.077E+08 years
8.01917E+15	5.40000E+42	Inner	0.900	has	0.9	5.10869E+20	3.82185E+15		1.211E+08 years
8.91019E+15	6E+42	MilkyWay	1	has	1	5.67632E+20	4.24650E+15		1.346E+08 years

The sun is located about 2.46731E+20 meters from the center of the galaxy and its orbital period around that black hole estimated to be about 7.57367E+15 seconds or about 2.40E+08 years. The chart below has been snipped (lines hidden) to fit onto the page, and the blue highlight, lines 23, 24, is the approximate location of the solar system.

Note that the blue highlighted section above in the chart above contains the correct radius distance but does not contain the appropriate orbital period of the sun. Mass and gravity alone calculations predicts a slower orbital period for stars the further away from

---

[66] Kepler's orbital period equation uses the mass of the sun and the distance of the semi-major axis to determine the planet's orbital period. However, adding the additional masses of the inner planets to the sun's mass and recalculating for the planet's orbital period gives us a better estimate, as though combining perturbations form the inner planets.

the center in the outer arms of galaxies. This chart is supposed to take care of all interactions between stars as we saw with the solar system, but it does not. As we will see in the next chapter, scientists tell us there must be some other force pushing the stars to orbit faster. For the sake of demonstration of this force, we are going to call it the slingshot effect, which we will learn later the true source, is the force responsible for this boost of acceleration. The slingshot effect occurs when multiple massive stars line up with the galactic center thereby combining their gravitational pull to tug and accelerate stars. It is something like an earth tidal bulge affecting the moon. We now incorporate this slingshot result added to the calculated Kepler speed in the spreadsheet below[67]:

Schwartzschild	kg		fraction	mass fraction	Radius = m	orbital period (sec)	x 31556952=	years	slingshot
1.38811E+10	9.3474E+36	Mass BH	2.45E-11	1.55789E-06	1.39E+10	411.4327697	1.304E-05	years	0.855
7.26849E+14	4.89450E+41	Inner	0.040 has	0.081575	2.27053E+19	1.18944E+14	3.769E+06	years	0.852
7.19832E+14	4.84725E+41	Inner	0.060 has	0.0807875	3.40579E+19	3.20560E+14	1.016E+07	years	0.849
8.05426E+14	5.42363E+41	Inner	0.080 has	0.09039375	4.54106E+19	5.90787E+14	1.872E+07	years	0.846
9.37324E+14	6.31181E+41	Inner	0.100 has	0.105196875	5.67632E+19	9.12061E+14	2.890E+07	years	0.843
1.78349E+15	1.20097E+42	Inner	0.200 has	0.200162402	1.13526E+20	2.89756E+15	9.182E+07	years	0.828
2.67310E+15	1.80003E+42	Inner	0.300 has	0.300005075	1.70290E+20	5.01152E+15	1.588E+08	years	0.813
3.56408E+15	2.40000E+42	Inner	0.400 has	0.400000159	2.27053E+20	6.95256E+15	2.203E+08	years	0.798
3.74228E+15	2.52000E+42	Inner	0.420 has	0.420000079	2.38406E+20	7.31082E+15	2.317E+08	years	0.795
3.92048E+15	2.64000E+42	Inner	0.440 has	0.44000004	2.49758E+20	7.65863E+15	2.427E+08	years	0.792
4.09869E+15	2.76000E+42	Inner	0.460 has	0.46000002	2.61111E+20	7.99605E+15	2.534E+08	years	0.789
4.27689E+15	2.88000E+42	Inner	0.480 has	0.48000001	2.72463E+20	8.32321E+15	2.638E+08	years	0.786
4.45510E+15	3.00000E+42	Inner	0.500 has	0.500000005	2.83816E+20	8.64033E+15	2.738E+08	years	0.783
5.34611E+15	3.60000E+42	Inner	0.600 has	0.6	3.40579E+20	1.00848E+16	3.196E+08	years	0.768
6.23713E+15	4.20000E+42	Inner	0.700 has	0.7	3.97343E+20	1.13236E+16	3.588E+08	years	0.753
7.12815E+15	4.80000E+42	Inner	0.800 has	0.8	4.54106E+20	1.23955E+16	3.928E+08	years	0.738
8.01917E+15	5.40000E+42	Inner	0.900 has	0.9	5.10869E+20	1.33328E+16	4.225E+08	years	0.723
8.19738E+15	5.52000E+42	Inner	0.920 has	0.92	5.22222E+20	1.35064E+16	4.280E+08	years	0.72
8.37558E+15	5.64000E+42	Inner	0.940 has	0.94	5.33574E+20	1.36758E+16	4.334E+08	years	0.717
8.55378E+15	5.76000E+42	Inner	0.960 has	0.96	5.44927E+20	1.38411E+16	4.386E+08	years	0.714
8.73199E+15	5.88000E+42	Inner	0.980 has	0.98	5.56280E+20	1.40026E+16	4.437E+08	years	0.711
8.91019E+15	6.00000E+42	Inner	1.000 has	1	5.67632E+20	1.41604E+16	4.487E+08	years	0.708
8.91019E+15	6E+42	MilkyWay	1 has	1	5.67632E+20	1.42296E+16	4.509E+08	years	0.705

With this adjustment, we note now that the green area within the highlighted section, lines 23 and 24, contains the correct distance from the galactic center and the correct estimated orbital period for the sun. Two things that greatly influence this slingshot effect or added acceleration are that the greater the mass of the inner galactic stars the greater it gravitationally tugs or propels the outer stars and the faster they move in their orbits around the galaxy. Each inner circle hub of larger stars and their combined masses accelerates the outer section concentric circle of stars, thereby causing some of the slingshot force, but not all of it. There is still an unknown force, which we will discuss later in this book, giving that additional boost. This slingshot transfer of energy action also slows down or decelerates the inner massive star and causes it to go inward, a corralling effect. Since the total mass of the galaxy is more homogenously distributed, the transfer of energy and speed from the inner stars to the outer stars flows much more rapidly than that within the solar system. The same type of energy transfer effect occurs on earth as the tide caused by the moon also causes the moon to increase speed to a higher orbit, and concurrently causes the earth's day-night rotational period to lengthen.

MILKY WAY GALAXY

Schwarzschild Radius Adjustments

Unlike the solar system where the sun possesses most of the mass of the system, the black hole at the center of the galaxy does not nearly contain close to the same ratio. As you can see from the two tables above, the adjusted Schwarzschild radii are ever increasing outward

---

[67] "Slingshot" effect also occurs within solar systems when planets line up. As a result, Jupiter and Saturn gravitational together have a significant impact on each other and the inner and outer planets lining up with them.

but still all within the inner .040 circle section of the galaxy and possess enough concentrated mass and gravitation attraction to hold the galaxy together. More specifically, the Schwarzschild radius entry in line 53 is less than the radius column entry in line 4. The depiction above shows the concentric circles of Schwarzschild radii in the hub:

We can roughly represent the movement of the galactic mass of normal gravity by a slight adjustment to Kepler's third law, $T^2 = (G_T+A_T)^2 = (4\pi^2/(G*M)) * a^3$, where G is Newton's gravitational constant, M is the combined mass of all inner matter, a is the length of the semi-major axis of orbit, G_T is gravity only orbital period, and A_T is that added boost of acceleration incorporated into the second galactic spreadsheet above. Hence, the star's orbital period is influenced by gravitational force and an additional acceleration force, more specifically orbital period $T = G_T + A_T$. We will further discuss, define, and present greater detailed explanation of this additional force in part 3 chapter 7 in the form of a universal equation, which will be submitted later with the intent to augment if not replace Einstein's famous equation $G_{\mu\nu} = 8\pi G/c^4 \, T_{\mu\nu}$.

For information (Verlinke, 2018) Erik gave a presentation at Delft, on March 8, 2018, titled "A new view on gravity and the cosmos" where he introduced his new equation defining the forces generated by Hubble Constant light-distance limit on galaxies. Reads:

$$\frac{1}{8\pi G}\int_{r\leq R} g_i^2 dV = \frac{M_B\, c\, R}{\hbar} \frac{\hbar H_o}{6}$$

where M_B is the galactic mass, c is the speed of light, R is the radius of the visible universe, $\hbar$ is Planck's constant, G is the gravity constant, and H_o is the Hubble constant. He claimed that the dark energy of the universe is responsible for causing galactic stars particularly the ones on the outer edge of the spiral arms to move faster than gravity alone can orbit them. The first part of the equation is from Einstein's GR equation applied at various distances from the center of the galaxy, while the other half of Erik's equation accurately resolves the rotational discrepancy but only for distances greater than r and less than R. He did not define dark energy. Let us set this aside to expand on Kepler's work.

A lecture found reflect Kepler's work under Astronomy 162 (Ryden, 2003) Professor Barbara Ryden had given on Wednesday, February 19, 2003. Ryden concluded that the high orbital speed of stars can be determined from Kepler's Third Law: by adding together all the masses of the inside stars orbit (in solar masses) M plus the mass of star M_*, to determine orbital speed around the galaxy[68]. Her equation was $M + M_* = a^3/P^2$, where a = radius of the stars orbit (in AU), and P = orbital period of star (in Years) (Ryden, 2003). My analysis is obviously slightly different, concluding that double counting the masses of the stars within the orbit provides a close approximate solution, but not good or accurate enough to be declared a universal equation. Unknown forces boost acceleration A_T leads us to the next topic, theories, and inexplicable elements in the universe. I will expand on this formula and produce an accurate mathematical equation defining galactic rotations and intergalactic interactions, applicable to solar periodic orbital periods, as well as two body gravity attractions. This universal formula reduces to Newton's gravity equation. Double counting the mass is an indicator that gravity is a push-pull force.

---

[68] In a sense, Ryden used the slingshot effect to justify double counting the mass within the orbital radius of the star being considered.

3.7 Solar-Planetary Rotational Torque Theory

"Pseudoscience speaks to powerful emotional needs that science often leaves unfulfilled. It caters to fantasies about personal powers we lack and long for."

— Carl Sagan

Abstract

Inertia and conservation of angular momentum keeps planets rotating about their axis and in their orbit around the sun. If no torque exists, then why is Earth's rotation not smooth? What factors cause a planet's spin to slow down or speed up at their aphelion or perihelion? What causes most planets to rotate west to east[69] or prograde? Is their rotation strictly a result of numerous angular impacts or bombardment in the solar system's early formation? A collision from large objects on orbiting planets or dwarf planets does influence their rotation, like Venus, Uranus, and Pluto. Can the sun's gravity apply torque to restore them to normal rotational direction? What causes non-rotating tidal lock? This paper uses the conservation of planetary extended angular momentum to analyze the possibility of stellar gravity torque on planets, and in doing so identifies one hidden aspect in the principle of equivalence; specifically, different parts of the planet fall at different rates toward the sun. Kepler's three laws, Newton's gravity laws, and Einstein's principle of equivalence apply to resolve the stellar effect or in our case solar torque on the planets, and other debris within the solar system.

Review of Laws and Equations

Johannes Kepler applied the concept of torque and polar coordinate system to the motion of planets orbiting around the sun and discovered three laws[70]:

0This is an illustration of Kepler's three laws with two planets P_1 and P_2 orbiting around the sun S_1. Planets follow elliptical orbits; equal time segments of orbital path have equal areas; and the square of orbital period is proportional to the cube of the semi major axis of its orbit. The angular momentum vector of a planet is a constant vector because the sun does not apply torque to the planet[71] per Kepler. As such, Kepler did not solve for planetary rotation. However, using Kepler's laws, Sir Isaac Newton discovered three laws: the law of inertia, the force law, and the action-reaction law. Particular pertinent to this paper is his gravity equation, $F = G (\{m_1 * m_2\}/r^2)$, where F is the force generated, G is the gravity constant 6.6734E-11 N $(m/kg)^2$, m_1 is object 1 and m_2 is object 2, and r is the distance between the centers of both.

Extended or point angular momentum (or rotational momentum) is the rotational analog of linear momentum. It is an important and conserved quantity – the angular

---

[69] West to East rotation is seen as counterclockwise motion for prograde orbits when viewed from the North Star perspective.

[70] Francis, E. M. (2016, 01 01). *A Proof of Kepler's laws*. Retrieved 01 24, 2016, from Kepler's laws: http://www.alcyone.com/max/physics/kepler/index.html

[71] Ibid

momentum of a system remains constant unless acted on by an external torque. Planets orbital path around the sun are point angular momentum, where angular momentum equals radius times the mass times velocity, L = r * m * v. Planet rotation is defined as extended angular momentum, which is equal to rotational inertia times the angular velocity, L = I * ω. Conservation of angular momentum requires both resulting vectors from the point angular momentum and that of the extended angular momentum to agree and go in the same direction. Physics defines that there is no solar torque applied to the planet orbital path to speed it up (around a point, the sun) or on its extended angular momentum to sustain, speed up, or slow it. Galilean Relativity defines that the laws of physics are the same for all observers in uniform motion relative to one another. In Galilean terms to gravity, a stationary object, or an object in uniform or constant motion equates to normal stable gravity. Specifically, in these instances, a spherical object's gravity is equal over the entire surface of the object as calculated per Newton's equation; there is no detected increase or decrease of gravity.

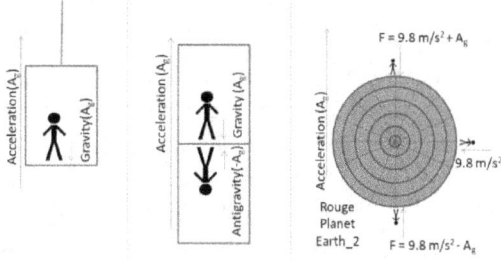

Einstein redefined Newton's laws in his Special and General Relativity Theories. Worthy of note here is his Strong Principle of Equivalence (see leftmost image above), where acceleration equates to gravity particularly in deep open space away from other massive objects[72]. An astronaut in a rocket lifting into space feels the added "G's" and a driver of a car pressing the gas pedal pushed back into his seat as he accelerates similarly feels the same effect.

A passenger in a bullet train facing in the direction of acceleration feels the same additional "gravity" while a passenger facing to the rear of an accelerating train feels "negative gravity." Obviously, passengers facing to the rear do not feel the additional "gravity"; instead, the acceleration may throw them from their seat if they do not

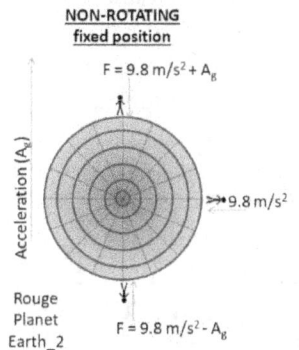 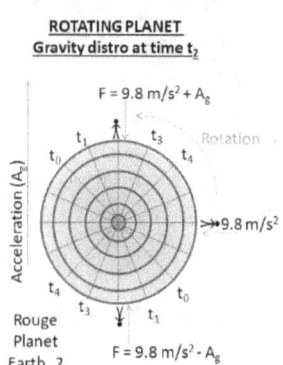

brace themselves or buckle in. A man standing on the lead edge of a non-rotating rogue planet (see rightmost image above) feels an increased gravity, while the same man feels less gravity on the trailing edge, and normal gravity in the position between both lead and trail edge. However, on a planet rotating about its North Pole axis, the man would feel a gradual increase in gravity from time t_1 through t_3, where the greatest gravity increase is when the man at the planet's equator is at time t_2 (shown on right image below). A man standing on either the planet's North or South Pole would feel normal gravity 9.8 m/s².

---

[72] Einstein/Golm/Potsdam. (2015). *Einstein online provided by Max Planck Institute for Gravitational Physics*. Retrieved 11 02, 2015, from E=mc^2: http://www.einstein-online.info/elementary/specialRT/emc

Acceleration Gravity Relationship

How does acceleration affect gravity distribution on the surface of massive objects? To visualize this, let us begin with a single object suspended in space. An ordinary massive object, in stationary state or in constant motion, experiences equal gravity over its entire surface. This is Galilean Relativity in action. In this state, the gravity constant G = 6.6734E-11 Newton (m/kg)^2 force applies equally to suspend that object in space.

Per Albert Einstein, acceleration equals gravity[73]. Accelerating a stationary or constant motion object will increase its gravity over the leading edge of the object in the direction of motion. Increasing the acceleration intensifies the gravity on its lead surface. Gravity intensity is proportional to acceleration. Spherical objects experience maximum gravity increase on the point of the surface intersecting the vector line drawn from the center of the object toward the direction of its acceleration. By the laws of symmetry, the spherical object also experiences decrease in gravity on the surface point intersecting the vector line directly opposite the maximum gravity increase, on the other side of the object.[74] We can call the decrease on trailing edge antigravity, as it is a negative of gravity increase on the lead edge. Comets develop tails as they gain speed on their approach to the sun, its trail edge gravity decrease releases debris off the back surface as its lead edge gains gravity and is heated up by the sunlight, similarly loosely bound pieces of an asteroid fall apart in the same situation. For an orbiting planet, this acceleration vector is tangent to the orbital path.[75]

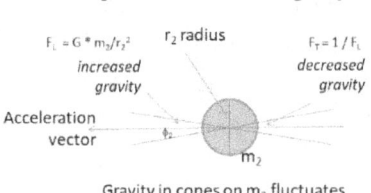

For two objects in space, object m$_2$ and object m$_1$, if the vector of motion of both objects point directly at each other, they will both experience mutual gravity acceleration toward each other, per Newton.[76] As one object increases acceleration toward another object, its gravitational attraction will increase in intensity on the object's side of the direction of movement[77] and develop an equal increase of antigravity, a weaker or negative gravity, on the opposite side, a zero-sum gain over entire object. As an object's acceleration increases, it creates an equal proportional concentration of gravity on its surface in the vector of motion, which of course is counterbalanced by an equal proportional decrease of gravity (antigravity) on the surface opposite vector.[78] The result of zero-sum-gain collision course is that the two objects mutually attract each other and accelerate equally

---

[73] Ibid

[74] For every action there is an equal and opposite reaction. In this case, an increase of gravity on one side must be counterbalanced by an inversely proportional decrease of gravity on the opposite side, a zero-sum gain Einstein neglected to announce, as it impacts the principle of equivalence.

[75] Also applies to objects not in direct collision course or in orbit.

[76] Newton, S. I. (1687). *Principia Mathematica*. Cambridge: Unknown.

[77] Per Einstein's theories.

[78] The zero-sum gain is required per Newton's laws. Similarly, increase in acceleration causes mass-energy increase at the leading edge while causing inversely proportional decrease at the trailing edge. Mass increase equates to gravity increase.

CHAPTER 3: GRAVITY AND OUR MILKY WAY GALAXY

proportional to product of their masses and inversely proportional to their distances, per Newton.

Let object m_2 and object m_1 be round spheres. The surface area of the spherical cone of object m_2 with the gravity increase drawn from the edge of object m_1 to the center of the object m_2 and vice versa, gives us the gravity force ratio. The area experiencing the decreased gravity is of an equal size spherical cone opposite that of the enhanced gravity[79]. The highest concentration of increased gravity during acceleration is the point on the object's surface intersecting the vector line in the direction of movement. Similarly, the highest concentration of decreased gravity during increase acceleration occurs on the point of surface opposite the vector of movement. The vector line drawn from the center of one object to the center of the other object is the gravity attraction at that moment in time and adjusts instantaneously as objects move. These are essentially the surface points at the center of the spherical cones on each side of the object or the center intersecting the vector line and the surface of the object.

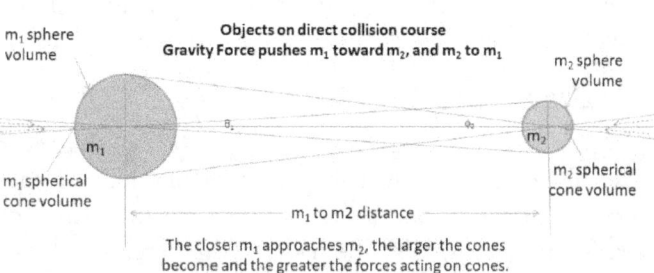

If the smaller object m_2's vector of movement is not pointing directly at the massive object m_1, object m_2 trajectory will curve or fall toward object m_1, as predicted by Newton gravity equation. As ordinary matter object m_2 accelerates toward object m_1, it experiences gravity increase on the side going toward the direction of motion, which in turn is counterbalanced by weaker gravity on the opposite side of the object. These enhanced gravity and decreased gravity provide torque for the object to rotate on its axis, enough to contribute increase speed and in some cases sustain or decreased angular momentum, like a ratchet turning a bolt. As object m_1 accelerates toward object m_2, it too experiences slight gravity increases on the side going toward the direction of acceleration, which is counterbalanced by a decreased gravity on the opposite side of the object.

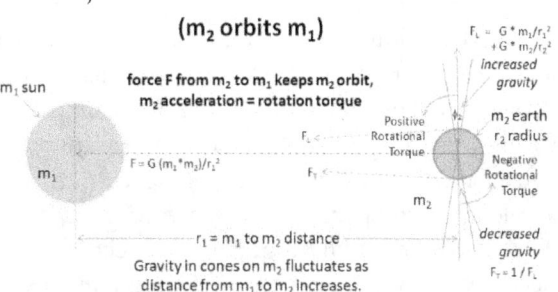

If acceleration is ignored in the image, the Earth has the same normal gravity over its entire surface. With acceleration, more gravity is added to the leading edge particularly at the point intersecting the vector running through the center of the world and simultaneously this gravity is subtracted from the trailing edge at the opposite side. Here orbital acceleration or point angular momentum transfers energy to the planetary spin or it extended angular momentum.

---

[79] Singular massive objects in space tend to form spheres, to include all stars and even black and white holes.

How can we determine the total gravity distribution? Starting from the centerline between the sun and the Earth that intersects the most amount of mass that slices the world in half, where the Earth's surface gravity is normal; here the sun's acceleration effects are zero, neither positive nor negative. As we move along the surface toward the leading-edge acceleration vector, Earth's gravity and sun's acceleration when summed together gradually increase in intensity until we get the full Earth's gravity plus acceleration at the point intersecting the vector direction (line through the center of the globe). The opposite occurs as we move toward the trailing vector line.

Note: When filmed in slow motion, the iron ball fired from the cannon reveals a similar rotational torque action as it rotates in the direction illustrated above. The lead edge of the ball rotates downward and the trail edge rotates upward as the ball travels to the left, a counterclockwise rotation direction. Is this solely the result of the ball rolling within the cannon barrel or from gravity torque? It does provide initial extended angular momentum. But a ball thrown by a catapult or slingshot without rotation begins to rotate in the same direction where the lead edge turns to gravity; this can only happen if Earth's gravity provides torque τ to begin its extended angular momentum motion, $\Delta L = \tau \Delta t$.

Analogy of Gravity Antigravity on Objects

Now let us take a moment to consider what acceleration does to the warping or curvature of space. Einstein predicted one-half of the results of an acceleration event, increased gravity[80], but did not address the other half, decreased gravity or antigravity. An increase in gravity is also the result of an increase in mass or the matter's energy (motion)[81]. As a massive object moves uniformly through space, space-matter, we will call *hygratium*, envelopes it equally in all directions as though that object was stationary. The resulting perfect curvature follows the object as it moves in space. We can visualize this effect easily in a soap bubble floating in the air; the atmosphere presses equally on all sides of the bubble and forms a perfect spherical shape. It moves with the flow of air, just as an object in space would move with the flow of space and its curvature and matter remain intact. Moreover, an object accelerating through space gains gravity at the forward edge, thereby compressing in that point to form denser curvature and through symmetry causes a pointed bulge or a protruding tip on the opposite end.

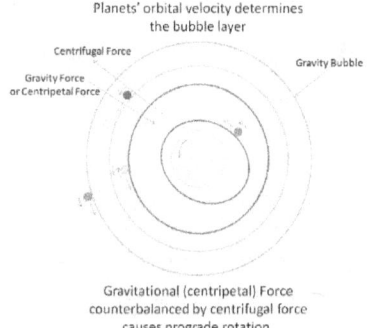

Acceleration is either a change in speed or a change in direction of motion[82]. We can see this effect in drops of water falling from the sky as rain and in comets; it forms a teardrop shape as it accelerates. Increase in mass equals increase in curvature of space[83]. So, an object accelerating in space causes space to compress or build up around the leading edge of the object (the same effect occurs when mass increases) forming a greater curvature

---

[80] Einstein, A. (1920). *Relativity, The Special and General Theory.* Kindle Direct Publishing Edition 2013.

[81] Ibid

[82] Definition of Acceleration.

[83] Einstein/Golm/Potsdam. (2015). *Einstein online provided by Max Planck Institute for Gravitational Physics.* Retrieved 11 02, 2015, from E=mc^2: http://www.einstein-online.info/elementary/specialRT/emc

CHAPTER 3: GRAVITY AND OUR MILKY WAY GALAXY

than otherwise detected around constant moving object or one that is stationary. Likewise, the tail end of the accelerating object will develop a trailing tip of less dense space-matter emanating opposite the vector of acceleration. The resulting space curvature is teardrop-shaped moving in the direction of acceleration; a comet is an example.

Gravity intensity is inversely proportional to the square of the distance[84] and best represented in concentric spherical layers or bubble layers of intensity surrounding the massive object or system[85]. A planet visualized orbiting the sun on its orbit path or such bubble layer follows prograde rotation and adherence to the law of conservation of angular momentum. A planet not following the natural prograde rotation would experience friction, increased temperatures, and gradual slowing of rotation speed as gravity tries to correct its direction.

Earth's Planetary Torque Equation

Mathematically, we can represent this invisible torque action or supplement to the conservation of angular momentum on planet Earth with a few simple calculations. We can solve for the gravity acceleration near the surface of the planet by multiplying the gravity constant G times the mass of the Earth divided by the square of its radius to get 9.818649 meters per second squared[86]. The Sun's average gravitational acceleration at Earth's average orbital radius equals G times the mass of the sun divided by Earth's orbital radius squared or 6.6734E-11 N (m/kg)^2 x 1.989E+30 kg / (1.495979E+11 m)^2 = 5.931044E-3 m/s^2. Earth's average velocity around the sun equals the circumference divided by the number of seconds in a year or 2*π*1.495979E+11 m / 31,557,600 seconds = 2.978525E+4 m/s. Earth experiences a slight increase of gravity at the surface of the vector direction tangent to the orbital path. This is calculated by adding Earth's gravitational acceleration plus the sun's gravitational acceleration divided by the Earth's gravity acceleration or 9.818649 m/s^2 + 5.931044E-3 m/s^2/ 9.818649 m/s^2 =1.000604059 G's, while the trail end opposite surface of the planet experiences a decrease in gravity of 0.9993963 G's obtained from 9.818649 m/s^2 - 5.931044E-3 m/s^2 / 9.818649 m/s^2.

This decrease gravity effect is not the same for spaceship artificial gravity creation. There is a limit when we consider how a spaceship travelling in space would create artificial gravity for its occupants. That spaceship would spin about its axis direction or vector of travel to generate artificial gravity. In doing so, an astronaut approaching the center axis experiences zero gravity, and the closer that person moves toward the inner surface of the outer hull the more gravity he or she feels. Speeding up the spaceship's spin increases gravity, but regardless of how fast it spins, the center axis remains at zero gravity[87]. This difference in torque force may not seem like much but if we multiply that lead-edge enhanced gravity times the gravitational attraction between the planet and the sun, we can

---

[84] Gao, S. (2014). *Understanding Gravity: Newton, Einstein, Verlinde?* Kindle Edition: Amazon Kindle Direct Publishing.

[85] NASA. (2012, 10 10). *Layers of the Sun*. Retrieved 09 02, 2015, from NASA Mission Pages: http://www.nasa.gov/mission_pages/iris/multimedia/layerzoo.html

[86] The 9.818649 m/s^2 acceleration is at or near the surface of the Earth.

[87] Artificial gravity is created with centrifugal force, is not true gravity, but a force perceived as gravity by space travelers. Einstein however equates it to gravity since acceleration (change of direction) is gravity.

calculate that torque force power. The average gravitation force between the Sun and Earth is about 3.54202E+22 Newton. The lead edge additional gravitational force of:

F_T = [(1.000604059 - 1) * 6.6734E-11 N (m/kg)^2 * 1.989E+30 kg* 5.972E+24 kg] / (1.495979E+11 m)^2 = 2.139589E+19 N

This provides the lead point a total force of 3.544159E+22 Newton.

Trail edge has decreased gravitational force of

F_P = [(1 - 0.999395941) * 6.6734E-11 N (m/kg)^2 * 1.989E+30 kg* 5.972E+24 kg] / (1.495979E+11 m)^2 = - 2.1389589E+19 N

This gives the trail edge a total force of 3.539880E+22 Newton, where 1 Newton Force = 1 kg * m/s^2, and the positive force F_T pushes the lead edge faster toward the sun, and the negative push force F_P can be seen as the results of centrifugal force on the planet making it rotate counterclockwise, west to east, contrary to equivalence principle[88]. The differences in land mass and ocean mass, and the position of the moon cause the Earth rotation to be not smooth[89].

We find that the best visualization of solar rotational torque effect on planet Earth is in the overall motion of the atmosphere and as the driving force behind the strong polar jet streams, at 9–12 km above sea level, that encircle the globe, and that of the weaker subtropical jet streams at 10–16 km. At this altitude and temperature differences, the air pressure below and above the polar jet stream are from 2.52 to 4.14 pounds per square inch (psi) (or at 0.17 to 0.28 atmospheres (atm)), while the air pressures of the subtropical jet streams are from 1.49 to 3.83 psi (0.1 to 0.26 atm), the pressures and angles impacted by solar torque. A jet stream forms high in the atmosphere between two air masses of difference temperatures. The greater the temperature differences the faster the wind in the stream, which could range from 161 km/h to 322 km/h and normally form during the winter. Scientists tell us that the Earth's rotation, friction, and heat energy transfers causes the airflow to go from West to East (to the right in Northern Hemisphere) and around the boundary between air masses, rather than directly from one air mass to the other. If this were the only factor, then the airflow should be slower than the rotation speed of the Earth, but the air moves faster over the land than the Earth rotates.

In this analysis, we now know that the solar planetary torque adds to planetary friction on the air molecules to accelerate it ahead of the rotational speed of the Earth, particularly at higher altitudes. Storms are an exception. As the season transitions from summer to late autumn in the Northern Hemisphere, easterlies or winds from the east develop between the two subtropical jets streams and travel 5 to 30 degrees latitude above the equator, hurricanes and typhoons that form, start out easterly, and then turn westerly as they move up toward the north. Similarly, cyclones form during the Southern Hemisphere's summer transition to autumn start moving easterly and then westerly as they turn toward the south. Southern Hemisphere's peak cyclone season typically runs from November to April (summer and autumn in the southern half). The gradual changes in the patterns of jet

---

[88] One instance where the equivalence principle fails since the entire object, planet Earth, does not fall equally toward another object, the sun.

[89] Time Systems and Dates – GPS Time: http://www.oc.nps.edu/oc2902w/gps/timsys.html

streams are also a result of the angle of solar planetary torque applied while the Earth transitions through the seasons. See image below provided by NASA.

What does this gravity difference mean in layperson terms? Well, if a 100 kg person or item were standing on a weight scale on the leading axis point of the Earth that person or item would weigh 100.0604 kg, while the same person or item standing on the opposite point on the other side of the world would weigh 99.9396 kg. On second thought, due to bodily functions replace the person completely with something more defined and consistent like a 100 kg dumbbell or free weights. We will find that anywhere within that small leading spherical cone surface of the Earth the weight scale result will deviate as mentioned above give or take a few grams. Again, the solar gravitational acceleration of 5.931044E-3 m/s² is enough to cause the planet lead-edge to accelerate 3.582700E-06 m/s² to rotate more toward the sun, and the trail end to fall slower (or push away) at -3.582700E-06 m/s² if and only if there is enough torque. When compared to the average rotational speed of the Earth at 460 meters per second, we can see that the additional solar gravitational acceleration at the Earth's distance from the sun might tend to add speed to the Earth's rotation. However, with the solar torque force ratio of just 1.186925E-02, it is insufficient to increase Earth's rotational speed at best it maintains it. Miniscule rotational increases occur when lead and trail points are both over large solid landmass. Lead and trail points over oceans lack traction due to the fluidity of the water. These actions contribute to the Earth's uneven rotation. Therefore, at the very least the solar torque force contributes and sustains the Earth's extended angular momentum.

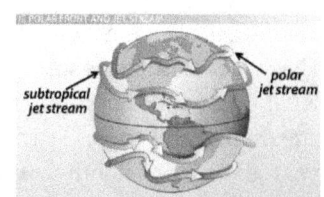

For clarification, solar-planetary torque is not a vertical force; it is tangential to the surface of the planet and therefore horizontal in nature. To calculate the solar gravitation acceleration effect on the rotational speed of the planet Earth we take 5.931044E-3 meters per second squared and divide it by the product of the radius of the planet times 2π. This amounts to an adjusted planetary rotational acceleration of 1.471845E-10 meters per second squared. When compared to the average rotational speed of the Earth is 460 meters per second, we can see that the additional solar gravitational acceleration at the Earth's distance from the sun tends to add speed to the Earth's rotation. However, with the solar torque force acceleration of just 1.471845E-10 meters per second squared, we can see that it is insufficient to increase Earth's rotational speed due to the presence of the moon. Miniscule rotational increases occur when lead and trail points are both over large solid and heavier landmasses. However, lead and trail points over oceans lack traction due to the fluidity of the water, where these forces translate to moving ocean currents and more evaporation. Moreover, this fluctuating gravity increase and decrease torqueing actions contribute to the Earth's uneven rotation. The solar torque forces contribute and sustain the extended angular momentum of the Earth.

Mutual Lunar and Terra Torque

The above solar torque analysis clearly shows that gravitational acceleration causes the planet Earth to sustain a counterclockwise prograde rotational speed, west to east, while the orbit of the moon causes that rotation of the Earth to slow down just a tiny bit every year. The Earth's torque on the lead edge of the moon equals 3.26846E+17 N and the trail edge of -3.26846E+17 N bringing about a lunar rotation speed matching its orbital time, and because they are in mutual orbit about each other, the moon's torque on the Earth equals to 6.65609E+14 N and trail edge of -6.65609E+14 N. When the moon is on the same side as the sun (front or back of Earth), both torques combine to affect the Earth's rotational speed. When the moon is on any other location, its miniscule torque on the Earth decreases the sun's torque force and therefore minutely slowing down the Earth's rotational speed. Incorporating tidal influences into the picture makes the moon a slow and steady braking force, at the same time causes the moon to incrementally drift away every year about four centimeters (or 4 cm/year) from this interaction. In contrast, the Earth's day will lengthen ever so slightly over a period of one hundred years amounting to only two milliseconds more time than today. The hollow moon's surface facing the Earth is internally thicker, a hidden bulk, causing tidal lock with Earth.

Venus' Planetary Torque Equation

Venus' invisible torque action is different as it is closer to the sun and smaller in mass than the Earth. We solve for the gravity acceleration near the surface of Venus by multiplying the gravity constant G times the mass of Venus divided by the square of its radius to get 8.867693 meters per second squared. The Sun's average gravitational acceleration at Venus's mean orbital radius equals G times the mass of the sun divided by Venus's orbital radius squared or 6.6734E-11 N (m/kg)² x 1.989E+30 kg / (1.082000E+11 m)² = 1.133776E-02 m/s². Venus's average velocity around the sun equals the circumference divided by the number of seconds in a year or 2*π* 1.082000E+11 m / 31,557,600 seconds = 2.154285E+04 m/s. Venus' slight increase of gravity at the surface of the vector is in the direction tangent to the orbital path. This is calculated by adding Venus' gravitational acceleration plus the sun's gravitational acceleration divided by the Venus' gravity acceleration or 8.867693 m/s² + 1.133776E-02 m/s² / 8.867693 m/s² = 1.001278547 G's, while the trail end opposite surface of the planet experiences a decrease in gravity of 0.998721453 G's obtained from 8.867693 m/s² - 1.133776E-02 m/s² / 8.867693 m/s².

When we multiply torque difference from enhanced gravity times the gravitational attraction between the planet and the sun, we can calculate that torque force power. The average gravitation force between the Sun and Earth is about 5.518090E+22 Newton.

The lead edge has an additional gravitational force of

F_T = [(1.001278547 - 1) * 6.6734E-11 N (m/kg)² * 1.989E+30 kg * 4.867E+24 kg] / (1.082000E+11 m)² = 7.055138E+19 N

This provides the lead point a total force of 5.525145E+22 Newton.

Trail edge has decreased gravitational force of

F_P = [(1 - 0.998721453) * 6.6734E-11 N (m/kg)² * 1.989E+30 kg * 4.867E+24 kg] / (1.082000E+11 m)² = - 7.055138E+19N

This gives the trail edge a total force of 5.511035E+22 Newton, where 1 Newton Force = 1 kg * m/s², and the positive force F_T pushes the lead edge faster toward the sun, and the negative push force F_P is the results of centrifugal force on the planet to rotate it counterclockwise, west to east, contrary to equivalence principle[90].

---

[90] One instance where the equivalence principle fails since the entire object, planet Earth, does not

A Venus 100 kg item set on a weight scale on the leading point of Venus would weigh 100.12785 kg if the vector force were vertical, while the object on the opposite point on the other side of the planet would weigh 99.8721 kg. On Venus, the solar gravitational acceleration of 1.133776E-02 m/s² is enough to cause the lead edge of the planet to fall 1.449587E-05 m/s² toward the sun and counter the retrograde rotation while the trail end pushes away at -1.449587E-05 m/s². In this case, the vector force is horizontal. We can see that the additional solar gravitational acceleration at Venus distance from the sun provides sufficient torque to slow down and eventually halt Venus' retrograde rotation; this resistance builds up friction and adds to Venus runaway temperatures. Temperatures will stabilize as Venus stops it retrograde spinning and begin to assume a prograde motion. The solar-planetary torque force is tangent to the surface of the planet. To calculate the solar gravitation acceleration effect on the rotational speed of the planet Venus, we take 1.133776E-02 meters per second squared and divide it by the product of the radius of the planet times 2π. This amounts to planetary rotational acceleration of 2.98159E-10 meters per second squared. When compared to the average rotational speed of the Venus is -18.05556 meters per second, we can see that the additional solar gravitational acceleration at the Venus' distance from the sun will tend to restore prograde motion. With the solar torque force acceleration of the planet at 2.98159E-10 meters per second squared, we can see that it is sufficient to stop and reverse Venus' retrograde rotation in time. The rotational increase occurs most when the lead and trail points are both over solid land. Since Venus has no oceans to impede traction, we can see that the solar gravitational acceleration at Venus distance from the sun provides sufficient torque to slow and halt its retrograde rotation and restore to prograde.

Mars' Planetary Torque Equation

Mars' torque action is different as it is further away from the sun but a lot smaller in mass than the Earth. Gravity acceleration near the surface at 3.7274 meters per second squared. The Sun's average gravitational acceleration at Mars is 6.387605E-04 m/s². Mars's average velocity around the sun equals 9.076070E+04 m/s. Mars' slight increase of gravity at the surface of the vector is in the direction tangent to the orbital path is 1.000171 G's, while the trail end opposite surface of the planet experiences a decrease in gravity of 0.999829 G's. The average gravitation force between the Sun and Earth is about 4.098990E+20 Newton.

The lead edge has an additional gravitational force of

F_T = [(1.000171365 - 1) * 6.6734E-11 N (m/kg)² * 1.989E+30 kg * 6.4171E+23 kg] / (4.558500E+11 m)² = 7.024253E+16 N

This provides the lead point a total force of 4.099692E+20 Newton.

Trail edge has decreased gravitational force of

F_P = [(1 - 0.999828635) * 6.6734E-11 N (m/kg)² * 1.989E+30 kg * 6.4171E+23 kg] / (4.558500E+11 m)² = - 7.024253E+16 N

This gives the trail edge a total force of 4.098287E+20 Newton.

---

fall equally toward another object, the sun.

where 1 Newton Force = 1 kg * m/s², and the positive force F_T pushes the lead edge faster toward the sun, and the negative push force F_P is the results of centrifugal force on the planet to rotate it counterclockwise, west to east, contrary to equivalence principle.

A Mars 100 kg item set on a weight scale on the leading point of Mars would weigh 100.0171365 kg if vector force were vertical, while the object on the opposite point on the other side of the planet would weigh 99.98286345 kg. On Mars, the solar gravitational acceleration of 6.387605E-04 m/s² is enough to cause the lead edge of the planet to fall 1.094615E-07 m/s² toward the sun and enhance rotation while the trail end pushes away at -1.094615E-07 m/s². This additional solar gravitational acceleration at Mars distance from the sun provides sufficient torque to sustain Mars 24.6597-hour day. Since the solar-planetary torque force is not a vertical force, to calculate the solar gravitation acceleration effect on the rotational speed of the planet Mars we take 6.387605E-04 meters per second squared and divide it by the product of the radius of the planet times 2π. This amounts to an adjusted acceleration of 2.99932E-11 meters per second squared. When compared to the average rotational speed of Mars is 239.8972 meters per second, we can see that the additional solar gravitational acceleration at Mars distance from the sun barely affects Mars' rotation speed. With the solar torque force acceleration of just 2.99932E-11 meters per second squared, we can see that it sustains Mars' rotational speed. Rotational increases occur when lead and trail points are over solid land.

Mercury's Planetary Torque Equation

Using the above equations for Earth, Venus, and Mars, we would anticipate that Mercury's torque action be greater than Venus' is, so why is it not spinning rapidly? Mercury orbits the sun in an elliptical orbit at an extremely close distance. At its closest approach it is about 46 million meters and at its furthest orbital point is at around 70 million meters from the Sun. This distance causes the side of Mercury facing the Sun to be scorching hot and molten, while the side away from the sun becomes fidget cold and solid. The molten face therefore develops a slow-moving wave or bulge several kilometers in height, almost sufficient to tidal lock it with the Sun's tidal bulge. At perihelion, Mercury's tidal wave is 43.58 km in height, and at aphelion, it is about 28.41 km. However, due to the extreme torque force applied by the Sun's gravity acceleration, it slightly wins over the tidal lock and causes Mercury to rotate once every 58 Earth days, while the planet takes approximately 88 Earth days to orbit the Sun.

Mercury orbit radius	Distance	Sun Gravity Accl	Trail push	-17034.8340	Torque Force		Tidal Bulge	Tidal Bulge
Perihelion 46 Mil meters	46000000 m	6.272870E+04	Lead pull	17036.8340	3.5105E+32	-3.5105E+32	43.5756	5767.6546
Aphelion 70 Mil meters	70000000 m	2.708856E+04	Lead pull	7357.6989	6.5464E+31	-6.5464E+31	28.4090	3780.1604
Average Orbital Radius =	58000000 m	m/s²	Trail push	-7355.6989	Force in Newton		Mercury(km)	Sun (km)

In the chart above, we see that at Mercury's perihelion, the Sun's gravitational acceleration on the planet is 6.27287E+04 m/s² generates the torque force on lead point is at 3.5107E+32 Newton, and the trail point is at -3.5103E+32 Newton. At Aphelion, the Sun's gravitation acceleration on Mercury is 2.70886E + 04 m/s² creates the torque force on the lead point is at 6.547E + 31 Newton, and the trail point is at -6.546E + 31 Newton. Both lead and trail points occur at or near the line between shadow and sunlit regions of Mercury, or semi-solid surface. This intense torque action is offset by the slow-moving tidal bulge partial lock between the planet and the sun, resulting in a slow turning rotation of about 58 Earth days to complete one Mercury "day."

CHAPTER 3: GRAVITY AND OUR MILKY WAY GALAXY

Again, solar-planetary torque force is not a vertical force; it is tangential to the planet's surface and therefore horizontal in nature. To calculate the solar gravitation acceleration effect on the rotational speed of the planet Mercury we take 3.958190E-02 meters per second squared and divide it by the product of the radius of the planet times 2π. This amounts to an adjusted acceleration of 2.58214E-09 meters per second squared. When compared to the average rotational speed of Mercury is 3.02556 meters per second, we can see that the additional solar gravitational acceleration at Mercury distance from the sun barely affects Mercury's rotation speed. With the solar torque force acceleration of just 2.58214E-09 meters per second squared, we can see that it sustains Mercury's rotational speed of which resisted by molten tidal lock. Rotational increases occur when lead and trail points are both over solid land. However, due to the planet's proximity to the sun, these points are half-molten, and the tidal bulge overcomes the torque force.

Extrapolation of Solar Torque on other Planets

The further the planets are away from the Sun the less gravitational acceleration are applied on them and therefore the weaker the torque received, which may appear relatively null when compared to the size of the gas giants of Jupiter, Saturn, Uranus, and Neptune, but they are not. If the gas giant rotated with none or little tilt about its axis, then their equatorial speed would be slightly faster than the rotational speed on the poles. From the Cool Cosmos website at CalTech we learn that "Jupiter is the fastest spinning planet in the Solar System rotating on average once in just under 10 hours… Jupiter's equator rotates a bit faster than its polar regions at a speed of 28,273 miles/hour (about 43,000 kilometers/hour). Jupiter's day varies from 9 hours and 56 minutes around the poles to 9 hours and 50 minutes close to the equator." How is this possible? We know from www.theplanets.org. "Jupiter does not experience seasons like other planets such as Earth and Mars. This is because the axis is only tilted by 3.13 degrees" It is because of the second fact showing almost no tilt that rotational torque is noticeable on Jupiter. Jupiter is the largest of all the planets. It has a radius of 71,492 kilometers at the equator and a radius of 66,854 kilometers at the poles.

Jupiter orbit radius	Distance	Sun Gravity Accl	Trail push	9.999907E-01	**Torque Force**		Tidal Bulge	Tidal Bulge
Perihelion 740.52 Mil km	7.4052E+11 m	2.420518E-04	Lead pull	1.000009E+00	4.2910E+18	-4.2910E+18	7.90596E+04	274.1691
Aphelion 816.62 Mil km	8.1662E+11 m	1.990407E-04	Lead pull	1.000008E+00	2.9015E+18	-2.9015E+18	6.50112E+04	225.4510
Average Orbital Radius =	7.7857E+11 m	m/s²	Trail push	9.999923E-01	**Force in Newton**		Jupiter(km)	Sun (km)

In the chart above, we see that at Jupiter's perihelion, the Sun's gravitational acceleration on the planet is 2.420518E-04 m/s² generates the torque force on lead point is at 4.2910E+18 Newton, and the trail point is at -4.2910E+18 Newton. At Aphelion, the Sun's gravitation acceleration on Jupiter is 1.990407E-04 m/s² creates the torque force on the lead point is at 2.9015E+18 Newton, and the trail point is at -2.9015E+18 Newton. Both lead and trail points occur at or near the equatorial line and between shadow and sunlit regions of Jupiter. This microscopic acceleration is enough to cause gas particles to increase speed at the equator. To calculate the solar gravitation acceleration effect on the rotational speed of the planet Jupiter we take 2.189710E-04 meters per second squared and divide it by the product of the radius of the planet times 2π. This amounts to an adjusted acceleration of 4.98495E-13 meters per second squared. When compared to the average rotational speed of Jupiter is 11,944.44 meters per second, we can see that the additional solar gravitational acceleration at Jupiter's distance from the sun barely affects its rotation speed. With the solar torque force acceleration of 4.98495E-13 meters per second squared, we see that it sustains the gas giant's rotational speed and the force behind its fast-moving atmosphere.

Saturn on the other hand has a tilt of about 26.7 degrees, and therefore the rotational torque is spread over a region around its equator like Earth. This causes this gas planet to have three separate rotational periods. Per NASA, System I, around the planet's equator, rotates once in about 10 hours and 14 minutes. Above and below the Equatorial Belt, System II rotates at about 10 hours and 39 minutes. System III, at the poles, rotates around 10 hours and 45 minutes.

Saturn orbit radius	Distance	Sun Gravity Accl	Trail push	9.999939E-01	Torque Force		Tidal Bulge	Tidal Bulge
Perihelion 740.52 Mil km	1.4E+12 m	6.772139E-05	Lead pull	1.000006E+00	2.3304E+17	-2.3304E+17	1.21951E+05	274.1691
Aphelion 816.62 Mil km	1.5E+12 m	5.899286E-05	Lead pull	1.000005E+00	1.7684E+17	-1.7684E+17	1.06233E+05	238.8318
Average Orbital Radius =	1.45E+12 m	m/s^2	Trail push	9.999947E-01	Force in Newton		Saturn(km)	Sun (km)

In the chart above, we see that at Jupiter's perihelion, the Sun's gravitational acceleration on the planet is 6.772139E-05 m/s^2 generates the torque force on lead point is at 2.3304E+17 Newton, and the trail point is at -2.3304E+17 Newton. At Aphelion, the Sun's gravitation acceleration on Jupiter is 5.899286E-05 m/s^2 creates the torque force on the lead point is at 1.7684E+17 Newton, and the trail point is at -1.7684E+17 Newton. In this case, Saturn's torque force is spread out over System I, an equatorial band, which in turn transfers energy to system II and then to system III.

Extreme Acceleration of Particles to Massive Object: Extreme acceleration makes the lead edge approach infinite gravity (and per Einstein) and creates infinite negative gravity or antigravity on the trail end. Particles of light, electron, and photon, travelling past a massive object at light speed develop temporary super "gravity" at the lead edge and simultaneously increase "antigravity" on the trail edge. These pulsating temporary gravitational increases and decrease causes the light wave-particles to turn sharply or bend, and follow the geodesies, Einstein defined as "additional gravitational energy".

Summary of Paper: Solar gravity planetary torque sustains planetary prograde rotation, counterclockwise motion, while orbiting the sun. Gravity torque enhances or at least sustains the conservation of angular momentum. The solar rotational torque is another way to explain conservation of planetary angular momentum. Newton third law, action reaction law, applies even within one object. Any temporary increase in gravity (or mass energy) on an object accelerated must be counterbalanced with the exact decrease of gravity (or mass energy) in the opposite side of the same object, to achieve a zero-sum gain over the entire object. This principle is required for Newton's laws and gravity equation to remain valid throughout the motion of all objects in space.

3.8 Stellar-Planetary Gravity Influence on Weather

Gravity Redistribution **Abstract**

Having survived two super typhoons, I often pondered what force could quickly intensify a storm from category 1 to 5+ in a matter of hours; surely, solar heat effect is insufficient. Where did the extra energy come from? The article, *Hurricane's atmospheric gravity waves help predict the storm's path* (Zhang, 2017), on May 16, 2017, announced that meteorologists (David Nolan and Jun Zhang) believe they have found a new way to track the intensity and trajectory of hurricanes by measuring atmospheric "gravity waves" emanating from the storms' centers. Storms do not create gravity; they respond to the gravity force. This article analyzes the sun's gravity, extended angular momentum, and the equivalence principle to define stellar gravity interaction on planets and in doing so identifies torque in gravity redistributions. Specifically, different portions of a planet tend to fall at different rates toward their star. Planetary inertia, tidal force, gravity acceleration and its negative, centrifugal, and centripetal, gyroscopic effects, reactions, and subsequent vectors all produce emergent forces. These variances consequently help us plot the locales and time of solar and lunar gravity disturbances on earth, and when applied to historical storm tracks reveals a pattern and cause behind weather systems, and a force that develop and intensify tropical storms and cyclones and influence their paths.

Introduction: Weather forecasting science has made tremendous strides and progress over the last several decades, particularly in accuracy and understanding, all due to the vast array network of deployed data collection devices, and the processing speed of modern-day computing power. Meteorologists typically present forecasts with confidence, regardless of the outcome, and at best alert the intended recipient for what might be. Forecasters periodically gather global data on water and air temperatures, directional flows, density, and pressures measurements, and apply fluid dynamic principles to develop weather prediction models. Complex models demand massive computer processing power to produce more granular resolution images, attributing heat energy rising motion, the Coriolis Effect, and fluid dynamics as the primary source of storms (NOAA, GFDL, 2016) while occasionally spotting and recording the peaks and troughs of illusive gravity wave flux (Fritts, 2016).

NOAA identifies gravity as a major force responsible for creating tides, where the moon's gravity bulges the closest water surface, and inertia counterbalances this with an opposite tidal bulge on the farthest water surface on Earth (Ross, D.A., 1995). This gives the planet two major bulges or high tides, and two areas of low tides (NOAA_Ross, 2017). Similarly, the sun's gravity causes two other tidal bulges on the planet, but because of its distance, are minor. The Earth's rotational speed makes the ocean high and low tides transition easterly nearly every six hours fifteen minutes at the equatorial band. Moreover, the earth's axial tilt, the sun's position in the sky and distance, the time of the year, plus the moon's orbital location and distance, influence the significance of high/low tides. Atmosphere fluidity responds similarly (Alexander J. , 2017), where low-pressure systems resemble the thicker bulge or "wave peak," and high-pressure is the lower bulge or "trough." The isobars or fronts are the neutral areas, where destabilizing effects of gravity waves appear in cloud patterns (Baumgarten, 2017). *Orbit* aired in 2012 takes one seasonal cycle in the Earth's journey around the sun, highlights its influence on weather, identifies earth's rotation as the reason air and ocean currents follow the Coriolis Effect, and ocean currents flow along the continental coastline (Colville, Nurmohamed, & Taylor, 2012). Gravity or more specifically, micro disturbances in gravity, or transitory redistribution of the gravity force power atmospheric vortexes. This paper calculates solar and lunar gravity torque (with average orbital paths), plots locale and times of the gravity disturbances, analyzes their effect on past weather systems, and draws a rudimentary correlation to weather predictions. It further integrates the gyroscopic effect and the vector force input output as an alternate way to view gravity energy redistribution that create tidal bulges and change atmospheric pressure.

Point and Angular Momentum: NASA defines that the most important invariants of the motion of a celestial body are the absolute values of spin angular momentum and orbital angular momentum. With noteworthy exceptions to this general rule, tidal effect, the absolute values of these vectors have remained essentially constant since the formation of the bodies, unless acted on by an external torque (NASA, Solar System Evolution, 2018). The orbital paths of planets around the star are point angular momentum, $L = r\, m\, v$, where the angular momentum L equals radius r times mass m times velocity v. Planetary rotation is extended angular momentum, $L = I\, \omega$, where L equals the rotational inertia, I, times the angular velocity ω. Hence, the conservation of angular momentum in solar system formation typically results in vectors aligning. Planets tend to rotate prograde in a counterclockwise path about a counterclockwise spinning star, else friction exists (i.e., Venus), and vice versa in a mirrored system. Massive collisions can cause vector misalignment.

In the modern equivalence principle, Einstein proposed that an occupant is unable to distinguish the effects felt or the light-bending observed (Marmet, 2012) in an elevator cab accelerating through some point in space where no gravity fields act, to that of a stationary elevator sitting in a planetary gravitational field (Guidry, 2011). In short, acceleration equals gravity, and the opposite is "negative acceleration equals antigravity." Acceleration alone does not create additional gravity or mass over the entire object in motion, it merely redistributes the object's gravity over the surface area between the lead and the trail vectors, mainly at two detectable key points on orbs. An individual who feels an increase gravity standing on an object accelerating in the vector direction would similarly profess an equal but decrease of gravity, or antigravity, from the negative acceleration on the opposite side (action-reaction law), giving a total zero-sum gain on the object's overall gravity force, fig 1.

Direct application of Newton's gravity equation on the surface of the ocean closest to the moon or the sun defines the inline-gravity upward vertical force from the earth's surface that causes tidal bulge, while the inertial force on the planet at its opposite side as the source of the tidal bulge there. Solar gravity force at earth's average orbital radius $F = G * (\{m_1 * m_2\}/R^2)$ is $6.6734E-11$ N $(m/kg)^2$ * $1.989E+30$ kg * $5.9724E+24$ kg / $(1.49598E+11\ m)^2 = 3.542255E+22$ N. So, the solar gravity force at the surface closest to the sun is $F = G * (m_1 * m_2) / (R - r)^2$ at orbital radius R minus the earth's radius r, equals $3.542557E+22$ N, and the difference between both forces gives us the solar tidal force of **$3.017312E+18$ N**. Similarly, the average lunar gravity force is $1.981437E+20$ N and the nearside force is 2.048787E+20 N, gives a lunar tidal force of **6.734999E+18 N**, twice more than the solar tidal force. Hidden between the tidal and inertial forces are solar and lunar gravity torque forces that augment planet angular momentum, and drive fluids toward tidal bulges, via gyroscopic input/release points, or gravity disturbances.

Solar-Planetary Rotational Torque: Gravity and torque mutually exist when an object orbits another in space. Pawet Zagorski's article, *Modeling Disturbances Influencing an Earth-orbiting Satellite*, discusses how the Earth's irregular surface and the uneven mass distribution of a non-spherical satellite create gravity torque, where its different mass parts fall at different rates... and that a slow spin about its center of mass and center of gravity adds stability (Zagorski, 2012). He defined this gravity torque as $T_G = (F_2 - F_1) \, l/2 \sin \alpha$, where F_1 and F_2 are gravity forces on each part; l is the line length between them, and α is the angle between line l and F_1 direction vector. Divide the torque force by the total mass to get the rotational acceleration force. The torque force τ on a gyroscope can be written out as $\tau = \omega_s \omega_r M R^2$, where ω_s is swiveling rate, ω_r is rolling rate, M is the mass, and R is the radius of the gyroscope (Cleonis, 2017). In planet terms $\tau = \omega_p \omega_s \omega_r M R^2$, the pitch rate ω_p equates to axis tilt change, swivel rate ω_s equals change in its precession, and roll rate ω_r is its rotational speed, where a pitch change causes swivel motion and vice versa.

To visualize solar and lunar torque, and planetary gravity disturbance effect on orbs, let us briefly imagine three individuals standing on a non-rotating rogue earth-size planet accelerating through intergalactic space as depicted in fig 1a. The equivalence principle gives these effects. The person on image top is at the lead acceleration vector, which passes through the planet's greatest mass, experiences normal gravity plus an additional amount by acceleration. Another at the opposite side notices a weakened gravity (minus acceleration), while the person standing at 90° longitude to both feels normal gravity. A liquid-covered planet would then conform into a teardrop shape, with "low-tide" "air-high-pressure" forming in the acceleration direction, and "high tide" "air-low-pressure" accumulating on the opposite, and a narrow band, mid-section of normal gravity, average sea levels, and atmospheric isobars. Take the same zero tilt planet accelerating through space, and rotate it at time interval t. At time $t2$, each person briefly feels the same effect as his or her colleague on the non-rotating planet, fig 1b. Similarly, a rotating liquid-covered planet initially assumes a similar teardrop shape, which changes as the planet rotates on time t. Due to the gyroscopic effect, a water-planet forms two low-tides air-high-pressure in each end of its acceleration direction, and two "high-tides" "air-low-pressure" on both sides perpendicular to its trajectory. Now, tilt the rotational axis by 23.5° and adjust spin to 460 km/h to get earth's gravity and antigravity points, and reflective points above below equator.

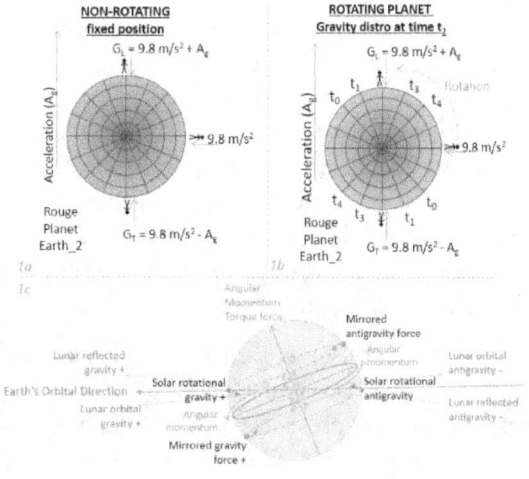

Figure 1: Acceleration and Gravity

How do these forces apply on an "anchored" spherical gyroscope? Obtain a model globe of the earth tilted at 23.5° to the left and anchored at both axes. You represent the sun. Touch a left finger to the globe at the left point parallel to the ground level at mid-section and pull it toward you (gravity +), while pushing the opposite right point on the globe with equal force (gravity -); the line between your two fingers is the acceleration vector through the planet's greatest mass. Your action makes gyroscopic forces travel around the globe to a 90° line, perpendicular to the acceleration vector, closest you, and concurrently to a 90° line opposite you and back again. The globe reacts to pull forces in two places on the left and push forces in two places on the right, equal-distant above and below the equator, and at points closest to sun and furthest at the equator during spring and autumn equinox, fig 1c. During the Summer Solstice and Winter Solstice, the left two, and right two points crossover each other at the equator and the closest center point diverges into two secondary points, and the opposite spot into two. These torque vectors transfer angular momentum forces and are a byproduct of the gravity forces that keep objects in orbit, as well as contributing energy to the forces that create the tidal bulges on Earth. Hence, a spinning sphere in space gyroscopically transfers vector forces to points along the longitude perpendicular to its acceleration direction and assists in the formation of tidal bulges along the "gravity line" between objects in orbit. With that said, we are now ready to discuss solar-lunar-earth torque forces.

Acceleration and Gravity: Using NASA's data estimates, we calculate the solar-planetary torque with these equations. The average gravity acceleration near Earth's surface is the product of the gravity constant G times the mass of the Earth 5.9724E+24 kg divided by its radius 6,371,008 m squared comes to 9.8192822 m/s². The Sun's gravitational acceleration at Earth's average orbital radius is gravity constant times the sun's mass divided by Earth's average orbital radius, which is 6.6734E-11 N (m/kg)² x 1.989E+30 kg / (1.49598E+11 m)² = 5.931041E-3 m/s². Thus, Earth's average orbital velocity about the sun equals the average circumference divided by the number of seconds in an average year or 2π*1.49598E+11 m / 31,557,600 s = 2.978526E+4 m/s. Earth experiences a slight increase of gravity at the surface of the vector direction tangent to the orbital path, and an opposing decrease on other side. By adding the Earth's gravity plus the sun's gravitational acceleration divided by the Earth's gravity, or 9.8192822 m/s² + 5.931041E-3 m/s² / 9.8192822 m/s² we get 1.00060402 G's. Simultaneously, at the opposite trail end the planet experiences a decrease in gravity of 0.99939598 G-forces found by the Earth gravity minus acceleration through space divided by normal gravity. Hence, 1 G normal gravity band exists between the lead and trail points. The G_L lead gravity is $1 + (M_S*r^2) / (R^2*M_P)$ and trail gravity G_T is $1 - (M_S*r^2) / (R^2*M_P)$, where G_L and G_T are the lead and trail gravity acceleration, M_S and M_P are the star and planet mass, R is the orbital radius, and r is the planet's radius.

At the earth's average orbital acceleration, the *Solar-Planetary Rotational Torque* generates both an increase and decrease of gravity, at 1.00060402 G's and 0.99939598 G's respectively, which sustains its prograde spin. Zagorsk's torque formula $T_G = (F_2 - F_1) \, l/2 \sin \alpha$, gives about (3.5444E+22N - 3.5401E+22N) * 6,371,008m/2 * 0.894006 = 1.218652E+26 kg (m/s)² Joule of the sun's gravity energy to spin earth, and lunar torque is (1.981443E+20N - 1.98143E+20N) * 6,371,008m/2 * 0.894006 = 3.81312E+21 kg (m/s)². Gyroscopic formula $\tau = \omega_p \omega_s \omega_r MR^2$ gives torque force of (2.52947E-11 m/s) * (4.28462E-05 m/s)*(463.83102 m/s)*(5.9724E+24 kg)*(6,371,008 m)² = 1.218614E+26 kg m² (m/s)³, leaving a difference of about 3.813122E+21 kg m (m/s)² accounted as lunar torque influence. The solar torque acceleration rate of the earth at 1.21865E+26 kg (m/s)² / (5.9724E+24 kg * 6,371,008/2 m) = 6.40549E-06 m/s² sustains rotation.

Substitute G_L and G_T above for F_2 and F_1 in Zagorsk's equation, we get the planetary torque acceleration equation $A_T = 2 * (G * M_S^2 * r^2) / (M_P * R^4) \sin \alpha$, or rotational acceleration where G is the gravitational constant, M_S is the star's mass, M_P is the planet's mass, r is the planet's radius, and R is the orbital radius of the planet. As the orbital distance increases, angle α approaches 90° at perihelion and aphelion, but varies less on inward and more on outward elliptic. Earth rotates faster than the moon orbits, so the moon's increase and decrease gravity 1.000003379 G's and 0.999996621 G's respectively, plus solar gravity, cause tides to rise and fall, and the equatorial currents to flow easterly in a band between 30° N and S of equator (Earth Science, 2018). Lunar torque at 3.81312E+21 kg (m/s)² or Joules has 0.0031% less energy than solar torque, affects earth's spin at 2.0043E-10 m/s² =3.81312E+21 kg (m/s)² / (5.9724E+24 kg * 6,371,008/2 m). Increased lead gravity forces $(G_L-1)*G*M_S*M_P/R^2$ and decreased trail radiate outward and dissipate uniformly in concentric circles forming peaks and troughs average to 9.8192822 m/s². The earth's acceleration around the sun creates a 9.8252133 m/s² gravity at its lead point and 9.8133512 m/s² at its trail, each with (+ or -) 2.139592E+19 N angled vector forces on the earth. The moon-earth mutual orbit creates an increased gravity of 9.81931541 m/s² at lead vector and 9.81924906 m/s² on opposite side, at angled vector forces of 6.694710E+14 N. The new moon and sun alignment provide a gravity increase of 9.826579 m/s² at the lead vector, and a decrease of 9.811986 m/s² on the trail, at momentary combined 2.265867E+19 N angled vector force. Since solar gravity torque are over 700% greater than tidal forces and concentrated on a relatively small area, they alter atmospheric conditions and can intensify or weaken storms.

Gravity Disturbance on Atmospheric Pressure: The solar-antigravity spot occurs at 6 pm local standard time LST derived from Greenwich Mean Time (GMT) plus/minus the longitude difference divided by 15, is a location where the earth's gravity changes from the normal 9.8192822 m/s² to 9.8133512 m/s². This antigravity pulse can decrease sea-level atmosphere millibars (mbar) at a rate of 0.0604% per second in low-pressure system over the area, and so two minutes within a solar antigravity area, over land on a non-rotating planet, can decrease atmospheric pressure from 1,013.2500 to 876.4763 mbar or from 14.696 to 13.650 lbs. per square inch. Solar torque at 6 am changes the earth's gravity from 9.8192822 m/s² to 9.825213 m/s² and heightens the local atmospheric high-pressure system from an average of 1013.2500 to 1169.8505 mbar in two minutes. These unnoticeable gravity fluxes gradually occur in time, can change tide and air pressure.

Similarly, the moon's antigravity point can lower atmospheric pressure from 1013.2500 to 1012.4287 mbar in two minutes, strengthen low-pressure systems, tropical depressions, and gradually intensify hurricanes. The moon's gravity+ torque enhances pressure from 1013.2500 to 1014.0720 mbar and weakens low-pressure systems and storms. Their combined solar-lunar negative torque reduces atmospheric mbar from 1013.25 to 869.0611, while the solar-lunar gravity+ point increases pressure from 1013.2500 to 1181.2459 mbar both influence weather systems. Since solar gravity+ and antigravity areas radiate concentrically outward and move easterly across the globe at approximately 1,000 mph or about 460 m/sec (NASA, NASA Image Education Center, 2017), it spreads the higher/lower pressure into a shallow trench/ridge and affects all atmospheric system it passes through. First and third quarter moons generate tidal bulges along earth's solar acceleration vector, a micro increased mass, and a minute increase in solar gravity and antigravity torque. But a full or new moon produces two tidal bulges aligned with sun, and low tides along earth vector path, with slightly less mass, but greater combined gravity and antigravity torque at new moon, and a decrease during full moon.

Gravity Disturbances Effect on Storms: Now the earth's and moon's acceleration through space in their mutual orbits around the sun creates gravity disturbances over specific areas on the planet and how these multiple forces increase or lower local air pressure, let us analyze effects on weather systems and ultimately their impact on storms.

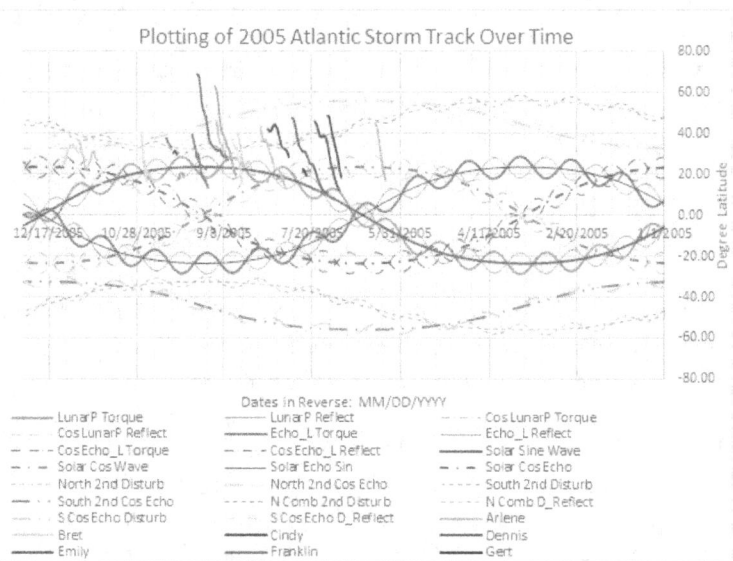

(Fig2: 2005 Atlantic Storm Track) The primary solar gravity and antigravity torque forces found at the LST of 6 am and 6 pm, with secondary at 12 noon and 12 midnight, echo at 3 am/pm and 9 am/pm, and reflective spots all at equal distance above and below the equator. These positive and negative gravity forces further disperse to additional equally spaced areas around the planet to form a gravity disturbance or energy grid. depicts the global path of the primary and secondary solar, lunar, and their combined torque points, and reflective spots during 2005, and plots the recorded tracks of 2005 N. Atlantic storms. The solid line sine-wave peaks represent earth's max tilt (23.5°) at the spring and autumn equinoxes. The intersections of the solid lines indicate the summer and winter solstices (Navy, 2016). The dashed lines show the secondary points, 90° off-phase, which possess reciprocal inline gravity disturbance aligned with the sun's

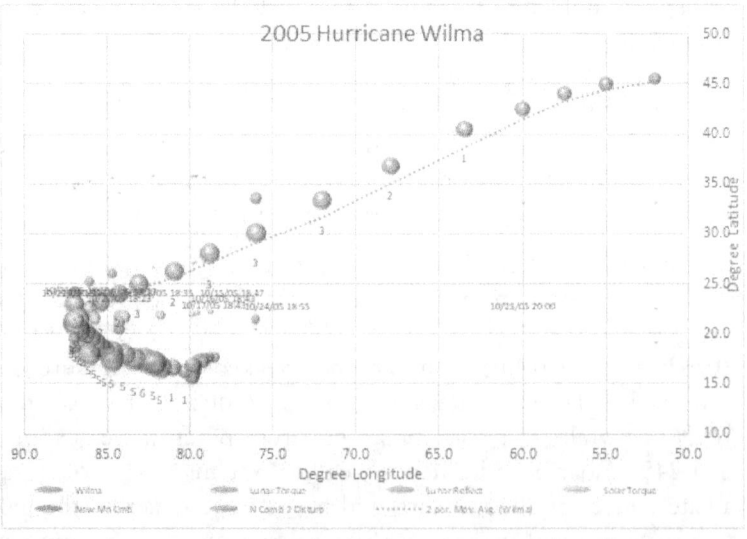

position in the sky. (Fig3: 2005 Hurricane Wilma – blue spheres) The wavy pattern, lunar cycle data (timeanddate.com, 2018), at 5.14° above and below solar torque lines, indicate the lunar torque points has enough gravity energy to start a low-pressure system, create tropical depressions, and storms.

In plotting tropical cyclones' position data for 2005, 2016, and 2017 (Figures 3, 4 and 5) from (NOAA, Weather Underground Storm Data, 2017) as blue orbs (size=wind speed; # = category), then overlaying the gravity disturbances spheres (size = energy transfer), we see the interactions between them. The intersection or close approach of these negative gravity disturbances intensify storms and or alter their direction. Similarly, the overlaying of positive gravity disturbances shows a weakening or steering of storms.

(Fig4: 2016 Super Typhoon Meranti -blue spheres)

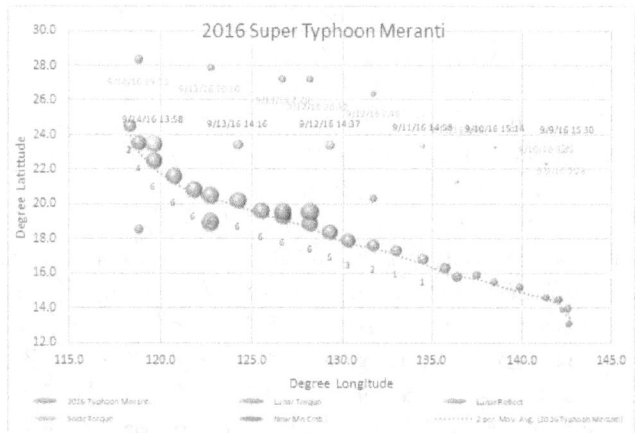

In this paper, the fictitious category "6" depicts super storms more than 175 mph, and the term "antigravity" refers to the negative aspect or decrease of gravity. Gravity redistribution, at its strongest gravity (or antigravity) "point," occurs within a 3-mile radius area, and dissipate concentrically outward at progressively weaker (or stronger) gravity at 3-mile increments until normal gravity is achieved. Its effects stretch outwardly to form a high (or low) pressure system front. Enhanced gravity strengthens local high pressure, and weakens low-pressure systems, while antigravity strengthen low-pressure areas and weaken high-pressure systems. Moreover, analyzing echo gyroscopic forces reveals additional energy behind storms like Hurricane Katrina in 2005, and others not near or directly over primary or secondary gravity disturbance points. Estimating a weather system's acceleration toward or away from a gravity disturbance can be complex. Take Newton's acceleration force $A = G*M/r^2$, disregard all other factors, and then determine the gravity disturbance "equivalent mass" M. Solar negative torque changes earth gravity from 9.8192822 m/s² to 9.8133512 m/s², amounting to 5.931041E-03 m/s² in a 3-mile radius. Here 5.931041E-3 m/s² represents the negative or positive acceleration superimposed onto rising warm or sinking cold air currents that triggers the central vortex or hub of low- and high-pressure systems respectfully. This energy is the power behind the low-pressure weather systems and intensification of depressions, storms, cyclones.

Solving Newton's formula $m_d = a_t r^2 / G$, given the acceleration a_t of 5.931041E-03 m/s², and radius r of 6,371 km, the "mass" m_d is 3.607448E+21 kg. The lunar gravity force at 3.31766E-05 m/s² acts as a "mass" of 2.0179E+19 kg, and a combined lunar-solar acceleration at 5.964218E-03 m/s² has a "mass" equivalent to 3.627627E+21 kg or about 0.06% of the earth's total mass behind it. The gravity disturbance "mass" has sufficient energy to influence weather system or cyclone within its grasp by initiating a vacuum when it enhances upward acceleration of warm air currents into the atmosphere or boosts downward motion of cold air toward the earth. As air currents flow toward the vacuum center of rising updraft, the Coriolis Effect turns them to the right, giving the cyclone a counterclockwise spin in the Northern Hemisphere, and to the left, giving a clockwise spin in the Southern Hemisphere. By applying Newton's gravity acceleration equation to a weather system, given its distance in meters from the gravity disturbance center, we correlate effects. The solar gravity torque acceleration rate is the product of the gravity constant 6.6734E-11 N (m/kg)² times "mass" of 3.607448E+21 kilograms divided by the distance squared between the system's and the gravity disturbance's center. This acceleration causes lateral movement of weather systems, where the "mass" either pulls or pushes the system, and or intensifies

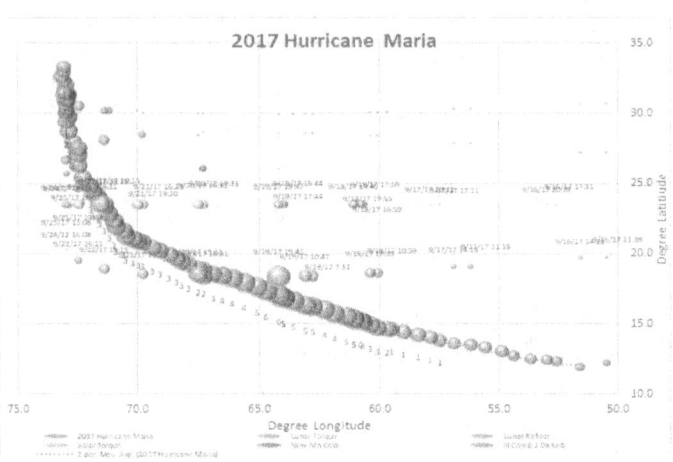

or weakens them. At about 20 kilometers (just over 12 miles), we get a solar gravity disturbance acceleration rate of 601.85 m/s² (or 0.374 mph/sec), a lunar gravity disturbance acceleration rate of 3.367 m /s² (or 0.002 mph/sec), and a combined lunar-solar acceleration rate of 605.22 m/s² (or 0.376 mph/sec). At 5000 km (3-miles), we get a solar disturbance acceleration of 9,629.58 m/s² (5.984 mph/ sec), the lunar torque rate of 53.865 m/s² (0.033 mph/sec), and combined rate of 9,683.44 m/s² (6.017 mph/sec), which all move along rotational geodesic lines at 1,000 mph or 460 m/s.

(Fig5: 2017 Hurricane Maria -blue spheres)

Equally, transient gravity disturbances affect storms caught within their range. Turn rate depends on the proximity of the storm's center to gravity disturbance's center and the angle of approach between both. To view this correlation, suppose a storm traveling northwest at a speed of 4.5 m/s (about 10 mph) passes just three miles to the north of an antigravity disturbance's center which is traveling west (along a geodesic path between lead and trail acceleration points).

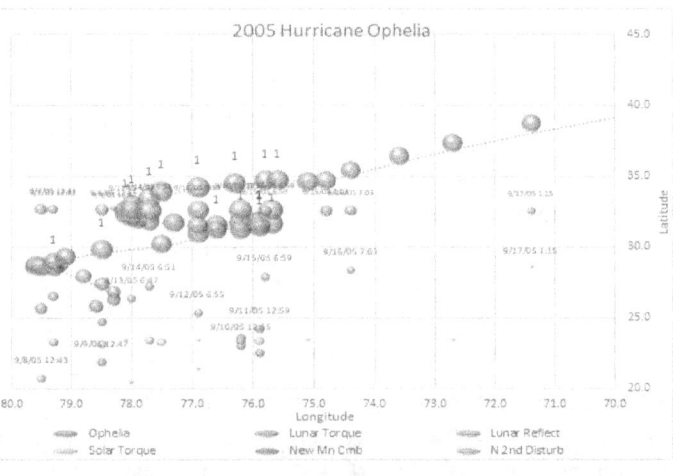

(Fig6: 2005 Hurricane Ophelia -blue spheres)

That storm will be pulled into a left turn and accelerated toward the west and may intensify. If that same storm was just three miles south of same antigravity disturbance, the storm will turn to the right into an intersecting path and intensify. Another pass by the same antigravity area will turn the storm to the right again and continue to make one complete loop; see fig 6 Hurricane Ophelia looped twice.

Three-year storms' data analysis revealed that when primary gravity disturbances traverse directly through the storm's center, they have the greatest effect on their intensity, either the negative gravity increases it, or gravity+ decrease it, and or simultaneously adjusts its path and speed. As antigravity disturbances traverse westward along the geodesics and south of the storm's center, they tend to pull or turn a storm's path southward depending on the proximity and strength of the disturbance; the storm reacts as though it is being pulled into a moving strong low-pressure trench or vacuum. Each pass the disturbance makes feeds more energy into the systems to either intensify or reduce its strength and or alter its path depending on proximity. Antigravity disturbances passing north of the storm's center pulls it northward, while positive gravity disturbances that traverse past and south of the storm's center at the lower latitudes tend to turn that storm north and gradually decrease its intensity depending on intersection proximity. The storm moves out of and away from developing high-pressure ridge. An increased gravity disturbance passing north of a storm pushes it south.

Conclusion: The solar-gravity acceleration of earth at 5.931041E-03 m/s^2 and lunar gravity acceleration at 3.31766E-05 m/s^2, and their combined alignment of solar lunar gravity disturbances rate of 5.964218E-03 m/s^2 has sufficient energy within the transient negative or positive gravity points to either trigger or dissipate the formation of weather systems, attract, or repel, and intensify or weaken them. In addition to generating the largest tidal bulges on earth, the alignment of the primary solar and lunar gravity disturbances, during a new moon, produces the greatest of intensity and the most rapid changes in weather systems within its clutch. A periodic alignment of *negative* solar and lunar gravity forces passing near or directly through a storm's center can significantly energize it from category one up to five plus in a short period of time or cause it to loop around once before continuing its original track, or both. Similarly, the combining of *positive* solar and lunar gravity forces can also weaken a category five storm to a category two or one cyclone or a tropical storm in a short period. Moreover, the secondary, tertiary, or other residual gravity disturbances provide weaker forces that power storms gradually.

Solar-planetary rotational torque and its gravity redistributions apply to all planets orbiting around stars, and to the stars themselves. Gravity disturbances generate a moving grid of peaks and troughs as the additional energy behind the ever changing and evolving high- and low-pressure systems around the world. Zero-tilt planets have horizontal atmospheric currents as their gravity disturbances are along its equator. The greater the planet tilt the more diverse is its weather patterns and global gravity redistribution grid, while the greater the star's distance the less its effect. When a gravity disturbance travels over land, it tugs solid mass, sustains rotation, and has less impact on atmospheric conditions, partially transferring its energy to and affecting low- and high-pressure systems, while disturbances over open water or fluidic surface, cause atmospheric changes, influence tides, and affect weather systems. The lead enhanced gravity formula is $G_L = 1 + (M_S * r^2) / (R^2 * M_P)$, and trail decreased gravity is $G_T = 1 - (M_S * r^2) / (R^2 * M_P)$. The stellar-planetary torque spin acceleration is $A_T = (G_L - G_T)\,l/2 \sin \alpha = 2(G * M_S^2 * r^2) / (M_P * R^4) \sin \alpha$, when the orbital distance goes to infinity the angle α approaches 90°. The stellar gravity acceleration is $a_\sigma = GM_S/R^2$. The solar gravity redistribution "mass" equation is $m_d = M_S * r^2 / R^2$. For planet gyroscopic effects, the torque equation is $\tau = \omega_p \omega_s \omega_r M_P r^2$. In all these equations, G_L and G_T are gravity lead and trail points tangent to the orbital path, M_S is the star's mass, M_P is the planet's mass, R is the orbital radius, r is the planet's radius, G is the gravitational constant, a_σ is stellar acceleration, and m_d is the "mass" that gravitationally influence weather systems. In the torque equation, the pitch rate ω_p is change in its axis tilt, swivel rate ω_s is change in its precession, and roll rate ω_r is rotational speed. A pitch change causes a swiveling motion and vice versa, while a torque acceleration change $F_t = A_T$ primarily affects the torque $\tau = F_t r$ and rotational rate ω_r.

Four stellar forces act on orbiting planets: gravity at lead half on acceleration vector, centrifugal at closest half to star, antigravity at trail half, and inertia at far half, while a planet's elastic gravity opposes each respectively with equal levels of antigravity in first two, and gravity plus in others. Then, due to gyroscopic effect, these response forces are released 90° out along their geodesic spin line, align to increase gravity and antigravity on acceleration vector as stellar torque, while the near and far forces fight to hold its spherical shape, or released in place for non-rotating planets. Lunar gravity has similar weaker forces and responses. Action reaction, positive-torque vector force is offset by negative-torque vector force, and vice versa, where both forces energize weather systems as they diverge. As an object orbits another, its acceleration through space creates torque, and its rotation develop gyroscopic interactions, and when combined with linear gravity form and move tidal bulges. Using earth's and moon's actual orbital tracks provide greater accuracy, and a more relevant and predictive correlation between gravity redistribution and weather systems; accounting for tilt variance, elliptical distance, and precession are beyond this paper. I used the average sun's position in the sky to derive solar lead and trail points, and the moon's average position for lunar lead and trail points, which are both, shifted ahead 90° off phased, then overlaid known storm tracks on them and deduced interactions.

PART 2: GRAVITY OF THE GRAVITON

Chapter 4: The Natural State of Balance

"The more clearly we can focus our attention on the wonders and realities of the universe about us, the less taste we shall have for destruction."

— Rachel Carson

Our universe is abundant with examples of balance or shift toward balance and symmetry, not chaos, as many so believe. We know this symmetry as the translational invariance or translational symmetry, where turning an object on a particular axis gives us the same object. A round ball has perfect symmetry. It does not matter which way you turn the ball; it is still the same ball. Nature tends to cause changes in things towards obtaining and staying in balance and symmetry. We have balance in the micro world, the macro universe, in physical and natural laws, formation of solar systems, galaxies, and in the universe. The structure of the universe is isotropic, almost self-similar[91]; therefore, computer-generated replication by fractal equations emulates it sufficiently. Natural and physical laws are constant and not autonomous to the motion and structure of the solar system, the galaxy, and the universe. Balance governs everything that goes on around us and symmetry translates to laws in physics. Nature prefers stability and balance to chaos. Some cultures have created a study centered on this relationship alone.

For this book, the most significant of these balances is the relationship between the four natural forces that control and govern all matter in the visible universe. How does the weak nuclear force affect the gravitational force, the electromagnetic force, and the strong nuclear force? How does the electromagnetic force affect the gravitational force, the weak nuclear force, and the strong force? How does the strong nuclear force affect the gravitation force, the weak nuclear force, and the electromagnetic force? Finally, how does the gravitation force affect the weak nuclear force, the electromagnetic force, and the strong nuclear force? The study of these relationships is the key to solving the mysteries of the past and unveiling our potential as a space-going race. There are four natural forces, which govern all visible matter, as we know it. One governs the macro world and the other three govern primarily the micro world. The three natural forces that govern the micro or inner universe are the weak nuclear force, the electromagnetic force, and the strong nuclear force. Each of these force characteristics are balanced with a negative, positive, and equal aspect. Without that balance, the forces would cause the atom to break apart or be crushed onto itself[92].

---

[91] Symmetry in the micro world and the macro universe defines physical laws. For instance, water molecules assume different symmetries as it passes through the different phases from solid ice, to liquid water, to gases and vapor; it symmetry defines its characteristic.

[92] The balance between atomic particles within an atom enables it to hold symmetry. The same balance and symmetry, found throughout the micro world as well as the macro world and the physics of the universe, defines gravity. Hence, the forces the produce gravity is expected to be positive, negative and neutral.

4.1 Balance in the Natural World

There are three types of matter in the universe: ordinary matter with gravity; antimatter with antigravity characteristics; and space-matter with neutral or zero gravity. From our perspective, gravity is hereby considered a positive force that pulls objects together; antigravity, a negative force that repels gravity objects away; and neutral matter (space-matter) that has zero gravity, which neither attracts nor repels the other two types of matter. Neutral space-matter is possible when the limited distance strong nuclear force is near equal to the electroweak force, where their quarks point-particles' motion outward is practically equal to their motion inward, but not enough to produce a steady "gravity" or "antigravity" field, symmetrical balance in a spherical shape. A result achieved by combining the neutral point particles of gravity matter with that of antigravity antimatter, we get the neutral quark particle contained within space-matter, which generates neither gravity nor antigravity, only the three other natural forces and their interactions. Quantum physics enhances this weak antigravity-gravity to gravity-antigravity outer repulsion and inner attraction mechanism with the quantum entangled force of 1 m/kg. Quantum mechanics or entanglement enables the mirroring of the movement of identical mass objects toward or away from each other. The larger the mass of an object the more quantum attraction it will exert on the smaller object. The next few sections should shed light on this age-old subject, the source of gravity, which still baffles brilliant physicists and scientists since the earliest development of science, the gravity of the graviton.

Nature also is teeming with examples that maintain symmetry and balance. Plants and animals alike develop with balance. The Fibonacci number pattern shows us that plants follow this pattern as they grow with balance in their leaves, buds, stems, or branches. This balance distributes weight evenly and allows the plant to grow to full height and strength. Balance also applies in the chemical world where molecules adhere to each other in patterns, adding strength and uniformity to the structure. For example, the carbon molecules pushed together with heat and pressure into diamonds bonded in perfectly symmetrical pattern, which makes the diamond super strong and almost transparent, a characteristic normal carbon does not have. Symmetrical balance is found everywhere in nature, whether inanimate objects or ones that are alive. Animals and human beings are balanced, left side and right side of the body, male, and female, tall and short, good, and evil, and so on and so forth. Physicists show us that deterministic equations produce chaotic results over long periods, and chaos operates to create a balanced structure; the phase space point-plot is an example of a balanced structure. Fractal geometry equation consisting of a real and imaginary part, developed by French Mathematician Benoit Mandlebrot, are another example of emulating structure found in nature, giving objects the natural jagged appearance, we are so used to seeing in landscapes, clouds, trees, coastlines, rocks, and other natural formations[93].

Gravity is the force the triggers nature to conform to balance. For instance, a tree grows both sides of the trunk uniformly to stay as vertical as possible depending on its environment and surroundings. Tree rings grow outward as the tree ages, oldest part at the center. Gravity acts on a plant's formation of new sections. As the plant grows, the segments of the growing new sections when counted just happen to conform to the

---

[93] Symmetry rules over the laws of nature whether inanimate or the living.

Fibonacci sequence. Gravity also affects how animals evolve, developing appropriate muscular and skeletal structure to equally provide the animal balance to move about. If you think about it, gravity works in balance with cell structure development and formation to create organism suitable to survive on the planet. Gravity influences evolution[94]. As we mentioned before, although gravity is the weakest of the four natural forces, it is the most dominant force over great distances. The three other forces balance each other; similarly, the other three work together to balance the effects of gravity. Gravity pulls object A to object B on the planet, and the other three natural forces within the object B pushes back on object A. Everything in existence in the visible universe is in some sort of balance. Negative electrical forces are balanced with positive electrical forces. Magnetic north is balanced with magnetic south. Protons in the nucleus are balanced with neutrons. Orbital levels of electrons are balanced by photons. Gluons hold protons and neutrons in balance. Negative electrical charges repel negative charges. Electrons repel electrons. Positively charge particles repel positive charges. Protons repel protons. Magnetic north repels magnetic north. Magnetic south repels magnetic south. Neutrons are buffers between protons holding them in place. It is miraculous that the atom and molecules in nature are held in balance in an intricate design between attractive forces and repulsive forces[95].

Gravitational forces or waves acts somewhat like magnetic waves. The closer two magnets come together, with north and south poles pointing toward each other, the stronger their pull of attraction become. The more powerful the magnets are the stronger their attraction. Similarly, the closer two objects come together in space, the stronger their gravitational attraction. Also, the more massive the objects are the stronger their mutual gravitational attraction. Gravitational attractions between any two objects are similar only to the attraction resulting when two opposite magnetic poles come together. In the visible universe, we see gravity in the attractive state pulling all objects together. Magnets have a repulsive nature imbedded into their structure. Specifically, if two like magnetic poles are pushed together, they will repel each other. The closer like magnetic poles are brought together the stronger their repulsion becomes. Likewise, the stronger the magnetic fields are the more you feel and realize the repulsive power between two like magnetic poles. Magnetic energy flows from magnetic north to magnetic south and vice-a-versa. Electrons attract protons, and electrical energy flows to positive. Negative ions flow toward positive ions. Positive particles repel positive particles. Neutron or neutral particles provide buffer between positively charged particles, and so on and so forth. Since all three of the four natural forces have a counterpart balancing them out. Then it is only natural to justify that there exists somewhere out there a counterbalance to gravity. We see gravity energy or a pull force toward objects; this force tugs the two objects together mechanically. This force must then flow from somewhere. That somewhere must be and is the counterpart of gravity; let us call that location the antigravity objects and galaxies. Just as two like poles of magnets experience a repulsive force, the gravity in this universe repels antigravity in the other part of this universe. "Uni" in universe means one.

---

[94] Gravity causes living organisms to produce left-handed organic molecules to move, breath, and live. These lefty molecules enable them to resist the force of gravity.

[95] These same attractive and repulsive forces are found throughout the universe, even in black holes and between galaxies, providing homogenous and isotropic.

What happens if you lay a large group of small bar magnets on a table? The magnets would align themselves north==south, north==south, north==south, and so on and so forth. Like poles will repel each other and unlike poles will attract each other. If necessary, the magnets will flip around to obtain the balance nature intended. If you had enough magnets on the table, the bar magnets can form a complete circle N==S, N==S, N==S, N==S, N==S, N==S, N==S, where the last South pole is reconnected to the first North pole in the lines curved around to make a circle. Circular formation does not violate the attraction and repulsive nature of magnets. Similarly, circular formation of objects bound by gravity does not violate gravitational attraction. In fact, gravitational attraction of sufficiently large objects in space molds that object into a spherical shape. This fact tells us that imbedded within each object is the potential repulsion energy associated with antigravity. A circle of bar magnets responds when a bar magnet is brought up to it, you might get an attraction or a repulsion force, depending on which part of the circle you approach, the potential is present.

Let us take this a step further, set the magnets in complete reverse. For example, S==N, S==N, S==N, S==N, S==N, where the last North pole is reconnected to the first South pole to make a circle. This configuration gives the same result as the one in the paragraph above. So, if you take a bar magnet with the south end pointing in toward the circle of magnets, you may get the same repulsive or attractive force; both potentials are present. Antigravity should therefore operate just as if we reversed all the magnets. In other words, objects with antigravity will attract other objects with antigravity and will repel objects with gravity; both potentials are present[96]. Let us look at this from another angle. Hurricane or Typhoon storms in the northern hemisphere rotate in a counterclockwise direction. Cyclone storms in the southern hemisphere rotate in a clockwise direction. The storm rotation is due to what is known as the Coriolis Effect, very much related to the conservation of angular momentum and rotation of the earth on its axis. Gravity, wind directions, and rotational spin of the planet trigger the storm's rotation direction. Antigravity forces cause antigravity planets to orbit retrograde rotation and orbit their star clockwise.

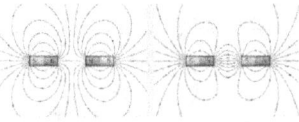

Now let us look at how this affects the micro world. Most of the molecules on earth have a chemical or atomic bond in the "right-handed" configuration. These bonds are influenced by the external gravitational force and internal pressure between objects. Since there is a potential of gravity and antigravity to exist on earth, we therefore have but a few "left-handed" molecules existing on earth, mainly found in life. Living organisms are composed primarily of left-handed or lefty molecules. This fact helps them smell and taste the difference between organic, and inorganic or inanimate objects. Plant life is unique in that it can absorb right-handed inorganic molecules and convert them into left-handed organic molecules with the aid of photosynthesis. These basic molecules are the exception to the rule caused by the antigravity potential; it gives us, and all other living organisms and plants, the rudimentary ability to resist gravity and thereby live. Life, and its ability to resist gravity, has the potential to occur everywhere in

---

[96] Einstein predicted that given the right conditions gravity too could become repulsive. Therefore, he created the cosmological constant, an "antigravity" nature that stabilizes the universe into a steady state.

the universe where all the necessary elements and conditions are exactly right, the "Goldie Locks" environment. We find organic type molecules floating in space, basic forms of alcohol are one example[97]. Any molecule with perfect symmetric configuration would, therefore, be identical to both parts of the universe, a world held together with gravity and a world held together by antigravity, for example the diamond or liquid water. We speculate that matter in the antigravity world would therefore consist primarily of molecules in the "left-handed" configuration, except for any of their living organisms composed of "righty" molecules resist antigravity, just as cyclones in the Southern hemisphere rotate in the opposite direction due to gravity.

Does a magnetic field push or pull? Well, it all depends on how you look at it. If you place two magnets with south poles facing each other, you get a pushing or repulsive force. If you place the magnetic north toward the magnetic south of another magnet, you get a pulling or attractive force. The strongest force of attractions is when both magnets touch each other. The same applies for gravity. All objects in the visible universe, with gravity, responses with the same attractive force as placing magnetic north up against or near magnetic south. The strongest attraction is when both objects touch each other. However, if you place magnetic north of one magnet against or near the magnetic north of another magnet, you get a repulsive or pushing force. The strongest repulsion occurs when both magnets are just short of touching each other. If we carry this same analogy forward to gravity antigravity relationship, we should get the same resulting repulsive force. It is only natural to say that the strongest repulsion is when the object possessing gravitational properties touches or is at the closest proximity to the object possessing anti-gravitational properties. This repulsive force or push instantaneously causes both objects to push away from one another, with the smaller antigravity object obtaining escape speed away from the larger gravity object without burning fuel; the antigravity object achieves escape speed regardless of how much mass and gravitational attraction the larger object has. Manmade forces cannot hold an antigravity object on a massive gravitational object.

By the way, if you place an electromagnet up against the circle of bar magnets, then turn on the electricity and alternate the current flow, we create a circular rotating motion with the permanent magnets or more specifically a functioning electric motor. Similarly, objects rapidly rotating somewhat generate a weak quasi "antigravity" force above and below the rotating objects caused by the centrifugal force outward. As an example, let us look at brief look at helicopters. Science tells us that the blades by themselves cannot possibly create enough lift through aerodynamics alone. Yet the helicopter is still capable of lifting off the ground and fly. In closer observations, the rotating blades on a helicopter develop a curved bend as they generate aerodynamic lift; this aerodynamic lift alone is insufficient to lift the whole weight of the helicopter plus the weight of occupants and cargo. Nevertheless, with the aid of tension caused by the outward centrifugal force of the rotating blades the aircraft obtains the additional lift necessary for it to become airborne. That centrifugal additional lift is analogous to a clothesline with wet clothes clipped to it or a tightrope walker, hanging by his hands, suspended over midair by a rope anchored at both ends to fixed objects like a building. Therefore, the faster the blades turn, the stronger

---

[97] Organic molecules form in intra-galactic space as appropriate elements come together by pull of gravity and antigravity, double tug, causes organic lefty molecules rather than inorganic right-handed molecules.

the centrifugal force, the tighter the tension becomes, and a type of "antigravity" effect achieved, suspension without anchoring support. In addition, the more weight or mass on the tips of the blades the more centrifugal force and the greater additional lift achieved. Another example is a spinning top. The faster the top spins, the more it appears to float over the surface. Rotating disc or blades closest to perpendicular or tangent to the massive planet surface achieves the greatest tension or "antigravity" effect[98]. Theoretically, an aircraft or spaceship with a downward concave curved disc or blades and heavy edge spinning at supersonic speeds or faster should be able to generate a stronger quasi "antigravity" or more weightlessness, greater tension, combined with generated aerodynamic lift or some other propulsion can enable the aircraft or ship to propel itself through the air or in space. Just like a spaceship rotates about its axis of direction to generate artificial gravity, the circular motion of rotor blades creates outward gravity just inside the edge and nullifies or near zero gravity at the center axle, through the center of the aircraft body, thereby lessens the total weight and enhances lift. This effect reveals why the water in Newton's relativity bucket experiment ride up the sides while the center concaves downward and is like what we get in a gyroscope, stability.

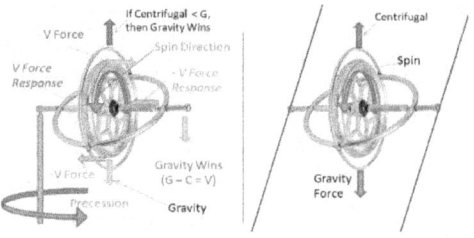

The closer to light speed objects rotate, the stronger the centrifugal force or "antigravity" force becomes, and the flatter the sphere or greater equatorial bulge of the object. Just as a skater increases rotational speed when they pull their arms inward during a stationary spin, the neutron star or black hole also increases rotational speed as it collapses on itself with gravity. Thus, a neutron star or black hole can have an equatorial bulge if they achieve a fast-enough spin. This quasi "antigravity" force results equatorial bulge that looks like the gravity object is being pushed from its North and South Pole axes, squeezed from both ends as if "antigravity" were present. In fact, the equatorial bulge results solely from centrifugal force produced by rotational speed, not antigravity, but the potential is there. The structure of life consists primarily of left-handed molecules. It enables us to resist gravity and live; an inherent attraction to the antigravity objects scattered throughout space including the white holes traversing within the wormhole tunnels all around us. Some scientists claim that these streaks in space are proof of "string theory". In a sense, they are right as we, as living organisms are attracted upward, thereby enabling us to resist gravity, stand up, and live, empowered by quasi partial antigravity quantum entanglement. This attraction speaks to our inner souls and encourages us to reach for the stars. The movement in the heavens fuels our resistance to gravity and is the basic source of motion of living organisms on earth, the breath of life wins over entropy. The heavens and all that is above created and evolved us, and the heavens within us has made humanity what it is and who we are, connected in the spirit of life[99]. It is internally the reason our ancestors and we desire to observe in study, document, and learn about the stars above us. We seek truth and understanding.

---

[98] The gyroscopic tension creating by centrifugal forces of a rotating narrow rotor blade adds to the lift capability of helicopters. Without this combined forces helicopters would not be able to lift of the ground.

[99] Humans connect with the heavens by spiritual and religious vocations. These heavenly bodies enable rudimentary inner organic molecules to move, resist gravity, breathe, and develop into life.

4.2 Balance in the Micro Elements

The weak nuclear force balances the interaction between atoms and molecules enabling each to adhere chemically or molecularly to each other while in some instances maintaining form and composition. The weak nuclear force prevents one group of atoms or molecules from passing through another when both are in solid form. At the same time, the weak force allows atoms and molecules to pass through or flow past another group of molecules if both are in the gaseous state or sometimes in liquid form. If one is in the solid form, then the gaseous or liquid form cannot pass through that solid object. Gaseous mater, however, can form bubbles and pass right through liquids, and liquids can form spheres or orbs and pass right through gaseous matter, just as rain passes through air when it falls to the ground. Likewise, solid objects can easily pass through gaseous and liquid matter with minimal resistance depending on the weight of the solid object. The push back force or friction force encountered by one object sitting over another object is a result of the weak nuclear force resistance. The weak force enables you and me to stand up, sit down, and move about in our daily lives. Without it, we would sink into the earth, and our molecules would flow right in between the molecules and atoms contained within the earth's surface. Without the weak force, there would be no earth's crust, oceans, and life, as we know it. This interaction of the weak nuclear force brings balance to bonding of atoms and molecules. Gases can bond to the outer layers of solid surfaces, causing a thin film of rust or discoloration due to chemical reactions. Liquids may be able to penetrate the first few molecules and atoms of a porous solid surface, like water on a paper towel or paint on a concrete block or wood, and in some cases change the structure or chemical composition of the solid object's surface, like paint thinner, dissolves and breaks away paint.

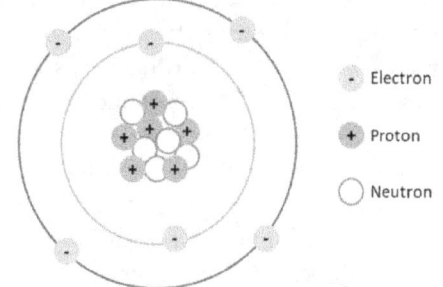

The electromagnetic force is the second natural force. It is stronger than the weak nuclear force. It is a combination of electric force and magnetic force. This force holds electrons within their orb level in an atom. Electron and photon waves work together to balance each other. Without the photon, the electron might plunge into the nucleus or fly off. For an electron to move to the next higher orb level in an atom, it would have to absorb an additional photon. The absorption of the photons and the energy that comes with it is balanced by additional mass and weight of the object being heated or energized. Electrons by natural law are allowed only to hold specific orb levels due to interaction with photons; they cannot create nor occupy a level halfway between these specific levels. The release of energy or photons enables the electron to drop from the high wave energy levels to lower energy orb levels, shedding off small amounts of mass in the process. The electrons orbiting the atom are also balanced by presence of protons within the nucleus of the atom; the negative electron balanced by positive particle proton. For light, the electric field of the electron is balanced by the magnetic field generated by its relationship with the photon. This interaction enables light to travel great distances in space without losing speed and energy. Similarly, the presence and movement of magnets, generates electrons and causes electrons to flow in a particular direction in relation to the movement of the magnet. Likewise, the flow of electrons in circular pattern over a core of certain materials like iron causes a magnetic field to form[100]. Electromagnetic force envelops and protects the earth from dangerous solar wind radiation and cosmic charges.

The strong nuclear force is the strongest of these three of the four natural forces. The strong nuclear force holds together the nucleus of an atom. It binds protons and neutrons together within the nucleus, in the form of gluons. Neutrons have no charge associated with it, and act as a buffer between protons contained within the nucleus. Without it, protons would repel each other, and atoms would fall apart. The more protons the nucleus has, the more neutrons are necessary to maintain balance and buffer the proton's electrically repulsion force. Hence, the more gluons are necessary to link the protons to these neutrons. This balance builds an invisible shell surrounding the nucleus something quite like an orb layer at the surface of the nucleus itself. The strong nuclear force holds quarks within protons, neutrons, and gluons, as well as within photons. Not to mention, the force that defines the quark as a sub-atomic particle is surely stronger than the strong nuclear force, possibly an unnamed fifth natural force, and a super-strong sub-nuclear force. This balance enables strong nuclear force, the electromagnetic force, and the weak nuclear force to assume and perform a variety of configurations to hold the atom and molecules together as they exist in nature and enables atoms and molecules to take on different characteristics, properties, and behaviors. This balance also enables atoms and molecules to absorb and release energy and/or sub-atomic particles, to react to one another differently in an infinite variety of different circumstances. ☐

(Wolff, 2008, p. 48) Milo on the other hand views all particles of matter differently and explains in wave structure of matter (WSM) theory those particles and their natural forces are in terms of energy wave interactions. "If two waves have identical amplitudes, they reinforce each other. If these two waves have opposite amplitudes, they cancel, and the waves annihilate each other. This according to Milo explains why charges attract or repel each other." This is resonance or wave balance, and the concept of anti-matter is just a form of matter that has opposite phases of the spherical waves of the particle resonances. (Wolff, 2008, p. 59) Milo states, "If two electrons are near one another, their identical waves add together producing maximum amplitude causing them to move apart seeking a minimum. If one is a positron, their waves will cancel each other producing a minimum amplitude they will be decreased as they move together." This WSM definition says that electrons always repel each other to seek minimum amplitude. There is a balance between discrete particle theory and the wave structure of matter theory, which we will explore later in this book. A unique way the natural laws interact with each other, one that gives us the unified theory science has been looking for such a long time.

---

[100] The circular flow of the molten iron core within the rotating earth generates a magnetic field around the earth. This field is responsible for protecting the earth and its life forms from harmful solar radiation.

4.3 Balance in the Macro Elements

The fourth weakest of the four natural forces is gravity. Gravity governs all matter in the visible universe at the macro level, the world of the large. Gravity is instantaneous and its influence is infinite, according to Newton, just as quantum entanglement too is instantaneous and infinite. It was Einstein's 1935 paper which was the first to point out "Quantum mechanics implies that something you do over here can be instantaneously linked to something happening over there, regardless of distance," in an effort to show that this concept was ludicrous, and its math needed much revision. However, the preponderance of detailed work in quantum mechanics experiments revealed, "Two objects can be far apart in space, but when it comes to quantum mechanics behavior, it's as if they're a single entity... and have temporal tentacles (Greene, The Fabric of the Cosmos: Space, Time, and the Texture of Reality, 2004, p. 12)." Gravity balances the other three natural forces. Gravity pushes us against other objects and the other objects push back in equal force. The push back employs the weak nuclear force, the electromagnetic force, and the strong nuclear force to counter the push of gravity.

An excellent example of this balance in the macro scale is the earth. The magnetic field of the earth gives the planet a magnetic north pole and a magnetic south pole. When the sun ejects solar flares at the earth, it sends with it a high concentrated and fluctuating magnetic disturbance. The interaction of the inbound magnetic fields and that of the earth's magnetic field can cause the earth to wobble in its orbit. Scientists conjectured that proto earth wobbled without a moon[101]. Additionally, the moon orbiting the earth adds stability and balance to the earth's rotational axis. The balance between the moon's orbit held by gravity and the suns influence on the earth's magnetic field, gives us here on the planet the protection we need from the potentially deadly radiation from the sun, while the earth orbits it. In the grand scheme of things, the other three natural forces do not influence orbits of large objects in space.

The total number of protons and electrons in essences zeros out the electrical charge of the macro object, and any internal magnetic field is overwhelmed quickly by gravity at great distances. However, at close distances gravity can be and in fact is overwhelmed by the other three natural forces, as we have already seen in stars, planets, and smaller objects. There is an exception to this rule. When the gravitational force is more than the weak force, the electromagnetic force, and strong nuclear force, then and only then will all three of these natural forces collapse, we have a black hole. The neutron star on the other hand is an example where the weak force and the electromagnetic collapse under the pressures of gravity, but the strong nuclear force is still able to maintain the structure of neutrons. In the neutron star, the electron's orbit is broken and the protons morph into neutrons, while the star releases gamma radiation and neutrinos[102].

---

[101] The earth also stabilizes the moon so much that the same side of the moon faces the earth. In other words, the moon's rotational period exactly matches its orbital period around the earth.

[102] The neutron star releases gamma radiation because with every jump a gamma photon takes on its

In the macro universe, gravity effects balanced by the orbital velocity of objects, as an object is drawn equally toward another more massive object with the greater effect on the smaller object. The smaller object will accelerate in speed at a given rate caused by a force inversely proportionate to the distance between the centers of both objects. There is a greater probability of the smaller object falling into some orbital path, decaying, ejecting, or stable, as long as the smaller object is not in a direct collision course with the larger object in space. Near misses can turn into stable elliptical orbits over time. Partial hits can cause the smaller object to assume orbit around the larger one. This is how the galaxy accumulates mass as it travels through deep space, adding stars and smaller galaxies to its size. Again, this delicate balance between gravity and acceleration keeps objects orbiting around larger objects in solar systems, and in galaxies, in space. This interaction between gravity and acceleration of objects creates an invisible bubble layer on which the smaller object travels over, at a west-to-east rotation or prograde motion, also known as a conservation of angular momentum. Matter and its mass generate gravity, which in turn generates acceleration due to attraction, which in turn generates the velocity to move the object to higher orbits, which in turn slows down the object and causes it to fall to a lower orbit. These changes of velocity cause the smaller object to follow an elliptical but stable orbit. Orbiting objects in a sense is a balanced exchange between matter and energy state.

The magnetron is another example of converting matter into energy. This time the energy is in the form of microwave radiation. The magnetron produces the microwave signal, which cooks the food within the microwave ovens' cooking chamber. Matter and energy are interchangeable. In this case, the chamber converts matter into energy or more specifically electromagnetic radiation. In other instances, as energy cools or loses strength, sub-atomic matter may be created as the energy dissipates, per mainstream scientists.

---

way to the surface, it is absorbed and re-released as the same gamma radiation without losing energy.

4.4 Balance in Formation of Galaxies and Universe

Our Milky Way Galaxy is composed of billions upon billions of stars orbiting around a central super massive black hole. Our scientists speculate that the galaxy is a barred spiral galaxy. Whether this is true or not is not pertinent for this chapter. What is pertinent is that the stars within the galaxy balanced between one-half and the other half, and between above and below the galactic plane, are held in constant struggle to obtain balance and symmetry. Symmetry in space is more commonly referred to as rotational symmetry or rotational invariance, where spatial direction is on equal footing with every other aspect. In other words, all positions are on a par equal to all other orientations. Clearly, symmetries are the foundations from which we can derive all laws.

This symmetry builds on itself. The galaxy starts with the formation of central stars igniting. The gravity of these stars then pulls gaseous material toward itself to create new orbiting stars. The central massive star then explodes to produce a black hole, which consume its closest companion stars. Galaxy formation is like that of tree rings, the oldest part is in the center. The additional mass causes the surrounding stars to accelerate and assume a disc like rotational orbit around the center black hole. Symmetry continues to build the galaxy into a flat rotating disc equal on both halves, as well as top and bottom. With the occasionally merging of two or more galaxies, the galaxy continues to pick up additional gaseous matter at its outer edges as it moves along in deep space. New stars ignite in the area where the solar system is to be born. These new stars are the first-generation stars in that region of the galaxy. First and second-generation stars die and provide the necessary building blocks for the solar system and its planets. Any imbalance of structure in celestial terms will correct itself over time. Collisions between proto planets and other proto planets if massive enough eventually form spherical orbs with time. Likewise, imbalances in the orbits of planets eventually correct themselves by moving toward either a closer lower orbit or a higher distance orbit. For example, Jupiter and Saturn tugged each other and changed orbits several times before settling in their current stable orbits.

Balance is found in stars. We all know that stars maintain their size and mass with a balance of nuclear fusion and gravity. Gravity pulls the matter inward, creates pressure, and heats it up. While nuclear fusion tries to expand and tear it apart. This balance provides for efficient burning of fuel for the life of the star. The densest of stars is the neutron star, which is of course a remnant of a super massive star that went super nova. The neutron star consists primarily of neutrons giving of intense radiation. The gravity in the neutron star is so intense that only neutrons are capable of existing, along with electrons, photons, neutrinos, and other sub-atomic particles. If it had any protons in it, they immediately would convert into neutrons due to extreme pressures and temperatures. If we carry this analogy forward to black holes, we will discover that there is also a balance maintain in them. What balance can we expect to find in black holes? Black holes are so massive and gravitationally intense that even neutrons are unable to hold their shape and existence. Specifically, the strong nuclear force that keeps the structure of neutrons intact, fails. The atomic particle collapses and leaves behind a "soup" of quarks, leptons, and force carriers, all pushing back with greater force to survive and retain their existence. As they push back against the intense gravity, they maintain balance by absorbing energy and increasing to medium and then the highest energy levels, with the occasional creation of anti-matter and further release of energy. This push back allows the black hole to retain its dense mass and volume. Without it, the inner core of the black hole would continually build up pure energy and cause the black hole to destroy itself in a colossal explosion more powerful and brighter than a hundred or even a thousand super novae[103].

How dense is the universe with normal matter? The galaxy's matter occupies about 0.01 % of the area of space where the galaxy sits. This relationship is also remarkably similar in the micro world. The atom's particles only occupy but a tiny fraction of the volume area of the atom. In both case, there is a lot of empty space between the actual particles and objects with mass.

All the matter in the visible universe is arranged primarily in a balanced configuration. This balance looks somewhat like sheets and filaments uniformly scattered throughout the universe, with what appears to us as voids of deep empty space with gaseous low-density clouds. In the grand scheme of everything, that is a whole lot of very empty space[104]. Cosmologists state that there appears to be some type of "dark matter" and "dark energy" causing the universe to accelerate in its expansion. Scientists believe that 90% of the matter in the universe is invisible to us. Black holes and white holes help hold balance in the universe, and how they affect where galaxies come. Image from NASA.

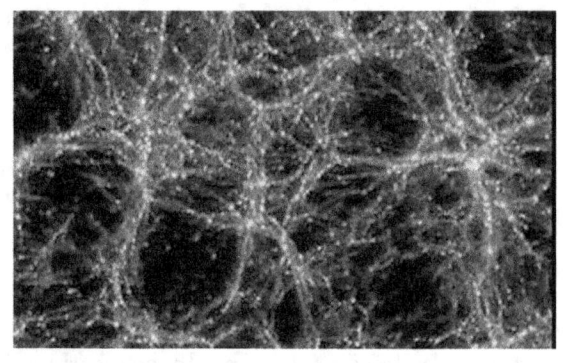

In the next two chapters, we will introduce and discuss the counterbalance to the visible universe in the Cosmological Balance Theory and provide the mathematical proof emulating how it functions. In doing so, we will discuss the reasons for this perceived "dark matter" and "dark energy" and how it affects us.

For now, it suffices to say that antigravity antimatter pushes on gravity matter and keeps the universe from collapsing on itself or from expanding exponentially outward. The Steady-State Universe for the moment has not been widely accepted by science and academia but is the most logical formation of the universe. The picture above depicts gravity matter forming intra-galactic web resembling strands and filaments. Hidden within that picture is also an intra-galactic antigravity strands and filaments formation.

---

[103] Without some type of release mechanism, chain-reaction antimatter-matter annihilation energy building up within a black hole could conceivably tear it apart. Since we have not recorded any exploding black holes, then there exists a release valve greater than just Hawking radiation.

[104] We refer to "empty" space as the "void" found within atoms and in between atoms. Actual particles are not part of "empty" space.

CHAPTER 4: THE NATURAL STATE OF BALANCE

4.5 Balance in Galilean and Einstein Relativity

There must be a balance between Galilean relativity and Einstein's special and general relativity. To start with, GR is based on SR, which is based on two assumptions below:

1. The laws of physics are the same for all observers in uniform motion relative to one another (Galileo's principle of relativity),

2. The speed of light in a vacuum is the same for all observers, regardless of their relative motion or of the motion of the source of the light.

Note that the second assumption is in direct contradiction to the first assumption. Light emitted within a uniformed motion ship or vessel should act and remain at the speed of light in the ship, in which observers are unable to detect any difference whether stationary or moving consistently. Why is this? Assuming all the windows and doors are closed, and shades drawn down to block all view to the outside, then all the matter, air atoms, molecules, and vacuum space between them within the confines of the ship are all moving at the same speed as the craft itself. Therefore, according to Galilean relativity, any light passing through these atoms will travel at the speed of light less effects from the air atoms, at approximately 3.00E+8 meters per second, whether stationary or moving at constant motion, regardless, if that motion is 10 meters per second or 100 million meters per second. For this to happen, the light's speed must be 3.00E+8 plus the speed of the vessel or system. Bottom line, assumption number 2 is erroneous and should have never been declared as a fact or principle regardless of what the physics of light math says. Remember after years of being lost in the tensors, Einstein had let go certain assumptions he initially made in the invariance theory quest to solve the general relativity theory.

Let us investigate Einstein's second assumption from two standpoints. First, we will look at light emitted from a constant moving object or ship. From there, we will conduct mental experiments in a ship at constant motion, and then from within a system that is in constant motion, and then from a system in constant accelerated motion. Finally, we will conclude with a modified second assumption for Einstein's SR, which of course puts both theories of SR and GR at risk. Einstein's principle of the constancy of the velocity of light is a delicate subject. As we said, it is based on an erroneous assumption for SR, and therefore should not be declared a principle. Einstein claimed that light has the same speed regardless of the comparison benchmark, which is a direct opposition to the Galilean relativity. Let us investigate this from an earth-bound perspective. A jet pilot moving at 100 m/s sends out a short beam of light emitted toward the direction of travel. Due to Galilean relativity, the beam of light initially departs the ship going at light speed plus the aircraft's speed. But because the air and space ahead of the jet is not moving in the direction or speed of the ship, the light almost immediately assumes a speed approximately 3.00E+8 m/s through the atmospheric air. The pilot is unaware of this and observes that the light beam transmitted from his lights to be going at the speed of light away from the aircraft. From a Galilean standpoint, we estimate the light to be traveling at 300E+8 m/s + 100 m/s at slightly more than 300 million meters per second, practically unnoticeable. But that light, on encountering air not moving with the ship, immediately assumes its normal speed approximately 3.00E+8 m/s through the air in front of it. Another observer on a fixed station on the mountaintop receives and sees that beam of light and clocks the arrival time of the jet airplane 30 seconds later. Results indicate that the jet traveling at 100 meters per second would have covered just a 3000-meter distance, calculated with the formula $t * v * (1 + v/c)$ equals 30 seconds times 100 m/s times $(1+100/299792458)$, like the Doppler Effect formula. The jet racing away formula is $t * v * (1 - v/c)$, where t is the time, v is the speed of the object, and c is the speed of light. In space, the light emitted from the spaceship would assume its normal speed within the vacuum of space, approximately 3.00E+8 m/s, this restriction is based on the fluidity of space, just as air molecules on earth governs the speed of light, the matter in space governs the light there.

Now let us take this into space travel at a constant speed of 100 million of meters per second. The spaceship pilot emits a short light beam in the direction of travel. According to Galilean relativity, the light emitted from the spaceship initially moves outward with its natural speed plus the speed of the spaceship. This would cause the light emitted from it to speed away at approximately 400 million meters per second. The pilot sees light emitted from his ship exactly as if it were stationary, going outward at about 3.00E+8 m/s. Another observer at a distant space station sees the flash of light after it has traversed the gap and clocks the arrival of the spaceship at about 30 seconds. Results indicate that the spaceship traveled a total of over 4 trillion-meter distance from the moment of sending out the light to it reaching the space station. Because of the extreme speed of the spaceship, it took light a total of 10 seconds to reach the space station (3 billion meters ahead) at normal vacuum speed before its observer started the stopwatch, $10 + 30 = 40$ seconds. In other words, the spaceship traveling at 100 million meters per second traveled a total of $1.0E+8$ m/s $* 40$ s $= 4.0E+9$ meters. Take it further, double the spaceship's speed. A spaceship going 200 million meters per second emits a flash of light which is picked up by another space station that clocks the spaceship's arrival to 30 seconds would have traveled a total of about 10 billion meters. This distance is the speed of light times 20 seconds plus the speed of the spaceship, 200 million m/s, times 30 seconds both equal to about 6.0E+9 m, giving us $20 + 30 = 50$ seconds of total travel time of the spaceship at 200 million m/s from emitting the light to arriving at the space station. Thereby, the total distance traveled from sending out the light signal to arriving is 10.00E+9 meters.

Objectively there are two possibilities for the speed of light emitted from a moving object, it is fixed at about 300 million meters per second or the speed of the ship is initially added to it. Light emitted from a stationary source traveling at normal speed can cover a 4-billion-meter distance requires 13.333 seconds, while light emitted from a 100 million m/s speeding spaceship per Galilean relativity would take 10 seconds to travel that same distance as calculated above. If we assume Einstein to be right, restricting light to its normal vacuum speed of about 3.0E+8 m/s, then 13.333 seconds would have passed before the observer in the space station spots the light from the spaceship and begins clocking the arrival of the ship around 30 seconds later. Implying that spaceship traveled a total of 43.333 seconds from emitting the light to it passing the space station. Hence, the ship traveled a total of 100 million m/s x 43.33 s = 4.333E+9 meters, which implies that light was traveling faster, or it needed more time to cover new distance of 4.333E+9 meters, at about 14.453 seconds. Nevertheless, this now changes the vessel's total time to 44.453 seconds and a new distance of 4.4453E+9 meters, where light now needs 14.828 seconds. Repetitive increase brings the spaceship refined travel time to 45.01 seconds and a distance to around 4.5015E+9 meters total. From the pilot's perspective, this restriction makes light move outward from the front of the ship at a much slower pace below 3.00E+8 m/s contrary to Einstein's claim that light travels at normal vacuum speed for all observers, including the pilot. Thereby showing assumption two has faults. For light emitted from the front of a spaceship or jet propulsion craft, that light emitted starts out with the additional boost of the ship's speed and after traveling for one second at that speed incrementally slows down, and eventually obtains, and sustains it natural vacuum speed, c within the surrounding air or space of the ship. Likewise, an observer will detect a spaceship racing past the space station to be emitting light that looks normal but after calculating the spaceship's total distance traveled will find that the light's speed normalized by subtracting the speed of the spaceship. In a reverse situation, the pilot sees light emitted from the rear of the ship to appear normal as if the craft were stationary. Due to the fluidity of space and its matter not moving with the ship traveling, the light either speeds up or slows down to eventually travel at the normal vacuum speed of about 3.0E+8 m/s.

What happens to light emitted within a ship or aircraft moving with constant motion? As with Einstein's mental experiment on a constant moving train of two observers, facing each other equally spaced from a light source set in the middle of a table of which they are sitting. These observers simultaneously detect the light flash from the light source within the moving train just as if they and the train were stationary. Einstein goes on to tell us that other observers on the platform adjacent to the moving train say otherwise; specifically, from their platform perspective, simultaneity did not occur. The second part we have already discussed, as light takes on a different speed upon leaving the moving object or system. As for the first part, Einstein clearly agrees and invokes Galilean relativity that anything within a constant moving vessel operates as if the whole system is stationary or at rest, including sound, heat, and light. Therefore, observers within a train, boat, cruise ship, jet airplane, or spaceship moving at a constant straight motion will experience the same Galilean relativity results as if they were stationary or at rest, to include observations involving light. That is light traveling equal distance toward the front of the vessel travels at the same speed as light traveling equal distance toward the back of the vessel. It does not matter if that distance is 1000 meters or 5 meters, both observers would simultaneously detect the flash of light emanating from the light source right smack in the middle between the two of them. It does not matter at what speed the vessel is moving if it remains at constant steady straight motion, whether it was 10 meters per second or 100 million meters per second; the results are the same. Keep this in mind.

Brian Greene tells us that Einstein predicts that the faster a "light clock" within a vessel moves toward the speed of light, the slower its clock runs (Greene, The Elegant Universe: Superstrings, Hidden Dimensions, and the Quest for the Ultimate Theory, 2003, p. 41). In that section of his book, he describes a simple photon particle bouncing between two parallel mirrors as a "light clock." When the light clock is functioning in a stationary vessel, the photon particle takes about a billion cycles per second. However, in a spaceship moving at extreme speeds closer to the speed of light each cycle the photon particle completes takes longer and longer as the photon must travel further between the moving top mirror and the advancing bottom mirror and therefore, Brian tells us Einstein correctly identified time slowing down the faster the ship moves. He even goes as far to tell us that mechanically clocks too within the same ship somehow slows down but avoids explaining how. Einstein's reasoning is in error based on the first assumption that Galilean relativity is accurate, implying that everything within the constant motion spaceship is moving as one system to include the light photon bouncing between the mirrors. As such, time does not slow down for the occupants of the spaceship moving at extreme speeds. Time functions exactly as expected normally, as if the entire ship was stationary. Hence, mechanical clocks also run normally as expected since all its parts are moving at the same speed along with the ship in constant motion. If we take the same mental experiment as the photon and mirrors above but change the orientation of the mirrors vertically instead of horizontally. This would give an outside observer the impression that the photon traveling toward the forward mirror takes longer than its movement toward the back mirror. In this case, if there was a time difference the photon moved between the mirrors, the complete cycle it takes would balance out and time remains consistent regardless of how fast the ship was moving. We know this perception is a mirage and is inaccurate. In this instance, Galilean relativity clearly tells us that both a photon clock and a mechanical clock operates exactly as expected, normally; there is no slowing of time. Bottom line, observations taken from outside the moving ship, does not affect the time within the ship; their perception of what is occurring within the ship is irrelevant.

Suppose we conducted a similar experiment within a jet airplane traveling between New York City to London (normally about 7.5 hours), where a young boy decides to bring his perfectly accurate bouncing ball clock to time the journey. He knows that two complete cycles the ball bounces between the top and the bottom paddle equates to one second. He sits there and diligently counts the bounces with the aid of a motion detector his dad bought him. As the jet reaches cruising altitude and speed at about 500 mile per hour, the boy detects no difference in his clock than when he was at home in his bedroom, it functions just as expect, in perfect rhythm, tick tock. However, to an x-ray observer on the ground the ball seems to take more time to travel between the upper paddle and the lower paddle as the jet streams across the sky, not because the distance between the paddles increased, but because the ground distance covered alters the paddle positions leapfrogging them forward as the ball moves up and down. That outside observer estimates that the two cycles the ball makes now take 1.05 seconds instead of 1 second, and that with the ball clock slowing down the jet time should read considerably less than 7.4 hours instead of 7.5 hours. When all is said and done, it does not matter what the outside observer says about his estimates, time within the plane functions exactly as expected, the trip took 7.5 hours and the boy's clock confirmed it. Point made; case closed. A ship's speed does not slow the clock within it, regardless of how fast it moves.

Now let us take the bouncing ball clock from within the cabin up onto the deck of a moving cruise ship out in the open seas. As the boy, opens the door to the upper deck, he immediately feels a wind hitting his face and interrupting his bouncing ball clock's rhythm. The wind encountered is the result of the ship moving through the atmosphere over the ocean surface. This visualization clearly shows that the environment between systems changes the speed of the ball. We know that sound initially travels from the ship's foghorn outward at the speed of sound plus the speed of the ship, and then reverts to just the speed of sound traveling through the surrounding atmosphere. Similarly, light emitted from the front of the cruise ship initial starts out with the combined motion of the ship and it normal vacuum speed $3.00E+8$ m/s, and then after a second reverts to the normal light speed of the new surrounding system, in this case the atmosphere of earth. The enclosed containment within the ship can be considered as one system, the upper deck is the layer between that system and that of the earth surface, which is another system. Departure from the earth into space is a transition from that system to the solar system, and departure from the solar system into deep space past the Oort cloud is a transition into the

CHAPTER 4: THE NATURAL STATE OF BALANCE

galactic system. Final departure from the galactic system past the galactic halo or the curvature of space around the galaxy into deep open space is a transition into the intra-galactic system of the universe. Each transition causes light to change speed as it enters another medium or system. Everything within the system moves with the system to include the "space" that is contained within it. Light, as it transitions between systems, therefore adjusts its actual speed to achieve and maintain the vacuum defined light speed of $2.9979E+8$ m/s within that new system. This is apparent when we observe light bending and refracting as it enters through a glass lens, and on exiting bends and refracts again and speeds up on the other side of the lens. Hence, light within the spaceship travels at exactly $3.00E+8$ m/s give or take the air reduction in it regardless of the speed of the vessel. As light leaks or leaves the ship, it assumes the speed of the system surrounding the ship. For a pilot moving at light speed, his ship's lights function normally accept the beam emitted seems to have limited outward shine as it transitions to match space system.

0The boundary of the solar system is the Oort cloud. Sunlight departing from the sun measures a speed of about $3.00E+8$ m/s on all sides of the sun. An observer in intra-galactic space, fully aware of the sun's speed through space, estimates the sunlight emitted in the direction of the sun and its system travels at about $3.00E+8$ m/s plus the velocity of the sun in the galaxy, and at $3.00E+8$ m/s minus the velocity of the sun system in the opposite direction. However, when the sunlight reaches that intra-galactic observer, he or she detects that it travels at exactly $2.9979E+8$ m/s contrary to estimates he took earlier, which is based on the sun's speed. This phenomenon occurs because light transitions in speed to match the intra-galactic system of open space it entered, of which its state of limited motion is different from the motion of the galaxy and of the solar system within it. Resulting speed measures exactly $2.9979E+8$ m/s within any system. Similarly, light changes speeds when moving through different mediums and each system is a medium.

Light adjusts speed as it transitions between mediums or systems

Airplane (100 km/hr)
Earth System (rotate 1020 km/hr)
Solar System (200 km/s)
Galactic System (402,000 km/s)

We conclude with the following modified Einstein's special relativity assumptions:

1. The laws of physics (in a system) are the same for all observers in uniform motion relative to one another (Galileo's principle of relativity),

2. "The speed of light in a vacuum is the same for all observers within the same system regardless of their relative motion or of the motion of the source of the light but is different for observers of different systems."

Light adheres to Galilean relativity and moves along with the system, increasing or decreasing total speed within that system to retain the measurements of motion observed in speed of light in a vacuum. Curvature of space affects light; clearly, high concentrations slow light enough to change its direction around massive objects, "gravity lensing".

4.6 Balance in Gravity and Antigravity

There exists a balance between gravity and antigravity in the universe. As identified in the previous sections above, there is a potential for antigravity on earth and it is responsible for the planet's west-to-east rotation or prograde motion, which we know also as the conservation of angular momentum. The discussion below reveals where to look to see it in action. Let us begin with what happens to a single object suspended in space. An ordinary massive object, in stationary motionless state or constant motion, experiences equal gravity over the entire surface of the object in all directions and no antigravity. This is Galilean relativity in action. In other words, space-matter applies 6.6734E-11 Newton force equally from all directions to suspend that object in space. If we take that same object and accelerate it, the gravity field will increase on the edge of the object in the direction of increased motion; this increase is proportional to the acceleration. Spherical objects will experience maximum gravity field increase on the point on the surface intersecting the vector line drawn from the center of the object toward the direction of its acceleration.

By the laws of symmetry and balance, the spherical object will also experience an antigravity field increase on the surface point intersecting the vector line directly opposite the maximum gravity field increase, on the other side of the object. This is the reason comets have tails as the comet gain's speed toward the sun, its antigravity increases which pushes debris away from the back surface of the object as it is heated up by the sunlight, all the while the front side gains gravity field and works to hold it together. Gravity decrease in the trailing surface of an accelerating object is inversely proportional to the gravity increase on the lead surface along the directional axis. For an orbiting planet, this axis is tangent to the orbital path. Science and the laws of physics tell us that the sun's gravity does not apply any torque on the planets. We think otherwise and mathematically discussed it in the Symmetrical Systems book and in Part 1 Fundamental Insights to Cosmological Balance. For now, it suffices to say that this gravitational difference generates a typically overlooked rotational torque on an orbiting object, such as a planet, comet, asteroid, and other debris.

Now let us investigate two objects in space, object A and object B. Space-matter, or "*hygratium*", provides the initial 6.6734E-11 Newton push force on object A toward object B and vice versa. If the vector of motion of both objects points directly at each other, they will both experience mutual gravity acceleration. As one object increases acceleration toward another object, its gravitational field will increase in intensity on the object's side of the direction of movement and develop an equal increase of antigravity field, seen as a weaker gravity, on the opposite side of direction of motion. As an object's acceleration increase it creates an equal but proportional concentration of gravitational field on its surface in vector of motion, which of course is counterbalanced by an equal proportional increase in antigravity field on the object's surface in the opposite vector. This increase of antigravity fields on the outside trailing edge enhances the space-matter's vacuum force and enables quantum entanglement to function in space as defined by quantum theory proven on particles with shared origin mirroring movement. The result is the two objects mutually attract each other and accelerate equally proportional to their masses and inversely proportional to their distances, per Newton.

Let us assume object A and object B are round spheres. The surface area of the spherical cone of object A with the gravitational field increase is drawn from the edge of object B to the center of the object A and vice versa. The area experiencing the anti-gravitational field is of an equal size spherical cone opposite that of the enhanced gravity. Hence, the highest concentration of increased gravity during acceleration is the point on the object's surface intersecting the vector line in the direction of movement. Similarly, the highest concentration of antigravity during increase acceleration occurs on the point of surface opposite the vector of movement. The vector line drawn from the center of one object to the center of the other object is the gravity attraction at that moment in time and is adjusted instantaneously as objects move. These are essentially the surface points at the center of the spherical cones (for round objects) on each side of the object or the center intersecting the vector line and the surface of the object.

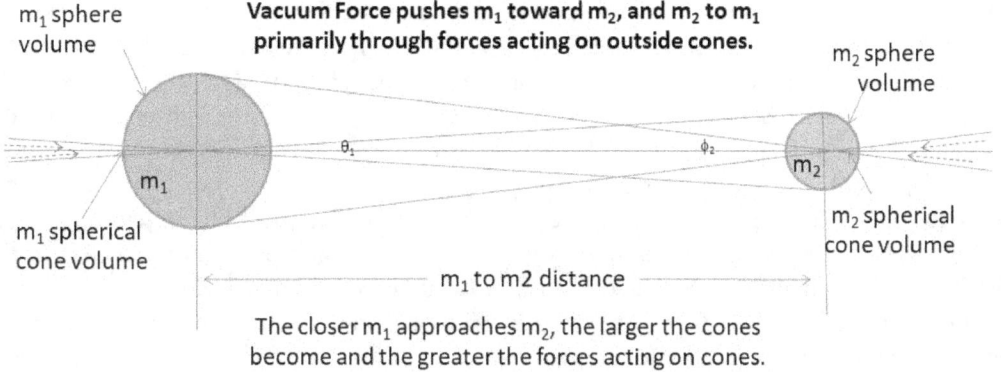

The closer m_1 approaches m_2, the larger the cones become and the greater the forces acting on cones.

If the smaller object A's vector of movement is not pointing directly at the massive object B, object A trajectory will curve toward object B, as predicted by Newton gravity equation. As ordinary matter object A accelerates toward ordinary matter object B, it experiences gravity on the side going toward the direction of motion, which is in turn counterbalance by antigravity field on the opposite side of the object. The enhanced gravity and lowered gravity (antigravity) provide torque for the object to rotate on its axis. As object B accelerates toward object A, it too experiences slight gravity increase on the side going toward the direction of motion, which is counterbalance by a slight antigravity field on the opposite side of the object.

Let the smaller object A be the earth and the larger object B the sun. Identifying such gravity antigravity circular area and center points on earth requires some work as one must consider the seasonal tilt of the earth on its axis, the rotational speed of the earth, the speed of the earth around the sun, the speed of the sun and local stars around the galaxy, and the galaxy movement through space toward Andromeda. These combined factors affect the earth's direction or vector of motion and continuously moves the gravity-antigravity points on its surface throughout the daily rotation period. Thanks to the Institute of physics earth and solar system facts, we know the following (Physics, 2015): The earth's rotational speed is about 1,000 km/h at 53 degrees latitude, and about 1670 km/h at the equator. The earth's speed around the sun is estimated at about 30 km/s. The sun's speed around the galaxy is estimated at about 200 km/s. In addition, the Milky Way galaxy's current speed toward Andromeda is estimated at 402,000 km/h. With all that said, the strongest effect on gravity antigravity is when the moon is lined directly above the vector motion gravity point caused by the speed of the earth around the sun, due to the proximity of the sun. Ancient cultures may have been able to harness these moving gravity antigravity points. In the 1940's, Edward Leedskalnin was able to tap into and take advantage of this anomaly when building the Coral Castle in Florida.

These temporary enhanced gravity and lowered gravity points (antigravity) are elusive. The closer you live to the equator particularly right between the summer solstice and the winter solstice latitudes, the more likely you can detect either the increase in gravity or the increase of antigravity fields during summer or winter solstice, where antigravity is seen as a decrease of the gravity field. Again, unless you live on the equator line, you will not be able to experience both anomalies within the same day, a 24-hour period. Enhanced gravity will come gradually, peak, and then dissipate and return to normal within a three-hour period around dawn. The same happens for the decrease in gravity, or antigravity; there will be a gradual decrease to a low around dusk then gradual increase to normal levels in a three-hour period. For example, those living at the northern summer solstice latitude will experience gravity increase on the dawn of the autumn equinox and will experience the weakening of gravity field or increase of antigravity on the dusk of the spring equinox for a few days peaking in the middle as the sun's path crosses over the equinox. The gravity antigravity circular area and their center points tend to move north south with the normal seasonal changes during the earth's orbit around the sun. The sun's movement around the galaxy also influences the strength of these points and may even move their position slightly. Antigravity fields combined with atmospheric conditions could easily initiate the low-pressure beginnings of tropical storms and other anomalies around the world, to include unexplained up drafts airplanes encounter.

Now let us take a review what acceleration does to the warping or curvature of space. Einstein predicted one-half of the results of an acceleration event but did not address the other half. As a massive object moves uniformly through space, space-matter envelopes it equally in all directions as though that object was stationary. The resulting perfect curvature follows the object as it moves in space. This effect is visualized easily in a soap bubble floating in the air; the atmosphere presses equally on all sides of the bubble and forms a perfect spherical shape. It moves with the flow of air, just as an object in space would move with the flow of space and its curvature and matter. However, an object accelerating through space gains gravity at the forward edge, thereby compressing in that point to form curvature and through symmetry causes a pointed bulge on the opposite end. Acceleration is either a change in speed or a change in direction of motion. The effect is seen in drops of water falling from the sky; it forms a teardrop shape. Essentially, an object accelerating in space causes space-matter to build up around the leading edge of the object forming a greater curvature than otherwise detected around constant moving object or one that is stationary. Likewise, the tail end of the accelerating object will develop a trailing tip of less dense space-matter emanating opposite the vector of acceleration. The resulting curvature of space looks something like a teardrop moving in the direction of acceleration; again, a comet is a perfect example of this phenomenon.

4.7 Balance in Quantum Mechanics is Gravity

In addition to the balance found within atoms discussed in the previous sections, quantum mechanics experiments have proven that point particles come in pairs or groups sustaining each other, known as super symmetry at the micro level. What causes gravity? In his book Understanding Gravity, Shan Gao reviewed Newton and Einstein gravity and then presented another idea for the cause of gravity. Gao looked at gravity as an emergent or an entropic force originating from interactions in a thermodynamic system (Gao, Understanding Gravity: Newton, Einstein, Verlinde?, 2014), but could not find a convincing analogy between gravity and entropic force in Verlinde's example. While string theory, particularly super symmetrical string theory, claimed several vibration patterns produces characteristics known to particles such as spin, charge, size, and mass.

String theorists even claimed that they have developed an equation and a specific pattern identifying the "graviton," which is only made possible by invoking the existence of nine spatial dimensions plus time as the tenth dimension, and some say its ten plus time. String theorists also tell us that the faster or more energetic the microscopic string vibrates the greater the mass of the particle containing the string. According to Brian Greene, "if strings were small enough, they would look like point particles and hence could be consistent with experimental observations (Greene, The Elegant Universe: Superstrings, Hidden Dimensions, and the Quest for the Ultimate Theory, 2003, p. 136)." Since no one has seen these Planck length strings nor are they or will they ever be detectable by our current technology, we hereby pass on the concept of strings and will go with the point particles, which are exactly what is "seen" in experimental observations. Like strings, point particles are as miniscule as or even smaller than electrons, and as undetectable and elusive. Subatomic particles, consisting of highly energetic pairs or groups of point particles, are detectable. Greene continues saying "Symmetry in physics has a very concrete and precise meaning: matter particles and messenger particles are far more closely intertwined than previously thought possible (Greene, The Elegant Universe: Superstrings, Hidden Dimensions, and the Quest for the Ultimate Theory, 2003, p. 167)." Pairs or groups of point particles produce patterns or shapes, which may look like or act like strings vibrating as identified in the super symmetrical string theory equations, not the other way around. In the quantum realm from Chapter 1, Milo Wolff used work from Schrodinger, Einstein, and Dirac to develop the Structure of Matter (WSM) theory, where spherical standing waves produced all the particles of the universe. Einstein proposed the points where wave-peak energy transfers occur in what he termed as "photons" as the "messenger" between all particles. However, unlike Einstein who envisioned only one-way photons, wave energy is a two-way symmetrical resonance signal between the source and the receiver. Wave resonance frequency determines the fixed levels electrons traverse around an atom. In- and out- wave interaction is the means of transferring energy and affecting particle behavior and motion. Milo also claims that resonating wave energy from electrons cause them to repel each other continuously. The WSM theory has it backwards; the motions of particles create waves, which affect other particles. Spherical standing waves do not create electrons or any other particles. Let us review how electrons repel.

Do electrons consistently repel at any distance? No, this is not the case since waves by its very nature have frequencies with cyclic amplitudes and troughs. Electrons have a quantum wave frequency, where the amplitude of the wave is represents the maximum peak energy and the trough represents the minimal low energy. Electrons exert maximum repulsion on each other at fixed distances predetermined by their out-wave frequencies, in other words, when their amplitudes match to add together, we get the highest repulsion energy and then it wanes at other distances in between these peaks. This explains the fixed electron spherical levels around any atom. An electron's out-wave frequency is the same as its duration cycle from merging to reemergence to maximum separation and back to merging of their - 1/3 charged sub-particle components, where merging is the peak energy (an electron particle) and the point of time of greatest point-particles separation is the lowest energy (appears as a cloud). The convergence of inward energy waves from nearby positively charged particles of matter overcomes the repulsion forces and pushes the three negative 1/3 sub-particles together and collide momentarily to form an electron charge of negative 1, which in turn repel and fly apart to give the electron the appearance of a spin characteristic and mass attribute. Perfect resonating in-and out-waves between the moving negative 1/3 charged particles maintains the elusive electron, -1 charge.

In his book, *The Particles of the Universe*, Jeff Yee suggests that everything in the universe is built from one fundamental particle, the neutrino (Yee, 2014) and various wave configurations thereof, even the aether of space. For instance, he claimed that an electron is a composite of ten neutrinos arranged to produce the appropriate standing waves to match the charge, mass, and spin of an electron, a concept like but different from string theory. We however view things slightly differently and want to expand the number of fundamental particles to three basic charges (+ - and 0) together they possess the grand unified force. Rather than using Yee's concept of wave structure matter, everything in the entire universe is composed of these three point particles, paired and grouped in different configurations, in unique motions and patterns to produce all subatomic particles, atoms, and all matter in existence, visible and invisible. As we eluded earlier there are three basic types of point particles: positive (unpaired positron), negative (unpaired electron) and neutral (neutrino or antineutrino). Technically, if we count each kind of neutral point particles then we have four types. An unpaired electron or positron has no spin.

The interaction between negative and positive point particles taps into the electromagnetic force. The interaction between neutral and another neutral particle accesses the elusive strength of the short-ranged strong nuclear force as a survival mechanism, which is about 100 times more powerful than the electromagnetic force. The perceived destructive contact between loose positron and loose electron point particles causes their energetic transition into loose neutrinos and antineutrinos. Similarly, neutrino and antineutrino transition on collision back into negative and positive point particles. However, the contact between a neutrino or antineutrino and a negative or positive point particle are mutual and beneficial for both, and non-destructive nor transitional. Of course, spin has an impact on kinetic energy interactions or physical collisions, like billiard balls on a pool table bouncing off each other. Spin in this context is not the actual rotation of the particle but provides the same angular momentum impact results. Spin aligned particles are more likely to bind together than opposite spin particles. Matching spins does not cause a binding only aid in the coupling of the particles for survivability. They bond securely while sharing energy and forces. Loose negative or positive point particles have no spin; they obtain spin when paired with a neutral point particle.

CHAPTER 4: THE NATURAL STATE OF BALANCE

In ordinary matter, a neutrino binds with a positive point particle to form a point particle pair with 1/3 positive charge, and an antineutrino couples with a negative point particle to form one of the three pairs in the center of quarks totaling negative 1 charge, or -1/3 times 3 = -1. This chain of point particles can be extended as long as the matter and antimatter point particles do not touch each other. In our analysis of an electron around a nucleus, we find three pairs of neutral point particles (antineutrino and neutrino) create three loose negative and three loose positive point particles, and it will take another six neutral particles to stabilize each of these loose charged particles. Hence, a group of three paired negative/neutral point particle has -1/3 charge each orbiting an atom gives that electron cloud the total charge of -1. These point particle patterns of motion have value to the Cosmological Balance Theory. Like those found in string theory, the patterns include Frisbee-like shapes or constituents, three-dimensional blob-like constituents, and many more possibilities. If string theorists are right, then they discovered the tiniest element that created everything. However, if string theorists are wrong, then these patterns are strictly a result of pairs or groups of point particles and their messenger point particles in a symmetrical dance around each other to produce the patterns identified, not from motions of strings. Messenger particles, normally neutral in charge, tell matter particles what to do even at the quantum level. As an example, electrons surrounded by photons repel by hurling photons at each other, resulting in a reaction force, which pushes them apart (Greene, The Elegant Universe: Superstrings, Hidden Dimensions, and the Quest for the Ultimate Theory, 2003, p. 124).

The photon messenger particle tells the electrons to move apart. Rather than the electrons throwing photons, the photons position themselves between the electrons and align their magnetic like poles toward each other so that their magnetic repulsion combined with repulsion of like charges, moves the electrons apart quicker. If the electrons have no photons, they align their neutral point particle pair to enhance their repulsive charge, which emulates wave interaction. Bottom line, the principle of symmetry in the laws of physics directly governs all four natural forces from the quantum level to the macro universe. Let us investigate this further. Quantum mechanics experiments prove that newly created point particles assume characteristics specifically aligned to complement existing particles within their surrounding vicinity, specifically to fill a relationship void. For example, if the newly created point particles are next to a free moving proton it conforms itself into an electron and photon, which begins to orbit the proton and thereby form an atom. Obviously, an atom is more stable than a free floating or moving proton. Per mainstream scientists, in some instances several point particles coming into existence creates counterpart particles or antimatter, such as an electron and its antimatter, the positron, shooting the positron outward and pulling electron into orbit around the proton.

Once created, these particles will remain that way unless they collide with something else, like their counterpart of matter antimatter. A positron colliding with an electron both annihilate each other, creating high intense energy output and probably a burst of light or photon/electron wave or neutrino/antineutrinos. Free positrons are positive, electrons are negative, and neutrinos or antineutrinos are neutral point particles. Mainstream science has experimentally proven that any particle colliding with its counterpart in matter antimatter annihilation releases high intensity energy and the simplest self-enduring of particle pairs, which may include the electron-photon wave we know and see as a light burst, as well as other bits of subatomic particles. Atom smashers experimentation, such as the Large Hadron Collider, has revealed to us the intricate internal constituents of nucleons. Despite these discoveries, scientists have not been able to explain why these quarks had these identified charges.

Let us look at the most basic and simplest of atoms, the hydrogen ^{1}H atom consisting of one electron and one proton and at least one photon. The photon is a massless particle of neutral charge probably constructed of two neutral, one negative, and one positive point particles in a tight mutual flat circular orbital pattern with the positive at the center orbited by one neutral and a negative point particle at the outside moving at light speed. All the photon's point particle adjacent surfaces touch to create a polarized magnetic field. The electron is also a massless negatively charged particle consisting of three pairs of one negative point particle and one neutral point particle grouped in a tight spin pattern, where each pair is generally separated in the cloud around the atom, which intersect in a tight loop every now and then, as they orbit. This gives the motion of the electron (when grouped the -1/3 * 3 = -1 charge) the quantum quality of being random and discontinuous, determined by probabilistic cause (Gao, Quantum Mechanics: A Comprehensible Introduction for Students, 2013, p. Loc 695). Shan Gao said, "It seems difficult to explain why the electron speeds up at the node and where the infinite energy required for this acceleration comes from." The strong force in their paired neutral point particles provides that energy. These are the point particle configurations and patterns for the photon, the electron, and the gluon. The photon and the gluon both include an antineutrino between the negative and positive point particles.

This concept also explains why electrons shot through the double slit experiment appears to function as "cloud" or wave while passing through the slits, then reconsolidate and hit the screen as a single particle (Gao, Quantum Mechanics: A Comprehensible Introduction for Students, 2013, p. Loc 1376). The electron and photon move together to maintain its predetermined distance from the nucleus given the energy level of the photon. In addition, at the nucleus we have one single proton for the ^{1}H hydrogen atom. For ^{2}H, we have one proton and one neutron at the atom's center.

POINT PARTICLE PATTERNS CREATE SUBATOMIC PARTICLES

Photon
Flat orbital shape = zero charge and polarized magnet

Electron
Spherical orbital shape = negative charge and repels negative

Gluon
Disc shape = zero charge, attracts and repels

N / S
Zero Charge

Negative Charge

Zero Charge

The proton as we learned in Chapter 1 consists of two up quarks and one down quark. The up quark has a charge of 2/3 positive and the down quark a charge of 1/3 negative. This structure consists of smaller point particles of different charges, two types of neutral, positive, and negative. I predict that the **up quark** composed of sixteen point particles has five positive/neutral pairs (1/3 * 5) orbiting one center consisting of three negative and a neutral pair (-1/3 * 3), where each of the five neutral point particle orbits their attached positive point particle partner (1/3 * 5 – 1/3 * 3 = 2/3). These five positive/neutral pairs then randomly orbit the three neutral and negative charged point particle pair at the center axis in a tight elliptical or most likely a hyperbolic orbit. The neutral or messenger point particles are buffers that prevent the positive point particle from contacting or colliding with the negative point particle. It does this by using 2/3 of the charged particle's energy to create the strong force, which is 100 times greater than that of the electromagnetic force. The neutral point particles control when the positive and negative repel by aligning themselves between the two (- n n +) and apply the repulsive strength of the strong nuclear force and allow the negative and positive to attract by placing themselves on opposite ends (n - + n) and turning off or halting the repulsion. The neutral point particle strength, the strong nuclear force, is stronger than charged point particles attraction, but shorter in range (about 100th that of Planck length) than that of the charged particles, the electromagnetic force. As a result, the neutral point particle slingshots the positive particle outward with a rapid acceleration for a short micro distance, then the charged particles attraction for each other overwhelms the outward motion and pulls the positive point particle inward while the neutral strong force cushions the descent, similar to freefalling inward. At the positive point particle's closest approach to the negative point particle, the neutral point particle's energetic strength then overwhelms the charged attraction and again slingshots the positive particle outward. Each outward swing, like that of a golf club, creates the jiggling edge appearance of the up quark and something amazing that we will discuss later.

The second component of the proton, the **down quark**, similarly constructed of ten point particles consists of two positive and neutral pairs (1/3 * 2), orbiting around a center group of three negative and three neutral point particles (-1/3 * 3) of which the center neutral particle is most likely antineutrinos. Reminder, the 1/3 charges are caused by the energy redistribution to enable the neutral point particle to apply the strong nuclear force. The two pairs of neutral and positive point particles randomly orbit the center three negative and three neutral point particles in a tight hyperbolic pattern (-1/3 * 3 + 1/3 * 2 = -1/3) give the down quark its -1/3 charge. Again, the neutral point particles prevent the positive point particle from colliding with the negative particle. Since the center has a greater charge than the orbiting pairs, it is bounded more loosely together, more energetic in movement, and hence the down quark is slightly greater in mass than the up quark. The sheer short-range strength of the neutral point particle, the strong nuclear force, slingshots or accelerates the positive point particles outward, and then halts them abruptly. Similarly, each outward elliptical swing, like that of pendulum, creates the jiggling surface of the down quark, and an effect to be discuss later.

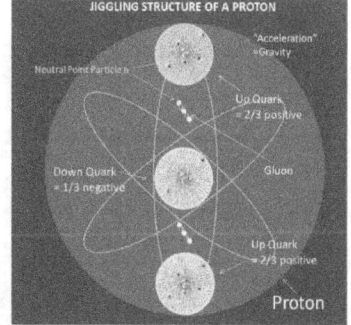

The centerpiece of the proton is the down quark, which is orbited randomly by two up quarks that are buffered by at least two gluons. This gives us the familiar form of a single proton structure. The up quarks travel around the center down quark in elliptical or hyperbolic orbits. Here again, each outward swing of the up quarks provides the jiggling surrounding surface of the proton. Each layer of the atomic structure adds another level to the overall energetic jiggling enhancement. Hence, the more energetic the point particle movement, the more massive the atom becomes. The three stronger forces of the four natural forces attract and repel to form configurations we see in nature. Given enough energy input, they can eventually form complex atoms and molecules; the interior of the sun is such a place to create new atoms through the fusion process.

Free-floating protons repel each other because of their like charges and electromagnetically attract three negative/neutral point particle pairs and a photon around them to form the simplest of atoms, the hydrogen-1 atom, ^1H. Since these completed hydrogen-1 atoms carry a total zero charge, their gravitational field enables them mutually to attract each other and given enough matter and time form stellar clouds. Depicted above is the point particle composition of a proton. As we mentioned at the beginning of this section, we accepted that the work showing shapes and patterns resolved by string theory equations; it provides us the baseline framework defining the shapes and patterns produced by the motion of point particles. It is anticipated that if the paired or groups of point particles form a rough spherical shape then additional magnifications would show multiple sporadic dimples and protrusions at the quantum level. Some paired or groups of point particles form disc-like or Frisbee-like shapes in order to produce the proper characteristics and properties desired, for example, the photon particle would be in this form in order to generate its magnetic polarity field. As mentioned earlier, the photon consists of one positive point particle shielded by a neutral point particle and orbited by a negative and a neutral point particle jointly traveling in a flat plane; this motion generates a north-south micro magnet while still having zero overall charge. The electron is three pairs of negative point particle orbited by a neutral point particle, giving it its spin characteristics and a 1/3 negative charge without adding mass. The photon and electron travel together as a light wave in a perpetual motion interweaving past each other as they move. These two groups of point particles cannot collide because their imbedded neutral point particle (or messenger point particle) repels them apart while the electromagnetic attraction pulls them back together, adding to the sustainability of its continuous perpetual motion, and provides the strong force energy needed to travel at the speed of light.

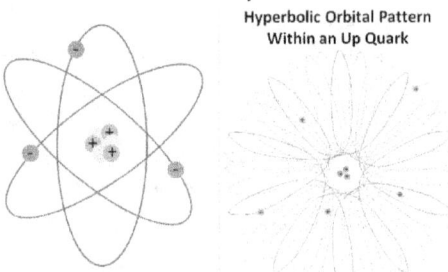

Typical Atomic Structure

Hyperbolic Orbital Pattern Within an Up Quark

In atoms consisting of several protons and neutrons in the nucleus, the gluons prevent the protons and neutrons from direct contact with each other. Gluons itself consist of two or three neutral point particles, a negative point particle and a positive point particle in a group with no charge, zero. Unlike the photon, which consists of one negative-neutral point particle pair in a flat orbit pattern around one positive-neutral point pair to create magnetic polarity while having overall neutral charge of zero. At the gluon center is the neutral point particle orbited by both the negative and positive point particles surrounded by neutral point particles. This gluon structure enables it to hold subatomic particles within the nucleus of an atom by repelling and attracting as needed. The electron, on the other hand, consists of three sets each consisting of one negative point particle orbited by one neutral point particle, overall charge -1 or negative one. The neutron consists of one up quark and two down quarks: $2/3 - 1/3 - 1/3 = 0$, which results in an overall zero or no charge. Since nuclear reactions conserve the number of leptons, the neutron cannot capture antineutrinos only neutrinos and the opposite are true for protons, they can only capture antineutrinos. The construct of the neutron, with the up quark at the center, is orbited randomly by two down quarks that are buffered by two gluons. The boundary or contact surfaces between the neutrons and protons buffered by gluons give us the appearance of a jiggling nucleus surface.

As discussed in Chapter 1, when protons are atomically smashed in colliders or decay naturally, they produce neutrons, quarks, electrons, photons, neutrinos, or other several other released subatomic particles. All these different variations are just the same source point particles (neutral, positive, and negative) reassembling into something else to survive, simply by forming new pairs or groups of point particles. The same applies when neutrons are atomically smashed or decay naturally, they breakdown or reconfigure into other subatomic particles. Since the weak nuclear force mainly moderates beta decay, we look to the strong force to differentiate between the anti-neutrino and neutrino in interactions with electrons and positrons. However, the strong force does not affect free electrons (or free positrons). Occasionally, the atom smasher creates antimatter particles, which annihilated one another on contact with their counterparts. You may be wondering, "What does all of this have to do with gravity?" Well, Greene explained, "With its 'graviton' pattern of vibration, string theory, is a quantum theory containing gravity (Greene, The Elegant Universe: Superstrings, Hidden Dimensions, and the Quest for the Ultimate Theory, 2003, p. 158)." We have an alternate explanation using point particles instead of strings that is more practical and conceivably more plausible. Note that the snapshot hyperbolic orbital pattern of point particles for an up quark on the right image shown below over time creates a Calabi–Yau manifold derived within string theory, compared to a typical atomic structure where electrons circularly orbit the protons and neutrons, which creates a spherical atom, although with a fuzzy surface.

Quantum physics has shown us that the faster point particles move the more massive they become. In other words, the more energetic the point particles are the more mass they exhibit. This can be shown using Einstein's energy-mass equation, $E=mc^2$. So how does more mass equate to more gravity? In Einstein's happiest moments, he discovered acceleration equals gravity, the equivalence principle. Now, let us analyze the structure of an up quark composed of sixteen point particles: five positives, three negative, and eight neutral point particles, where five pairs of neutral and positive point particles orbit randomly the three pairs of neutral and negative charged point particle in the center axis. Every time the positive point particles moves inward toward the negative point particle, the neutral point particle buffers or cushions its inward movement making it gradually slowdown, freefall, and then sends that positive point particle outward with acceleration reaching near the speeds of light, which of course is halted by the negative point particle pulling it back toward itself. Acceleration provided by the strong force of the neutral point particle, and the halting and returns provided by the electromagnetic force, which is cushioned by the strong force. The resulting motion is an elliptical or hyperbolic orbit with rapid outward acceleration and slower inward movement. There are no strings here.

Imagine for a moment that the positive point particle was an individual jumping on a trampoline and the neutral point particle was the trampoline cushioning and catapulting the individual upward, while the central negative point particle operates like the earth pulling the jumper down, not with gravity, but with electromagnetic force or electrostatic charge. Both positive point particles' rapid repeated acceleration outward and their slow "freefall" descent inward create a gravitational field around the entire up quark. The same effect occurs within a down quark, where the rapid outward accelerating positive point particles and slower descent create a gravitational field around it. Therefore, the movement speed of the point particles within each quark determines its mass, while the rapid outward acceleration causes its gravitational field, the "graviton" for which scientists are searching. The looser bound and more energetic negative and neutral pairs within the core of the down quark give it its additional mass over that of the up quark.

What is a graviton and why have scientists not been able to find it?

To reiterate, the "graviton" is not a unique spin-2 massless point particle but a result primarily of the motion of point particles within quarks and enhanced by the motion of quarks within nucleons. The "graviton" is therefore within every atom's nucleus. Graviton can be associated particularly with the effect of the neutral point particles' action within the quarks. They are the messenger point particles that turn on and turn off strong nuclear repulsion force to avoid collisions of positive and negative point particles. The same type of motions occurs within protons where the two up-quarks rapidly accelerate outward away from and then are cushioned in descent to the center down-quark thereby increases that gravitational field and enhances the "graviton's" reach. This simple "graviton" movement also occurs within neutrons and between neutrons and protons buffered by gluons, we observe as jiggling. Again, subatomic particles constructed from four types of point particles: two types of neutral, a positive, and a negative, are assembled in various patterns and shapes. Below are three possible spherical configurations that causes gravity field around an up quark. The image on the right produces the most dominate and long-range macro universe force, gravity, yet it is weakest among the four natural forces. Specifically, the three natural forces working together create gravity, and as such clearly explain why it is the weakest of the four. The "graviton" is the result of point particle motions within quarks, and among quarks in nucleons, and among subatomic particles within the nucleus. These three layers of "gravitational field" enhance their internal layer's graviton range. The more atoms the greater its mass and the greater its gravitational field.

In summary, we note that these simple motions or jiggling within and of quarks and nucleons all explains why larger atoms have both more mass and more gravitational pull, and why electron levels around atoms are fuzzy layer in quantum flux. All this is possible based on one simple reason, rapid outward point particle acceleration and slow inward return movement equates to the creation of the gravity field. The neutral point particle pairs slow buffering and rapid outward spring motion is the 'graviton' scientists seek.

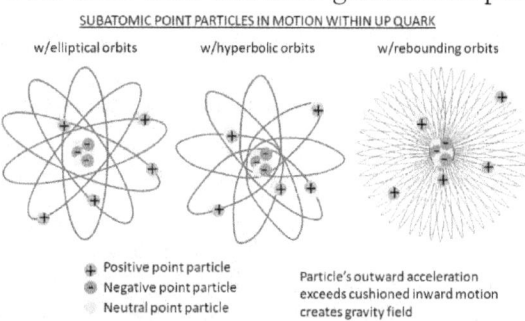

Mathematical Justification for Quantum Mechanical Graviton

To analyze this "graviton" effect mathematically, we need to compare it to a real-life situation. Imagine for a moment that you are an astronaut aboard a rocket near the Cape Canaveral, Florida, 0.003 kilometers above sea level. By your calculations, your rocket engines produce a thrust 100 times greater than the gravity of the earth, so whenever your engines are blasting your rocket speeds up and away with tremendous acceleration while being pulled downward by gravity. The upward acceleration generates a gravitational field in the direction of motion, as discovered by Einstein. However, due to a malfunction there is a limit on the rocket's acceleration range. When your rocket reaches above a height of 10 kilometers, the engines turn off and the rocket losses momentum, abruptly stops, and begins to fall back to earth slower than its upward journey. On the freefall way down, you panic and flip every possible switch; no matter what you do the rocket engines refuse to reignite until you have descended just below 5 kilometers. At which point, the engines somehow reignite and send you and the rocket rapidly racing upward with renewed acceleration. The fast-upward motion and slow descent generate an oscillating gravitational field we will call the "graviton" effect.

Mathematically, the rocket zooms upward from five kilometers to past ten kilometers in 3.2104756 seconds with total acceleration thrust of 970.2 m/s², derived from the thrust of 100 minus the gravitational pull of one multiplied by 9.81896 m/s². Acceleration time is calculated by square root of two times distance divided by acceleration minus acceleration divided by the distance squared, or SQRT($2r/(A - A/r^2)$). Upon reaching 10 kilometers, the engines shut off, and the rocket takes 31.9438289 seconds to fall 5 kilometers. This repeated rapid acceleration upward and relatively slower descent to earth generates an overall oscillating "gravitational field" on each upward motion at 100 G's or 100 times earth's normal gravity. We know that the strong nuclear force is 1E+38 times more than gravity, with a strength approximately 6.6734E+27 Newton (kilogram/meter)². We also know that the strong nuclear force is 100 times greater than the electromagnetic force 1E+36 which is at about 6.6734E+25 N (kg/m)². If we use the strength of the electromagnetic force instead of the earth's gravity field, then it yields this equation:

Gravitational field generated by rocket's acceleration is roughly:

= [6.6734E+27 N (kg/m)² * 3.2104756 s * 99 kg/s² /1E+38

+ 6.6734E+25 N (kg/m)² * 31.9438289 s * (-1 kg/s²)/1E+36]

/ (99 kg/s² * 3.2104756 s – 1 kg/s² * 31.9438289 s)

= 6.6734E-11 N (kg/m)²

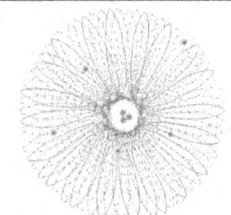

GRAVITON EFFECT WITHIN UP QUARK

Point particle's outward acceleration of 2.9222E-38 seconds exceeds cushioned inward motion of 2.9222E-37 seconds creates gravity field, the "Graviton" of 6.6734E-11 N (m/kg)²

Note that the rocket's total 99:1 up and down motion ratio does not change the overall gravitational field of the earth. It creates an oscillating "gravity" field augmenting earth's own. Now imagine taking this motion into the quantum world within an up quark, where we have five positive point particles with their companion neutral point particles (messenger switch for the acceleration) strong nuclear force outward acceleration and down cushioned descent to central three negative and three neutral point particles. The neutral particle senses proximity and turns on and off the repulsion as needed to prevent the positive particle from colliding with the negative particle and prevents the two particles from getting too far away, holding it in orbit like a stretchable glue. The on-switch, maximum strong nuclear force of 100 times, occurs by aligning the neutral point particles on the same side. At the maximum range of the strong force, the off switch halts the outward motion. This off switch, which activates the electromagnetic force of one, occurs when the neutral point particles are on opposite sides, and the negative and positive pull on each other, a weaker attraction. The "graviton" action envelopes the entire quark and generates a very tiny amount of gravitational field in every direction amounting to a miniscule quantum contribution to the overall 6.6734E-11 N (kg/m)². Note that the central axis of the quark by itself does not have, create, or exert gravity. This "graviton" effect in one quark symmetrical system universe, combined with those of all the quarks and nucleons within every atom in the earth creates its significant gravitational field.

Using the Planck length (1.6E-35) divided by 100 of 1.6E-37 and the equation for travel time equals square root of $(2r/(A - A/r^2))$ where r is distance and A is acceleration, the outward motion time of an accelerating point particles at light speed is

$= \text{SQRT}(2*1.0000001\text{m}/(299792458 \text{ m/s}^2 - 299792459 \text{ m/s}^2/1.00001\text{m}^2)*1.6\text{E-}37$

$= 2.99222\text{E-}38 \text{ s}$

and the inward motion time is

$= \text{SQRT}(2*1.0000001\text{m}/(2997924.58 \text{ m/s}^2 - 2997924.59 \text{ m/s}^2/1.00001\text{m}^2)*1.6\text{E-}37$

$= 2.99222\text{E-}37 \text{ s}$

Using the outward and inward time to estimate the "Graviton" effect within a quark:

$= [6.6734\text{E+}27 \text{ N (kg/m)}^2 * 2.9222\text{E-}38 \text{ s} * 99 \text{ kg/s}^2 /1\text{E+}38$

$+ 6.6734\text{E+}25 \text{ N (kg/m)}^2 * 2.9222\text{E-}37 \text{ s} * (-1 \text{ kg/s}^2) /1\text{E+}36]$

$/ (99 \text{ kg/s}^2 * 2.9222\text{E-}36 \text{ s} - 1 \text{ kg/s}^2 * 2.9222\text{E-}35 \text{ s})$

$= 6.6734\text{E-}11 \text{ N (kg/m)}^2$

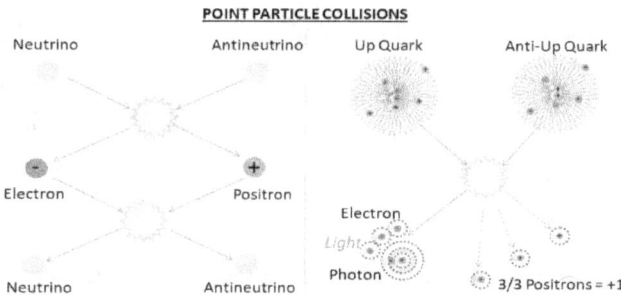

Like a rectifier smoothing out an alternating current into a steady direct current, this "graviton effect" equal to 6.6734E-11 N (kg/m)² occurs within every quark in every atom in the visible universe all contributing a miniscule amount to its overall gravitation field. According to Brian Greene, quantum fluctuations in space spontaneously produce electrons and positrons, which move about and then annihilate each other. We alter this in that matter and antimatter particles come into existence from the exceedingly rare collisions of other particles, such as neutrinos.

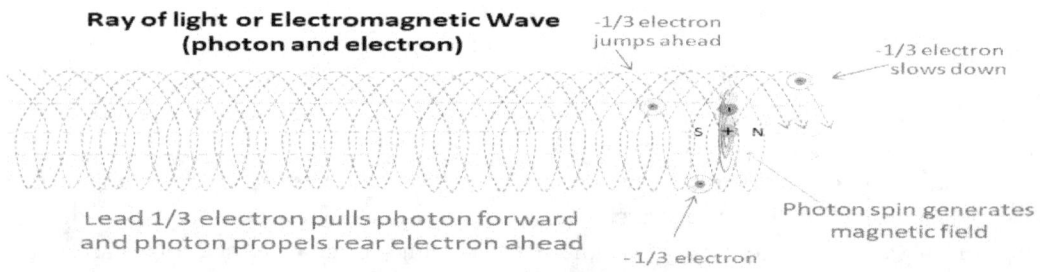

Above shows an up-quark colliding with an anti-up-quark.

Chapter 2 presented science typical electromagnetic (EM) wave. The light emitted below has an electron consisting of three 1/3 negative point pairs, and a photon spinning to generate a magnetic field. The three negative charged pairs spiral orbit around the photon alternating leapfrog jumps and creates a magnetic field to propel the photon forward, which in turn combines fields to slingshot the separate negative charges forward in perpetual motion. The resulting wave interaction differs slightly from the typical science diagram where the electron moves up and down while the magnetic wave moves left and right in the forward motion. More details of how these particles move and interact will be discussed in chapter 5.

Before we leave this subject, we should mention why neutrons survive within a neutron star, and why quarks survive within black holes. The mass of a proton is approximately: 1.6727×10^{-24} grams. The mass of a neutron is about 1.6750×10^{-24} grams; it contains two down quarks, which has slightly more mass than an up quark. And the mass of an electron is approximately 9.110×10^{-28} grams. The neutron structured much like that of a helium atom, consists of an up quark orbited by two down quarks, is very stable. The down quarks constructed with three negative particles and three neutral particles within its center and randomly orbited by two pairs of positive and neutral point particles are more energetic and provide the strength to withstand the crushing gravity in a neutron star. The extra strength comes from the loosely bounded and more energetic negative-neutral point particles (strong force) at its center, the negative charge of the two down quarks surrounding the neutron, and the fact that neutrons are zero in overall charge. Their extra motion is more energetic, gives it a slight increase in mass, without an increase in their "gravity" field. Any protons within a neutron star reconfigures into neutrons, by emitting a positron and neutrino. Similarly, quarks can withstand the gravitational forces within a black hole because they are the source of the black hole's gravity, without their existence there would be no gravity field generated within the black hole. Dissolving the atom into its basic quarks maintains the overall gravity field and allows the black hole to be denser. We predict that the quarks within a black hole might realign themselves into zero charge quarks, three pairs of negative-neutral point particles at its center hyperbolically orbited by three pairs of positive-neutral point particles. It does this by transferring positive-neutral pairs between high-energy top and bottom quarks. Such zero charge quarks soup resembles the neutrons in a neutron star. The super symmetrical structure of quarks is like miniscule quantum forms of "atoms" complete with positive, negative, and neutral point particles all working together to maintain existence and survive; exactly how is beyond the purpose of this book. As such, these quarks transform to produce neutral non-graviton matter and anti-graviton matter to relieve some pressures within the black hole, thereby enabling "excretion" of antimatter.

Super symmetry demands the existence of a mirrored pair to matter with gravity. Antigravity antimatter repels gravity matter and vice versa. In antigravity part of the universe, neutral point particles operate differently than in the gravity universe; their atoms built primarily with antineutrino point particles generate antigravity. The swapping of neutral point particles within a quark and the reverse spin of those particles could be the cause of antigravity versus gravity. While a sustained strong force that neither turn off nor on could cause zero gravity in space-matter, where outer positive point particle pairs of the quark orbit in spherical pattern at a fixed distance rather than a hyperbolic motion.

Chapter 5: Theory of Cosmological Balance

"Theories crumble, but good observations never fade."

— Harlow Shapey, *Director of the Harvard College Observatory*.

The timely introduction of an enlightened, compelling, and logical unified theory will cause the inevitable and decisive rejection of the Big Bang Theory and select others. Scientists and many of us laypersons have been searching and waiting for this plausible solution to come forward. What is this theory? Are we ready for this revolutionary change of such magnitude? Maybe not at first, but eventually it will come to be a reality for all. The Cosmological Balance Theory could be this theory, concise and detailed in process development, while identifying the defects of the Big Bang Theory, and pointing the way to future fruitful fields of investigation, research, and experimentation. Cosmological Balance predicts: There exists matter that defines space; time is a universal linear motion based on the average vibrations internal to quarks at the quantum level. Gravity is both a push force over a layer known as the curvature of space[105], not pull force; and light is a particle influenced by gravity and lensing as well as a wave dispersing with distance. The theory also predicts that galactic red shift is due primarily to photon energy absorption as it passes through matter in space; the universe is Euclidean and not expanding exponentially. And the ages of galaxies do not directly tie to the distance from us; empty space is not empty and solid matter is not solid; and matter exists three dimensionally. The interaction of all three types of matter is a nine-dimensional relationship, Cosmological Balance Universe!

Summary of Theory and Assumptions:

In Steady-State Theory (Bondi, 1948), the application of the natural and physical laws to the cosmos is assumed, as all four natural laws influences cosmology are examined to be true in the motion of galaxies, solar system, and in the planets and debris that orbit. Planetary or terrestrial physics can be used explicitly only in a homogeneous and infinite universe, implying a strict logical basis for cosmological balance exists in such a universe. Newtonian definitions, laws, axioms, and gravity equations are also assumed valid in determining temperatures and pressures in stars as well as in black holes, and that gravity is instantaneous, as is quantum entanglement. Cosmological Balance assumes and predicts the existence of three types of matter: matter with gravity; matter with antigravity in the invisible hidden part of the universe; and matter with no gravity or antigravity characteristics, neutral space-matter, *hygratium*, yet pulses gravity-antigravity. All objects have innate attraction initiated by their entangled quantum particles' connection to like

---

[105] Einstein curvature of space-time describes how ordinary matter when massive enough curves space and time simultaneously. Physicists view space as completely empty and void, nor filled with aether, massless, and filled with the Higgs particle ocean. Einstein describes space as being flexible, expandable, able to curve, twist, ripple, and has the inherent ability to tell matter what to do. Cosmological Balance predict space to have mass, vacuum pressure, energy, and force; space-matter forms positive curvature around objects, concave shaped. Negative curvatures exist in gravity and antigravity repulsion and are saddle seat shaped.

types to merge fueled by vacuum energy of space, instantaneous infinite intensity layers, not limited to the speed of light.

The First Law of Thermodynamics: Conservation assumes that matter and its motion cannot be created or destroyed. The Cosmological Balance adheres to this first law. The Big Bang Theory, however, clearly presents the worst violation of conservation ever formulated and a concept full of holes. Its creation of everything from nothing is based more on a religious assumption, not a scientific one, ignoring the laws of nature. It presented that there was a "beginning" involving something, a tiny "singularity," from which energy dispersed and all matter came into existence. If such a "singularity" existed "before" the beginning, then it really was not the beginning, as something happened before then to create this "singularity." This law of conservation cannot be violated because fundamental physical laws and its basic assumptions always existed before the visible universe came to be and will continue to remain valid for all time. Thermodynamics considerations and the alternate proper reading of the red shift of distant galaxies, identified through astronomical observations that the universe is not expanding, but relatively stable, with occasional slightly contracting and expanding motions as matter and energy interchange in an infinite symmetrical and balance universe. The galactic red shift, not solely Doppler Effect, results from photon energy absorption. Space is not void of matter. Hence, there is a continuous conversion of space-matter to energized sub-atomic particles and into another form of matter with gravity or antigravity and then back to original form of matter, space-matter, at a rate unobservable to us. The observable properties of such a balance are obtained, and all the observational tests are found to give good agreement to the theory. There is no motion without matter, and there is no matter without motion. All phenomena can be classified as either matter, which exists, or the motion of it, energy, which occurs; particles "jiggle" in perpetual motion to exist.

Mathematics, however eloquent, can never represent a completely accurate depiction of reality. Applied math must conform to science; science should not bend or be compliant to mathematics. Calculations is just that an idealism; a representation of what mathematicians believe is reality, not reality itself. Einstein for instance had a choice when he his equations told him that mass would increase as its velocity approached the speed of light (Einstein/Golm/Potsdam, 2015); he kept the math and discarded the physics that told him otherwise that only the energy of the moving mass is what increased. Obviously, Newton's gravity equation $G \times m_1 \times m_2/r^2$ do not have objects exponentially increase in mass as m_1 and m_2 increase speed toward each other or decrease in mass as they move apart. Time is universal, it is measured by the motion of an object in relation to other all other things around it. Plotting time does not make it a fourth dimension. Einstein idea that matter could be converted completely to energy, $E=mc^2$, is misleading claim the universe at one time contained only pure energy and therefore can be completely void of matter, impossible since energy is the motion or potential motion of matter. He claimed mass "curved space-time", or warped fabric of space-time caused gravity. Cosmological Balance is not going to discard proven physics. It accepts the concept and equations of the curvature of space Einstein proposed, but redefined in a unified theory; where space "curvature" is dense space-matter accumulated around a massive object causes gravity[106], and defines the creation process with the formulation of a field theory.

---

[106] Nesting of gravity layers or curvature of space around massive objects conforms to the shape of

5.1 The Basic Theory

The basic theory for Cosmological Balance uses the baseline Machian theory[107] of gravity first proposed by Hoyle and Narlikar in which the origin of inertial is linked with a long-range scalar interaction between matter and other matter. Specifically, the theory is derivable from an action principle with the simple action: $A = -\sum_a \int m_a \, ds_a$, where the summation is over all the particles in the universe, labeled by a, the mass of the ath particle being m_a. The integral is over the existence of the particle, ds_a representing the element of proper moment in time of the ath particle. The Cosmological Balance takes the simple action and goes a step further:

$$A = -\sum_a \int m_a \, ds_a = \sum_g \int gm_g \, dg_g + \sum_n \int nm_n \, dn_n + \sum_s \int sm_s \, dp_s,$$

where the summation is over all three types of particles in the universe: those with gravity, labeled g, which we are all familiar with; those with antigravity, labeled n, which we will learn about; and those without gravity or antigravity called space-matter, labeled s, which we will discuss in great detail. The mass of the gth, n^{th}, and s^{th} particle are labeled gm_g, nm_n, and sm_s. Their integral is over the existence of the particle, dg_g, dn_n, and dp_s, representing the element of proper moment in time of the gth, n^{th}, and s^{th} particle. The mass itself arises from interaction with other particles, to include interactions between all three types of matter: space-matter, gravity matter, and antigravity antimatter, where the neutral graviton in space-matter is synonymous with the Higgs particle. Thus, the mass of particle a $m_a = \sum_{b \neq a} m^{(b)}(A)$, at the point A on its existence arises from all other particles b in the universe: where $m^{(b)}(X)$ is the contribution of inertial mass from particle b to any particle situated at a general space-time point X. The long-range effect is Machian in nature and is communicated by the scalar mass function $m^{(b)}(X)$, which satisfies conformably invariant wave equation. $\lambda m^{(b)} + \frac{1}{6} R m^{(b)} + [m^{(b)}]^3 = N^{(b)}$,

Here the wave operator is with respect to the general space-time point X. R is the scalar curvature of space-time, as defined by Einstein, and the right-hand side gives the number density of particle b. The field equations are obtained by varying the action with respect to the space-time metric g_{ik}. It is important to note that the above formalism is conformably invariant. One can choose a conformal frame in which the particle masses are constant. If the constant mass is denoted by m_p, the field equations reduce to

$$R^{ik} - \frac{1}{2} g^{ik} R + \lambda g^{ik} = -\frac{8\pi G}{c^4} [T^{ik} - \frac{2}{3}(c^i c^k - \frac{1}{4} g^{ik} c^l c_l)],$$

where c is a scalar field, which arises explicitly from the ends of broken world lines or changes of existence, that is when there is a transition viewed as creation (or annihilations) of particles in the universe. So, divergence of matter tensor T^{ik} does not always need to be zero, as creation or annihilation of particles is compensated by the non-zero divergence of the c-field tensor, equation above. Quantities G (gravity constant) and λ (cosmological constant) are related to the large-scale distribution of particles in the universe. Thus, where N is number of particles in a $G = \frac{3\hbar c}{4\pi m_p^2}$, $\lambda = -\frac{3}{N^2 m_p^2}$, cosmic horizon.

Luminiferous aether was a sought-after invisible medium thought to be required for light to propagate in space, sometimes referred to as just aether,

---

the object or group and is not necessarily spherical but always wraps around the object like a "halo."

[107] Machian Theory was first announced in Mach, E., *The Science of Mechanics*, London, 1893, pp. 229-238.

also spelled ether. Space-matter, similar to ether, but an exotic form or type of matter without gravity or antigravity properties, occupies space at one atom per cubic meter generating pressure at 3.80379E-63 N/m² and vacuum energy at 8.83919E+15 kg*m²/sec² = 8.83919E+15 N*m = 8.83919E+15 Joules, and a vacuum force of 6.6734E-11 kg*m/s² = 6.6734E-11 N. Detailed explanation of space-matter is discussed in the chapter section titled "What makes outer space appear void?" This vacuum force when transferred via collisions of space-matter and gravity matter provides the push force of gravity[108]. In addition, this force when transferred with collisions between space-matter and antigravity antimatter provides the force for the anti-gravitational attraction of antigravity antimatter. Fueled by vacuum pressure and energy, the vacuum force of 6.6734E-11 N within one kilogram of unique high-energy space-matter atom (an exotic matter) at one meter per kilogram colliding with normal gravity matter at one meter per kilogram transfers the total force of 6.6734E-11 N m²/kg², which is equivalent to Newton's gravitational constant G used on normal matter equations.

1_{sm} 6.6734E-11 N m/kg x 1_{gm} m/kg = 6.6734E-11 N*m²/kg² = G

These transfers of energy and force make gravity a push force from space, not totally a pull force as seen from earth. The closer two objects come together the less space-matter between them can keep them apart and the stronger the surrounding outer space-matter is able to push them together. Since normal gravity matter, due to its low density and lower electromagnetic repulsion than space-matter, its particles are less likely to push apart and more likely to clump together through natural weak chemical, molecular, and "gravitational" bounds, at the atomic and quantum level. This push force is then transferred continuously toward the center of each object (believed to possess gravity) making them appear like the gravity force pulls the two centers together per Newton's gravitational equation. The more massive the gravity matter becomes the further out its gravitational attraction appears to reach, hence massive black holes and massive white holes gravitational influences all the stars within their entire galaxy. When, all of the space-matter surrounding these bodies provides the force to push them to each other while the remaining three natural forces keep them together, defined as quantum mechanical graviton or simply *quantum gravity*. The Cosmological Balance Theory therefore defines the interaction of all three types of matter in the universe. The gravitational constant, as known in Newton's gravity equation, is here thereafter redefined as the product of vacuum force and natural gravitational attraction of like objects, where vacuum force is 6.6734E-11 Newton and the natural quantum entanglement or gravitation of two normal objects in the visible universe is one meter per kilogram. The revised equation reads 6.6734E-11 N x (1 m/kg x m_1) x (1 m/kg x m_2) / r^2 = Gm_1m_2/r^2, where 1 m/kg is the inherent attraction of sub-atomic particles in what scientists call the quantum mechanics entanglement of objects. Quantum entanglement of particles with a common origin possesses an innate ability to affect each other over infinite distances and share a tendency to function or merge as one. In essence, we can call this portion of the gravity equation (1 m/kg), *quantum entangled attraction*[109]. The same applies within the invisible part of the universe, antigravity

---

[108] Space-matter existed long before any matter or antigravity antimatter; space-matter is the source of the other two types of matter.

[109] All ordinary matter shares a common origin, they all come from the decomposition of space-matter and are therefore quantum entangled and attracted. The same applies to all antigravity antimatter attraction and a connection between gravity matter and antigravity antimatter repulsion as well.

antimatter, and their galaxies, where vacuum force is 6.6734E-11 Newton and the natural anti-gravitational attraction of negative one meter per kilogram. The revised antigravity equation reads 6.6734E-11 N x (-1 m/kg x m_1) x (-1 m/kg x m_2) / r^2 = 6.6734E-11 N m^2/kg^2 x m_1 x m_2/r^2 = Gm_1m_2/r^2.

However, when it comes to space-matter on itself, gravitational constant as defined per Newton's gravity equation has no impact on it because the natural attraction of two space-matter particles or atoms is zero meters per kilogram. The revised equation reads 6.6734E-11 N x (0 m/kg x m_1) x (0 m/kg x m_2) / r^2 = zero. The interaction equation between space-matter and normal matter with gravity is 6.6734E-11 N x (1 m/kg x m_1) x (0 m/kg x m_2) / r^2 = zero; and between space-matter and antigravity antimatter it reads 6.6734E-11 N x (-1 m/kg x m_1) x (0 m/kg x m_2) / r^2 = zero. Finally, the equation interaction between matter with gravity and matter with antigravity is written as 6.6734E-11 N x (1 m/kg x m_1) x (-1 m/kg x m_2) / r^2 = -Gm_1m_2/r^2. Note: the signs of the various constants are determined by the theory, and not manually inserted by hand. For example, the constant of gravitation is positive, the cosmological constant or antigravity negative and the coupling of the c-field energy tensor to space-time is negative, as described above. A more complete picture of creation of matter, which incorporates inputs from quantum theory, is needed to determine the coupling of the c-field to matter and to determine the rate of creation. Matter creation is possible at a given point in space-time provided the ambient c-field satisfies the equality c = m_p at that point. In normal circumstances, the background level of the c-field will be below this level. However, in the strong gravity of compact massive objects, or in intense space low-pressure rotating systems, the value of the field can be locally raised. This leads to creation of matter, as we know it, along with the creation of negative c-field energy and antigravity antimatter. The latter also has negative stresses, which have the effect of pushing the space-time outward as the antigravity antimatter mutually repels away normal matter with gravity, as well all know. Antigravity antimatter mutually attracts other antigravity antimatter just as the matter with gravity attract each other, both forming own gaseous cloud nebulas.

We shall refer to such pockets of creation, which primarily result from low-pressure space "storm systems" and spontaneous stellar cloud nebula formations, as *gravity-matter creation* or *antigravity antimatter creation*. We shall refer to such pockets of normal matter or antigravity antimatter-antimatter annihilation and conversion, which primarily occur within high-pressure high-energy dense massive black holes or massive white holes through excretion into wormhole tunnels, as *space-matter restoral*. A spherical (Schwarzschild type) compact matter distribution will lead to a spherically symmetric excretion whereas an axis-symmetric (Kerr type) distribution would lead to a jet like excretion along the symmetric axis. These are two types of wormhole tunnel excretion or ejection pattern determines the shape of the galaxy hosting the massive black hole, or massive white hole; makes elliptical or barred galaxy. The third major type of pattern occurs from rapid rotating symmetric excretion, makes spiral galaxy. There is no state of infinite curvature and terminating world lines, as in the standard big bang theory, nor is there a black hole type horizon where nothing is ejected, as the presence of the c-field causes the collapsing object to return into space the building materials necessary to sustain space, *space-matter or* **hygratium** (which is without gravity or antigravity characteristics, yet pulses both). Details discussed in the physics of creation below.

5.2 The Cosmological Balance Defined

Physical science calls for the unrestricted repeatability of all experiments as a fundamental axiom, implying that the outcome of an experiment be unaffected by the position and time at which it is completed. Cosmology must be primarily concerned with this fundamental assumption and, therefore, a suitable cosmology is required for its justification. Physics in the laboratory require us to distinguish between conditions, which can be by "chance" at will and those "inherent" laws, which are absolute[110]. Such distinctions between the "chance" and the "inherent" laws and constants of nature is justifiable so long as we can control the "chance" conditions and can test the validity of the distinction through multiple experiments. In astrophysics, it is very difficult to control and hard to prove which are "chance" and "inherent" laws, but we need not get concerned when it comes to interpretation of the dynamics of the solar system, and the galaxy in the universe. The "inherent" laws, gravity, for example, satisfactorily and accurately permit the solar system and all its numerous orbits. However, when we consider the actions of the entire universe, then the logical basis for the distinction of the "chance" conditions and the "inherent" laws are questionable. Observations of the structure of the universe give us unique results, and if interpreted in error, cause confusion, particularly in the velocity of light and or the constant of gravitation. Regardless, we are required to accept and assume some of such observations to represent the "chance" conditions and "inherent" laws to contemplate the universe. Such assumptions are in fact implied in all theories of the universe, as a necessity to specify a problem and recommend a solution and define rules and justification to extrapolate into the future or into the past. The presentation and acceptability of these assumptions will determine adoptability by the scientific community.

Any interdependence of natural or physical laws and large-scale structure of the universe might lead to a fundamental difficulty in interpreting observations of light emitted by distant objects. For if the universe were seen from those objects, presented a different image, then we cannot be justified in assuming the same processes responsible for the emission of light analyzed. This difficulty is removed by the Cosmological Balance universe theory. According to this theory, all large-scale averages of quantities from astronomical observations (specifically the mean density of space to include "dark matter" and "dark energy," the average size and shapes of galaxies, the ratio of condensed and uncondensed matter, the luminosity distances, etc.) would tend statistically to have a similar value independent of the observer's position, as their distance range increases, if the observations are performed at similar equivalent times, near simultaneity. This widely recognized principle implies that the universe on a large scale and is homogenous, and that any differences, which exist, are only of local significance. The universe presumed to be capable of altering its large-scale structure, but only in a way as not to upset its homogeneity, such measurement is of universal time.

We might therefore look to the "Cosmological Balance" Theory for justification of the assumption of the general validity of physical laws; but while the theory supplies the justification with respect to changes in place, it still leaves the possibility of a change of physical laws with universal time or maybe not. Any system of cosmology must involve a

---

[110] Outcomes considered "chance" are sporadic and accidental, like the rolling of several dices at once.

speculation about this dependence as one of its basic assumptions. These assumptions provide us the baseline to make interpretations of observations of light emitted from very distant objects, even those from a different scale of universal time, and accordingly the processes causing the emission may be unique and unfamiliar to us or a mirage misinterpreted. For that matter, cosmology systems might as well be categorized by their assumptions, implied or not. Cosmological Balance theory regards all if not most of laboratory physics to be applicable, without regard to the state of the universe, and draws its own conclusions from careful analyses of observed data. We shall pursue the possibility that the universe is in balance, self-recycling perpetuating state, without making any assumptions regarding the circumstances leading to the creation of space in the universe. We regard this pursuit as compelling, a universe where the laws of physics are constant. Without such assumption, our knowledge, obtained virtually at one instance in time, becomes inadequate for an interpretation of the universe and the dependence of its laws on its structure, hence becoming unreliable for any extrapolation or prediction into the future or of the past. Cosmological Balance stays the course, defined not only by the usual physical laws, but also by that extension of it, which is obtained on assuming the universe to be not only homogeneous, but also relatively constant on a large scale. We do not claim that this principle must be true, but we say that the evidence itself shows convincing argument to its validity, thereby confirming that terrestrial physical laws are applicable in cosmology as a science. If the Cosmological Balance Theory is satisfied in the universe, then we can base ourselves confidently on the permanent validity of all our experiments and observations and explore the consequences of this theory and principle.

Gravitation is constructed from both a pull and a push force. As a push force, it moves two objects toward each another. However, Newton's laws interpret it as a pull or mutual "gravitational attraction" between two objects, disregarding the force or mechanism pushing the two objects together, space and its energy, pressure, and force to move matter[111]. As an object moves from point A to point B, it will push any other matter or particles in its path out of the way to get to point B. As such, create less "push" apart between the objects and thereby allowing the "push" toward each other to be stronger and increase with proximity, defining gravitational effect as a force that is directly proportional to the product of their masses and inversely proportional to the square of the distance between them. This is Newton's Gravity Law written as a "push" force. It is the same effect, we get when air flows over a forward moving wing causing the airplane to achieve lift, which of course could be misinterpreted as "antigravity" by uneducated people. The question is do you see the air pushing up the airplane, or do you see the vacuum above the wing pulling up the wing; most see the compressed air below the wing pushing up or lifting the airplane. Gravity is a matter of interpretation and visualization. The "push" of gravity is necessary to counterbalance the outward "push" of centrifugal force as a body orbits another, instead of visualizing it as the pull of gravity or centripetal force counterbalanced by the push of centrifugal or tangent speed. The push force of space on massive objects toward each other gives gravity its instantaneous and infinite range, while quantum generated part of gravity contributes to its pull aspect.

---

[111] Einstein's general relativity equation shows us that space tells matter what to do while matter too tells space what to do. In other words, space makes matter move, and moving matter makes space respond accordingly.

5.3 Application of Cosmological Balance

"The universe is then one, infinite, immobile.... It is not capable of comprehension and therefore is endless and limitless, and to that extent infinite and indeterminable, and consequently immobile."

— Giordano Bruno, *Cause, Principle, and Unity*, 1588

The theoretical framework presented is solid; however, the mathematics representing it may need supplementary development and refinement. Several physics disciplines may need further investigation in relation to Cosmological Balance theory. These include electromagnetism, fluid dynamics, cryophysics or low temperature physics, high-energy physics, high-pressure physics, light physics, molecular physics, nuclear physics (atomic nucleus), particle physics, and acoustics. Each one of these disciplines' analysis in near vacuum environment will confirm our interpretation and understanding of the existence of space-matter and its properties and affect the acceptance of the theory. Physical laws are imbedded into the structure of the Cosmological Balance, and conversely the universe's structure depends upon the physical laws.

Cosmological Balance is homogeneous, renewable self-perpetuating balance, and on a large scale, somewhat stationary or "breathing," and not expanding at an accelerated rate. These three principles are vital to the definition of the theory. We must now apply the principle of Cosmological Balance to laboratory physics and to astronomical observations. We regard this theory as of such fundamental importance that we shall be willing if necessary, to be critical and possibly reject theoretical extrapolations from experimental results if they conflict with the Cosmological Balance Theory. We must be critical, as "chance" conditions could have influenced experimental results. For instance, the cesium clock counting (ticking) mechanism or the rate of decay of the radioactive material in the clock due to changes of environment could affect the time measurements, thereby giving us two distinct times, one in orbit around the earth that is different from the one on the ground. Such "chance" conditions would thereby cause scientists to conclude inadvertently that Einstein was right about his predictions of Special and General Theory of Relativity, concerning space and time. Of course, we shall never purposely disregard any direct observational or experimental evidence and we shall make every effort to see that we can easily satisfy all such requirements.

For the Cosmological Balance Theory to apply, one might at first glance expect that the universe would be unchanging and possessing no consistent large-scale detectable motion. This however, conflicts with observations of nearby and distant galaxies, and would conflict with the observed thermodynamic state. A Cosmological Balance however would clearly reach that thermodynamic equilibrium after some time, as it is an infinitely old universe, with complete equilibrium between energy and matter, and large-scale motions, with limited expansion and contraction. The creation of matter with gravity and antigravity leads to perceived expansion of the universe. In addition, the conversion of matter with gravity and antigravity back into the sustainment of ocean of space between galaxies leads to perceived contraction. In this case, hydrodynamic continuity, which has been regarded as unqualified truth and not as an approximation of physical laws, applies to Cosmological Balance expansion and contraction.

The universe is not expanding at an accelerated rate, despite the now more rigorous interpretation, as cosmologists are still baffled by "dark matter" and "dark energy" influences. Edwin Hubble's declaration of the run-away accelerated expansion of the universe[112] may have been made in haste when he observed red shift light from distant galaxies and concluded the motions of those galaxies speeding away from us produced these results. He ignored two important clues pointing to error in judgment. First, that all the galaxies were showing red shift, and second, that the larger the galaxies' distance the greater the red shift. These clues indicated something else must be causing the red shift, but he, and all other scientists after him, did not look seriously into those clues, instead they relied strictly on the "Doppler Effect" of objects speeding away as the reasoning for the red shifted light.

The creation of gravity matter or antigravity antimatter discussed above can make the space-time appear to expand, particularly if significant new antigravity antimatter is created between two or more normal gravity matter galaxies. Its presence will cause gravity galaxies to drift apart; the deSitter metric represents such an example in space-time. However, the creation of gravity matter or antigravity antimatter from space-matter and its return into space-matter through epochs of ups and downs, shows an oscillation over a long-term evolution, requires us to Sachs et al metric[113]. Sachs et al have computed the simplest such solution with the line element given by

$$ds^2 = c^2 dt^2 - S^2(t)[dr^2 + r^2(d\theta^2 + \sin^2\theta d\varphi^2)],$$

where c stands for the speed of light and the scale factor is given by

$$S(t) = e^{t/P} \left[1 + \eta \cos \frac{2\pi\tau(t)}{Q}\right],$$

The constant P and Q are related to the constants in the field equations, while $\tau(t)$ is a function $\sim t$, which is also determined by the field equations. For details, see Sachs *et al.* The parameter η may be positive and is less than unity. Thus, the scale factor never becomes zero: the cosmological solution is without a space-time singularity.

We can now examine the requirements, which the Cosmological Balance places on the evolution of stars and galaxies. The mean ratio of condensed to uncondensed matter with gravity has to stay constant, and for this reason, new stars and galaxies have to be formed as older ones move away or dissipate and return back to sustaining the ocean of space and space-matter. Cosmological Balance imply that no feature of the universe is subject to any consistent change, and that no observer capable of any unique definition of a universal time besides quark vibrations, and that the ages of a group of galaxies in a sufficiently large volume of space follow a uniformed statistical distribution evenly spread between older, larger and younger smaller galaxies. This fact has been confirmed by observation in the diversity of galaxies in a region of space.

The next point to discuss is the thermodynamics state of the universe as cosmologists see it. It is believed that this state is far from any equilibrium. Very much more energy is in the form of visible matter than in radiation; and that very much more energy is radiated away than is absorbed by visible matter. This disequilibrium is significant and is marked

---

[112] Edwin Hubble, 1936, announced that the universe was expanding at an accelerated rate. This caused Einstein to withdraw the cosmological constant.

[113] The Kantowski-Sachs et al metric, in Einstein's general relativity, describes a homogeneous but anisotropic universe in terms of ds^2 for massless scalar field equations.

and characteristic feature of the visible universe that it must be one of the main tasks of cosmology. Any photon emitted by a star has a large probable free path, owing to the tenuous distribution of opaque matter in the universe, and most likely travel uninterrupted in areas where there is less matter. Photon energy absorbed by intercepting matter will be subject to a loss of energy and red shifted by a miniscule fraction, with each encounter repeatedly, before continuing its journey to the observer. This sink for radiation energy is commonly known and available to most photons emanating from the surface of stars. The process of getting rid of both matter and radiation from stars of any fixed volume is by pushing both across the surface bounding this volume; and then both replenished from within. It is of course important for any theory of cosmology that it should not predict thermodynamic equilibrium, and this is possible only by supposing either that insufficient time has been available since the beginning of radiation by stars, or, by suggesting that the bulk of the energy radiated does not again become available as heat. In the case of Cosmological Balance, the bulk of radiated energy is absorbed by space-matter for the maintenance of the structure of space itself and is not available as heat. Cosmologists attribute this absorption to "dark matter and dark energy" characteristic feature. Without this additional radiated energy, space would not be able to provide the force behind the gravity we see in the visible universe, nor the antigravity in the invisible part of the universe. Cosmological Balance Theory brings us remarkably close if not into thermodynamic equilibrium, a balance between all matter, all radiation, and all energy, known and unknown by cosmologists.

Now let us investigate the conservation of mass and energy in the Cosmological Balance. An observer attempting to estimate the quantity of matter in the universe will find that his result depends on the threshold intensities and wavelengths of his instruments. He observes a distribution of about 15% visible matter with gravity and 85% dark matter with unknown characteristic to cosmologists. In the Cosmological Balance, dark matter or as we will get to know it as space-matter, which is a gaseous unique high-energy matter (an exotic atom) with no gravitational or antigravity properties, on excessive collisions with itself is converted to visible matter with gravity and invisible matter with antigravity characteristics. The energy, pressure, and force of the space-matter become the force of gravity and antigravity. The light photon energy emanating from stars with gravity or antigravity become the energy that sustains space-matter and maintains space as it is, a vacuum. Black holes and White holes excretion provide the essential building blocks for replenishing space-matter. Cosmological Balance provides and is the proper conduit for matter and energy conservation and balance.

5.4 Observational Tests

Our observations of the universe are influenced by what is happening between the rest of the universe and us, whether viewed through dynamic or electromagnetic interactions. Dynamically, the rest of the universe affects us primarily through inertia's long-range force gravity, defining frames of references in this region of the universe. It is strong and comes almost certainly from the whole universe. The other view of the universe is electromagnetic interactions, which are readily visible and interpreted frequencies in the spectrum. Light and all other EM spectrum reaches us in observable intensities from many extra-galactic objects, and a whole lot of information that comes with it collected. Especially, the distribution over the night sky, the magnitude, and the radial velocities determined for each object seen. We must look at the whole region of cosmology to interpret the data properly, determining if the observation is universal or purely local. In observing nearby isotropy and homogeneity nebulae, we find their distribution and magnitude of the radial velocities and assume the observation applies to a considerable amount of the universe, and relevant to cosmology. The work of Robertson and Walker is applicable to our theory, except that we have specified a balance universe, not expanding exponentially. Light paths are the null geodesics of the metric:

$$dx^2 = dt^2 - (dx^2 + dy^2 + dz^2) \exp(t/T),$$

where the (x, y, z) coordinates of any particle in general motion in the universe is constant and t measures the time of any observer travelling with the particle. T is a universal constant and the velocity of light set to unity. It is the simplest of Robertson metrics, primarily used to describe empty space.

The de Sitter universe geometrical properties are also quite simple and applicable to the theory. Changing the Cartesians (x, y, z) to spherical polar coordinates (r, θ, φ), then the observer at the origin would find that the light's source spectral lines' luminosity L at (r, θ, φ) received by him at t = 0 would show a red shift

$$I + \delta = \lambda_{received} / \lambda_{emitted}$$
$$= \exp(-t_{em}/T) = I + r/T,$$

While the intensity of light received would be

$$L/4\pi r^2 \exp(2t_{em}/T) = L / (4\pi r^2 (I+r/T)^2).$$

Now let us incorporate the null geodesics metric. Accordingly, if n is the number of nebulae per unit volume, the number of nebulae between r and r+dr is

$$4\pi r^2 n dr \exp(3t_{em}/T) = 4\pi r^2 n dr (I + r/T)^{-3}.$$

The difference between the two quantities depends on the nebular spectrum and is a function of the red shift. This red shift is not due to vector velocity of distant objects but due to absorption of photon energy by gaseous or transparent material between the object and us, very much like the atmosphere on earth causes a red shift of the Sun's light as it sets or as it rises. This gaseous material defined in the Cosmological Balance as space-matter, evenly spread across space at about one substantial energetic atom per cubic meter. This space-matter, an exotic atom, can absorb photon energy directly from the light passing through it, briefly slow down the light before reemitting it with minor change to its color[114]. While matter with gravity, resulting from colliding and decaying of space-matter, accumulates in low-pressure areas, absorb light photon energy from distant galaxies and causes the red shifts in their light. Matter with antigravity, similarly created by excessive colliding and decaying of space-matter, absorbs antigravity light photon energy from distant antigravity galaxies and causes red shifts in their light.

Through analysis and observation, we discovered and located several black holes in the galaxy, including the massive black hole at the galactic center in the Constellation of Sagittarius, estimated their event horizon and conjectured radius size. Cosmologists still have not agreed as to what is going on within the black hole. Some speculate that quantum physics affects black holes; some apply Einstein's Theories, while others apply Newtonian physics to the matter within a black hole. Cosmological Balance applies Newtonian equations to conjecture the internal collisions within the black hole. In our analysis, we speculate the continuous creation and annihilation of anti-matter within black holes. In our estimates, the black hole somehow either releases energy and mass, or blows itself up. Since we have yet to see a black hole explode despite from the creation of anti-matter and their annihilations within them, nor find the large wormholes leading away from the black hole to white holes, as predicted by Einstein and Hawking, then there must be some other energy and matter release mechanism, a realization of the grand unification theory. Cosmological Balance Theory predicts that release to be trillions upon trillions of micro-wormhole tunnels releasing micro white hole or black hole orbs into space, as the sub-atomic building block source of space-matter, where the four natural forces are one[115]. The collapse of these wormhole tunnels at the end of their paths also releases a short microwave burst at 160.2 GHz, without significantly raising the temperature of the surrounding space, keeping it around 2.7-Kelvin degrees. The 2.7 Kevin temperature of the universe obtained by reading the microwave background signal from space was analyzed to fluctuate randomly from 2.7249 to 2.7251 in different pocket areas. Quantum mechanics makes all things jittery and turbulent within the micro universe (Greene, The Fabric of the Cosmos: Space, Time, and the Texture of Reality, 2004, p. 308).

Einstein introduced his Special and General Theory of Relativity, predictions, and proposed experiments backed up with supporting tenants for acceptance by mainstream science and academia. Considered by many as a genius in his time, he paved the way of

---

[114] Space-matter moving toward the massive object provides the force of gravity on the object and contains it in space. This moving exotic space-matter contained within the densely packed layer around the massive object, known as the curvature of space, repeatedly briefly absorbs and slows down the starlight photons and then reemits them at noticeable angle attributed to space-matter's optical density light refraction. Einstein captures this repeated light deflection or "gravity lensing" as the result of additional "gravitational energy" cause by the curvature of space, where his general relativity equation accurately calculates that starlight's geodesies path around the massive object.

[115] The Cosmological Balance Theory defines that there is no single wormhole tunnel connection between a massive black hole and massive white hole, and that massive black holes are capable of creating and ejecting trillions upon trillions of minuscule white hole orbs (known as evaporation), which travel outward in almost invisible wormhole tunnels like tentacles in all directions. Massive white holes too can create and ejecting trillions upon trillions of minuscule black hole orbs (known as evaporation), which travel outward in almost invisible wormhole tunnels at the speed of light squared like tentacles in all directions. These paths if detected would look just like strings from the galactic center to the galaxy edges.

modern scientific thinking and experimentation. He took advantage of the limitations of his time and in his own words the "stupidity" of humanity to popularize his theories and childhood dreams and spent the rest of his life defending them. What becomes of Einstein when his tenants are stripped away? We modify and update Einstein's achievements and get a corrected refined unified theory advancing humanity. The Cosmological Balance Theory analyzes each of his tenants and proposes adjustment to some. Einstein's solution to Mercury's orbital anomaly was unnecessary, as there is a Newtonian solution to Mercury's perturbations. It certainly is not a result of bending of space and time as predicted by Einstein. Moreover, Einstein's prediction of the bending of light around the sun, star, or any object of significant size and mass, has nothing to do with the curvature of space time. It is a result of the special characteristic of exotic space-matter and its intense gravity or curvature of space that bends the path of the minuscule mass the photon particle takes as predicted by Sir Isaac Newton coupled with refraction due to density changes in space surrounding the sun's clear colorless outer atmosphere.

Einstein's perception of space-time tied the two elements together. He declared the speed of light as the constant, relative to anything and everything. In other words, the total motion in space added to the total motion in time must equal the speed of light, in an absolute space-time. Zero motion in space means all the motion is forward in time. Motion of a vehicle in space close to the speed of light means that motion in time is close to zero. Therefore, according to Einstein an increase of speed decreases one's motion in time. Using Einstein's logic, an individual traveling at 2/3 the speed of light tells us that the clock in their spaceship is running 2/3 slower than their "stationary" sibling's clock on earth. Two hours' time in the spaceship equates to three hours on earth, thereby the space traveler ages more slowly. However, if we unequivocally apply Einstein's perception of space-time to the orbit of planets and their true combined speeds (rotation plus paths around the sun, around the galaxy, and galaxy movement through space), we immediately get an unexpected result, gravity equations fail and orbital paths decline, completely oppose to viewed and known observations per Kepler, Newton, and others. Science tells us that any theory that opposes observations and confirmed data is hence questionable.

Einstein's prediction of time slowing down as gravity increases, means that particles falling into a black hole would steadily build up just within the event horizon where Einstein predicts time stands still, thus never reaching the black hole surface itself, a conundrum in itself. By the same prediction, Einstein says that as the vehicle's or object's velocity increases then their clock slows down, thereby, again causing particles falling at ever increasing speed into the black hole to build up just within the event horizon. Although, our current interpretation of clocks in orbit around the earth, compared to clocks on the earth may seem to prove Einstein is right in his principle, it does not seem to make sense when it comes to the sun, other stars, and especially, a black hole. It does not even pass the logical test for an object in a stationary position just outside the Milky Way, as such, an object would have infinitely fast clock speed due to near zero velocity and near zero gravity, an impossibility, as oppose to zero time and space just before the Big Bang, another impossibility. Newton's universal time must therefore be correct.

Einstein also used a flat 3-D graphic representation to explain the fabric of space-time, which has inherent problems. He used gravity in space to define gravity on objects in space, which of course is not typically a valid approach to defining phenomena. In other

CHAPTER 5: THEORY OF COSMOLOGICAL BALANCE

words, he used an unknown force that mimics gravity to explain why the fabric of space-time is bent or depressed downward[116]. He states that massive objects like the sun or a star bend the space-time fabric, and its resulting curvature makes the planets or smaller objects follow a circular or elliptical orbit around it. If this were true and we followed an orbiting object as portrayed, then the planet would be orbiting in retrograde direction to follow the curvature he depicted, but it is not. We also know that space has no gravity force within it that can bend space-time, let alone create a curvature that Einstein claims. Nonetheless, Einstein has the support of the scientific community working his theories and predictions. The imagination of the early 1900s, although still valid and acceptable, does not do us justice today; he was a genius that paved the pursuits of the scientific community and took advantage of the stupidity of humanity. If Einstein's famous equations of General and Special Theory of Relativity and the view of the curvature of space-time are so elaborate and accurate, then why do they always try to simplify them by setting c or the speed of light to one, more precisely ct represented by 1 meter = 2.99792E-8 second, or use some other short-cut of these equations? They are just too complex for most of our simple minds, and occasionally deemed unnecessary when Newton's equations are simply fine.

Cosmological Balance Theory, however, presents a more simplified, comprehensible and believable picture of the force of gravity and the interactions between objects, stars like the sun and their solar system, justifying the normal prograde motion of planets because of this depiction and explaining the circular or elliptical speeding up and slowing down motions of these objects. The Cosmological Balance Theory also provides a viable explanation of why galaxies stars orbit in unison and why galaxies have certain shapes and patterns. The black hole and white hole excretion process through micro wormhole tunnels, not just Hawking radiation, provides part of the necessary force to make all stars orbiting the galaxy hub, its center black hole or white hole, appear to move as one huge, interconnected wheel. The excretion process also sustains space-matter and in turn space between galaxies, as well as provides essential sub-atomic particles and gaseous materials for galaxies to develop new stellar nebulas and new stars, consisting of hydrogen, helium, and other light elements. Observation shows us that the age of the universe according to the Big Bang Theory may not be accurate. The observable objects in NASA's Hubble Deep Field Images show us fully formed galaxies several billions of light years away. These galaxies according to the galaxy's stage of development surely must be at least ten to thirteen billion years old themselves, while humongous galaxies are at least 100 billion years old. Homogeneous universe projects that if we were sitting in one of those distance galaxies within the deep field image looking further away from our current location, we would also be looking at fully formed galaxies, which are an additional 12 billion light years away[117]. The distance from us and their apparent age in development make them older than age of the so-called Big Bang, which is a drastic miscalculation in extrapolating one singularity destruction as the creation of the universe, practically an impossibility, breaking all laws of physics, even claiming physics did not exist before hand.

---

[116] Einstein used the phrase "the fabric of space-time," which gives us the impression the space consists of some type of matter. However, nowhere in his description does he define what that matter is.

[117] Cosmological Balance Theory interpretation of the images provided by Hubble space telescope speaks volumes of the true age of the universe.

5.5 The Physics of Creation

The Cosmological Balance does not claim an age of the universe, for all we know the universe is infinite in age, or at least unknown to us. We see therefore that our Cosmological Balance theory agrees with observations at least as well as the relativistic models do when their numerous free parameters have been adjusted. Edwin Hubble's original deduction that his data suggested a large infinite universe is also in agreement with our theory, which however disagrees with Edwin's perceived run-away expansion. We shall now attempt to consider and elaborate some of the details of the physical process of creation. The discussion must of course be speculative, but some deductions can be made which limit the range of possibilities.

The Cosmological Balance Theory does not claim a specific age of the universe, and therefore, does not identify when creation began. It does assume, however, that space and time always existed even at the beginning of the universe, as we know it, that calm state and "clear skies before the storm" in the tranquility of flat barren open space. Given such a state, there was no visible light, just complete darkness of space throughout the universe. The Cosmological Balance claims that this space was not void, but already occupied by a high-energy unique space-matter called *hygratium*, which contained only .00015 % of actual particles with mass, a mass without gravity or antigravity characteristics for the sole purpose of sustaining space itself; it has a neutral graviton, also known as the Higgs particle[118]. This completely alien form of matter to us is the true normal state of space-matter that repels or bounces off each other, like hydrogen and helium atoms and molecules creating the gaseous upper atmosphere on the planet or in a balloon, but instead of atmosphere space-matter spreads out and creates the vacuum of space. In the very beginning, God created angels and formed space for them, a gaseous grid of slow moving almost liquefied gas evenly spread at about one to 1.00015 atoms per cubic meter over a finite area. The space-matter and space came into existence simultaneously and resulted in vacuum perceived as empty space, exactly how is beyond the purpose of this book, suffice it to say it was not a big bang. The original coexistence gave space its pressure, force, energy, and finite volume. The energy of space, also sometimes called, vacuum energy and its force are the source of the phenomena we call gravity and the source of antigravity force as it transfers this force via collisions of space-matter with normal gravity matter, as we know it, and antigravity antimatter, discussed later in detail. However, we are jumping ahead of ourselves, as there was no normal matter with gravity nor antimatter with antigravity at the beginning, the creation, the calm before the storm.

Space, like all gaseous environments, regardless of the how vacuumed it may be, has pressure even if miniscule. Space-matter flows in such a vacuum environment leads to pockets of higher-pressure and lower-pressures within space itself. Along with the high-pressure, we see clear blackness or perceived void empty space. While low-pressure regions of space-matter, leads to the consolidation of space-matter, increased collisions, the beginnings of a rotation, and potential decay of space-matter into medium and the less energized sub-atomic particles, appearing like a storm brewing in space. These new sub-

---

[118] The Higgs particle discovered using the Large Hadron Collider, gives ordinary matter its mass. Exotic space-mater contains the Higgs particle or neutral graviton, and gravitationally pushes ordinary matter together, thereby giving it mass.

atomic particles, primarily consisting of low-energy quarks, leptons, and force carriers, reassemble into normal protons, neutrons, electrons, and photons, which become the normal hydrogen and some helium atoms we are all quite familiar with, to include gravity and its antigravity counterparts. Yes, the rare sporadic decay of hygratium yields both gravity and antigravity particles, which reassemble as the lightest elements of normal matter and antigravity antimatter, and of course repel away from each other, while like matter consolidate with mutual attraction. Space-matter most likely shatters or decomposes on impact with itself directly into new hydrogen atoms with gravity and their counterpart antihydrogen with antigravity, which simultaneously repel each other in opposite directions. During the split, the neutral graviton or Higgs particle yields an ordinary graviton particle and antigravity graviton particle. Today, we detect these highly concentrated sections of low-pressure accumulations or turbulence as areas of gravitational lensing caused by the newly formed matter and growing rotating but practically invisible stellar nebulae, the identified presences of the "dark matter" cosmologists' conjecture. Weak low-pressure systems of space-matter are more likely to dissipate just as low-pressure depressions do on earth. However, the stronger the space-matter low-pressure system the more likely it will give birth to stars and a globular cluster galaxy, and even progress through further gravitational compression into fully developed elliptical, spiral, or barred galaxy at various stages of progress, something like a hurricane, typhoon, or cyclone on earth, at various stages of categories all possessing an "eye." In the hub of a well-developed galaxy resides a massive black hole, which becomes the source of high-energy dense sub-atomic particles or building blocks for space-matter atoms, and the intergalactic background microwave signal of 160.2 GHz. These massive black holes excrete by ejecting sub-atomic orbs in the form of micro white holes tunneling through the galaxy into deep intergalactic space. The massive black hole appears black as empty space much like the center of the cyclone is calm and feels and looks something like normal atmosphere on earth[119]. Do not be fooled by its appearance as its edges provide clues of its strength and force, like the event horizon of a black or white hole.

Super Typhoon Nina approaching the Philippines on November 25 at 0702 UTC. This image was produced from data from NOAA-9, provided by NOAA.

Every created space-matter rotation of low-pressure system is balanced by the concurrent creation of another low-pressure rotation in the opposite direction, means that for every particle of matter with gravity created, there is most likely an opposite, antimatter, with antigravity. Rotations in the antigravity galaxies are clockwise when seen from the North Star point of view, and stars and their planets in retrograde orbital motions. These low-pressure systems of antigravity antimatter are lensing object by intelligent beings residing, if they existed, in the invisible part of the universe, antigravity galaxies. Weak rotations of antigravity antimatter led to dissipation and dispersing of material. While stronger rotations of

---

[119] The eye of the cyclone, hurricane, or typhoon although appearing calm and quite actually sends air currents upward and outward into the atmosphere. Massive black or white holes do the same they send particles back into space to replenish space-matter.

antigravity antimatter grow into large stellar nebulas to give birth to anti-matter stars, solar systems, and anti-matter globular cluster galaxies, and eventually through space-matter forced anti-gravitational compression become fully developed elliptical, spiral, or barred galaxies at various stages of progress. At the hub of all fully developed antigravity galaxies resides a massive white hole. These massive white holes also excrete and eject high-energy dense particles, which become the building blocks for space-matter atoms and the maintenance of space, as well as the source of the intergalactic antigravity microwave signal of 160.2 GHz. The return of material back to space-matter completes the Cosmological Balance's endless cycle.

Excessive collisions between space-matter hygratium atoms with itself sometimes lead to atomic decay say about .014 % convert into lower-energy particles and matter with gravity. However, collisions between space-matter atoms and matter with gravity transfer energy, we see as the force of gravity[120]. Likewise, the collisions between space-matter atoms and matter with antigravity similarly transfer energy, which is seen as the force of antigravity by intelligent beings, if they existed, within antigravity galaxies. Because space-matter is extremely stable and constructed of high-energy particles, any collision with other matter other than itself does not cause decay or instability of atomic structure, it remains stable. The structure of space-matter gives it the energy to sustain the void found in space. Depicted below, picture provided by NASA, a compressed gaseous cloud is just beginning to spawn new stars:

Uniformly dispersed space-matter in the current universe, estimated at one atom per cubic meter, at about 71.30% of all matter in the universe is balanced by matter with gravity at 14.35% and matter with antigravity at about 14.35 %. Any more matter with gravity leads to larger galaxy and a more massive black hole, which produces more building blocks for space-matter hygratium to counterbalance and return to norm. The same applies if there were more matter with antigravity it would lead to larger massive antigravity galaxies and more massive white holes, which would produce more building blocks for space-matter. The current ratio of 71.30% space-matter, 14.35% matter with gravity, and 14.35% matter with antigravity is as ideal a Cosmological Balance as possible. At this current ratio, new low-pressure systems produced by space-matter pressure turbulences are mostly weak and not self-sustaining systems they disperse almost as quickly as they assemble in a few millennia. Any less gravity matter or antigravity antimatter than this ideal balance, can potentially lead to stronger low-pressure systems of space-matter turbulences and spawn new stellar nebulas, star systems and galaxies to counterbalance the ratio shortfall.

---

[120] Newton envisioned some force providing the mechanical aspect of gravity. However, due to limited information he was not able to identify it, so he wrote his mathematical equations as a pull force generated by the mass of objects from a distance.

5.6 Energy, Mass, and Momentum

Einstein's famous equation $E=mc^2$ has inspired the world and lead countless scientists and physicists in academia believing in it. In today's modern scientific view, we need to apply a slight modification Einstein's famous equation $E=mc^2$, because this equation lacks the simple fact that we cannot have energy without the presence of matter for which the energy moves or gets absorbed into[121]. Specifically, energy is defined as the motion or potential motion of matter, and as such, energy does not stand alone.

Paul Adrien Maurice Dirac (1902-1984) provided insights to this clarification through what he defined as the Dirac Sea, where an infinite sea of particles with negative energy exists in a vacuum. Paul Dirac, a British physicist, first postulated it in 1930 to explain the anomalous negative-energy quantum in his equation for relativistic electrons, which is based upon Schrödinger's formulation of quantum mechanics according to paper on *The Dirac Equation and the Prediction of Antimatter*, by David Vidmar (Vidmar, 2011). For the Schrödinger Equation, the Hamiltonian used is the non-relativistic classical Hamiltonian of $p^2/2m + V$, and then squared for convenience, by $H^2 = p^2c^2 + m^2c^4$ before it can be used within Schrödinger's equations. Klein-Gordon equation carried this step further and yielded two energies given by $E = \pm \sqrt{(p^2c^2 + m^2c^4)}$. Therefore, this equation, uniquely relating energy, mass, and momentum, uses the sides of a right triangle to graphically represent this $E^2 = p^2c^2 + m^2c^4$ relationship below:

For the case where a particle is considered at rest (p = 0), the above equation, reduces to $E^2 = m^2c^4$, which is usually quoted as the familiar $E=mc^2$. However, this is an oversimplification because, while $x * x = x^2$, we can also see that $(-x) * (-x) = x^2$. Concluding that:

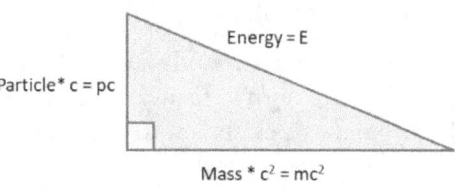

$E = \pm mc^2$.

Although the above equation, enabled Dirac to claim the negative solution predicted anti-matter, which Carl Anderson eventually discovered as the positron in 1932, his assumption that a particle can be at rest is in error. A particle cannot be at rest, for if it was it no longer exists. All particles vibrate or "jiggle" to exist and therefore are never completely at "rest." Declaring that the particle must travel at speed of light c is also presumptuous.

Assume for the moment that $E=mc^2$ is a good relation defining that energy is the product of mass (at rest as defined by Einstein) and the speed of light squared for one side of the triangle, the adjacent side. However, let us modify the other side of the triangle, the opposite side, to read pv^2, based on the action caused by waves, vibrations, or "jiggle" of particles. For instance, in light a photon and electron travel together, the photon moves with outward speed v and companion electron moves in the opposite direction perpendicular to that plane with same speed v, therefore light electromagnetically vibrate at pv^2. Similarly, the electron orbiting around an atom, held in place by a photon, vibrates at

---

[121] $E = mc^2$, where E stands for energy, m is mass of the object or system, and c is the speed of light.

pv^2. Likewise, a quark vibrates inside a proton with the other two quarks at pv^2 energy. We thereby realign the triangle relationship between energy, mass, and momentum as:

Energy = ± square root ($Mass^2 * c^4 + Particle^2 * v^4$)

= ± √ ($m^2c^4 + p^2v^4$)

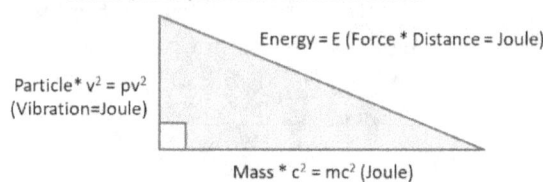

For particles at super cold temperatures close to zero Kelvin degree and supposedly near what most call the "rest" state, their velocity v goes close to zero, to give us an equation approaching $E = \pm mc^2$ per Dirac. However, we know that particle vibrations will never get to zero; if it did, the particle ceases to exist and the triangle collapses with all its sides become non-existent, resulting in zero mass, producing zero energy. For this reason, intergalactic space or deep open space has an average conjectured temperature of 2.7-Kelvin degrees; it contains matter, which moves, vibrates, and produces vacuum energy, pressure, and force, and sustains space exactly as we perceive it, appearing empty but not exactly. Hence, the proper energy, mass, and momentum equation, slightly different from Klein-Gordon yet compatible, reads:

$E = \sqrt{[(-m)(-m) c^4 + (-p)(-p)v^4]}$,

or

$E = \sqrt{[(m)(m) c^4 + (p)(p)v^4]}$,

where mass –m and particle –p can be concluded as anti-matter, while mass m and particle p are normal matter. Therefore, as the particle's velocity v approaches zero, we get $E \approx \pm mc^2$ as a closer approximation.

As we said earlier, Dirac conjectured that the negative mass –m represents the existence of anti-matter and positive mass m represents normal matter[122]. For anti-matter and matter collisions, scientists' conjecture that both annihilate each other completely, and yet they show this annihilation with in a spectacular bright burst of energy. What scientist failed to associate with that burst is the release of light and heat, self-perpetuating light in the form of photons and electrons electromagnetically travelling outward in all directions from the collision, with the product of its multidirectional burst and light speed of c^2, which carry heat with it. More specifically, we have the situation where the Energy E obtained from the annihilation causes the velocity of the particles p to approach the speed of light, v=c. So, the equation now reads:

$E = \sqrt{[(-m)(-m) c^4 + (-p)(-p)c^4]}$,

or

$E = \sqrt{[(m)(m) c^4 + (p)(p)c^4]}$,

where anti-matter –m annihilates matter m as they encounter each other resulting as mass equals zero in the following equation:

$E = c^2 \sqrt{[(0^2 + p^2)}$

---

[122] Dirac read the + and – mass to be the prediction for normal matter and its opposite anti-matter. It can also read to be ordinary gravity matter and its opposite antigravity antimatter.

$$= c^2 \sqrt{p}^2$$
$$= \pm pc^2$$

where particle p, ejected at light speed in opposite directions from each other, results in repulsion motion that achieves the speed of light squared from the point of annihilation occurred; mathematically represented as

$$E = pc^2 = (\sqrt{p}*c) * (\sqrt{p}*c),$$

or

$$E = - pc^2 = (-\sqrt{p}*c) * (\sqrt{p}*c),$$

Where $\sqrt{p}$ is always a photon and companion electron, or $\sqrt{-p}$ is an anti-photon and positron, travelling as light (EM) pulse away from each other in opposite directions, regardless of what was the original particle p. For example, if the colliding masses were a proton and anti-proton annihilation, then the resulting burst would produce normal photon and electron light pulses, most likely in the visible range. However, if the original colliding masses were an electron and positron, then the photon and electron resulting from the annihilation would be a gamma ray photon and a higher-energy electron (possibly a tauon).

Energy of Antigravity Gravity Contact

In the proof above, the negative mass –m not only represents the existence of anti-matter, it also predicts the existence of antigravity antimatter produced within massive black holes, as an opposite of normal gravity matter m, but with anti-gravitational properties. These antigravity antimatter particles mutually attract and bond to each other to form micro white hole orbs deep within the black hole that created them, and slowly rise by mutual repulsion between gravity and antigravity particles toward the surface.

We can essentially say that the additional energy, resulting when antigravity is repelled by normal gravity matter, as we know it, forms wormhole tunnels, powered by a grand unification of three natural forces and the presence of a negative graviton (opposite of ordinary matter). For antigravity antimatter and normal gravity matter contacts, the Cosmological Balance predicts that neither one gets annihilate in a spectacular bright burst, but rather gets repelled from each other at the speed of light squared fueled by the ejection power of the strong nuclear force, the electroweak nuclear force, and antigravity/gravity combined[123]. If we consider a microscopic object (sub-atomic micro white hole orb) being repelled from the surface of a super massive black hole, it is obvious that the microscopic object will receive the greater energy repulsion and thereby move outwardly at the speed of light squared, c^2, while the black hole itself receives the minimal nudge (practically zero movement). This sudden burst of energy forces the white hole orb into a wormhole tunnel from the black hole's surface to at least 100 thousand light year distance beyond the edge of the galaxy containing that massive black hole. Upon leaving the surface of the black hole, the orb's initial ejection speed of light squared will receive an additional acceleration from the force $G*M_{BH}/r^2$, where the distance r is from the center of the black hole to the

---

[123] Prediction of the grand unification force occurs within the black hole. Such a force enables black holes to expel miniscule white hole orbs from the surface of the black hole to past the outer edges of the galaxy.

center of the white hole orb. The acceleration speed is augmented by the additional mass of each star the white hole orb passes on its journey outward. Upon passing the edge of the galaxy, the white hole orb's speed gradually slows down and eventually the wormhole collapses as the white hole orb's speed drops below the speed of light. More specifically, we have the situation where the Energy E obtained from the repulsion causes the velocity of the particles p to approach the speed of light squared, c^2. Since all matter within a supermassive black hole is broken down into their basic subatomic particles: quarks, leptons, and force carriers, the equation now reads:

$$E = \sqrt{[(-m)(-m) c^4 + (-p)(-p)c^4]}$$

or

$$E = \sqrt{[(m)(m) c^4 + (p)(p)c^4]}$$

where normal matter m repels antigravity, $-m$, from the surface of the black hole, of which all its matter was broken down into sub-atomic particles p, i.e.... m=p. The energy released is:

$$E = c^2 \sqrt{[(p^2 + p^2)}$$
$$= c^2 \sqrt{(2p^2)}$$
$$= \pm pc^2 \sqrt{2},$$

where the particles p, in this case grouped into a white hole orb is ejected at light speed squared, travels away from the black hole's surface within a wormhole tunnel. Mathematically we represented this as $E = pc^2 \sqrt{2}$, where the square root of two ($\sqrt{2}$) represents the energy creating the wormhole tunnel and the $\pm$ represents repulsion from either the surface of a black hole or a white hole.

For example, if p= 1 kg, then the Energy value of E is:

$$E = \sqrt{[(8.98755E+16)^2 + (8.98755E+16)^2]}$$
$$= \sqrt{(1.61552E+34)}$$
$$= 1.27103E+17 \text{ kg m}^2/\text{sec}^2$$

where the additional energy $\sqrt{2} = 1.414213562$ initially creates the wormhole tunnel and sustains it, as the particles, white hole orb, travel at the speed of light squared, c^2, within it[124]. Similarly, black hole orbs ejected from massive white holes gets repelled with the same energy and speed. This wormhole tunnel ejection is a brief explanation of how black holes and white holes effectively excrete matter; detailed descriptions provided in a later chapter.

---

[124] This little bit of extra energy enables the orbs inside the wormhole tunnels to slip right through space-matter without colliding with them. It creates an electrostatic field around the tunnels, which repels space-matter similar to a force field.

5.7 Planetary and Galactic Orbital Periods

Orbital periods of planets around their stars, and stars around their galaxies are application of and along the same lines as energy, mass, and momentum. This relationship provides the basis and existence of galaxies. The dilemma here is that the orbital period of planets around their stars are calculated by a different formula than that estimating stars orbiting around their galaxy[125]. The Cosmological Balance Theory provides the mechanism causing stars to orbit faster than solely with just the force of gravity, and then captures that momentum in a universal equation applicable to all galaxies. In this section, we continue from where we left off with the discussion of Kepler's third law in chapter one and the partial analysis of the rotational speed of the galaxy in chapter three. The equations given below in this section are a preview of results produced in later chapters of this book.

Again, the presence and motion of space-matter produces the vacuum-force of 6.6734E-11 Newton, which when combined with the inherent quantum mechanical attraction of 1 m/kg for gravity matter or -1 m/kg for antigravity antimatter, provides the force behind Kepler's and Newton's laws. We have the Cosmological Balance orbital period equation below,

$$T_c = [\sqrt{(4\pi^2/(GM_C)*a_c^3)}] * [(c^2 * \sqrt{(1/(2GM_C))} * e/\pi],$$

where the first part is basically a modified Kepler's third law, and the second part represents acceleration caused by gravity-antigravity particle or object interactions. The second part can also be viewed as the rate of angular momentum transfer of energy between stellar objects within a galaxy. This equation unifies Newton and Kepler laws, the Schwarzschild radius and mass, quantum mechanics, and the principles of geometric and polar coordinate systems.

The Cosmological orbital period for shielded systems, like the solar system, is

$$T_c = [\sqrt{(4\pi^2/(GM_C)*a_c^3)}],$$

where M_C is the accumulated mass of the concentric inner part of the system, where there is practically no interaction between antigravity and gravity matter. This modified Kepler's third law roughly accounts for planetary interactions and perturbations.

Likewise, the resulting Cosmological Balance galactic orbital period for antigravity systems is:

$$T_A = [\sqrt{(4\pi^2/(GM_A)*a_A^3)}] * [(c^2 * \sqrt{(1/(2GM_A))} * e/\pi], \qquad (5.7.1)$$

and its shielded solar system orbital period is:

$$T_A = [\sqrt{(4\pi^2/(GM_A)*a_A^3)}], \qquad (5.7.2)$$

where T_A is the orbital period of an object in the antigravity system with semi-major distance a_a, M_A is the accumulated mass of the inner part of the system with the center massive object, G is the gravitational constant 6.6734E-11 N m²/kg², c is the speed of light, e is Euler's number, π is the constant pi.

---

[125] Einstein's general relativity equation solves gravity lensing and Mercury's orbital anomaly but does not provide and acceptable solution to why stars orbit around the galaxy in excessive speeds. The Cosmological Balanced Universe Theory presents a universal cosmological equation for all situations.

The above galactic orbital equations (5.7.1) and (5.7.2) are applicable to the universe, and thereby called the Cosmological Balance Universal Law.

Since space-matter, as defined, does not attract each other and makes space, what it is, appear empty and void, a vacuum. Systems where there are interactions between gravity matter and antigravity antimatter, the constant $[c^4 e^2/2G\pi^2]$ times mass M represents the acceleration resulting from that action-reaction. We have the Cosmological Balance relational equation below,

$$S_m * [4\pi^2/G] * [c^4 e^2/2GM_A\pi^2] * A_m = [4\pi^2/G] * [c^4 e^2/2GM_C\pi^2] * C_m,$$

and

$$S_m * [4\pi^2/G] * [c^4 e^2/2GM_C\pi^2] * C_m = [4\pi^2/G] * [c^4 e^2/2GM_A\pi^2] * A_m,$$

where e is Euler's constant, G is the gravity constant, S_m represents space-matter and its motions; A_m represents antigravity antimatter and its motions; and C_m represents normal gravity matter and its motions. As antigravity particles traverse through the gravitation system, they significantly accelerate the rotation of the system under observation, moving stars faster than with gravity or antigravity alone. In these equations, space-matter and antigravity tells normal matter how to move, and likewise space-matter and gravity matter tells antigravity antimatter how to move. Detailed proof of these equations is discussed later in this book.

Cosmological Balance is a unified theory. So, to complement the gravitational forces of the macro universe, we need to resolve how gravity works in the quantum world. In chapter 4 we learned that the ratio of strong force to electromagnetic force generates a 99:1 up and down motion, which creates a gravitational field internal to all nuclei. This oscillating "gravity" field occurs within all quarks. In the up quark, we have five positive point particles and their companion neutral point particles' strong nuclear force outward acceleration (gravity) and slow down cushioned descent to central three negative and three neutral point particles (zero gravity). The neutral particle senses proximity and turns on and off the repulsion as needed to prevent the positive particle from colliding with the negative particle, and also prevents the two particles from getting too far away, holding it in "orbit" like a stretchable glue. At the "on-switch," the maximum strong nuclear force of 100 times occurs by aligning the neutral point particles on the same side. At the "off switch," only the electromagnetic force of 1, occurs when the neutral point particles are on opposite sides halting outward acceleration, and the negative and positive forces pull on each other, a weaker attraction, free-falling the return. In the next section, I present a standalone Quantum Gravity Unified Theory paper for both micro and macro universe.

5.8 Quantum Graviton Unified Theory

"Nothing happens until something moves."

- Albert Einstein

Abstract

What causes gravity? Sir Isaac Newton discovered and expressed gravity in his equations as a force that is instantaneous and of infinite range[126], while Einstein redefined it as a force resulting from the curvature of space-time limited by the speed of light[127]. Since then, modern physicists search for the "graviton" massless spin-2 particle has been fruitless[128]. Could they be looking for a particle that does not exist? Quantum mechanics experiments have proven that point particles come in pairs or groups sustaining each other, known as super symmetry at the micro level[129]. In his book *Understanding Gravity*, Shan Gao discusses Newton and Einstein gravity, and then presents gravity as an emergent or an entropic force originating from interactions in a thermodynamic system[130], as in Verlinde's example, concluding that entropy plays an important part of the gravity attraction. But according to Brian Greene, string theory, particularly super-symmetrical string theory, claims its vibration patterns produces characteristics known in particles such as spin, charge, size, and mass[131] and define a specific equation and pattern identifying the "graviton." Moreover, Weinberg identified the electroweak force[132]. This paper defines Quantum Graviton as an effect resulting from a precisely fine-tuned and motion generated by the other three natural forces, which create an entropic force, not limited by the speed of light constancy, known as "gravity," which truly operates exactly as defined by Newton, instantaneous and infinite in range. Graviton is not a particle; it is a result.

Assumptions:

1. The laws of physics are the same for all observers in uniform motion relative to one another in absolute time and space (Galilean Galileo's relativity principle).

2. Newton's gravity equation, $F = G(m_1 * m_2/r^2)$ is fractal, where m_1 and m_2 are the objects mass in kg, and r is the distance between centers of both bodies in meters.

3. Einstein Equivalence Principle, without time dilation, declares that equivalent accelerated motion is indistinguishable from an actual gravitational field.[133]

---

[126] Newton, S. I. (1687). *Principia Mathematica*. Cambridge: Unknown, Newton's law of universal gravitation

[127] Einstein, A. (1920). *Relativity, The Special and General Theory*. Kindle Direct Publishing Edition 2013, Constancy of the Speed of Light Principle

[128] Carlip, S. (2011). *Does Gravity Travel at the Speed of Light?* Retrieved 11 2, 2015, from Physics Frequently Asked Questions: http://math.ucr.edu/home/baez/physics/Relativity/GR/grav_speed.html

[129] Collier, P. (2014). *A Most Incomprehensible Thing*. Kindle Edition: Incomprehensible Books.

[130] Gao, *Understanding Gravity: Newton, Einstein, Verlinde?*, 2014

[131] Greene, *The Elegant Universe: Superstrings, Hidden Dimensions, and the Quest for the Ultimate Theory*, 2003

[132] Weinberg, S. (2008). *Cosmology*. Oxford: Oxford University Press.

[133] Einstein, A. (1920). *Relativity, The Special and General Theory*. Kindle Direct Pubishing Edition 2013.

4. Gravity is the weakest of the four fundamental forces of nature[134] at approximately 10^{-38} times the strength of the strong nuclear force (i.e. 38 orders of magnitude weaker), 10^{-36} times the strength of the electromagnetic force, and 10^{-29} times the strength of the weak nuclear force.

5. The most common basic and simplest of atoms is the hydrogen, ^1H protium, consists of a proton, electron, and a photon[135]. Its proton has two up quarks and one down quark.

6. All matter is assembled from four basic point particles: positive, negative, and two types of neutral, estimated at 9.1095E-43 kg, in various structures, patterns, and bonds sharing energy.

Point Particle (point particle)

Quantum physics has shown us that the more energetic point particles move the more mass they achieve, per Einstein's energy-mass equation, $E=mc^2$, and that acceleration equals gravity in his equivalence principle.[136] Rapid acceleration outward provided by the strong nuclear force of the neutral point particles results in generated gravity, and the slow return inward motion provided by the electromagnetic force results in zero gravity, or the total combined "graviton" effect at 6.6734E-11 N (m/kg)^2. The total energy of the point particles within each quark determines its mass, while the generated rapid outward acceleration and slow inward motion causes its gravitational field, the "graviton." Explicitly, the three natural forces working together create the fourth natural force, gravity. The "graviton" is not a single spin-2 massless particle, but the result of point particle motions within quarks. No strings attached.

Point Particle Configurations

The hydrogen 1 atom, ^1H, consists of one electron and at least one photon, and one proton[137]. The **electron** is a massless negatively charged particle consisting of three pairs each with one negative and one neutral point particle grouped in a tight spin pattern and a negative 1/3 charge, where pairs generally separated in the cloud around the atom intersect in a tight loop every now and then, as they orbit. This gives the motion of the electron, when grouped, periodically the **negative one charge,** from negative one-third times three, or (-1/3 * 3 = -1), a quantum quality of being random and discontinuous, determined by probabilistic cause[138]. This concept also explains why electrons shot through the double slit experiment appears to function as "cloud" or wave while passing through the slits, then reconsolidate and hit the screen as a single particle.[139]

The **photon** is a massless particle of neutral charge probably constructed of one or two neutral, one negative, and one positive point particles in a tight mutual flat circular orbital pattern with the positive-neutral pair at the center orbited by one neutral and a

---

[134] Weinberg, S. (2008). *Cosmology*. Oxford: Oxford University Press.

[135] Wikipedia. (2015, June 21). Protium Definition.

[136] Einstein, A. (1920). *Relativity, The Special and General Theory*. Kindle Direct Publishing Edition 2013.

[137] Wikipedia. (2015, June 12). Atom Composition

[138] Gao, *Quantum Mechanics: A Comprehensible Introduction for Students*, 2013, p. Loc 695

[139] Ibid, p. Loc 1376

negative pair point particle at the outside moving in a circle. All the photon's linear point particle adjacent surfaces touch to create a polarized magnetic field at **zero charge**, while the speed of the orbiting negative particle determines its frequency. The electron and photon move together to maintain its predetermined distance from the nucleus given the energy level of the photon, and at the center sits one proton to form the ^{1}H hydrogen atom. The **proton** consists of two up quarks and one down quark. The up quark has a charge of 2/3 positive and the down quark a charge of 1/3 negative. This structure consists of smaller point particles of different charges, two types of neutral, positive, and negative. We predict that the **up quark** composed of sixteen point particles has five positive-neutral pairs (1/3 * 5) orbiting one center consisting of three negative and a neutral pair (-1/3 * 3), where each of the five neutral point particle orbits their attached positive point particle partner (1/3 * 5 - 1/3 * 3 = 2/3). These five positive-neutral pairs then randomly orbit just outside the center three neutral-negative particle pairs in a tight elliptical or most likely a hyperbolic pattern, a quantum uncertainty possibility.[140]

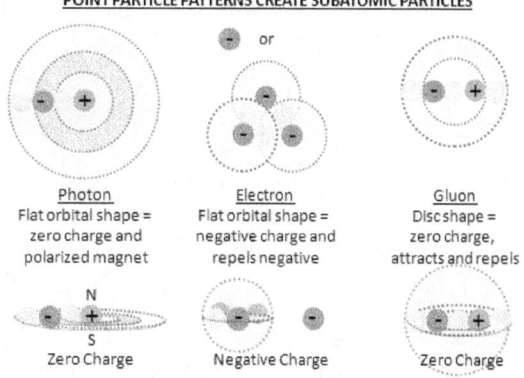

The **neutral** or "messenger" point particles are buffers that prevent the positive point particle from contacting or colliding with the negative point particle, thus avoiding annihilation. It does this by using 2/3 of the charged particle's energy to invoke the strong nuclear force, which is one hundred times greater than that of the electromagnetic force.[141] The neutral point particles control when the positive and negative repel by aligning themselves between the two (- an n +) and apply the repulsive strength of the strong nuclear force but shorter in range, about 100th that of Planck length. The neutral particles also allow the negative and positive to attract each other by placing themselves on opposite ends (an - + n) and turning off the repulsion, where "an" is the antineutrino and "n" is a neutrino. This motion displays a hint of "intelligence," a survival instinct[142]. As a result, the neutral point particle slingshots the positive particle outward with a rapid acceleration for a short micro distance, then the charged particle's attraction for each other overwhelms the outward motion and pulls the positive point particle similar to freefalling inward[143]. At the positive point particle's closest approach to the negative point particle, the neutral point particle's energetic strength then overwhelms the charged attraction and again slingshots the positive particle outward. Consecutive hyperbolic iterations of these pairs create a jiggling **hexahedron pattern**, which eventually cover the entire "spherical" surface of the quark.

---

[140] Rouse, M. (2015, January). *Quantum Theory Definition*. Retrieved June 15, 2015, from whatis.techtarget.com: http://whatis.techtarget.com/definition/quantum-theory

[141] Wikipedia. (2015, June 20). Strength comparison between the four natural forces.

[142] Bailey, A. A. (2013). *The Consciousness of the Atom*. Kindle Edition: Kindle Direct Publishing.

[143] Einstein defined free fall as the absence of gravity, hence weightlessness or zero-gravity.

The second component of the proton, the **down quark**, similarly constructed of ten point particles consists of two positive and neutral pairs (1/3 * 2), orbiting around a center group of three negative and three neutral point particles (-1/3 * 3) of which the center neutral particle is most likely antineutrinos. Reminder, the 1/3 charges are caused by the energy redistribution to enable the neutral point particle to apply the strong nuclear force. The two pairs of neutral-positive point particles randomly orbit the center three negative-neutral point particles in a tight hyperbolic pattern (-1/3 * 3 + 1/3 * 2 = -1/3) give the down quark its -1/3 charge. Again, the neutral point particles prevent the positive point particle from colliding with the negative particle. Since the center has a greater charge than the orbiting pairs, it is bounded more loosely together, more energetic in movement, and hence the down quark is slightly greater in mass than the up quark. The sheer short-range strength of the neutral point particle, the strong nuclear force, slingshots or accelerates the positive point particles outward, and then halts them abruptly. The more energetic the point particle movement, the more massive they become. The three stronger forces of the four natural forces attract and repel to form various configurations we see in nature[144]. Given enough energy input, they can form complex atoms and molecules; the interior of a star is such a place to create new atoms through the fusion process.

Free-floating protons repel each other because of their like charges, while each proton electromagnetically attracts three negative-neutral point particle pairs and a photon around them to form the simplest of atoms, the hydrogen 1 atom, ^1H. Since these completed hydrogen-1 atoms carry a total zero charge, their gravitational field enables them mutually to attract each other and given enough matter and time form stellar clouds[145]. The nucleus of a hydrogen 2 atom, ^2H, also contains a neutron with zero charge. The centerpiece of the **neutron** is the up quark, randomly orbited by two down quarks buffered by at least two gluons[146]. The down quarks orbit the center up quark in elliptical patterns. The hyperbolic orbital pattern of point particles in a quark over time creates a Calabi–Yau manifold derived within string theory[147], which is different from the typical atomic structure where electrons spherically orbit the nucleus with a fuzzy "cloud-like" surface[148].

These are typical atomic structures, and that of the quark, and particle composition within a proton:

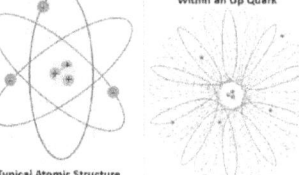
Typical Atomic Structure

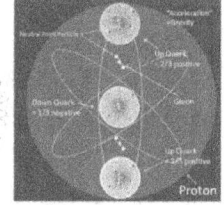
Hyperbolic Orbital Pattern Within an Up Quark

Gluons, the buffer between protons and neutrons, consist of two or three neutral point particles, a negative point particle and a positive point particle in a group with no charge, zero. Unlike the photon, which consists of one negative-neutral point particle pair in a flat orbit pattern around one positive-neutral point pair to create magnetic polarity while having overall neutral charge of zero, the gluon's center is the neutral point particle surrounded by both the negative and positive point particles and neutral point particles at each end. The gluon enables it to hold

---

[144] Wikipedia. (2015, June 20). Strength comparison between the four natural forces.

[145] Weinberg, S. (2008). Cosmology. Oxford: Oxford University Press.

[146] Wikipedia (2015, August 15). Structure of Quarks and Gluons within a proton.

[147] Ibid

[148] Ibid

subatomic particles within the nucleus of an atom by repelling and attracting as needed. Since nuclear reactions conserve the number of leptons, the neutron cannot capture antineutrinos only neutrinos and the opposite are true for protons, only capture antineutrinos.[149]

Quantum Graviton Effect

Quantum physics has shown us that the faster point particles move the more massive they become or exhibit, as defined by Einstein's energy-mass equation, $E=mc^2$. So how does more mass equate to more gravity? Using Einstein's equivalence principle[150], let us analyze the structure of an up quark composed of sixteen point particles: five positives, three negative, and eight neutral point particles, where five pairs of neutral and positive point particles orbit randomly the three pairs of neutral and negative charged point particle at the center axis. Every time the positive point particles moves inward toward the negative point particle, the neutral point particle buffers or cushions its inward movement making it gradually slow down, freefall, and then expels that positive-neutral point particle pair outward with acceleration exceeding the speed of light at a hundred-fold[151]. Then the outward motion of the positive pair halted abruptly by the negative point particle's pull freefalls inward to repeat another cycle. The resulting motion is a hyperbolic orbit with rapid outward acceleration and "freefall" inward movement, illustrated above, forms a vibrant hexahedron shape. There are no "strings" in this structure. Therefore, the movement speed of the point particles within each quark determines its mass, while the rapid outward acceleration causes its gravitational field, the "graviton" for which scientists are searching. The looser bound and more energetic negative and neutral pairs within the core of the down quark give it its additional mass over that of the up quark.

To reiterate, the "graviton" is not a unique spin-2 massless point particle but a result primarily of the motion of point particles within quarks and enhanced by the motion of quarks within nucleons. The quantum "graviton" effect spikes out of every quark and atom's nucleus in an infinite omnidirectional infinitesimal

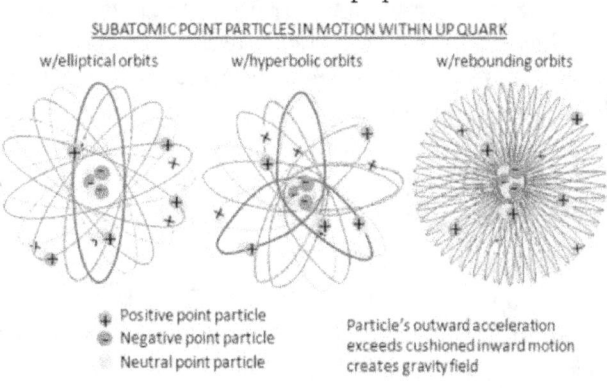

pinpoint of pure energy, an out-wave cresting instantly to infinity at an infinite frequency, liken to a "flash two-way hotline" or instant transmission. The graviton attribute can be associated particularly with the effect of the neutral point particles' action within the quarks. They are the messenger point particles that turn on and turn off strong nuclear repulsion force to avoid collisions of positive and negative point particles, and the

---

[149] Wikipedia. (2015, June 27). Radioactive Decay Definition.

[150] Einstein, A. (1920). *Relativity, The Special and General Theory*. Kindle Direct Publishing Edition 2013.

[151] Strength comparison of strong nuclear force to the electromagnetic force.

collisions of opposite types of neutral point particles. This "graviton effect" occurs within neutrons and protons buffered by gluons.

Again, subatomic particles constructed from four types of point particles: two types of neutral, a positive, and a negative, are assembled in various patterns and shapes. Here are three possible spherical configurations that cause gravity around an up quark. The image on the right produces the most dominate and long-range macro universe force, gravity, yet it is weakest among the four natural forces. The "graviton" is not a single massless particle, but the result of point particle motions within quarks in nucleons and atoms; it is an entropic force, where the total point particle motions within any two objects cause the attraction motion between both objects. This entropic force instantaneously projects in- and out- spherical wave energy at infinite distance, an inherent effect of the wave generated by the structure of matter[152]. So, any two quarks, even that of "neutral" quarks, within close proximity to each other are gravitationally bound and develop an unbreakable entropic force, motion begets motion, pulling each toward the other in what we call gravity, in precisely the same formula Newton defined; this extreme is evident within black holes. The more atoms are in an object the greater its mass and the greater its gravitational field[153]. Let us investigate this in the real world.

Graviton Mathematical Proof

To show the math, imagine for a moment that you are an astronaut aboard a rocket at Cape Canaveral, Florida, 0.003 kilometers above sea level. By your calculations, your rocket engines produce a thrust 100 times greater than the gravity of the earth, so whenever your engines are blasting your rocket speeds up and away with tremendous acceleration while being tugged downward by gravity. The upward acceleration generates a gravitational field[154], per Einstein. When your rocket reaches a height of 10 kilometers, the engines turn off and the rocket loses momentum, abruptly stops, and begins to freefall back to earth, much slower than its upward journey. You continue on this freefall in "zero gravity" until you have descended just below 5 kilometers, at which point, the engines reignite and send you and the rocket rapidly racing upward with renewed acceleration, only to repeat action again. The fast-moving upward motion and slow descent generate an oscillating gravitational field we call the "graviton effect."

Assuming the earth's mass is $5.9723 *10^{24}$ kg, and the radius is 6371.008 kilometers[155], we use the formula acceleration = Gravity Constant times Earth's mass divided by radius squared or $a=G*M_E/r^2$ to obtain the gravitational acceleration of 9.819117821 m/s². Mathematically, the rocket zooms upward from the altitude of five kilometers to ten kilometers in 3.2073 seconds with total acceleration thrust of 972.0927 m/s², derived from the thrust of 100 minus the gravitational pull of one times 9.8191178 m/s² or 99 * 9.8191178 m/s². Acceleration time is calculated by square root of two times distance

---

[152] Wolff, M. (2008) *Schrodinger's Universe: Einstein, Waves & the Origin of the Natural Laws*, Outskirts Press. Quantum Graviton "wave structure of matter" reverses Wolff's WSM (in-out-waves structure create matter).

[153] Gao, S. (2014). *Understanding Gravity: Newton, Einstein, and Verlinde?* Kindle Edition: Amazon Kindle Direct Publishing.

[154] Ibid.

[155] https://nssdc.gsfc.nasa.gov/planetary/factsheet/earthfact.html

divided by acceleration minus acceleration divided by the distance squared, or SQRT(2r/(A - A/r²)). Upon reaching 10 kilometers, the engines shut off, and the rocket takes 31.91271642 seconds to fall 5 kilometers.

Upward acceleration time is:

= SQRT(2*5000/(972.0926642- 972.0926642/5000²))

= 3.207348679 seconds

Freefall time is:

= SQRT(2*5000/(9.819117821 - 9.819117821/5000²))

= 31.91271642 seconds

This repeated rapid acceleration upward and relatively slower descent to earth generates an overall oscillating "gravitational field." We know that the strong nuclear force is 1E+38 times more than gravity, with a strength approximately 6.6734E+27 Newton (meter/kilogram)². We also know that the strong nuclear force is 100 times greater than the electromagnetic force 1E+36 which is at about 6.6734E+25 N (m/kg)². Using the strength of the electromagnetic force instead of the earth's gravity field, we get the following equation:

Gravitational field generated by rocket's acceleration is roughly:

= [6.6734E+27 N (m/kg)² * 3.207348679 s * 99 kg/s² /1E+38

+ 6.6734E+25 N (m/kg)² * 31.91271642 s * (-1 kg/s²)/1E+36]

/ (99 kg/s² * 3.207348679 s − 1 kg/s² * 31.91271642 s)

= 6.6734E-11 N (m/kg)²

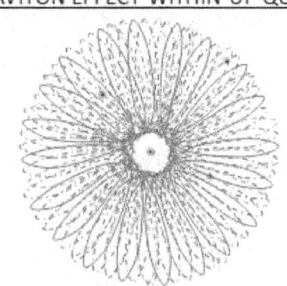

GRAVITON EFFECT WITHIN UP QUARK

Point particle's outward acceleration of 2.9222E-38 seconds exceeds cushioned inward motion of 2.9222E-37 seconds creates gravity field, the "Graviton" of 6.6734E-11 N (m/kg)²

Now let us take this motion into the quantum world of an up quark, where we have five positive point particles each paired with their companion neutral point particles which act as the messenger switch for acceleration invoke the strong nuclear force outward acceleration and cushion their descent toward the central three negative-neutral point particle pairs. The neutral particle senses proximity and turns on and off the repulsion as needed to prevent the positive particle from colliding with the negative particle and prevents the two particles from getting too far away, holding it in orbit like a stretchable glue[156]. The on-switch, maximum strong nuclear force of 100 times, occurs by aligning the neutral point particles on the same side. At the maximum range of the strong force, the off switch halts the outward motion. The off switch, which activates the electromagnetic force of one, occurs when the neutral point particles are on opposite sides, and the negative and positive pull on each other, a weaker attraction. The "graviton effect" action envelopes the entire quark

---

[156] Wikipedia. (2015, June 25). Gluon Definition.

and generates a very tiny amount of gravitational field in every direction amounting to a miniscule quantum contribution to the 6.6734E-11 N (m/kg)². This "graviton effect" in one quark symmetrical system universe, combined with those fractal forces from all the quarks in the nucleons within every single atom in the world creates the earth's total gravitational field. Note that the central axis of the quark by itself does not have, create, or exert gravity; the quark's center has zero gravity[157]. This fractal notion is applicable to all objects in the known visible universe.

Using the Planck length (1.6E-35) divided by 100 of 1.6E-37 and the equation for travel time equals square root of $(2r/(A - A/r^2))$ where r is distance and A is acceleration, the outward travel time of an accelerating point particles is

=SQRT(2*1.0000001m/(299792458m/s² - 299792458m/s²/(1.0000001m)²))*1.6E-37

= 2.99222E-38 s

And the inward travel time is

= SQRT(2*1.0000001m/(2997924.58m/s² - 2997924.58m/s² /(1.0000001m)²))*1.6E-37

= 2.99222E-37 s

Using the outward and inward travel times to estimate the "Graviton" effect within a quark (see **Tab 1** for Coulomb Law and Lorentz Force justification), we get

= [6.6734E+27 N (m/kg)² * 2.9222E-38 s * 99 kg/s² /1E+38

+ 6.6734E+25 N (m/kg)² * 2.9222E-37 s * (-1 kg/s²) /1E+36]

/ (99 kg/s² * 2.9222E-38 s – 1 kg/s² * 2.9222E-37 s)

= 6.6734E-11 N (m/kg)²

Like a rectifier smoothing out an alternating current into a steady direct current, the "graviton effect" levels out to 6.6734E-11 N (m/kg)² within every quark in every atom in the visible universe all contributing a miniscule amount to its overall steady gravity[158].

Quantum Gravity in Stellar Fusion

According to mainstream science, fusion is a hydrogen burning process within a star[159]. Hence, the first test of the point-particle "Quantum Graviton" concept in quarks is to evaluate what happens to their structure within a star, such as the sun. In the diagram, two hydrogen-1 atoms are fused together to form hydrogen-2 and release a positron and neutrino.

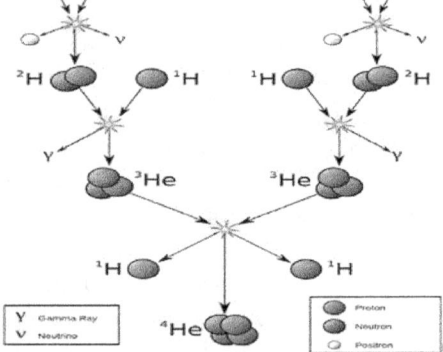

---

[157] Note that a spaceship rotating about an axis to create artificial gravity has zero gravity at the center axis.

[158] Newton, S. I. (1687). *Principia Mathematica*. Cambridge: Unknown

[159] https://en.wikipedia.org/wiki/Stellar_nucleosynthesis#Hydrogen_burning

Left image in this structure with point particles (note: the -1 electron orbiting the proton consists of three one-third negative charge pairs and a photon). The next image to the right demonstrates that when two outer up quarks of protons collide one releases three positron-neutrino pairs equating to one positron and a neutrino. This collision converts the up quark into a down quark, goes from 2/3 positive subatomic particle into a 1/3 negative charge.

Why are three pairs of particles released? The loss of one pair would change the quark to 1/3 positive making it still repulsive, the loss of two pairs would make it zero charge and still resist merger, and the loss of exactly three pairs changes the quark to 1/3 negative, a down quark, making it receptive to attach to the inbound up quark. The ejected three pairs of positive-neutral point particles rearrange themselves into a positron and a neutrino.

Thus, this fusion step transitions one proton into a neutron with a total of zero charge and creates a hydrogen-2 atom with an extra electron orbiting the

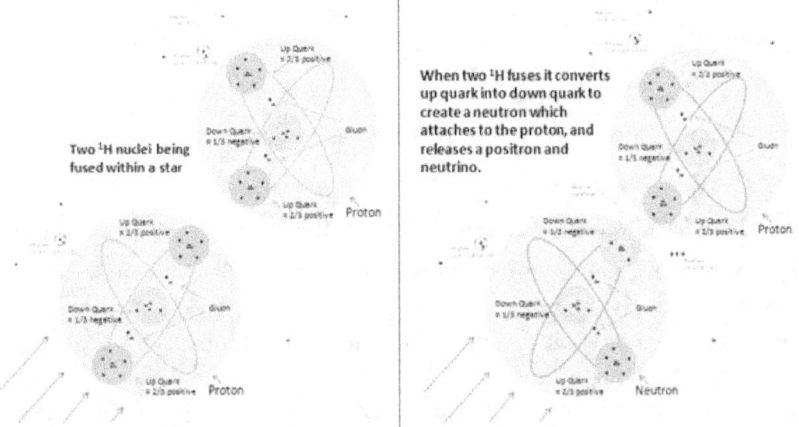

new atom. In the above hydrogen atom, the down quarks attract electro-magnetically toward the up quarks in a mutual bond to form an almost stable atom, hydrogen-2 but with an extra electron loosely orbiting it.

In the next step, hydrogen 1 atom fuses with the new hydrogen-2 atom above. The proton of the hydrogen-1 atom attaches itself to the neutron within the hydrogen-2

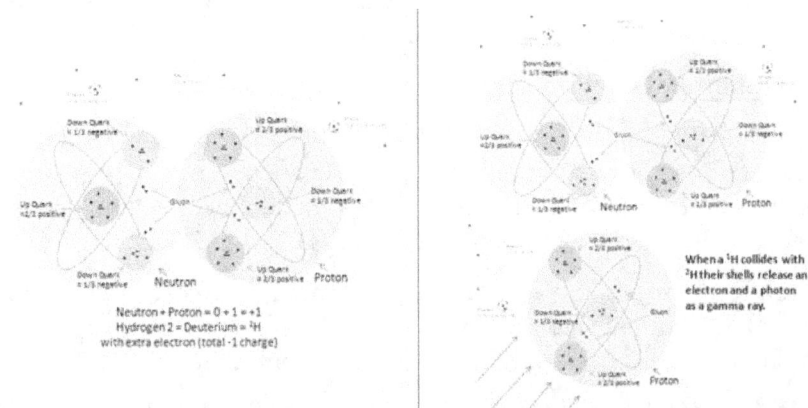

nuclei to form a new helium-3 atom. This fusion process releases the extra electron and photon as high-energy gamma ray radiation resulting in a helium atom with two electrons and photons orbiting it.

In the next step of the fusion process, when two helium-3 atoms collide one nucleus rips the neutron from the other helium nucleus and ejects two protons with orbiting electrons/photons as two new hydrogen-1 atoms. The above analysis concludes that the initial point-particle structure, which produces "gravity" as proposed within quarks, holds valid during the fusion process within the sun and therefore can be extrapolated as the process within all stars in the visible universe, consisting of ordinary matter with gravity. See **Tab 2** for discussion of which subatomic structures came first, the neutron or proton, the up quark or down quark.

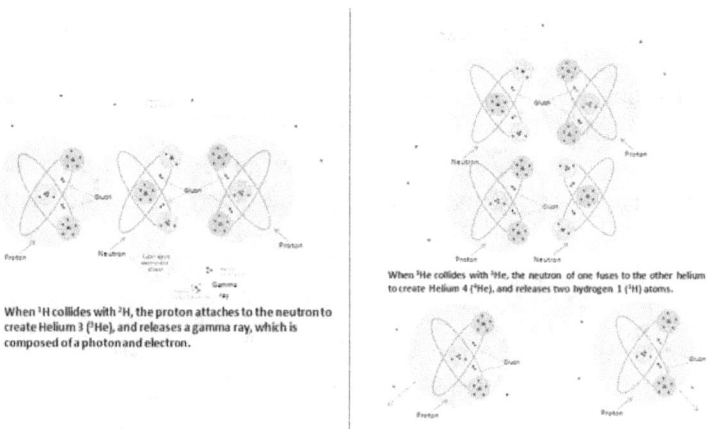

When ¹H collides with ²H, the proton attaches to the neutron to create Helium 3 (³He), and releases a gamma ray, which is composed of a photon and electron.

When ³He collides with ³He, the neutron of one fuses to the other helium to create Helium 4 (⁴He), and releases two hydrogen 1 (¹H) atoms.

Summary of Quantum Graviton Theory

The "graviton effect" within every atom in the visible universe results from the neutral point particle's ability to turn on and off the strong nuclear force to avoid collisions of positive and negative point particle to survive. We observed this motion as jiggling of quarks, which explains why larger atoms have both more mass and gravitational pull, and why electron levels around atoms are fuzzy layers in quantum flux. The "graviton" action envelopes the entire quark and generates a very tiny amount of gravitational field in every direction amounting to a miniscule quantum contribution to the overall $6.6734E-11$ N $(m/kg)^2$. This graviton effect combined with those of all the quarks within every atom in the world creates the earth's total gravitational field, a source for Newton's gravity based on Einstein's equivalence principle insight.

Quantum Graviton Math:

Using the distance of Planck length (1.6E-35) divided by 100 or 1.6E-37, and the equation for accelerated travel time equals square root of $(2r/(A - A/r^2))$ where r is distance and A is acceleration. The outward motion time of accelerating point particles is equal to $2.99222E-38$ s, and the inward motion time is equal to $2.99222E-36$ s

Since the strong nuclear force is 100 times greater than the electromagnetic force, we obtain a ratio of outward acceleration at 99 kg/s^2 to inward at -1 kg/s^2. Using the outward and inward time above, we estimate the "graviton" effect within a quark

$= [6.6734E+27$ N $(m/kg)^2 * 2.9222E-38$ s $* 99$ kg/s^2 $/1E+38$

$+ 6.6734E+25$ N $(m/kg)^2 * 2.9222E-36$ s $* (-1$ $kg/s^2) $ $/1E+36]$

$/ (99$ $kg/s^2 * 2.9222E-38$ s $- 1$ $kg/s^2 * 2.9222E-36$ s$)$

$= 6.6734E-11$ N $(m/kg)^2$

The "graviton effect" occurs within every quark in every atom in the visible universe. In gravity matter, the internal core of the quark consists of antineutrino and negative point particle pairs making it an "antigravity" core to counterbalance the "gravity" field developed by the neutrino/positive pair generating the "graviton" motion effect. Therefore, reversing of neutrino and antineutrino pairing with charged particles may be one way to create antigravity accelerated anti-graviton effect for the invisible part of the universe, while equal speeds of inward and outward motion results in neutral zero gravity space-matter. The most significant of the fusion process test is when one hydrogen-1 atom (single proton) fuses with another. During the fusion, three positive-neutral point particle pairs are broken away from one of the up quarks colliding. This converts that up quark into a down quark making its core more loosely bound and energetic. The released point particles reassemble into one positron (a group of three positive and two neutral point particles) and one neutrino (neutral point particle).

(Electroweak + Strong Force) * (Quarks' Energy) = Gravity

The graviton effect employs both the repulsion between an antineutrino and a neutrino particle made possible by borrowing energy from their paired charged particle, and the additional Lorenz Force created when a negative point particle curves near a positive point particle. This motion generates a burst magnetic field, and a counter recoil system, to eject the positive-neutrino pair with the "strong nuclear force." These combined motions and forces within the quark generate an omnidirectional 6.6734 E-11 N (m/kg)^2 gravity constant, which is fractal in nature, an energy wave operating at the quantum level as it does in the macro universe. The quantum graviton's ability to project a common center of mass between any two or more quarks and or the objects that contain them, make their mutual attraction to each other instantaneous and infinite in range.

The combined charge of all atoms in each star or object in space balances out to zero. But their quantum graviton force does not zero out. The larger object's total negative charge of all quark cores is more than the smaller object and therefore moves slower toward the common center of mass than the smaller body. The differences in quantum charge and neutral particle attraction translates into a difference in mass that exerts an entropic force equal to the gravitational force acting on both objects. The quantum energy wave emanating from each subatomic particle with the quark is released with each cyclic motion, as out-waves at a rate commensurate to the medium in which it traverses. Quantum gravity is hence an instantaneous force and of infinite range, directly proportional to the product of their masses and inversely proportional to the square of the distance between their centers, per Newton. This "Quantum graviton" force applies in the quantum realm and the macro universe, precisely from a fine-tuned motion generated by the other three natural forces create an entropic force, not limited by the speed of light per Einstein, a force known as "gravity"; see **Tab 3** for detailed analysis.

Quantum Graviton Way Ahead

Symmetry demands the existence of antigravity antimatter to counterbalance the existence of gravity matter, as well as the existence of "neutral" matter also known as "dark" matter, the type which neither generates antigravity nor gravity. The application of super symmetry predicts the existence of a mirrored pair to matter with gravity, in this case an entire counterpart to the visible gravity matter universe. Antigravity is the reverse of gravity, a negative gravity; its existence could be invisible to us. The Dirac antimatter prediction can also be interpreted to allude to the existence of antigravity, counterbalancing ordinary matter with gravity, in as much as antimatter is the opposite of matter[160]. Antigravity antimatter repels gravity matter and vice versa. Physics allows for this mirrored antigravity universe[161].

Black holes exist within gravitational bound galaxies, and their counterpart white holes in antigravity galaxies. See **Tab 4** for the composition of black holes. In addition to Hawking radiation, black holes are predicted to release pressure by emitting innumerable amounts of miniscule antigravity-orbs through micro-wormholes tunnels and seed the antigravity part of the universe. White holes similarly release its pressure by emitting the micro orbs of "ordinary matter" that seed the abundant amounts of lighter elements detected and suspended in intergalactic open space[162]. The universe is a symmetrical system in a delicate cosmological balance, homogenous and isotropic, and ruled by quantum gravity and antigravity.

Tab 1: Natural and Perpetual Particle Motion

Up to this point, we discussed and developed a specific motion within the quark that generates a unique pattern, which creates quantum gravity. We then tested the graviton effect structure and motion within the stellar fusion process currently known to science and discovered that it met the stringent requirements. Several factors enable the quantum graviton effect to generate force at the micro level and evidently exert gravity at the macro universe to operate the way it does, in a word its fractal. To answer how quantum gravity is instantaneous and of infinite range, we need to take a step back to re-explain the process of how point particles assemble from "scratch." The constituents of quarks, which are positive-neutral pairs and negative-neutral pairs, do not assemble magically into an up quark and down quark configuration by themselves.

To bring the process into perspective, let empty space exist teaming with loose random point particles that collide and transform neutrinos and antineutrinos into positive and negative point particles and back again. Over time, a neutral and charged point particles pair up for survivability reasons. The key here is that these particles seek stability, primarily in neutral non-destructive groups. As three negative-neutral pairs combine into an electron, assuming all neutral point particles are the same, antineutrino, they experience stability around one neutrino at the center. This happens when one pair's negative point particle connects with that neutrino, then it attracts another pair's negative point particle to that neutrino, and so forth until three pairs of negative point particles are recoiling from, jiggling off, or sharing that center neutrino to form a stable configuration.

The Quantum Gravity "Engine"

The central neutrino pulls the other three antineutrinos toward itself into a tight group, while the negative point particle of the pair becomes the buffer, forming an extremely

---

[160] Ibid

[161] Weinberg, S. (2008). *Cosmology*. Oxford: Oxford University Press.

[162] www.NASA.gov, abundant amounts of lighter elements discovered in intergalactic space.

active recoiling hexahedron shape, like pistons moving in and out within an invisible engine, the beating "heart of the quark." Shown are the particle motions within a quark:

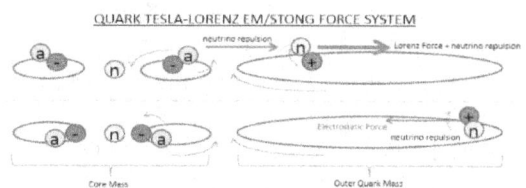

All quarks have within them this built-in recoil mechanism, analogous to a field artillery gun employing a "gas-piston" counter-recoil system and muzzle brakes, which adds stability. Since the center neutrino is not permanently paired, it does not have full access to the strong force. If for some reason, the negative point particle touches the center neutrino as it attracts the antineutrino inward, it will not bind; the center neutrino particle at that moment may borrow a bit of energy from the negative point particle and eject the paired negative-antineutrino outward. Likewise, the negative-antineutrino pair taps into the strong nuclear force just enough to prevent its paired antineutrino from contacting the neutrino. This short burst and partial quantum plank distance is just enough for the center core to have mass and act as a counter-recoil mechanism or shock absorber. This recoil mechanism stabilizes the quark and enables the core to move in the direction of the quark's quantum attraction toward a common center of mass between quarks.

The neutrino to negative-antineutrino counter recoil mechanism and the electromagnetic Lorenz Force provides 33.3334 of the "Strong Nuclear Force," which transfers completely to the positive-neutrino pair (see image on right). The negative-antineutrino in ejecting the positive-neutrino exerts another 33.3333, and the positive-neutrino pushes back onto the negative-antineutrino adds another 33.3333, while the electroweak positive negative particle interaction subtracts 1.000, giving us a 99:1 force ratio. From Nicolas Tesla's point of view, the electromagnetic force plays an important part in catapulting the positive-neutrino particle pair outward. The additional force also known as the Lorenz Force provides this extra boost. As the negative-antineutrino pair passes the positive-neutrino pair, the near collision creates in a fraction of a nanosecond an intense electromagnetic field, sufficient to multiply the positive-neutrino outward motion tenfold. In essence, the positive-neutrino pair acts like a quantum rail gun's armature and is expelled outward at stupendous speeds, but without the heat or friction. Now combine this motion with the repulsion between the antineutrino and the neutrino, both of which borrows energy from their charged partner. This short-ranged rapid acceleration, known as the "strong nuclear force," becomes 100 times that of the electrostatic force alone.

Using the Coulomb's law Force equals $K_e * q_1 * q_2 / r^2$, where K_e = 8987551 / 8.98755E+16 and q_1 and q_2 represent the quantities of charge on the two interacting objects, down to the Plank Length of $1.616\ 229(38) \times 10^{-35}$ divided by 100, we can calculate the internal functions of a quark. Let the size of a quark radius be one-hundredth that of the Plank. This will make the internal recoil hub 0.3333 times 1/100th that of Plank Length, and the outer projectile be 0.6666 times 1/100th that of Plank Length. We will increment the internal hub at 0.0005 starting at 0.0025 and decrement the external orbit from 1 at rate of 0.0010, taking length measurements and calculating force for each of the 644 increments, converging around .3333. All these fractional measurements when multiplied by 1/100th Plank Length produce an accurate predicted length between point particles internal to the quark and complies with the range of the strong force.

The outward force of outer hyperbolic orbit is 10 times 3 and 1/3 stronger than the inward force of 1/3, where antineutrino-negative pair generates 5/3 and neutrino-positive pair generates 5/3, and the multiplier of 10 generated by the Lorenz Force "rail gun" push. Outward force of 10 times 10/3 or 10 * 3.3333 equals 33.3333 times the electromagnetic force. While the inward force is 1/3 or 0.3333, this gives us a ratio of 99:1 of the outward strong nuclear force to inward electroweak force. The Lorenz Force generated by a brief pass of the recoiling core negative-antineutrino pair's closest approach to the positive-neutrino point-particle pair projectile provides the multiplier to the outward force of ten. Together the average outward force resolves to about 6.673400E+27 N $(m/kg)^2$ and time of 2.922240352E-38 seconds squared, and the inward force to 6.673400E+25 N $(m/kg)^2$ and the "free fall" time of 2.922240353E-37 seconds squared. In addition, the recoil mechanism estimated to affect the entire system with a force of 1.000000002E+00 N $(m/kg)^2$ provides maneuverability to the quark and stability to the overall quantum graviton force and is therefore multiplied to the average or rectified result, gives us the gravity constant. See 3-D image below:

The math reads: Using Force Equation = $((F_1*T_1*99/1E+38 + F_2*T_2*-1/1E+36) / (99*T_1 - 1*T_2))$ times the R_F Recoil Force, where F_1 is outward force, F_2 is inward force, T_1 is outward time, T_2 is inward time, and 1E+38 is strength of strong nuclear force and 1E+36 is electromagnetic force compared to gravity; quantum graviton force in a quark resolves to:

= [6.673400E+27 N $(m/kg)^2$ * 2.922240352E-38 s^2 * 99 kg/s^2 /1E+38

+ 6.673400E+25 N $(m/kg)^2$ * 2.922240353E-37 s^2 * (-1 kg/s^2) /1E+36]

/ (99 kg/s^2 * 2.922240353E-38 s^2 − 1 kg/s^2 * 2.922240353E-37 s^2)

* 1.000000002E+00 $(m/kg)^2$

= 6.673400E-11 N $(m/kg)^2$

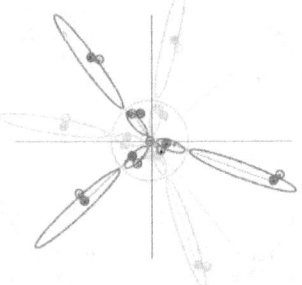

QUARKS INTERNAL COUNTER RECOIL SYSTEM

From Coulomb's law and Lorenz Force perspective, the graviton effect reconfirms Newton's gravity and agrees with Tesla's view of nature; electromagnetic forces affect everything. Depicted above is the quark's counter-recoil system. The core's resulting strong negative one charge attracts three positive-neutrino pairs. As one positive pair (+1/3) moves toward the negative one hexahedron electron group, it taps into the strong nuclear force to halt the inward motion and expels the paired particle abruptly outward for a very short distance, where it turns around into an inward motion because of the electrostatic pull. This motion forms a hyperbolic orbital pattern just outside the negative one charged group. The "core" then attracts two more positive-neutrino pairs to form a stable, zero-charge neutral quark. The three positive-neutrino pairs' hyperbolic orbital paths would then develop a pattern comparable to the Calabi–Yau manifold. Over time, we end up with space filling up with these "neutral" quarks, each of course attracting others. Exactly how will be discussed below, suffice to say Newton calls it gravity.

Tab 2: Particle Interactions and Dark Gravity

When a negative point particle collides with a positive point particle, they do not necessarily "annihilate" each other per science, they transition into two opposite neutral point particles we call neutrino and antineutrino. Likewise, when a neutrino collides with an antineutrino they transition into negative and positive point particles. However, when a negative point particle hits with a neutrino the two will bond into a pair sharing energy and spin, with a 1/3 negative charge. Similarly, when a positive point particle intersects with an antineutrino, they bond into a pair with 1/3 positive charge. When three positive-neutrino pairs grouped together with one antineutrino in the center, known as a positron, collide with an electron consisting of three negative-neutral pairs grouped together, they are doomed to "annihilate" each other on contact, forming free or loose point particles or new pairs or groups of various types. In addition, the process starts again, pairing up to form either a stable neutral quark or their opposites. The configurations formed by switching the neutrino and antineutrino places results in zero-charged "anti-neutral quark" that will move away or repel from normal matter, as we know it and combine with like particle groups, suffice to say we will call the attraction antigravity. Model shown.

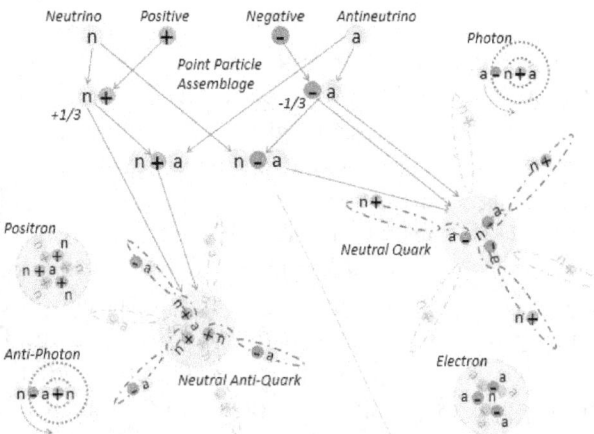

In another possibility, the tetrahedron (consists of four positive-neutrino pairs with a center antineutrino) attracts four individual negative-antineutrino pairs to create a "neutral anti-quark." Four particle pairs give us a three-dimensional shape; however, it does not resolve easily into -1/3 and +1/3 charges of down and up quarks. Three pairs or six pairs both resolve into the -1/3 and +1/3 charges. However, nature has chosen the simplest of forms, three. There are three natural forces; the fourth is just a byproduct of the first three. Neutral anti-quarks will attract and annihilate their antimatter counterpart, the anti-neutral quark, on contact. This transition action resets the field of space with loose or smaller pairs of point particles to reassemble repeatedly until they form either normal neutral quarks or anti-neutral quarks with gravity or antigravity respectively, which eventually group into two separate pockets of matter in space, that with gravity and that with anti-gravity.

Dark Matter and Dark Gravity Transition

As large groups of neutral quarks consolidate to form into a "micro black hole," their energy level elevates with the increases in motion instigated by their "quantum gravity" mutual attraction. The collision of two high energy-level neutral quarks causes one to transfer a positive-neutrino point particle pair to the other quark, transforming the donor quark into a negative one-third charge and the receiver quark into a positive one-third charge, which hold together electrostatically as a pair of quarks, neutral-charged subatomic particle (-1/3, +1/3). When two of these zero-charged paired particles collide, one of three things happen. Either each pair returns to original state of all neutral quarks (0, 0, 0, 0), or we get one of two other possible structures. When the sides containing the negative charges collides, (+1/3, -1/3) hits (-1/3, +1/3), one quark becomes negative two-thirds charge, attracts the two pairs of positive one-third charge, and releases a neutral quark, (+1/3, -2/3, +1/3) (0). Or when the side containing the two positive charges collides, one quark becomes positive two-thirds charge and attracts the two negative one-third charges, and releases one neutral quark, (-1/3, +2/3, -1/3) (0). Both subatomic particles are neutral and zero-charge. The continued transitioning of neutral quarks into neutrons changes the "micro black hole" into a miniscule neutron sphere. Science refers to this unknown invisible matter in space as "dark matter," which clearly exerts "dark gravity:"

What happens when these two possible neutral tri-quark particles collide with each other? Specifically, the outside edge of one has negative one-third charge and the outside edge of the other has positive one-third charge would bind; centers bind; and other edges bind. Together they form stable neutral bond. However, the situation begins to change as more and more groups that are neutral assemble into a small sphere massive enough to develop higher pressures, and its overall heat energy increases. If a smaller group of neutral (-1/3, +1/3) quarks collide with the **neutron** we get two possible outcomes. First possibility: the two -1/3 collide and push over a positive-neutron pair toward the larger group destroying its integrity and transitioning both, (-1/3, +2/3, -2/3) and (0, +1/3) resulting in (-1/3, 0, 0) and (0, +1/3) to become five neutral quarks splitting apart and restarting the assemblage process over. Some of these neutral quarks may fall apart and create photons groups and loose negative-neutrino point particle pairs. In the second possibility: the outside -1/3 collides with one of the +1/3 and sustains the **neutron** subatomic particle; (-1/3, +2/3, 0) and (0, -1/3) and releases two neutral quarks and or all its point particles as photons and electron constituents, which flow around the antineutrons positive exterior but resist the neutron's negative exterior.

If the same smaller group of neutral (-1/3, +1/3) quarks collide with the **antineutron** we get two possible outcomes. First possibility: the two +1/3 collide and push over one pair of positive-neutron toward the larger group transitioning both; (-1/3, +2/3, +2/3) and (0, -1/3) resulting in the rearrangement of the antineutron into a **proton** (+2/3, -1/3, +2/3) and rips apart the small group into a neutral quark and a down quark. In the second possibility: the outside +1/3 collides with the one of the -1/3 and sustains the antineutron subatomic particle; (+1/3, -2/3, 0) and (0, +1/3) and releases two neutral quarks and or all its point particles as gluons, photons, and electrons constituents. These electrons flow around the positive charged exterior of antineutrons and protons but resist the neutron's negative exterior and are suspended in a level determined by the accompanying photon. The end-result is the creation of the hydrogen ^{1}H atom with a single proton surrounded by an electron and its photon; or a hydrogen ^{2}H atom with a proton and a neutron at its core surrounded by an electron and its photon. The electron shell of each newly created atom absorbs energy and resists gravitational pressure, making the whole sphere less dense, and given enough accumulation of matter expands the orb into a proto star. Note that in this particle assembly sequence, the neutral quark forms before the up or down quark, which leads to the neutron and then the proton. The proton's positive charge enables the

CHAPTER 5: THEORY OF COSMOLOGICAL BALANCE

formation of the atom. A group of neutral quarks is a micro black hole while a grouping of neutrons is a micro neutron orb detected as "dark matter", "dark gravity."

Tab 3: Instant Infinite Range of Quantum Gravity

How can one quark attract another? Newton defines gravity as the attraction between two objects[163]. Greene tells us that in quantum entangled particles simultaneously react, regardless of distance, as if inextricably linked[164]. When one particle moves toward another particle, its movement is mirrored by the second particle causing it to move toward the first particle. He also tells us the electrons throw photons at each other to gain energy to repel each other, similar to the quantum entangle and electromagnetic principle combined[165]. Science explains that electrostatic force is not a player when it comes to attracting objects in space as the charges of all the atoms within each object total to practically zero and attributes all attraction to gravity alone. Science also tells us that matter is fractal, where events that occur in the micro world also occur in the macro world at an exponential size. Gravity is fractal by definition; so, quantum gravity created within a quark should also function in the same way it does in the macro scale of the universe.

These ideas combined explain how one quark instantaneously attracts another quark regardless of their distance. Using Newton's gravity concept, let us investigate how this applies between two quarks in the quantum realm to determine the smallest fractal scale of gravity. Assume for a moment that we have two neutral quarks each with zero charge. Just as two objects in space having zero charges are attracted gravitationally per Newton, we expect these two neutral quarks to attract each other. On analyzing the internal motion of each neutral quark, we see a negative one (-1) at its core and positive one (+1) at the surface of its fluctuating or jiggling "shell." If each quarks' point particles were motionless or have flat spherical orbit motions, then they would have no effect on each other.

However, the quark's outward hyperbolic motion of their positive point particle pairs extends its quantum entanglement connection to the other quark's outward motion of the positive particles, creating in effect a "target" point midway between both quarks. The "common center of mass" midpoint between both neutral quarks, like the barycenter between two objects in space that gravitationally attract each other. This is the point in space to which both objects are drawn towards, a center of gravity. There is no need for a messenger particle to travel between them. Within the quarks common center of mass is the projected combined total of both shells' charges, a positive two charge or (+2), which draws both negative one or (-1) cores' charge toward it. Similarly, if the positive charged particles were outwardly thrust in the opposing directions, we get the negative two (-2) charge in the center of mass and the positive one (+1) then nudges the core toward the common center of mass. In both instances, each quark reacts as though "quantum graviton" attracted toward each other, powered by the natural "electromagnetic and weak" force and extended infinitely by

---

[163] Newton, S. I. (1687). *Principia Mathematica*. Cambridge: Unknown

[164] Greene, B. (2003). *The Elegant Universe: Superstrings, Hidden Dimensions, and the Quest for the Ultimate Theory*. New York: W.W. Norton & Company.

[165] Ibid

the "strong nuclear" force. This is an entropic force in action, mirrored as both objects come together by the quantum graviton effect, which is instantaneous; it does not require a messenger particle to traverse the space between them.

Does the graviton effect work between neutrons and protons? For simplicity sake, we will assume that the quark core has three negative-neutral pairs. In a neutron, we have two down quarks and one up quark. This gives us a three negative one or (-3) in the quark's cores and a combined total of positive-three or (+3) in the outer shells. The overall graviton effect between two neutrons when their positive particles are moving toward each other is -3, +6, -3, where each core's negative three or (-3) is attracted toward the common center of mass containing positive six or (+6). When the positive particles are moving away, we get +3, -6, +3, where each core moves toward each other. Similarly, in a proton, we have two up quarks and one down quark. This gives us a three negative one or (-3) in the quarks' cores and a combined of positive four (+4) in the outer shells. The overall graviton effect between two protons when their positive-neutral particle pairs are moving toward each other is -3, +8, -3, where each core's negative three or (-3) is attracted toward the center of mass containing the positive eight or (+8). When the positive particles are moving away, we get (+4, -6, +4), where each core moves toward each other. All other directions not aligned yield a negative core equal the positive surface; for example, -3, +3, provides no movement in that direction. Moreover, the alignment of the neutral particles plays a more significant part in which direction the quarks (and all the atoms in the entire object) move. The alignment of the charges in the above paragraph helps the layman to understand electrostatic attraction. Nevertheless, the neutral particles' strong force causes the quark to move. In two single neutral quarks, we also get 3*an, 6*n, 3*an, where each quark's antineutrinos (an) are drawn to the perceived larger group of neutrinos, their common center of mass, at 100 times more force than the electrostatic difference. Neutral particles overwhelm charged motion as they control the strong force by using energy from their charged pair. Neutral alignment primarily determines if we see attraction or repulsion. An alignment between ordinary matter and antigravity antimatter would appear to be 3*an, 3*n: 3*an, 3*n, where each neutral particle. 'an' is an antineutrino and 'n' is a neutrino, repels their counterpart in the vector line.

Introducing electron clouds around the atom does not change this quantum attraction force, as these electrons are massless and move along with the atom, thereby does not change the center of mass between two atoms. Carrying this quantum "graviton effect" concept from the micro to macro world, we can see that this attractive entropic force too is fractal and applies to the universe. Two objects each with a total of zero charge attracts each other; visually one object's negative googolplex quark cores always move toward the two times positive googolplex common center of mass, while the other object's negative googolplex core moves in the mirror direction toward the same common center of mass. This "graviton effect" then converts into an attraction force that is directly proportional to the product of their masses and inversely proportional to the square of the distance between them; said another way, Newton's prediction applies to the quantum level of and within quarks just as much as it applies to the macro universe.

STELLAR OBJECTS
(Zero Charge Attraction = Quantum Gravity)

Center of Mass

CHAPTER 5: THEORY OF COSMOLOGICAL BALANCE

Above is a depiction of a common center of mass between two stars, the attraction point. Each stars' overall combined atoms balance out to a charge totaling to near zero. However, their quantum graviton force does not. The large star's total negative charge of all its quarks' core is twice as much as that of the smaller star. Let the smaller star have googolplex quarks and the larger star have two times googolplex quarks. The smaller star would have one googolplex negative charge and the larger star a negative charge of twice googolplex, while the center of mass would then emulate a total positive charge of three times googolplex. This difference in quantum charge attraction equates into a proportional difference in the mass exerting an entropic force equal to the gravitational force acting on both objects, which of course pulls the smaller star faster toward the center of mass than it tugs on the large one. Image below shows center of mass between two stars and a planet, attraction point.

Here all three objects move toward the common center of mass, exactly as in the quantum world between three neutral quarks or three neutrons in space. The quantum graviton solution, in essence, is an electromagnetic based action with strong force extension, infinite range, and instantaneous quantum entangled property. Quantum gravity exactly as Newton predicted is not limited by the speed of light per Einstein, because it does not require a "messenger particle" to work.

THREE OBJECTS IN SPACE
(Zero Charge Attraction = Quantum Gravity)

Tab 4: Quantum Gravity in Black Holes

Now that we understand quantum gravity, we should discuss why neutrons survive within a neutron star, and why neutral quarks are expected to survive within black holes. The mass of a proton is approximately: 1.6726×10^{-27} kilograms. The mass of a neutron is about 1.6749×10^{-27} kilograms[166]; it contains two down quarks, which has slightly more mass than an up quark. An electron's mass is approximately 9.110×10^{-31} kilograms. The neutron, structured much like that of a helium atom

Neutrons within a Neutron Star

consists of an up quark orbited by two down quarks, is very stable and zero charge. The down quarks constructed with three negative particles and three neutral particles within its center and randomly orbited by two pairs of positive and neutral point particles are more energetic, coupled with the neutron's total neutral charge; provide it the additional strength to withstand the crushing gravity in a neutron star. Depicted are tightly packed neutrons in a neutron star. The extra strength comes from the loosely bound and more energetic negative-neutral point particle pairs at its center, the negative charge of the two down quarks within the neutron, and the fact that neutrons are zero in overall charge. Its more energetic motion gives it a slight increase in mass, without an increase in gravity. Any

---

[166] http://www.citycollegiate.com/atomic_structureXIg.htm

protons within a neutron star reconfigures into neutrons, by emitting a positron and neutrino[167]. Pictured is a tightly packed quark "soup" in a black hole. Similarly, the "neutral" quarks can withstand the gravitational forces within a black hole because they are the source of the black hole's gravity, without their existence there would be no gravity field generated within the black hole. Dissolving the atom into its basic quarks maintains the overall gravity field and allows the black hole to be densest of all known gravitationally bound matter. We predict that all quarks within a black hole realign themselves into zero-charge neutral quarks, containing three pairs of negative-neutral point particles at its center hyperbolically orbited by three pairs of positive-neutral point particles. It does this by transferring positive-neutral pairs between high-energy top and bottom quarks. Neutral quark soup resembles neutrons in a neutron star.

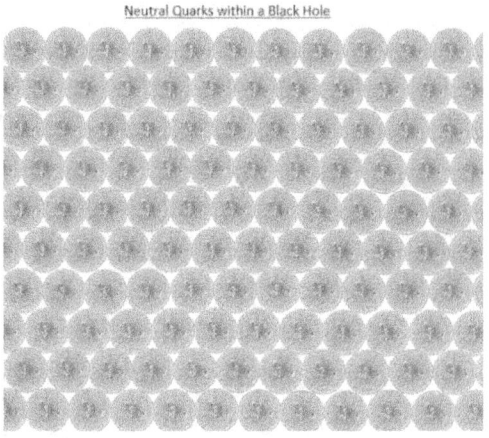

Neutral Quarks within a Black Hole

The super symmetrical structure of quarks is like miniscule quantum forms of "atoms" complete with positive, negative, and neutral point particles all working together to maintain their existence and survive. Any up or down quark within a black hole is quickly converted back into a neutral quark, just as a proton transforms back into a neutron within a neutron star. Given enough pressures and energy, these quarks may transform to produce neutral non-graviton matter and anti-graviton matter to relieve pressures within the black hole, thereby enabling another form of "excretion," miniscule matter expelled outward. The tightly packed quark "soup" structure within a black hole has significantly less "empty space" between each point particles than that of the neutrons within the neutron star; clearly showing us why black holes are the densest matter known[168] yet fluid. Black holes constituents are made up of free-flowing neutral quarks, and as such, contain extremely intense temperatures and exuberant amounts of pressure[169]. By this definition, black holes are black stars, not cold dead singularities, but vibrant with significantly intense motion. Since super symmetrical structure of quarks are like miniscule quantum forms of "atoms," the smallest space-matter particle in space, the hygratium, a unique neutral quark, primarily neutral and yet pulses both gravity-antigravity at the rate of one-five-hundredth times the speed of light, approximately 600,000 cycles per second. Hygratium exists in its purest form, one quark per every cubic meter in deep open space or intergalactic space and becomes more tightly packed as the form the curvature of space around massive objects whether they are gravity matter or antigravity antimatter in nature, to include black holes and white holes. Hygratium can also generate a negative electrostatic shell to prevent collisions with normal matter and a positive electrostatic shell near antigravity antimatter. These shells are non-existent when around black holes and white holes.

---

[167] Wikipedia. (2015, June 12). Neutron Star Definition.

[168] Wikipedia. (2015, June 12). Black Hole Definition.

[169] Weinberg, S. (2008). *Cosmology*. Oxford: Oxford University Press.

5.9 The Function of a Field Theory

We have now completed that part of the development of our theory, which deals with the motion and stability of the universe, with the new physical character of the new concepts introduced. The next step is to make a more detailed formulation of the theory in terms of which it is possible to discuss the motion of individual particles. Two methods are possible. One can follow kinematical relativity to examine the consequences of the Cosmological Balance, or one can try to create a field theory. Milne, Walker, and their school explored the first possibility for less stringent Cosmological Balance, making our progress easier and additional assumptions less necessary. However, it is known from kinematical relativity that the application of this technique is by no means easy unless the Cosmological Balance is assumed to apply to every detail of world and universe structure, purely a physical and statistical law in the universe, applying to all averages.

The development of a field theory is therefore of particular importance. We wish to stress that in our view general relativity could not be satisfactorily adapted to our theory, at least not in totality. Though general relativity is presumably correct in stating that the light cone, together with space and time measurements, defines a tensor field throughout space-time, we disagree with the claim that this is the only fundamental covariant field universally defined. An equally fundamental vector field is defined by the state of motion in which an observer must be to see isotropy in the universe around him or her. This universally defined vector field exists in every cosmology, but it is of particular importance in the theory presented here.

A field theory is only required to deal with deviations the Cosmological Balance Theory. For example, planets in the solar system that orbit retrograde are the exception to the Cosmological Balance, where the norm is the prograde planet rotation. So, exoplanets discovered in other than this solar system within the visible galaxy, the Milky Way, are expected to also orbit normally prograde, by the laws of conservation of energy, predicted by the Cosmological Balance. Careful examination of any planets with retrograde motion in the visible universe will reveal that these planets rotations will be or are slowing down and will eventually come to a stop before reversing its rotation direction to prograde. The action of slowing down and reversing a planet's rotational direction causes internal friction between the motion of the molten core within the planet, if it had one, and its crust, thereby overheating the planetary atmosphere[170].

Stars orbiting faster or slower than the rest of the galaxy's arms are the exceptions to the Cosmological Balance. Larger nearby stars either catapulted these stars forward or yanked them behind in the galactic arm, per Newton's gravitational attraction. Although these stars may be currently excessively fast or slow, the Cosmological Balance states that in the end these stars will eventually return to balance orbital harmony with the rest of the galaxy and when counted among the whole galaxy are already in balance. Similarly, planets, given enough time will eventually clear their own orbital path within their solar system, provided nothing else causes them to fail.

---

[170] Exoplanets found within the "goldilocks" zone of a star to be rotating in retrograde direction could be too hot for life to exist. This excessive heat is produced by rotational friction.

The density of space-matter hygratium atoms on occasion will slow light emitted by distant galaxies with gravity or antigravity to a crawl without changing its color or red shifting it. This effect may distort images of distant visible galaxies with gravity or the invisible ones with antigravity. Concentrations of newly formed gaseous matter with gravity within low-pressure space-matter systems will however, significantly red shift only normal light from visible distant galaxies and if combined with the gravity bending caused by curvature of space, the galactic halo, be seen as lensing based on the density and volume of the accumulated matter and layered space-matter respectively[171]. Similarly, concentrations of new gaseous antigravity antimatter formed by opposite rotating low-pressure space-matter systems will likewise only red shift antigravity light from antigravity distant galaxies or be seen as lensing based on the density of the antigravity antimatter accumulated by their intelligent inhabitants, should they exist.

Antigravity galaxies are invisible to us. Why? Because the light emitted by antigravity consist of particles possessing antigravity characteristics and therefore turn or curve away from any object with gravity, thereby avoiding our instruments and us. Conceivably, antigravity light could also travel in a different configuration unlike the normal light waves we all know. The antigravity light photon/electron pattern could be a self-perpetuating spiraling weave, like a corkscrew. In any case or by chance these antigravity photons come into our realm, their different patterns of travel are currently highly unlikely to be picked up by our eyes or our instruments simply because they were not designed to detect such obscure or abnormal patterns. Likewise, light from the galaxies and stars are invisible to any intelligent life, assuming they existed, within an antigravity galaxy. On May 24, 2016, physicists from Trinity College Dublin announced the discovery of such a new form of light not normally observed (Patel, 2016). In their discovery, the light photon spirally rotates around the axis of travel with its counterpart electron. This discovery was made possible by studying angular momentum.

The Cosmological Balance Theory fully describes the universe, as it is, infinite, ageless, constantly creating gravity matter, and antigravity antimatter to form stars, galaxies, and planets capable of supporting life, as we know it, or some related version of life, and in turn constantly restoring space-matter hygratium. These galaxies' super massive black holes and white holes are the gateways that in turn eject and provide the necessary high-energy dense sub-atomic particles to restore space-matter exotic atoms, which sustain space. We owe our material existence to this self-perpetuating renewable balance of the universe, the Cosmological Balance. The rest of this book will present in excruciating detail the arguments and illustrations supporting this theory.

---

[171] The movement of exotic space-matter toward the massive object in space (force of gravity) has the ability to redirect starlight in the direction of motion is what Einstein referred to as additional "gravitational energy." We simply call it light refraction without changing the frequency and color of the starlight passing through it. Both are the curvature of space surrounding the massive object. Space-matter has a high optical density similar to ordinary glass lens.

Chapter 6: Theory and Proof of Balance

"The universe is big, its vast and complicated, and ridiculous. And sometimes, very rarely, impossible things just happen, and we call them miracles. And that's the theory. Nine hundred years, never seen one yet, but this would do me."

— Steven Moffat

In the first two chapters, we have reviewed to the best of our abilities and knowledge what we have come to know of the visible micro and macro universe, to include the building blocks from the smallest sub-atomic quarks to the largest galaxies and the awesome black holes within them, and the physics defining each. We discussed two of the most inexplicable elements that cosmologists have yet to find, the missing "dark matter" and "dark energy," and existence of white holes, to explain why the universe is expanding in an accelerating rate. Then in chapter three, we investigated where galaxies come from and conducted an analysis of the timeline for the Milky Way to assemble. The reviews of these three initial chapters were necessary to provide the required background and mathematics for the rest of the book. Moreover, chapter 4 in this part summarized the balance found throughout the universe. Concerned over an expanding universe, Stephen Hawking stated; "There might be some other form of matter which we have not yet detected and which might still raise the average density of the universe up to the critical value needed to halt the expansion (Hawking, The Theory of Everything: The Origin and Fate of the Universe, 2002, p. 34)." He was alluding to something more than "dark matter" and "dark energy." In another discussion about black hole emissions, Hawking stated, "However, in the end most people, including John Taylor, have come to the conclusion that black holes must radiate like hot bodies if our other ideas about general relativity and quantum mechanics are correct." He goes on to say, "The existence of radiation from black holes seems to imply that gravitational collapse is not as final and irreversible as we once thought[172] (Hawking, The Theory of Everything: The Origin and Fate of the Universe, 2002, p. 92)." He then continues, "Even the types of particle that were eventually emitted by the black hole would in general be different from those that made up the (object entering the black hole) (Hawking, The Theory of Everything: The Origin and Fate of the Universe, 2002, p. 93)."

Chapter 5 in this part defined the Cosmological Balance Theory, generalized mathematics outlining this cosmology, and laid down some assumptions. The next few chapters ahead will present the details necessary to visualize this theory. We propose that the universe consists of the visible part with the stars and galaxies that light up the dark outer space, which we all are just beginning to see, document, know and understand, and the invisible part, which encompasses what scientists call dark matter and dark energy, which is still a mystery to mainstream cosmologists and astrophysicists. Some even say that dark matter and dark energy could be leaking matter and energy from another dimension or another parallel universe, or even a multi-verse. These other concepts are beyond the purpose of this book and will not be discussed further here.

---

[172] Professor Stephen Hawking predicted that radiation (evaporation) was sufficient from keeping the black hole from crushing itself into a singularity, and yet he described the same phenomena as a singularity. Other physicists disagree and see the black hole itself not as a singularity but a homogenous sphere with a "soup" of subatomic particles capable of pushing outward against the forces of gravity.

If the predictions are correct that there is that much "dark matter" scattered throughout the universe, then we can extrapolate that this unknown matter is actually the composition of space itself. It consists of an almost undetectable type of substantial high-energy matter separated by extreme gaps between molecules and atoms. Space-matter is a type of exotic matter that does not "gravitationally or anti-gravitationally" attract each other but electromagnetically repels each other apart like most gases on earth, giving us the perception of "void" empty space with lots of energy, "dark energy" or as we labeled it, anti-gravity. The push force hygratium in space provides is the gravity we recognize. It is everywhere and comes in every direction all at the same time, also known as vacuum pressure, energy, and force[173]. As such, objects moving in and through space, will respond to the presence and push of vacuum force as if "gravity" is instantaneously pulling them together, when in reality the normal matter objects are being pushed together incrementally as they move through space, rectilinearly responding exactly as Newton explained. In this scenario, the gravity waves Einstein predicted do not travel at the speed of light. If it did, then this makes the Earth-Sun gravity speed per Einstein at least an eight-minute delayed response, a concept hard to visualize by most of us, but apparently acceptable by those who understand and know how to apply Einstein's General Relativity equation. The Einstein predicted gravity speed limitation is completely opposed to the principle of consistency set forth by Newton. Yes, the Cosmological Balance Theory is centered on one explanation of what scientists perceive as "dark matter" and "dark energy." The concept literally leads us to the development of a cosmological balance equation. An equation that determines the orbital period of stars around galaxies given the distance from the galactic center (and vice versa), and the same equation when in simple form becomes applicable to orbital periods of solar systems (it is a slight modification to what we know as Kepler's Third Law). The same equation also reduces to solve the interaction of galaxies with each other. It becomes Newton's gravity equation when ordinary gravity matter galaxies interact with each other. The cosmological balance equation explains why some galaxies oppose or do not follow predicted gravitational behavior, as if something is going against their gravity, scientists have call it "dark energy;" cosmological balance theory calls it repulsion of galaxies.

Cosmologists conjecture that super massive stars go supernova at the end of their life span and when large enough sometimes collapse to form a neutron star or even a black hole[174]. We also speculate that at the center of every massive galaxy is at least one super massive black hole if not two or three in a downward spiral toward merger. We also know that all the stars within the galaxy orbit around this super massive black hole at its center. The more massive the black hole is, the more likely it would create and eject subatomic antigravity particles. This black hole pressure release leads us to the solution of next dilemma, why the outer stars of a galaxy orbit at a uniform rate.

---

[173] Space-matter is an ocean of a unique matter with equal pressure, energy, and force throughout. It is a perfect "fluid" and "gaseous" substance.

[174] The neutron star and the black hole are the two of the densest objects we have discovered in the universe. They are both remnants from the death of super massive stars.

CHAPTER 6: THEORY AND PROOF OF BALANCE

6.1 Two Sides of Balance Explained

The following definition concerning Yin Yang in Taoism is a summary from the website: http://personaltao.com/teachings/questions/what-is-yin-yang/

Yin Yang is a well-known and documented concept commonly used within Taoism. Yin and Yang are two halves, which are necessary to complete wholeness. It defines the nature of change, a balance held within two halves of a whole. The word Yin means the "shady side" and Yang represents the "sunny side". Yin Yang together is a duality concept and struggle forming a whole. We all see examples of Yin and Yang in our daily routine in life whether we know it or not. These are some examples: night (Yin) and day (Yang), shadow (Yin) and light (Yang), female (Yin) and male (Yang), evil (Yin) and good (Yang), sick (Yin) and well (Yang), and so on and so forth. Generations upon generations over thousands of years have compiled quite a collection of documented examples, logically separated, sorted, and grouped into their designated Yin Yang classification systems.

In Taoism, the symbol for Yin Yang is known as Taijitu. In the western world, we most commonly know it as the yin yang symbol. The Taijitu symbol appears in several cultures and over the years has come to represent Taoism itself. The deep meaning of Yin Yang is illustrated seen in the following:

Yin Yang illustrated from the Tao Te Ching

When people see things as beautiful,

ugliness is created.

When people see things as good,

evil is created.

Being and non-being produce each other.

Difficult and easy complement each other.

Long and short define each other.

High and low oppose each other.

Fore and aft follow each other.

Basic Concepts Defining the Nature of Yin Yang

Each side of Ying and Yang is neither absolute by itself. In other words, nothing is completely Yin or completely Yang. Each aspect contains the beginning point for the other aspect and vice versa. For example, as the earth rotates in its orbit around the sun, day becomes night and then night gives way to daytime. Yin and Yang halves are interdependent upon each other so that the definition of one requires the existence of the other to be complete.

Yin Yang relationship is not completely static and absolute. The nature of Yin and Yang flows and changes with time and space, from generation to generation. A simple example is thinking about how the day gradually flows into night as the sun sets. However, the length of day and night are changing as the earth orbits the sun. Even as the earth ages, its spin is gradually slowing down causing the length of day and night to get longer, in due time our 24-hour clock will replaced with a 25-hour clock and so forth[175]. Day and night halves are not static entities itself. Changes in the relationship between Yin and Yang can sometimes become dramatic where one aspect can quickly and literally just transform into the other. Some species of fish and other unique asexual animals have a portion of their females transform quickly into males to compensate for low count of male fish population. In summation, Yin and Yang halves together form a complete whole representation of nature. As a result, an increase on one side causes the other to decrease to maintain overall balance of the whole. Of course, the balance of Yin Yang sometimes skewed due to outside influences, is represented through the following four possible imbalances:

1. Deficiency in Yang
2. Deficiency in Yin
3. Excess in Yang
4. Excess in Yin

These four imbalances are usually paired: so, an excess of Yin can also stimulate a Yang deficiency and vice versa. The Yin Yang concept is especially important for traditional Chinese healing practices. An excess of Yang interpreted as a fever. An excess of Yin could depict the accumulation of unnecessary fluids in the body. Traditional Chinese healing examines a person's health is in terms of these eight principles: Internal and External stimuli, Deficiency and Excesses, Cold and Heat, and Yin and Yang. Yin Yang halves can be further subdivided into additional Yin and Yang aspects. For example, a Yang aspect of Heat subdivided into a Yin warm or Yang burning hot.

Examples of Yin Yang in cosmology could be categorized like this: earth (Yin) and sun (Yang); moon (Yin) and earth (Yang); potential energy (Yin) and kinetic energy (Yang); invisible universe (Yin) and visible universe (Yang); destruction (Yin) and creation (Yang); nucleus (Yin) and electron (Yang); electron in light (Yin) and photon in light (Yang); and so, on and so forth. Every opposite half of a balanced relationship in physics could also be listed as either Yin or Yang, even magnetic north (Yin) and south (Yang); anti-matter (Yin) and normal matter (Yang); antigravity (Yin) and gravity (Yang); death (Yin) and life (Yang); and negative (Yin) and positive (Yang) to name a few.

Additional principles defining Yin and Yang qualities exist in several other cultures. These concepts listed above are only a starting point in illustrating the nature of and struggle within Yin and Yang principle. Taoism does a good job of not codifying life, which is ironic since many Taoist's readily list what is Yin and what is Yang in their view. Normally, Taoist texts will list a few examples of Yin and Yang and then move on to the next important topic on their agenda. This makes perfect sense from a Taoist perspective for it is up to the reader to reveal and see life from his or her own perspective. The Taoist passage quoted above from the Tao Te Ching is an excellent example of this idea, where the reader must discover on which side he or she sits. In conducting your own analysis, you will discover a few additional aspects to Yin and Yang realizing that the passage is not a complete definition either. To learn more about Yin Yang, you as the reader are encouraged to go out and to explore these basic ideas on your own (Julie, 2005).

---

[175] Rest assure, the changing to a 25-hour per day clock will not occur within our lifetime.

6.2 Visible Side of The Universe

Now let us apply the Yin Yang principle to the visible universe. Everything we see and know in the visible universe exists within the realm of the Sunny or Yang side of the universe. The sun is a star just massive enough and with enough gravity to compress itself into a ball, that fuses hydrogen atoms into helium atoms and gives off heat and light. The photons emitted from such fusion takes up to 100,000 years of moving between atoms and losing energy to go from within the star to its surface and then once at its surface races at the speed of light outward, as sunshine and radiation. The sun shines and bathes the earth and the other planets orbiting around it with sunlight, which sustains life, as we know it, on the planet earth. The light from the sun that does not encounter matter continues through space for millions and billions of years. Everything we have discussed in the previous chapters all occur within the Sunny or Yang side of the universe of ordinary matter or from our view, the visible part of the universe. The microelements and the macro-elements exist within our realm, along with the mathematics, physical laws, and gravitational effects that all occur in the Yang side of the universe, the visible part. Scientists study reality, everything else is speculated, or either unknown or undiscovered.

Matter, energy, momentum, and time are all interconnected. From this we know that time is universal outside the black hole event horizon and possibly as well as within it. Without the steady ticking vibrations of time embedded within all quarks shielded by a super strong sub-particle force and the energy they give us; the universe would not exist. In other words, if all quarks or the smallest sub-atomic particles in the universe suddenly stop vibrating or jiggling, they would cease to exist and with their disappearance, all atoms and matter they occupy would cease to exist as well. Time would fail to flow at the speed of light, no energy and matter produced, and the universe would vanish. This is obviously impossible for not even the shear gravitational force of a black hole in the visible universe can stop their quarks from vibrating or their leptons and force carriers from functioning. Hence, time exists even within the most massive of black holes; time is truly universal and absolute. Conceivably, such an environment gives rise to the grand unification force found in space and its matter. What can withstand a black hole's gravitational forces? First, the force maintaining the quark's sub-atomic particle shape is stronger than the strong nuclear force and is due primarily to the quantum mechanics quality of entanglement they possess where they all function as one and in time. Second, together their combined energy can generate the necessary repulsive force to maintain a homogenous "soup" like consistency in the black hole. Thomas (Thomas, Hidden in Plain Sight 2: The equation of the universe, 2013) argued that a repulsive force exists within objects fully enclosed completely within their Schwarzschild radius. The black hole is such an object, and the presence of repulsion equates to existence of pressure, which of course leads to increase temperatures. This repulsive pressure force is the result of all quarks vibrating to sustain their existence. The vibrations of the quarks clock time; therefore, time exists within the black hole[176].

We introduced the electroweak force, which is a unification of the electromagnetic force and the weak nuclear force discovered by experiments with higher temperatures and energies. The grand unification theory expands on this and suggests that even higher

---

[176] Newton equation demands that time continue to function normally within black or white holes without slowing down or speeding up. Time is universal in all parts of the universe.

energies and spikes in temperatures at or above 10^{27}-Celsius degrees can merge all four forces of nature into one, where repulsion prevails within a gravitation field and gives way to a homogenous state of particles or of "soup" like consistency. The gravitational pressure, energy, and temperature within the black hole caldron can create such a single unified force and antigravity, which provides the steady release valve or "excretion" means for the black hole itself. This unified force also provides the extensive energy of repulsion between gravity matter and antigravity antimatter and stabilizes into the space-matter *hygratium* atom. Space-matter, composed of the original natural fundamental force, through several phase transitions due to temperature, pressure, and energy changes, becomes the four natural forces we know today in ordinary matter. Georgi, Helen Quinn, and Weinberg also suggested that three of the four forces (strong and weak nuclear forces, and the electromagnetic force) are unified at temperatures above 10^{28}-Celsius degrees to obtain complete symmetry among non-gravitational particles (Greene, The Fabric of the Cosmos: Space, Time, and the Texture of Reality, 2004, p. 267). Moreover, at slightly higher temperatures and energy all four forces form perfect symmetry, and conceivably antigravity and the force to create wormhole tunnels for escaping microscopic white hole orbs[177]. On the wormholes collapse, released perfectly symmetrical particles reconfigure into space-matter *hygratium* atoms, with neutral graviton (aka Higgs particle) and take their place in the space grid or Higgs Ocean.

In the visible universe everything is linked in a balance of natural laws and symmetry, the yin yang principle. An electron's negative charge is counterbalanced with a proton's positive charge in an atom. The neutron, although neutral, counterbalances and holds protons together in the nucleus. The photon and electron energize and balance each other electromagnetically. Magnets have two poles, North, and South. An electron has an opposite, the positron, or anti-electron. Matter has an opposite anti-matter. Everything in nature has an opposite, so why not gravity. Without antigravity, the visible universe could potentially collapse on itself under its gravitational pull. Without antigravity, black holes would blow up from excessive anti-matter matter explosion pressure. The infinitesimal chance creation of antigravity within a black hole releases pressure and a minuscule amount of mass, and still allows the black hole to maintain its super dense mass and extremely small, compressed size. Any existence of antigravity particles on the world on earth would immediately be repelled to fly away at speeds faster than the escape speed before we even notice it was there. Therefore, let us assume that when a black hole creates antigravity particles, they will begin to "float" to the surface and be ejected outward at escape speed or, in this case, at the speed of light squared[178]. As Stephen Hawking said, "Black holes aren't as black as they are painted (Hawking, The Theory of Everything: The Origin and Fate of the Universe, 2002, p. xii)."

How can a black hole expel an object at the speed of light squared and what passage method can allow this? Recall Einstein's equation $E = mc^2$. We know that light itself (or

---

[177] The pressures, temperature, and energy internal to massive black and white holes are predicted to be the right environments for the "grand unification" to occur.

[178] The pressure, temperature, and energy within the black hole create antigravity particles along with the grand unification of all four natural forces into orbs. The newly formed orb float upward and is ejected from the surface at the speed of light squared, powered by mass, energy, and momentum relationship.

CHAPTER 6: THEORY AND PROOF OF BALANCE

any other EM radiation), cannot escape from within the event horizon of the black hole or from the black hole surface itself, as defined. Therefore, any object or particle if it were to escape from the black hole surface must have anti-gravitational properties to achieve an escape velocity of the speed of light squared, defined in chapter 5 section 5.6 Energy, Mass, and Momentum, it must travel as practically pure energy with fluid properties. It generates its own wormhole tunnel, implying that the antigravity mass is expelled with the energy of all four natural forces, unified into one, repelling it outward to escape the intense gravity of the black hole. The wormhole tunnel is the barrier between the forces of antigravity and the forces of gravity. The antigravity antimatter or *exotic* matter (Ebrahimi, 2014) is ejected in a thin shell micro non-traversable wormhole tunnel or throat, without violating standard energy conditions. This wormhole tunnel is not traversable for humans but perfectly suited for fluid micro white holes.

Let us take a moment to determine the escape speed from the black hole at the center of the galaxy by using the power of gravity alone. The standard escape speed equation for most objects from mass M with radius r is the square root of $(2 * G * M / r)$. Using this equation, the black hole escape speed should be equal to $2.99792E+8$ m/s or the speed of light. However, from the definition of the black hole, cosmologists tell us that this is not enough for even the photon particle traveling, as light cannot escape such a massive gravitational force. However, using the adjusted Schwarzschild radius of the black hole, we get the square root of $G \times M \times \lambda/2\pi = 8.03E+9$ m, where λ equals $6.5E-7$, and with this adjusted radius, the escape speed should be at least, square root of $2 \times G \times M / r = 3.94086E+8$ m/s. Supposing we divide λ by the constant e 2.518281828 to make it smaller, we get a greater adjusted escape speed of $5.06E+8$ meters per second. And if we square π in the Schwarzschild radius we get square root of $G \times M \times \lambda/2\pi^2 = 4.53E+9$ meters, and a much greater required escape speed of $5.2466E+8$ meters per second. In order to ensure a sub-atomic orb is able to escape the surface of the black hole it must therefore have an escape velocity, in theory, to be at least twice the speed of light, $5.99585E+8$ meters per second. Therefore, in order to ensure a sub-atomic orb is able to escape the surface of the black hole it must therefore employ the power of the other three natural forces, in antigravity quantum mechanical configuration, to eject it to the escape velocity, per Cosmological Balance Theory, of the speed of light squared, $8.98755E+16$ meters squared per second squared. The super grand unification of all four natural forces integrated into one generates this gravity-antigravity repulsive escape speed. This black hole escape velocity was defined in section 5.6 in chapter 5. Remember that gravity is the unification of the strong, the electromagnetic, and the weak forces working together at the quantum level. We use the term super grand unification to differentiate it from the String Theorist's definition of grand unification (strong plus electroweak force).

If you think, "the speed of light squared" velocity violates physical laws or breaks Einstein's law, which declares that nothing can exceed the speed of light, you are mistaken. To show that this is in error, we should recall the explanation of Hawking radiation at the edge of the black hole's event horizon. There Hawking explains that two gravitational objects, in this case, particles, at the event horizon are about to fall into the black hole. One is ejected outward via slingshot effect and the other falls inward toward the black hole. The same effect applies to binary stars or binary black holes (A and B) too close to the center massive black hole; one (A) is flung outward at great speeds and the other (B) plunges into the black hole. We also saw that the escape speed from just within the black hole event horizon at the center of the galaxy must be at least the speed of light. Therefore, as the small black hole (A) flung outward from the mutually orbiting binary black holes would have to been moving already near, at, or greater the speed of light for this to occur. Now, let us then suppose that the escaping black hole (A), say about one-fourth the mass of the center black hole, speeding away from the galaxy, just happens to be on its way directly toward Andromeda Galaxy. For arguments sake, Andromeda has a black hole at least equivalent to the mass of the black hole, and an acceleration attraction of $G*M/r^2 = 3.24E+6$ meters/second squared (m/s²) at its surface, and a mutual gravitational force of $G*M*(0.25*M) / r^2 = 2.5949E+27$ Newton. Let us add another factor into this equation, the speed the Milky Way is moving toward Andromeda, 402,000 kilometers per hour, which equates to $1.1167E+5$ meters per second (m/s), each galaxy of about equal size, traveling half that speed $5.5833E+04$ m/s. Therefore, this ejected black hole (A), 1/4 the size of the center black hole, travelling at $2.99848E+8$ m/s would continue to gradually accelerate toward the closest galaxy, and obviously exceed the speed of light multiple times before it gets to Andromeda.

Because of the great distance between galaxies, the acceleration rate on the small black hole (A) on its first 200 hundred years is around $1.1104E-9$ m/s², just enough to accelerate the rogue black hole (A) to $3.00E+8$ m/s, slightly more than the speed of light. If we fast-forward to 4560 years, the acceleration on the rogue black hole (A) would be $1.1107E-9$ m/s² and the black hole speed increases to $3.1E+8$ m/s. At 21,035 years, the acceleration on the rogue black hole (A) would be about $1.111E-9$ m/s² and the black hole speed increases to $4.0E+8$ m/s. At 33,382 years, the acceleration rate equals $1.1114E-9$ m/s² and the black hole (A) speed increases to $5.0E+8$ m/s. At 43,738 years, the acceleration rate equals $1.1117E-9$ m/s² and the rogue black hole achieves the speed of $6.0E+8$ m/s, over twice the speed of light. At 250,000 years, the acceleration rate equals $1.1215E-9$ m/s² and the rogue black hole (A) achieves $4.4E+9$ m/s. At 311,575 years, the acceleration rate becomes $1.1268E-9$ m/s2 and the rogue black hole (A) achieves $6.06E+9$ m/s and with every year thereafter increases acceleration rate as it moves to Andromeda Galaxy.

Now let us extrapolate this information onto this galaxy, the Milky Way itself. Cosmologists tell us that the galaxy is moving toward an Andromeda Galaxy at about $4.02E+5$ Kilometers per hour, which equates to about $1.1167E+5$ m/sec. Similarly, the acceleration rate between both galaxies is about $1.1104E-9$ m/sec². At that initial speed and initial acceleration rate, it will take about 43,700 years for the Milky Way Galaxy to achieve and exceed the speed of light, $2.9980E+8$ m/s, and acceleration rate of about $1.1106E-9$ m/s². At 70,000 years, the acceleration rate increases to $1.11109E-9$ m/s² and the galaxy's speed increases to $6.0772E+8$ m/s, twice the speed of light. If we fast-forward to 150,438 years, the Milky Way Galaxy will be feeling the acceleration effects of $1.1135E-9$ m/s² and achieve the speed of $1.9144E+9$ m/s. In addition, at 300,000 years, the acceleration rate increases to $1.1228E-9$ m/s² and the galaxy's speed hits $5.391E+9$ m/s. These speeds are all possible based on one simple fact, mass does not increase to infinity, as it approaches the speed of light, a grave error made by Einstein. Why is this possible? The simple answer is gravity does not fail; gravitational acceleration continues even way beyond the speed of light unless interrupted by another force. Observation of the universe shows us that visible galaxies collide at great speeds beyond the speed of light, and because of their extreme distance from us, seem to merge in a slow dance (Einstein calls this effect time dilation). We disagree with Einstein on time dilation. We attribute this effect to speed-to-distance ratio (not time dilation); just as an airplane at a great distance, flying across the horizon on earth, seems to be barely moving along its journey. In addition to Andromeda and Milky Way Galaxies moving toward each other, the entire local galaxy group is moving at about 6.43736E+8 m/s toward the Virgo local galaxy group, which apparently does not have sufficient gravity to cause this attraction. Scientists attribute this unknown attraction or motion to an energy they call "dark flow", to be discussed later. Another example of objects exceeding the speed of light is the simple yet elegant merger of two black holes spiraling downward into each other. Imagine two black holes that initially gravitationally capture each other starting with the typical speed around the galaxy they orbit. As the two objects approach each other they begin their spiraling pattern. With each new orbit around their barycenter, both black holes proportional increase in acceleration as the distance between them decreases; they fall faster and faster inward toward each other. Eventually, both black holes break the speed of light barrier multiple times long before colliding and merging into one.

Let us return to black hole excretion. What makes the black hole eject antigravity antimatter? In the case of antigravity particle ejected from the black hole, the attractive property of the mass of the antigravity particle is negative its graviton is negative, and the attractive property of the mass of the black hole is positive. In the case of gravity particle ejected from a massive white hole made of antigravity antimatter, the mass graviton of the ejected particle is positive, and the mass graviton of the white hole is negative. In both cases the gravitation or anti-gravitation force resulting is negative or repulsive. This ejection or excretion process is like results you get from a permanent bar magnet with south pole downward in a tube sitting directly over an active electromagnet with north pole upward. If the current is reversed, then electromagnet immediately repels the permanent magnet upward and away. In the case of antigravity sitting on a gravity object's surface, the escape speed is achieved instantly at the surface because that is where the greatest gravity antigravity repulsion exists. Escape speed depends on the mass of the larger object. From the Earth surface, it would be at a minimal of 11.2 km/s or 25,000 miles per sec. However, due to the significant mass differences between the Earth and the small object say 235 kilograms or less, its escape speed will be less than the speed of light squared beginning at the surface and continues at that speed throughout its trajectory away from the earth. From the black hole surface, that antigravity escape speed is the speed of light squared. Note that the speed of light squared is valid escape speed regardless of how massive the black hole, and is thereby defining a new law, the speed internal to wormhole tunnels, c^2. The ejected black hole or white hole orbs' estimated composition and size will be discussed later. If the Large Hadron Collider accidently created an antigravity particle of quark size or smaller not bounded by electromagnets, it would instantly achieve escape speed upward and away from the earth at speeds greater than 11.2 km/sec and maximum of the speed of light squared, tunneling its way through all matter, missing every normal particle due to the vast emptiness within atoms. This speed and avoidance of our instruments makes it exceedingly difficult to find and document such an event. Additionally, upon leaving the Earth, that antigravity particle would curve outward and continue directly away from the sun and the solar system into deep space beyond the edge of the galaxy. No gravity object in the galaxy will be able to slow it down, not even, super massive stars.

6.3 How Black Holes Excrete vs Evaporate

As we alluded to in the previous section, we believe there is more to black hole evaporation than that defined by Stephen Hawking radiation theory. In it, Stephen Hawking radiation is a black body radiation that is predicted to be released by black holes, due to quantum effects near the event horizon. This type of "evaporation" occurs at the event horizon and not from the black hole itself, thereby making it difficult but not impossible for the black hole to lose a tiny amount of mass[179]. Hawking radiation only limits what the black hole "eats," thereby losing some mass due to the laws of thermal dynamics but practically a negligible amount from the surface of the black hole.

I predict that black holes "excrete" in a different way than that described by Stephen Hawking. Black holes, especially the one at the center of each galaxy, are so massive and dense that they overwhelm the other three natural stronger nuclear forces to crush normal matter into its smaller components of individual protons, neutrons, and electrons, but without the empty space, ordinary atoms have between particles. Then at the outer core they breakdown into a "soup" of still smaller individual sub-atomic particles: quarks, leptons, and force carriers, etc. Over time, sheer pressure, heat, friction, and high energy causes waves of particles to gain energy to medium and then to high-energy particles until their imbedded graviton characteristic flip to neutral first, and then become anti-graviton, which of course gives the resulting particles "antigravity" characteristics. All massive black holes eventually create and eject these modified high-energy quarks, leptons, and force carriers with antigravity properties. Antigravity sub-atomic particles, held together by the grand unification force, must travel in curved paths around normal particles to either rise or *float* to the surface of the black hole or *sink* to the inner core, depending on where they were created in relation to the black hole's adjusted center of gravity.

The center of gravity within the black hole moves upward as more antigravity particles sink to and accumulates at the center of the black hole. The diagram depicts two paths of antiquarks with antigravity within the Black Hole.

Normal particles gravitationally push antigravity quark sub-particles upward and outward to the surface of the black hole, a process that could take a long time depending on the mass of the black hole. On their way to the surface, they will bump into other antigravity sub-particles, grow in density, and emerge as a "white hole" of a specific size with sufficient density and antigravity to hold itself together as they are ejected outward. In another perspective, it is like the bubbles consolidating at the bottom of glass of beer reaching a specific size and

---

[179] Hawking radiation emission is produced from particles at the event horizon where one particle is pulled inward while its companion particle is flung outward. Hawking tells us that this evaporation process takes a long time and continues after the black hole has eaten everything around it. How can this be so when there is nothing at the event horizon? A single lonely black hole with no stars or large amount of gaseous matter around it in intergalactic space would have little or no Hawking evaporation.

then begins to float toward the surface. Each bubble sometime bumps into other bubbles but do not grow and when they hit the surface of the beer, they create foam consisting of hundreds of micro bubbles not one large bubble. Likewise, once the micro white hole orbs begin to float or sink, they will not grow further in size.

The mass of the black hole at the center of the galaxy is estimated at 4.3 million times the solar mass of the sun. The movement of an antigravity sub-particle within that black hole could take anywhere from a hundred to a thousand years to reach the surface of the black hole far below the "event horizon" but actually takes a short amount of time for observers outside the "event horizon," possibly as short as a couple of decades to a few hundred years. The size of the black hole influences how long the antigravity sub-particles take to reach the surface. The time difference or time reversal is per Einstein's Theory of General Relativity not ours, which may or may not actually happen[180]. Based on the structures of stars and planets we have seen thus far, we predict that the composition of a black hole is as follows: On entering the event horizon, normal matter is "spaghettified" by the shear gravitational friction, stretched, pulled apart, and streamed onto the surface of the black hole. On impact these atoms are packed as tightly as possible with other atoms all at the verge of collapse; this is the outer crust of black hole. At each level down within the black hole, matter is further crushed, squeezed, and broken down into its sub-atomic components of protons, neutrons, and electrons, and then further down smashed and ripped into its smallest particles: quarks, leptons, and force carriers (outer core), and the inner core consists of antigravity sub-atomic particles which pushes outward with antigravity force. The inner core pushes upward against the outer core, which in turn pushes inward to counterbalance. The gravity at the outer core overwhelms and obliterates the" strong" nuclear forces internal to normal matter, and the antigravity in the inner core does the same. Super grand unification of all four natural forces enables gravity antigravity homogeneity within the black hole.

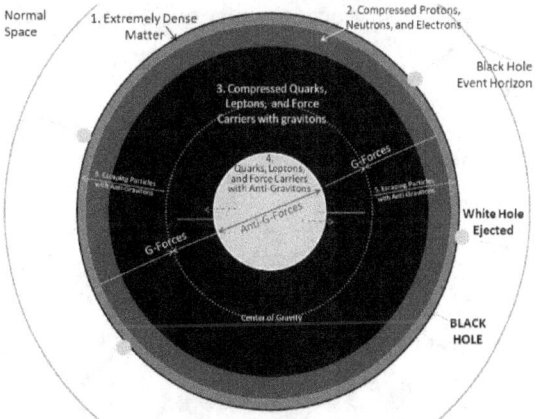

The outer core of normal sub-particles and inner core of antigravity "soup" of sub-particles are held in a delicate balance in this configuration, opposing each other. When the compressed outer core produced too much antigravity sub-particles that *sank* to the core, then antigravity inner core will enlarge and begin to produce more of the normal sub-atomic particles with gravity that of course will rise or float to the outer core. The sub-particles moving upward to the outer core expands it again and causes antigravity core to shrink to original configuration. This continuous flow of antigravity and gravity particles will eventually balance out. If the black hole "eats" nearby stars or matter, its outer-core and inner core will grow as well. If the black hole has not consumed matter for some time, then its inner core shrinks as well until it balances with the outer core.

---

[180] The Cosmological Balance Theory predicts the time functions normally within the black hole or white hole, the same as it does throughout the universe.

Unlike starlight leaving a star with self-sustaining electromagnetic waves, the black holes emissions are elusive and invisible to us. As antigravity particles grouped into micro "white hole" orbs break through the surface, they will be ejected with a massive gravitational antigravity repulsion, powered by the super grand unification of the four natural forces, at the speed of light squared through wormhole tunnels, and carry a very minuscule amount of the mass of the black hole with them into space. Their almost pure energy ejection will be quite similar to that of speed skier springing out of the starting gate and racing rapidly down the steepest "mountain" one can imagine, at each passing second the skier picks up more and more speed until they reach escape speed and enter into a wormhole. The orb immediately achieves escape speed as it bubbles up and leaves the surface of the black hole. Using Einstein's imaging, the "mountain" is the peak of the black hole from the "antigravity" perspective. In actual physical terms, these wormhole tunnels shoot directly straight out from the surface of the black hole in all directions, like starlight leaving the star's surface. Since the speed of light in a vacuum is about 299,792,458 meters per second (m/s), then the antigravity orb also gets ejected within a wormhole tunnel at the speed of light squared escapes the black hole, about $8.98755E+16$ m^2/s^2. The white hole orb's speed will instantly create wormhole tunnel that weave around stars and planets to distances of over 0.5 million light years beyond the galaxy, accelerated by the force GM_{BH}/r^2. The acceleration speed will then be augmented by the additional mass of each star the white hole orb passes on its journey outward, an incremental Schwarzschild mass increase. The wormhole path will follow the "valley" between "mountains" produced by star's gravity. Below is a depiction of the "mountain" seen by an antigravity white hole orb travelling outward in a wormhole tunnel.[181]

Hundreds of millions of these particles with antigravity properties will emerge from the black hole surface as compressed micro "white hole" orbs with at least eight or nine quarks, three leptons, and associated force carriers (all with antigravity properties) and tunnel away from the massive black hole in all directions inside the wormholes they created. Typically, they will follow straight lines outward unless they encounter gravity from normal matter or stars, at which point they will curve away and follow the "valley" between the "mountains" of the star's gravity. The angle of departure of these orbs from the black hole are not affected by the black hole's rotational spin even if it is very close to the speed of light; therefore, the angle should be exactly 90 degrees perpendicularly outward, the same angle light takes when leaving a rotating star. The stronger the gravity of the object in its path, the sharper the convex curve away the orb will take. After passing the object, the wormhole tunnel would gradually curve back to its original orientation directly away from the center of the galaxy, due to gravitational forces of everything it passed since leaving the black hole surface. As the antigravity particles curve away from an object, it will cause that object to deviate either speeding it up or slowing it down ever so slightly, while moving in the opposite direction - a result caused by the transfer of energy. Due to laws of physics, the wormhole path outward causes all the stars orbiting around the galaxy to experience a "galactic wind" which would be enough to maintain

---

[181] The path these wormhole tunnels make in space if detected appear like faint images of "strings" tossed in complete disorder.

unison orbits of all stars as though they were all connected. Granted some of the "galactic wind" is deflected above and below the galactic plane. Stars at the galactic edge would orbit around the galaxy faster than expected.

Antigravity wormhole tunnels extending outward and above the galactic plane will exert a "downward" and accelerating anti-gravitational force to keep stars within the galactic plane orbiting in unison and from travelling too far above the plane. Wormhole tunnels extending outward and below the galactic plane will also keep stars orbiting in unison and from travelling too far below the galactic plane with an upward force. This action is somewhat like the above the door fans at the entrance of a grocery store trying to keep the temperature-controlled air in the store from flowing out the front entrance of the store. Like the store fan, this galactic corral system is not perfect. Sometimes a star or two are ejected outward above or below the galactic plane, while another star plunges in the opposite direction. In any case, the wormhole tunnels travel outward in all directions much as sunlight travels away from the sun in all directions. See depiction:

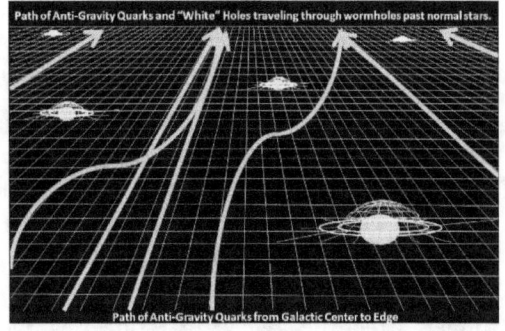

In addition to keeping stars orbiting in unison around the galaxy, the galactic wind produced by the wormhole tunnels also affects objects in deep space. It pushes objects toward other normal objects and allows gravity to draw the objects closer to each other. The galactic wind in this case behaves like "density waves." We just need to look for these clues. Here is another clue of the existence of wormholes containing antigravity "white holes orbs." We know that the pioneer spacecraft is slowing down in contrary to NASA's space propulsion and gravity calculations. Our scientists call that fact the "Pioneer Anomaly" as they are still trying to figure out what causes it. I predict that the pioneer spacecraft that is currently leaving the solar system is encountering the repulsive energy of nearby antigravity wormholes[182]. That interaction is what is causing that spacecraft to slow down. The further out our spacecraft goes into deep space the more galactic wind it will encounter, and the stronger these antigravity forces will push against the spacecraft.

As the antigravity micro "white hole" orbs reach the outer edge of the galaxy, it straightens its path directly outward and gradually begins to slow down the further it moves away from the galaxy into open space, losing energy, and possibly expanding and falling apart when they emerge from the wormhole tunnel. The wormhole tunnel that held the antigravity particles together gives way and collapses. This antigravity micro white hole orb ejection process is how black holes effectively "excrete" material. Each antigravity sub-particle group leaving the black hole takes with it a minuscule amount of mass of the black hole with it, a quantity much more than the Hawking radiation, and thereby evaporating the super massive black hole faster over time into a smaller one until balance is reached. In this theory, smaller less massive or micro black holes do not have the mass necessary to

---

[182] In addition to encountering the forces of antigravity, the pioneer spacecraft moves closer to the edge of the "halo" around the solar system, also known as the outer curvature of space surrounding entire system.

create antigravity quark particles and therefore do not excrete this way, but only through the Hawking evaporation method. Unlike Einstein's view of the connection between black hole and white holes with one massive wormhole tunnel, we will have one massive black hole with billions upon billions of micro-wormhole tunnels each carrying a micro white hole. The pattern of ejection from the massive black holes will be like that of water spewed out from a water sprinkler in unlimited number of streams as it turns, as long as the black hole remains massive enough to produce these micro white hole orbs. The streams or wormhole tunnels all go straight out but appear to be forming a spiral shape because the center black hole or hub of the sprinkler is spinning. When the streams appear to be forming a barred shape, then the center black hole is pulsating in two main streams opposite of each other. The pattern of the galaxy is dependent on the movement or rotation of the black hole at its center and the pattern of the wormhole tunnels flowing outward. The three main types of galaxies: elliptical (slow or no spin with ejection in all directions), barred spiral (slow spin with periodic pulsating ejections), and spiral (rapid spin with continuous steady ejection). If the galaxy is a barred spiral shaped, then it is in the middle of transition between its former spiral configurations to the barred shape[183]. This explains why the Milky Way is most likely a barred spiral galaxy. Black hole excretion and the wormhole tunnels are invisible to us and occur whether the super massive black hole is or is not *eating* nearby stars or other material. Because these ejected micro white holes have antigravity properties, neither our instruments nor our eyes will ever get the chance to see them. They are first traveling in a wormhole at the speed of light squared and second, they will curve away from any stars, the planet, and from any of our instruments. Therefore, undetectable and are essentially "invisible" to us. As they pass this star, it transfers energy to it and causes it to either speed up or slow down. The presence of these wormhole tunnels affect the spacecraft Pioneer as it leaves the solar system, and occasionally sends a new comet from the Oort cloud towards the sun.

Antigravity micro white hole orbs ejected from the black hole will bend space not time and look like the micro thin spikes coming out from a sea urchin. These spikes are atomic, sub-atomic, or micro *wormholes* leading away from the super massive black hole in curved paths around normal stars and other large matter directly to at least 1/2 million light years beyond the edge of the galaxy containing that super massive black hole, or a distance far enough away from the galaxy's gravitation influence. These billions and billions of spikes are threads or strings with intense concentrated mass all work together to gradually nudge and keep all the stars orbiting the galaxy to move generally in unison. String theorists believe these faint trails are what they have been looking for to move their agenda forward. Hawking says, "A black hole would settle down to a stationary state. It was generally supposed that this final stationary state would depend on the details of the body that had collapsed to form the black hole. The black hole might have any shape or size, and its shape might not even be fixed, but instead be pulsating (Hawking, The Theory of Everything: The Origin and Fate of the Universe, 2002, p. 59)." Roger Penrose and John Wheeler clarify Werner Israel's work on black hole shape, showing that a black hole should

---

[183] Galaxy patterns are dependent on two factors, the rotational speed, and the pattern of ejection from the center black hole.

behave like a ball of fluid, and therefore become spherical. Imaged is the result of various black hole spins and how its antigravity white hole wormhole tunnels affect the type of galaxy around it. The spikes depict the microscopic wormhole tunnels. A Barred Spiral Galaxy with spiral shaped edges is the result of the black hole in the center of a spiral galaxy changing into a pulsating black hole. The change takes time to propagate outward, leaving spiral arms last to change.

As these antigravity micro white hole orbs reach the edge of the galaxy, their combined antigravity force will repel the outer edge of the galaxy and cause those stars and their solar systems closest to the outer edge the galaxy to get an additional boost and pick up more speed. Thus, explaining why the entire spiral arm moves the way it does, faster than calculated by gravity alone, hence not lagging as predicted. Our cosmologists view the unexplained push or influence acting on the galaxy's outer stars as "Dark Energy." The Dark Energy they are searching for is the resulting energy transfer as the antigravity white hole orbs weaving their way through stars leave the galaxy and from the surrounding space vacuum anti-gravitational force. In each curved path away from a star or planet, the wormhole transfers energy to that star or planet. The micro wormhole spikes or *strings* moves the stars like broom bristles moving dirt away on the floor that is being swept. As the antigravity micro white hole orbs lose momentum, energy, and decelerate below the speed of light, they will emerge from their wormhole and expand to mostly neutral space-matter sub-atomic particles and a small fraction will retain antigravity properties. Wormhole exits are anticipated to be hundreds if not a thousand of light years distance away from the galaxy from which they came. On exiting the wormhole, the expansion and energy loss of the antigravity micro white hole orb will make neutral space-matter sub-atomic particles and forced carriers within it transform into atoms and neutral space-matter, as well as some atoms with antigravity properties. The wormhole collapse releases a short 160.2 GHz microwave burst, at the temperature of about 2.7 Kelvin, in all directions[184]. This microwave radiation pattern in space was originally thought to be the remnants of the "Big Bang." However, it occurs every day by collapsing wormhole tunnels and will not be dissipating as originally predicted by Hawking.

These wormholes will release particles that develop into neutral space-matter and antigravity quarks, leptons, and associated force carriers, to produce new neutral space-matter *hygratium*, like hydrogen atoms, and some antimatter atoms with antigravity properties. The neutral space-matter atoms and molecules will join the rest of their type in space. The antigravity hydrogen atoms and molecules will continue movement outward away from normal galaxies and accumulate together and move toward concentrated "antigravity" matter or galaxies with "antigravity" properties. The "light" emitted from the antigravity stars of these "antigravity" galaxies are invisible to us, only because they never reach us. The galaxies' gravitation waves, combined with the white wormhole tunnels leaving the galaxy, pushes or bends the "antigravity light" away, a lensing effect. From another perspective using Einstein's fabric of space and time, the paths these wormholes make through the galaxy appears to us as crests of anti-gravitational waves like those that surfers wait for on the open waters to ride onto the shore. Repeated antigravity waves can and does influence the orbital velocity of even the larger stars. Much like the constant

---

[184] An explanation of how these frequencies is produced is provided later in the book. Suffice is to say that the wormhole shape emulates what happens within a microwave oven.

repeated action of small waves hitting the east coast shore of the United States does to erode miles of it out to sea over decades and hundreds of years, leaving buildings that were once on the shores, stranded out in waist deep ocean water in South Carolina for example. Likewise, constant pounding of small waves against the side of a large ship can and does eventually move it. High concentrations of wormhole tunnels provide antigravity "density waves" which provide the necessary energy to compress the interstellar medium gases into dense spinning mass, which potentially ignite into new stars. Wormhole tunnels travelling above the highest stars above the galactic plane are responsible for keeping those stars from flying off into empty space. Similarly, wormhole tunnels below the lowest stars below the galactic plane push those stars back into the galactic plane. The interaction between gravity and antigravity keeps the galaxy together. This antigravity wormhole also delays the smaller globular galaxies around the Milky Way Galaxy from plunging right into the galactic plane. Pictured here is but a few examples of the antigravity waves that move toward and hit the stars. They are as numerous and as constant as sunlight hitting the earth. The combined effects of all these waves affect and moves even the largest of stars orbiting the galaxy.

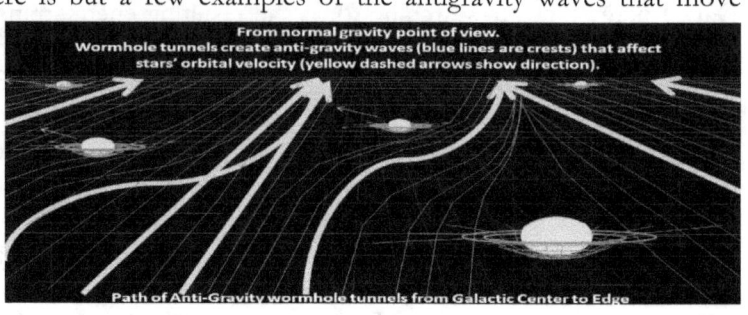

According to the Cosmological Balance Theory, the rapid growth of massive black holes or the merger of two massive black holes briefly releases a tremendous amount of antigravity antimatter particles into massive amounts of wormhole tunnels racing outward in all directions[185]. Such massive release (in one short intense black hole excretion) can be visualized as an "antigravity wave" emanating outward; this is a slight adjustment to Einstein's prediction of a "gravitational wave." Technically, gravitational waves move inward toward the massive object (not outward) to seem to pull other objects to itself. The Laser Interferometer Gravitational-Wave Observatory (LIGO) device used to detect this "ripple" is the same type of instrument that can also detect the sub-atomic interference of two merging "space-matter waves" moving through the solar system at the speed of light, where the space-matter wave emanating outward from the galactic center collides with an inbound space-matter wave. Such inbound space-matter wave could come from as close as one or two billion light year distance away such as the black hole merger event or rapid feeding of one massive black hole when two galaxies collide. Einstein saw this event as the "ripple" of space-time. I see it as a ripple in the curvature of space only, not time. They are identical phenomena. I see the same event as the overlapping of two space-matter particle "waves" traveling toward us at the speed of light, with some sections of increased intensity or peaks, and some of cancellation or collisions, causing the same type of momentary bending and warping of ordinary matter on earth as Einstein predicted. Since the space-matter vacuum force is the source of the force of gravity, then the space-matter wave or

---

[185] Merging black holes breaks the surface tension of both black holes and stirs up the contents held within them thereby release one quick and tremendous amounts of antigravity suspended within each one of them. Once the merger is complete, the surface tension returns to the new massive black hole and effectively slows down evaporation to normal pace.

ripple LIGO detected can also be perceived as a "gravity wave" in Einstein's mind. However, they are not the same. Gravity is a constant force analogous to the air pressure you feel day in and day out, while the space-matter "ripple" that LIGO detected is liken to a sound wave or more accurately a lightning bolt traveling through the air or electricity running through a wire, the conduit is space. Without space-matter, we would not exist. Space-matter provides the uniform constant force, which supply the push force combined with the quantum atomic attraction of all ordinary matter to each other, which provide the pull, give us the gravity force we all know[186].

Before we leave this subject on LIGO, let us review how the detection device works. LIGO is an L-shaped device built with two 1-meter diameter tubes of exactly four kilometers long, where lasers sent down both tubes simultaneously to mirrors returns and cancels each other out at the detector. Any deviation on the lengths of either tube by a tiny fraction of less than a millimeter will cause the detector to emit an alarm. This alarm signifies the detection of the "ripple in space." Scientific theoretical analysis has proven that waves are real in space and can ripple at the speed of light weakening as the move outward. However, they are space-matter waves, not gravitational waves, which travel via transmission of electrostatic charge from space-matter atom to atom, as electrons move through wire. Remember that in the definition of a Black Hole no particle, EM wave, or any other wave traveling at the speed of light can escape the gravitational pull within the event horizon of a black hole, let alone that of two massive black holes merging into one. Therefore, Einstein's predicted "gravitational wave" traveling at that speed cannot be emitted from the surface of the colliding black holes. However, antigravity particles traveling in wormhole tunnels can make that leap outward. Massive amounts of these antigravity wormhole tunnels affect ordinary matter like an "antigravity galactic or intergalactic wind" pushes a star to orbit faster or acts like an "antigravity wave" but drop below the speed of light and are released once they get far enough away from the black hole and the galaxy. On the collapsing of this intense wave of wormhole tunnels, we get a sudden burst of the creation of billions upon billions of newly formed space-matter atoms trying to occupy their place in the grid for themselves. This quick creation releases a jolt of electrostatic charge, which sends an outward wave traveling at the speed of light that eventually reaches us as a "ripple" in curvature of space. The merger of two black holes into one released extensive amount of anti-gravity and sent a shock wave into the space-matter immediately around it. In doing so, space-matter itself sent out vibrations in the form of an electrostatic wave traveling at the speed of light carried through grid of space-matter atoms at maximum speed electrons travel through a copper wire or speed light travels in space. The vibration passed from one atom to another via electrostatic repulsion, analogous to a sound wave moving through air. When this wave reached the galaxy, was but a "ripple", compared to what originated it. The ripple collided with space-matter around earth created peaks and troughs, which caused ordinary matter on the planet to warp slightly just as Einstein predicted. These overlapping opposing space-matter waves intersected at an angle, bent, and warped matter as it continued to distort space-matter. LIGO is a quantum microphone detecting the vibrations of space.

---

[186] Gravity is the result of two forces working together to achieve one result, the consolidation of ordinary matter with others of its kind, and that of antigravity antimatter with its type (antigravity attraction).

6.4 Invisible Side of the Universe

In the Yin Yang system, there exists an opposite part to the known visible universe, the invisible part. As the Black Hole at the center of the galaxy excretes, it releases antigravity micro white hole orbs at the speed of light squared, which generates wormholes. These antigravity orbs will race outward away from the galaxy at the speed of light squared and will not show minimal signs of slowing down until it has traveled a great distance away from the galaxy, a distance far enough that the galaxy's gravitation force is negligent or until it encounters and is attracted to another antigravity particle or objects. When significant amounts antigravity particles accumulate and come together by antigravity attraction, they will eventually form antigravity stellar cloud nurseries[187].

Given enough antigravity antimatter particles and compression hydrogen atoms will be produced, which will eventually accumulate under their own antigravity force to form antigravity stars, invisible to us as their "antigravity light" are made of photons possessing antigravity properties and characteristics. The reason antigravity galaxy's photon light is invisible to us, is that their beams never reach our eyes nor instruments, as we are made of matter that repels antigravity photon light particles. Gravity repels antigravity photons, causing it to weave through or curve away from normal visible galaxies. Eventually, these newly created antigravity stars will accumulate to form antigravity galaxies and white holes at their centers, which are also invisible to us, but not to any evolved intelligent life, if they existed on those habitable antigravity planet surfaces. To add clarity, the antimatter white hole at the center of an antigravity galaxy operates like any other black hole; it devours everything antigravity that falls within its event horizon. It is unlike the white hole Einstein and Hawking defined or imagined that spews out inordinate amounts of matter. Yet, Einstein's white hole is overlaid right on top of the antigravity white hole, where its excretion process spews out inordinate amounts of normal matter with gravity. Similarly, the mirror excretion process of a black hole in the visible galaxy spews out massive amounts of antigravity antimatter as a release valve, thus, appears as the sought after "white hole" or opposite to the "black hole" at the center of antigravity galaxies.

What this means is that every pocket of what we perceive as "void" empty space between galaxies, could potentially contain antigravity galaxies, invisible to us. These invisible galaxies therefore have anti-gravitational effects pushing on the visible galaxies. Antigravity galaxies will also mutually attract other antigravity galaxies to form a more massive antigravity galaxy. In doing so will push smaller normal galaxies out of the way. Under passing galaxies' antigravity forces, some small ordinary galaxies that are not able to get out of the way might collapse into a massive black hole, surrounded by a few surviving stars. Massive antigravity galaxy wedged in between two smaller normal matter galaxies will push the smaller normal galaxies apart, thereby, appearing to us as an expansion of that area of the universe. However, two massive normal visible galaxies like, the Milky Way and Andromeda Galaxies, that are relatively close enough to exert gravitational pull on each other, will push any smaller antigravity galaxy, located within what we perceive as *empty* space in between, out of the way as the two massive galaxies come together. Antigravity galaxies not able to move out of the way might collapse into one massive white hole, which

---

[187] We already learned that such gaseous consolidation process in intergalactic or open space could take billions of years.

may or may not be surrounded by a few surviving stars. The "void" pockets in space that cosmologists claim is there between visible galaxies may not actually be a void after all. We need to look for clues showing the galaxies being pushed apart to infer that an antigravity galaxy of significant size could be right in between the two visible galaxies that are moving apart. The speed the two visible galaxies are moving apart can be used to determine the size of the antigravity galaxy in the middle. Moreover, the direction of their movement can be used to determine the direction the antigravity galaxy is also moving. This result is similar to moving a magnet under the table with north side up, which is not in plain view, to cause the motion of two wide visible flat magnets sitting on top of the table, with their north side down.

From aliens' point of view within the antigravity galaxy, the normal galaxy is invisible to them, just as much as their antigravity galaxy is invisible to us. In this scenario, antigravity antimatter and normal matter will not be able to come together due to their mutual repulsive force. Lurking at the center of each massive antigravity galaxy is a massive "*white hole*," an area in space with enough mass that even antigravity light or antigravity photon beams cannot escape due to its massive anti-gravitational pull. These massive white holes are not directly connected via one wormhole to a massive black hole, but in the grand scheme and balance of the universe are somewhat connected as Hawking predicted with his concept of Hawking evaporation occurring in both. Both have Hawking Radiation, and both eject miniscule orbs through wormhole tunnels in their excretion process.

6.5 How White Holes Excrete vs Evaporate

Much like visible galaxies, we predict that super massive *white holes* exist at the center of each antigravity galaxy in the Yin part of the universe, the invisible part. These white holes are so massive and dense that they crush antigravity atomic particles into their smallest building blocks, quarks, leptons, and force carriers, all with antigravity properties. The force that gives these sub-atomic particles shape is stronger than the strong nuclear force and the quantum mechanics quality to function as one and in time, together with their combined energy enables them to remain intact and form a homogenous "soup" like structure within the white hole. In the final compression, some of these individual antigravity antimatter sub-atomic particles anti-gravitons are flipped to neutral first, and then take on gravitational properties. If any anti-matter particles (with antigravity) are created in the process, they will instantaneously be annihilated when they contact their counterparts in the antigravity white hole. The release of energy provides additional fuel for quarks, leptons, other sub-atomic particles, and associated force carriers to become medium and higher energy sub-atomic particles and finally transition to neutral and then gravity particles, which of course is not destroyed when they pass near antigravity particles. These tiny particles now with internal gravitons, therefore, repel and curve away from antigravity antimatter sub-particles within the white hole and rise to its surface or sink to the inner core. Particles with antigravity push gravity particles away. Particles with gravity gradually make their way by weaving in curves from within the inside of the White Hole toward the outer edge much like a photon in a star but without being absorbed as it makes its way from the inside of a star to its outer edge to escape as starlight. Much like starlight, these sub-particles of matter with gravity rise to the surface they will combine with other quarks, leptons, and force carriers all with gravity into a *micro black hole orb* of a specific size and break through the surface of the white hole at tremendous speeds, the speed of light squared escape speed propelled by gravity antigravity repulsion. From Einstein perspective, these black hole orbs will be "rolling down" or descending the side of a sharp-peaked "mountain" in the fabric of the space-time continuum. They will be ejected almost as pure energy at the speed of light squared, powered by the super grand unification of the four natural forces into one, and instantaneously create a wormhole tunnel through their galaxy and beyond the edge.

Before they begin their rise to the surface of the white hole, these sub-atomic particles will accumulate into compressed micro black hole orbs, groups of at least eight or nine quarks, three leptons, and associated force carriers all with gravitational properties. These micro orbs will remain that specific size regardless of how many times they bump into other black hole orbs due to the force of the grand unification bubble that surrounds them. This action is like bubbles of air molecules bumping into each other as they rise within a beer bottle without enlarging. Once breaking the surface, billions and billions of these micro black hole orbs will race away from the white hole in their own wormhole tunnels, sustained by the force of the grand unification shield. They will follow an outward curved path away from antigravity stars, antigravity planets, and any objects with anti-gravitation exertion. Each pass will transfer small amounts of energy and thereby affect the antigravity stars' orbit around the antigravity galaxy. Possible structure of a White Hole depicted below.

Due to laws of physics, the path outward causes each antigravity star orbiting around the antigravity galaxy to experience a "galactic wind" which causes the majority of antigravity stars to appear to orbit in unison as though all connected. The process of micro black hole orb ejection is how white holes excretes, each gravity matter sub-particle orb leaving the white hole takes with it a very minuscule mass of the white hole and thereby *evaporating* it over time. Excretion continues until the white hole reaches a size that it cannot produce any more gravity particles and black hole orbs. Again, the pattern of micro wormholes emitted from the white hole will be very similar to that of a water sprinkler ejecting billions and billions of jet streams of water as it turns and creating something in the shape of this barred spiral galaxy, but instead of fine jet streams, we have micro wormhole tunnels. White hole excretion remains steady if the white hole is not "eating" any nearby antigravity stars or other antigravity material. The more the white hole consumes nearby stars, the more micro black hole orbs it will produce and eject. Again, white holes excretions slow down and stop when the white hole does not have sufficient mass to flip the anti-graviton properties of the particles in it.

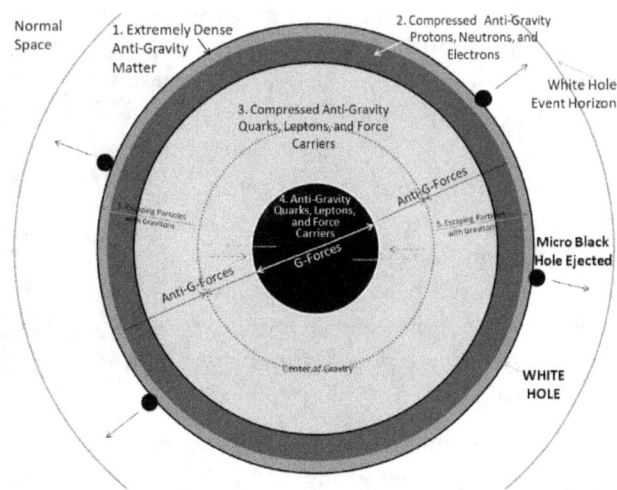

Gravity Wormhole tunnels extending outward and above the galactic plane will exert a "downward" and accelerating gravitational force to keep stars within the galactic plane orbiting in unison and from travelling too far above the plane. Gravity wormhole tunnels extending outward and below the galactic plane will also keep stars orbiting in unison and from travelling too far below the galactic plane with an upward force. This action is somewhat like the above the door fans at the entrance of grocery store trying to keep the temperature-controlled air in the store from flowing out the front entrance of the store. Like the store fan, this galactic corral system is not perfect. Sometimes an antigravity star or two are ejected outward above or below the galactic plane, while another antigravity star plunges in the opposite direction. In any case, the wormhole tunnels travel outward in all directions much as sunlight travels away from the sun in all directions, billions, and billions at a time. Depicted below is how white holes excrete.

Micro black hole orb ejected from the white hole will bend space not time and look like the spikes coming out of a sea urchin. They bend space in that these orbs punch right through space and slip right through in their tunnels in between the gaps of space-matter. Their streaks leave behind billions upon billions of faint images of strings in space, something like those jets leave behind as the move across the sky[188]. These spikes fit the

---

[188] String theorists might even say that these faint images of streaks in space are the evidence they believe they have found to support their theory.

exact definition of sub-atomic or microscopic wormholes leading directly away from the white hole directly to a point about ½ million light years beyond the edge of its originating galaxy. These wormhole spikes also act as threads or strings. Millions if not billions of wormhole spikes emanating from the galaxy center, work together to gradually nudge and keep all the stars orbiting in unison. Pictured here is the result of various white hole spins, and how each affects the type of antigravity galaxy formed:

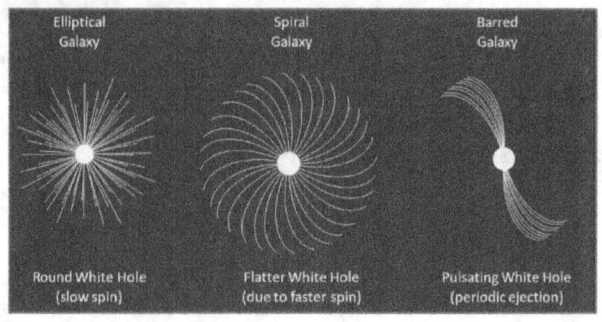

The accumulation of black hole orb with gravity ejected from antigravity galaxies appear in deep space to us as "dark matter" of which we can detect its mass. We in the visible galaxy see the bending of light around what we perceive as massive "Dark Matter." In fact, the light that is reportedly being bent around so-called empty space is actually being repelled around an antigravity galaxy and then curved by the gravity orbs leaving the antigravity galaxy; the "Dark Matter" source is the invisible antigravity galaxy in the middle of our view. It possesses a mass and halo of densely packed space-matter so great that it curves or bends light around it from extremely distant galaxies. We see the effect as lensing (looking through a lens). The "dark energy" which causes galaxies to drift away from each other is the result of massive invisible "antigravity galaxies" moving in between two normal smaller visible galaxies, and space-matter. Only super massive white holes provide the building blocks for the galaxies and likewise the super massive black holes provide the building blocks for the invisible "antigravity galaxies", never the two shall meet. This is a continuous process and an endless supply of particles and gaseous material for each other. One cannot exist without the other. Each complements the other, the Yin and the Yang.

Any imbalance between Yin and Yang parts of the universe will restore itself over time. Excessive number of invisible antigravity galaxies will cause them to anti-gravitationally attract each other and collide to form still larger galaxies. These super large antigravity galaxies, in turn produce greater numbers of micro black hole orbs with gravitation properties. When these micro black hole orbs drop out of their wormholes below the speed of light, the shock wave and sudden decompression will release neutral space-matter quarks, lepton, and force carriers, and some particles with graviton properties (normal gravity) sending out a burst of 160.2 GHz microwave signal of about 2.7 Kelvin degrees in all directions. The release sub-atomic particles eventually form protons, neutrons, and electrons, to make neutral space-matter atoms of primarily hydrogen, and some normal matter like hydrogen, helium, and Lithium. When significant numbers of normal hydrogen atoms and stellar clouds are developed and consolidated, they make new visible stars and galaxies. The same holds true if we had excessive numbers of visible galaxies, they would merge more often into larger galaxies capable of producing more building blocks for new Yin invisible antigravity galaxies[189].

---

[189] Symmetry governs the production of gravity and antigravity material.

6.6 Depiction of the Cosmological Balance

As we have said earlier, each part of the visible and invisible universe supplies the building blocks for the other part of the universe, and endless supply of Hydrogen and antigravity hydrogen[190], created from the death of massive stars and excretion of white and black holes. Only super massive black holes and super massive white holes have sufficient mass and can flip the graviton and anti-graviton characteristics internal to their particles. The collapse of the black hole wormhole tunnels can also create neutral space-matter mostly hydrogen and some normal hydrogen and helium and might create some other light elements like lithium likewise white hole wormhole tunnels collapsing can create the same neutral space-matter and some antigravity counterparts of these elements. Therefore, the interstellar medium is teeming with these elements, the local "fluff." The local bubble stretches 300 or more light years long. Some believe that it is the remnant of an ancient super nova; we think not. It is the result of space-matter and antigravity galaxies creating and pushing normal matter together in one long gaseous cloud.

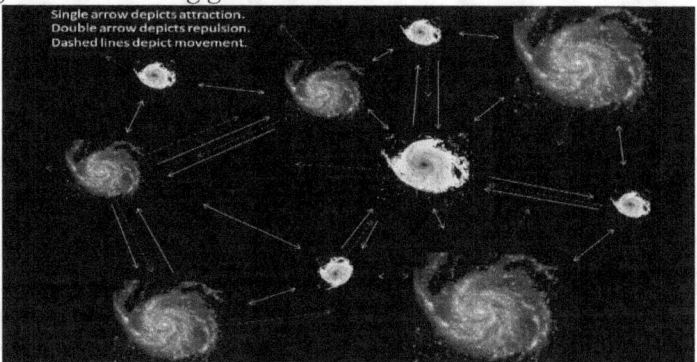

The "white galaxies" in picture represents the "antigravity galaxies" and are invisible to us when we look directly at where they should be from the normal visible galaxy. The area where such "antigravity galaxy" sits appears to us as empty space or voids with anomalies of "dark matter" and "dark energy." The same applies in reverse. This galaxy is invisible to intelligent beings, if they existed, in the antigravity galaxy or "White Galaxy" as seen in the picture above. Invisible larger antigravity galaxies moving toward each other repel normal smaller visible galaxies away. Smaller visible galaxies unable to get out of the way might be compressed into rouge black holes with possibly a few surviving stars orbiting them. Likewise, larger normal visible galaxies on a collision course push away smaller antigravity galaxies residing between the two of them. Again, the locations of these antigravity (white) galaxies appear to us as "voids" or empty space. It is, therefore predicted that a group of antigravity galaxies on one side of the local group of galaxies, containing the Milky Way and the Andromeda Galaxies, could conceivably be pushing the local group toward the Virgo local group of galaxies. This prediction explains why the Virgo group, which lacks the gravitational mass to attract the local group, appears to be pulling the local group toward itself at a speed of about 6.437E+8 m/s.

---

[190] Rogue black holes wandering in open space will easily pick up these gaseous hydrogen and helium within its gravitational reach and accumulate them, compress them into stars to add into a developing galaxy.

We also predict that molecules formed in the antigravity part of the universe will consist primarily of "left-handed" molecules, flipped when compared to their counterparts in the visible world, like mirror images of the molecules on earth. However, when compared to the right-handed molecules, they possess the same characteristics and chemical properties. In other words, the Vitamin C molecule in the world is the same the Vitamin C in all 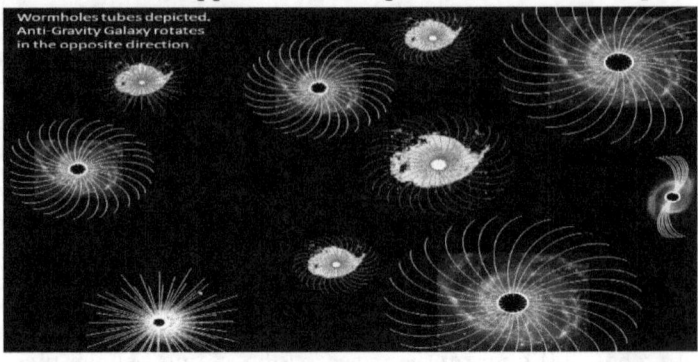 aspects except ours is a right-handed molecule and the one in the antigravity world is a left-handed molecule, except for the living organisms in both. Depicted are the galaxies and their respective general wormhole paths and patterns.

"The laws of physics do not distinguish between the past and the future. More precisely, the laws of physics are unchanged under the combination of operations know as C, P, and T. C means changing particles for anti-(gravity) particles. P means taking the mirror image so left and right are swapped for each other. In addition, T means reversing the direction of the motion of all particles – in effect running the motion backward. The laws of physics that govern the behavior of matter under normal situations are unchanged under the operations of C and P on their own. In other words, life would be just the same for the inhabitants of another planet who were the mirror images and who were made of anti-(gravity) matter (Hawking, The Theory of Everything: The Origin and Fate of the Universe, 2002, p. 129)." Antigravity antimatter fits the descriptions of C. Left-handed atoms and molecules fit the description of P. Moreover, reverse direction of orbits, and rotation of planets and stars fit the description of T. Therefore, life in the antigravity galaxies, solar systems, and their planets would be just the same as the inhabitants of earth. If you meet someone from the antigravity galaxy and he or she holds out a hand to shake yours, do not reach for it. If you do, both of you would immediately feel the strong repulsion of gravity and anti-gravitational forces, and your hands would fly apart, and bodies repel each other[191].

---

[191] Theoretically, entities of negative gravitations would never be able to come close enough to each other to "see" each other let alone close enough to touch.

6.7 Mathematics and Proof of Cosmological Balance

"Nothing happens until something moves."

— Albert Einstein

In part 1 chapter 2, Inexplicable Elements in the Universe, we briefly discussed the topics of Dark Matter and Dark Energy. Observation has shown us the existence of something, which mainstream cosmologists have named as "Dark Matter" and "Dark Energy." The observed "lensing" effect of distant galaxies by what appears to be "void" or empty space, tells us that there might be massive amounts of matter there that we cannot see, thereby naming it Dark matter. The observed expansion of the universe has also been attributed to "Dark Energy," a source that has eluded cosmologist for years. Both previously inexplicable characteristics have been explained with the existence of invisible antigravity galaxies in the Cosmological Balance, the invisible and the visible parts coexisting, the Yin and Yang.

Thomas in his second book (Thomas, Hidden in Plain Sight 2: The equation of the universe, 2013) goes on to argue that gravity as defined by Einstein needed modifications in order to fit his theory, so that gravity is both attractive and repulsive depending on how much of the mass of the object is within the Schwarzschild radius. For instance, Thomas states that since the majority of the earth's mass is outside that radius hence, its gravity is attractive; exception to this is its core within that Schwarzschild radius where the mass above the radius pulls matter outward to the radius edge like a repulsive force going from the center outward. This outward motion is not the result of time in reversal. Specifically, the center of gravity of the earth is its Schwarzschild radius. The same thing happens with the sun, the majority of its mass is outside its Schwarzschild radius hence its gravity, as seen from the outside, is attractive, with the exception of that portion of the mass within the Schwarzschild radius, which acts somewhat repulsive, according to Thomas. More specifically, the core of a star has slightly less pressure than originally estimated by gravity equations alone. Thomas then goes on to say that since a black hole has all its mass within its Schwarzschild radius, the gravity within that radius acts as a repulsion force on mass within the black hole, pushing much of the mass toward the event horizon, while its interior remains loosely bound and less dense than the edge or crust. The black hole crust surface is below its Schwarzschild radius, and grows outward as the black hole consumes more and more mass in conjunction with the radius expansion. With his adjusted gravity crossover scale and predicted modification to gravity, he defined gravity as attractive for most of the universe, but repulsive within black hole event horizons, and within the universe, thereby declaring that this is the reason the universe is expanding, from a singularity like a white hole exploding with a big bang. This of course is contradictory to his claim that there is no singularity within a black hole. Thomas discussion of repulsive gravity within Schwarzschild radius gives credibility for the existence of antigravity within a black hole[192].

We concur with Thomas' prediction of antigravity, or repulsive gravity as he calls it within the black hole, inadvertently confirming that high-energy pressure and of course,

---

[192] In support of Thomas' claim of antigravity is that of Einstein and of scientists studying the negative Higgs field, the source they say produced the "bang" for the big bang theory.

extremely hot temperatures exist within such a structure. Yes, repulsive force equates to particle pressure, and pressure yields heat energy and raises temperatures. We do not agree with Thomas' analysis that all the mass of the black hole is repelled toward the event horizon to form a thick super dense crust and less dense interior. We do however believe that the dense mass of the black hole being more evenly spread and a homogeneous "soup" of particles. From our perspective, the repulsive force predicted within the Schwarzschild radius also produces a homogeneous and isotropic distribution; this result applies to both the universe and within the black hole where attraction and repulsion compete. As for the universe itself, we also agree with Thomas' prediction for a stable, isotropic, and spatially flat universe all of which resides within the boundaries of itself, its own Schwarzschild radius, and that gravity equations needed some modification to show the effect of antigravity, gravity, and non-gravity. Cosmological Balance also agrees with Thomas that black holes are not infinitely dense singularities, but orbs with various layers of densities, and capable of producing antigravity particles. We will elaborate more on these points within this chapter.

Now let us investigate the mathematics to support the creation of and the ejection of antigravity particles from a super massive black hole in the Cosmological Balance Theory.

$$F = -G \frac{m_1 m_2}{r^2}$$

To make Newton's Gravitational Equation into an antigravity repulsion equation, either we make the smaller mass a negative number, or we change G to $-G$ (negative Gravity). In each case, the Gravitation Force F is a negative number.

Again, Newton's gravity equations are simple and reliable when it comes in very handy to defining the interactions of objects in relation to other objects. Think of gravity in terms of waves traveling toward and coming onto the "shore" of that object. Where the two waves meet is where the two objects feel the mechanical tug on each other. If their tug on each other is sufficiently strong to pull the two objects together then and only then will the two objects move toward each other, otherwise their trajectory or paths will only experience a slight deviation, or perturbation. While antigravity force in the realm would appear to us as a wave moving outward from the object, whether that is a wormhole tunnel or a white hole orb. The antigravity wave will then add to and enhance any normal gravity wave it encounters, intensifying its tugging effect on normal matter. We will further discuss this effect later in this chapter.

6.8 Newton Gravity Effects in Black Holes

As we have said in the previous sections of this chapter, we predict that at the depth within the black hole near the black hole's center of gravity that produces antigravity quarks, leptons, and force carriers, also produces anti-quarks, anti-leptons, and anti-force carriers. The difference is that when anti-quarks, anti-leptons, and anti-force carriers are created, they are immediately destroyed when they contact their counterparts. That release of energy surges intense pressure, friction, and heat, enabling the elevation of quarks, leptons, and force carriers to medium and higher levels, and the eventual creation of antigravity quarks, lepton, and force carriers, which of course stay intact, consolidate, and rise to the surface of the black hole, or sink to the center core. The creation and release of antigravity particles, leptons and force carriers allows the Black Hole to remain gravitationally bound and intact. The center of gravity of the black hole is not the center of the black hole. It is above the inner core. Gravity experienced at the center of the black hole is negligent as the entire black hole's mass pulls outward equally in all directions. Within the prediction of the Cosmological Balance, the black hole center, or inner core, is occupied with antigravity antimatter that sank to the center since they were created just below the black hole's center of gravity line, or the Schwarzschild radius.

How does this happen? Objects (to include photons and all electromagnetic radiation) passing the event horizon gets "spaghettified" and plunges to the surface of the black hole and immediately begins to get crushed and broken down. The further down it goes within the Black Hole the more gravitational forces it encounters. Any atom or particle approaching the black hole's center of gravity experiences gravitational forces capable of breaking down the strong force, the electromagnetic force, and the weak force of normal matter. The gravitational force is approximately 10^{-38} times the strength of the strong force (i.e. gravity is 38 orders of magnitude weaker), 10^{-36} times the strength of the electromagnetic force, and 10^{-29} times the strength of the weak force. We also know that astrophysicists have estimated the mass of the black hole at the center of the galaxy to be about (4.7×10^6) times the mass of the sun or the maximum estimated mass of the black hole at the center of the galaxy is approximately $((1.98855+0.00025)*10^{30}) \times (4.7 \times 10^6) =$ 9.34736e+36 kg = m_2. The Gravitational Constant G= 6.673 x 10-11 N (m/kg)². Set r = .000001 for the slightest movement of any 1 kg group of particles within the black hole. Since time, in opposition to Einstein, does exist within a black hole and less empty space resides between particles (unlike the vast empty space we see within atoms on earth), we can use Newton's equation to see what G-Force does to particles based on what we found within the Large Hadron Collider but at a googol times more frequently. The sheer gravitational force within the black hole causes all particles to move about like a homogenous soup[193]. Now using Newton's gravity equations $F = G((m_1 \times m_2)/(r^2))$, we get the results for any 1 kg object, m_1, within that black hole:

F_{BH} = (6.673 x 10-11 N (m/kg)²) x ((1kg x 9.34736e+36 kg) / (.000001m)²)

  = (0.00000000006673 N (m/kg)²) x (9.34736e+48 (kg/m)²)

  = 6.237493328e+38 N

---

[193] Extreme pressure, temperature, and energy beget movement even in the most densest of circumstances within the massive black hole or white hole.

$F_{BH} > 10^{38} > 10^{36} > 10^{29}$

In the example above, the Gravitational Forces, F_{BH}, within the Black Hole are greater than that of the Strong Nuclear Force, Electromagnetic Force and the Weak Nuclear Force within the atoms and matter inside the same black hole. The overwhelming Gravitational force makes all known sub-atomic particles to either collide with each other or flow freely around each other. In any case, energy is constantly being transferred from sub-atomic particle to sub-atomic particle. If you believe the equations above to be wrong, then let us look at another in terms of pressure and temperature within a black hole.

We know that the densest of stars, the neutron star consists of plasma soup of particles and atoms, and still manages to give off some light and tremendous amount of radiation, namely high-energy gamma rays. If these densest of stars were extremely hot because of their pressure, then why would anyone imagine that the black hole is solid and cold, especially when it "eats" hot plasma from nearby orbiting stars. Now, let us apply Newton's gravitational equations to see how pressures and temperatures build up within a black hole, namely the one at the center of the galaxy, as an example. The chart below depicts the two methods.

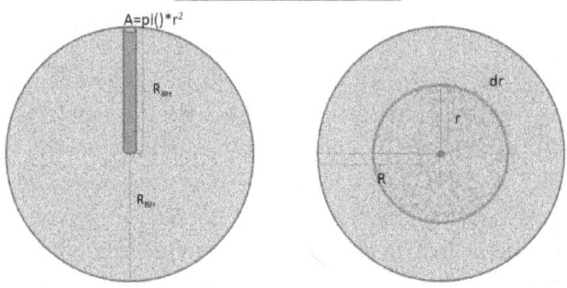

PRESSURES WITHIN A BLACK HOLE

In the left image of the black hole, we have drawn a cylinder with area $A = \pi r^2$ with length R_{BH} from surface of the black hole to its center. The volume of that cylinder is $V = \pi r^2 \times R_{BH}$. Assuming the black hole has a constant homogenous density ϱ and Newton's law of gravity $F = G \times m_q \times M_{BH} / R_{BH}^2$. Where gravity constant $G = 6.673 \times 10^{-11}$ N (m/kg)^2, the mass of the black hole $M_{BH} = 9.3474 \times 10^{36}$ kg, and the mass of a quark $m_q = 3.0$ to 5.5×10^{-30} kg (averaged at 4.25×10^{-30}). Then the density of black hole is $\varrho = 1.61 \times 10^{26}$ kg/m^3, the radius of the black hole $R_{BH} = 1.39 \times 10^{10}$ km, and set r=1.1 then the area $A = \pi r^2 = 3.801327111$ m.

Pressure p(r) at the center of the black hole can also be roughly estimated using:

$p(r) = \varrho \times G \times M_{BH} / R_{BH}$

$= 1.61 \times 10^{26}$ kg/m^3 $\times 6.673 \times 10^{-11}$ N (m/kg)^2 $\times 9.3474 \times 10^{36}$ kg / 1.39×10^{13} m

$= 7.2246E+42$ N/m^2

We can also look at the force of gravity shown in the right side of the figure above as a layer of the black hole with a thickness "dr" at radius r with mass m caused by the inner sphere of mass M_{BH}:

$m = \varrho \times V = \varrho \times 4\pi r^2 \times dr$

$M_{BH} = \varrho \times (4/3)\pi r^3$

We get:

$P(r) = (4/3) \pi \varrho^2 * G *$ integral from r to R_{BH} of r^2 dr $= G *(4/3) \pi \varrho^2 (R_{BH}^2 - r^2)$

$$P(r) = G *(4/3) \pi p^2 \int_r^{R_{BH}} r^2 \, dr = G *(4/3) \pi p^2 (R_{BH}^2 - r^2)$$

So the pressure at $p(0) = G * (4/3) \pi p^2 (R_{BH}^2 - 0^2)$,

And with $M_{BH} = \varrho \times V = \varrho (4/3) \pi R_{BH}^3$

Applying force $F = G \times M_{BH} \times m / R_{BH}^2$, we get the same pressure result as before:

$P = \varrho \times G \times M_{BH} / R_{BH} = 7.2246E+42 \text{ N/m}^2$

This pressure generates extreme temperatures within the black hole. Using Botzmann constant $k=1.4 \times 10^{-23}$ J/K, the temperature within the black hole can be approximated with the formula

$T = G * m_q * M_{BH} / (k * R_{BH})$

And m_q is the mass of an average quark (4.25×10^{-30} kg).

$= 6.673 \times 10^{-11}$ N (m/kg)^2 × (4.25×10^{-30} kg) × (9.3474×10^{36} kg) / (1.4×10^{-23} J/K × 1.39×10^{10} km)

$= 13,833,023,682.89$ K

$= 1.38 \times 10^{10}$ Kelvin degrees

Clearly, these extreme pressures and exuberantly high temperatures do not conform to the hard-cold desolate objects that some mainstream cosmologists believe of black holes. How can something that consumes nearby matter, stars, and other black holes with such ferociousness be anything but hot or plasma like? Black holes are most likely a "soup" of particles constantly colliding with each other, emitting, and absorbing tremendous energy.

Gravitational Forces within the Black Hole (F_{BH}) are sufficient to cause particles to either be transformed or be shattered into smaller pieces, just as particles do on impact within the Hadron Collider experiments here on earth. When anti-matter particles like anti-quarks, anti-leptons, and anti-force carriers are created, they will immediately be destroyed on encountering their counterparts, and release tremendous energy (but insufficient to rip the black hole apart as we have yet to see a black hole explode). We predict that the excess energy released affects gravitons within normal quarks, leptons, and force carriers, makes them neutral, and then gives them antigravity properties (which are not destroyed). We also predict that the release of these antigravity particles from the surface of the black hole, ejected via wormholes from the center of the galaxy and beyond the outer edges of the galaxy) explains the unison movement of the stars orbiting all galaxies. These wormhole tubes curve around stars (much like a ball travels the valleys (open space) between the hills (the stars)) are the "dark energy" cosmologists are looking for, which of course are invisible to us due to their extreme speeds, at the speed of light squared.

This concept also defines another "excretion" process of black holes different from that given by Stephen Hawking. Note: The center of gravity of the black hole is not the center of the black hole. Gravity experienced at the center of the black hole is negligent as the entire black hole's mass pulls outward on it equally in all directions, cancelling all forces. We predict that the core of the BH consists of antigravity particles pushing outward and in balance with the inner core of the black hole. The Black Hole core consisting of antigravity particles could likewise create normal gravity particles (quarks, leptons, and force carriers) which "float" upward to the inner core of the black hole, thereby balancing "antigravity" and normal gravity particles[194]. We have already drawn up a power point slide showing the predicted structure within a black hole.

We predict that these white hole orbs and their antigravity particles emerging from the wormholes at extremely great distances away from the galaxy have already gravitationally or should we say anti-gravitationally attracted each other to form "antigravity" galaxies, which push normal galaxies apart, giving us on earth measurements of an expanding universe. Normal galaxies like Andromeda and the Milky Way galaxy are gravitationally attracted and moving closer together. This movement forces any smaller "antigravity" galaxy in between to move out of the way. Antigravity galaxies not able to get out of the way might collapse into one massive white hole, which may or may not be surrounded by a few surviving stars. Similarly, two moving antigravity galaxies could crush and collapse ordinary matter into massive black holes, which may or may not be surrounded by a few surviving or new stars. We call this the "Cosmological Balance" universe, the visible and the invisible parts of the universe. This concept explains "Dark Matter." At the present time, there are slightly more "antigravity" galaxies in the universe. They reside within the "voids" between normal galaxies in the visible universe. Given time, these "antigravity" galaxies will produce enough "normal" matter for the visible universe to grow, obeying "Yin Yang" principles, to balance.

By the way, normal light gets repelled away from "antigravity" galaxies, long before they enter that galaxy ("mountain" in Einstein's space time perception but upside-down). This gives us the "lensing" effect of distance galaxies, see image above. The presence of the antigravity galaxy causes normal light to curve away, while the galaxy halo, consisting of ordinary matter, refracts the light around the galaxy; the total effect produced is what we see as lensing[195]. The same applies to "anti-light" or light with "antigravity" characteristics emitted from antigravity stars and antigravity galaxies, this anti-light does not reach earth because the galaxy's gravitational force or waves pushes or curves it away from us, and our instruments never get the opportunity to capture or "see" it, and therefore is "invisible" to us. Part 3 will discuss another hidden aspect. We see that part of space as "voids."

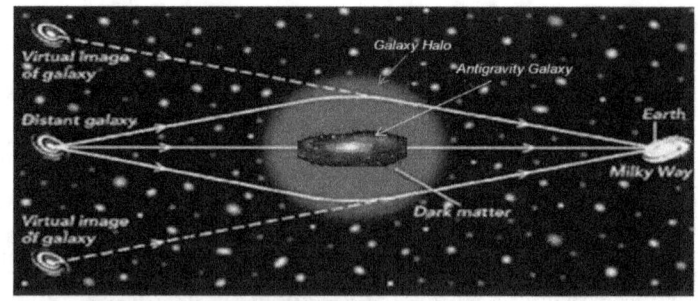

Antigravity galaxy bends light away, while the galaxy halo, consisting of ordinary matter, refracts light around galaxy.

---

[194] The grand unification of all four of the natural forces into one enables the orb to survive, float to the surface, and eject through and beyond the galaxy.

[195] The "halo" surrounding ordinary and antigravity galaxies are densely packed space-matter wrapped around the galaxy in order to "contain" it in space. Einstein called this formation the curvature of space. Space-matter has the unique ability to bend light within that dense halo due to its optical density; this effect is equivalent to "gravity lensing."

We almost forgot to mention that the antigravity particles travelling within the wormhole experience no loss in time as they move through the tunnel. According to Einstein's Theory of General Relativity, particles travelling at the speed of light see no passage in time. We do not necessarily agree with Einstein's perception of time when it comes to speed. Why is this so? If the photons of light travelling at light speed experiences no passage of time, then how can the photon waves move up and down and their companion electron move side to side or vice versa in an electromagnetic pulse that perpetually self-propels the photon in light through empty space. No movement due to stand still or lack of time means we have no EM pulse or wave to propagate the beam of light, truly an impossible situation or as some say, "catch 22." In any case, if Einstein is correct, then when the particles emerge out of the wormhole, they will come out as they were at the time of their creation. Either way, this gives them the necessary time to accumulate under mutual antigravity to form gaseous stellar clouds, antigravity stars, and even antigravity galaxies. The same thing happens to black hole orbs ejected from massive white holes. They travel through similar wormhole tunnels and come out as they were at the time of their creation. Whether this is true or not, this time paradox tells us that there are endless supply of normal particles waiting to become hydrogen and particles containing antigravity to produce antigravity hydrogen to replenish the visible and the invisible parts of the universe, respectively.

Above is the result of gravitational (or should we say antigravity galaxy) lensing, we observe through the Hubble telescope images. Antigravity galaxy and its halo, curves light away and then refracts light around a galaxy.

6.9 Antigravity Interaction Theory

Now let us talk about the predicted rate of production of antigravity particles and rate of ejection of these particles from the black hole surface into wormholes going out in all directions. We predict that for around every hundred anti-matter particles created and annihilated, one group of antigravity particles is created. The creation process occurs through four steps. In the first step, about twenty anti-matter matter energy releases move quarks, leptons, and other particles with their force carriers to medium energy levels. In the second step, about forty more anti-matter matter energy releases takes the medium energy particles to their highest energy level. In addition, in the third step, the next forty anti-matter matter energy releases flips the internal graviton to neutral in a group of quarks, leptons, other particles, and their force carriers, and in the fourth step, the orb absorbs the energy to become particles and force carriers with antigravity characteristics. Creating this extremely high-energy level orb enables the super grand unification of all four natural forces into one and allows this group of subatomic particles to slip right through space and its space-matter because it possesses both the neutral space-matter building blocks and the antigravity antimatter within it. Estimated total rate of antigravity creation is approximately 1.41E+52 particles per second with the associated Higgs Boson particles assuming negative mass, a negative graviton. The Higgs particle is a neutral graviton[196]. Approximately 2/3 of these particles with antigravity characteristics will "float" to surface or about 1/3 will sink to black hole inner core. As particles rise to the surface, they will merge with other antigravity particles and grow to a specific size of approximately eight to nine quarks, leptons, and force carriers and remain that size throughout their journey until they exit the wormhole tunnel. Again, once these antigravity particles reach that specific size the grand unification of the forces unites them into one shielded negative graviton orb capable of rising upward passing all other gravity particles and escaping outward away from the black hole surface.

We know that the sun is estimated to emit 1×10^{45} photons per second and the black hole in the center of the galaxy is 4.7×10^6 times the solar mass. Using the formula $R=(2GM)/(c^2)$, we predict that the black hole will emit white hole orbs consisting of between two billion to four billion sub-atomic particles, at a rate of 1.175E+42 to 2.35E+42 orbs per second in all directions. The mass of each set of orbs estimated between 7.37E-19 kg to 1.47E-18 kg, and a radius between 1.09449E-45 mm and 4.36881E-42 mm. The white hole orbs ejected from the black hole surface will instantaneously reach escape speed of the speed of light squared and enter into wormhole tunnels primarily traveling outward in lines, unless they encounter stars, causing their path to curve away. The angle of departure of these orbs from the black hole will be minimally affected toward rotation of the black hole's rotational spin if it was rotating near or at the speed of light; otherwise, its angle is exactly 90 degrees perpendicularly to the surface of the black hole and directly outward. The stronger the gravity of the object in its path, the sharper the convex curve away the orb will take. Stars orbiting in close proximity to the black hole at the center of the galaxy acts like shutters on a ship's Morse code light device, letting groups of white hole orbs pass through in segments of time toward the sun and nearby stars, like a lighthouse beacon matching the orbiting star's cycle.

---

[196] The Cosmological Balance Theory solely predicts this idea.

Each massive group of wormholes carries with it enough antigravity "galactic wind" to nudge the sun due to its size and mass, while minimally affect the planets and the Oort cloud[197]. Newton's Gravitational law tells us that normal gravity waves travel toward the sun, with increasing strength as they approach the surface of the sun. When significant groups of antigravity wormholes bend away from the sun, they send their antigravity waves toward the sun, which combine with and strengthen the intensity of the normal gravity waves going to the sun. This action distorts the gravity force of the sun and nudges it in the opposite direction. In the paragraph above, the black hole emits 1.175E+42 to 2.35E+42 orbs per second in all directions, implying that the total mass ejected from the black hole into antigravity white hole wormholes per second is 1.175E+42 x 1.47104E-18 kg = 1.73E+24 kg to a maximum of 2.35E+42 x 1.47E-18 kg = 3.45693E+24 kg. Solving for 1-degree cube in space over the entire sphere, we get 1.73E+24 / 360^2 = 1.33E+19 kg per second to about 3.45693E+24 / 360^2 = 2.66739E+19 kg per second. Below is an illustration that shows impact of wormhole tunnels passing by the sun during window of "galactic wind:"

Image on the left is the Cosmological Balance equivalent of increasing strength of gravity waves proportional to the distance from the surface of the massive object also means an increasing density of space-matter layers that overlap the outer edges of the massive object, and is what Einstein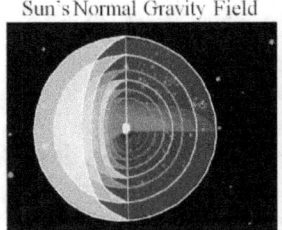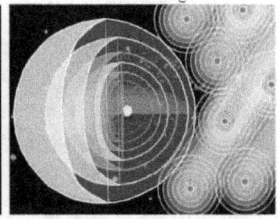

Sun's Normal Gravity Field | Wormholes Nudge Sun
Planets Orbit Consistent | Perturbs Mercury's Orbit?

calls the curvature of space. The more massive the object is in space, the greater the density of the layers of space-matter surrounding it, the greater the gravitation force acts on it and the greater the curvature of space blankets it. Close contact between space-matter and normal matter is the transfer force of gravity. The sun is about 26,000 light years from the center of the galaxy, and travels about 782,000 km/hour in generally a circular orbit with an upward and downward motion above and below the galactic plane. We are lucky that the solar system is not in the center of the galaxy and to have a helio-sheath or a shield around the solar system to deflect some of the cosmic radiation heading our way. This turbulent magnetic barrier is created at the edge of solar wind and appears more like a "froth" of bubbles, likened to the foam on the surface of beer or that of shaving cream. In intergalactic space, John Wheeler calls this effect around galaxies the "space-time foam", solar "halo" or curvature of space around the outer edge of the sun's gravitational reach. When in the window of channeled "galactic wind," the total antigravity wormhole tunnels over an hour period is sufficient to nudge the sun and increase its speed ever so slightly, say by a 1 to 2 m/sec^2 depending on the number that came through that window. When the sun is not in the window of the "galactic wind," the other stars within the local group gravitationally pull the sun and slows it down just a tad, let us say by about 0.5 to 1 m/sec^2 or speeds it up by that amount. These slight speed increases maintain the unison motion of all the stars orbiting the galaxy, gravity alone is not sufficient.

---

[197] Galactic wind thought to be "density waves" provide movement and sometimes compression of normal ordinary matter.

6.10 Mercury's Orbital Anomaly Solved

Physicists have been looking for the missing piece that allows Newton's gravitational law to work fine on Mercury's orbit. They looked intently for another object between the Sun and Mercury able to cause the anomaly but could not find one. All they had to do was apply the earth-moon precession relationship to that of the Sun and Mercury[198]. Let us take a closer look at mathematically solving Mercury's elliptical orbit discrepancy from another angle, from the perspective of tidal influences from the planet Mercury causing tidal waves just below the surface of the Sun and vice versa the sun causing Mercury's hot fluid surface to bulge as the planet passes during perihelion. We know that the moon causes tides to rise and fall on Earth. We also know that the moon slows down the rotational spin of Earth at a rate of a couple of milliseconds every hundred years. Similarly, the moon's orbit is increasing in speed and therefore drifting away at a rate of 3.78 centimeters per year. Mercury's close orbit around the sun has the same tidal effect on the plasma fluid within the sun, which cause tides to rise and fall, solar mass ejections waves, and we suspect impacts the eleven-year annual sunspot cycle movement. On Mercury's approach when heading toward its perihelion, its gravitational pull causes the sun to react with a small tidal wave that begins to build within the sun's plasma outer core and travel along with Mercury's approach. Since there is no "solid land mass" on the sun, with each passing day, the solar tidal wave grows more and more in height and increases in speed until Mercury just reaches its perihelion. At Mercury's perihelion, the now tsunami tidal wave on the sun is at its greatest peak, just underneath its "gaseous atmosphere" or photosphere, surpasses Mercury's orbital speed, gravitationally attracts the planet, and accelerates it forward. This action holds the planet's perihelion at that distance for a bit longer while tugging it forward, resulting in the formerly unaccounted orbital precession astronomers observed. In review, the moon causes the earth's tide, which in turn causes the moon's own elliptical orbital precession and its drifting away. Hence, the invisible solar tide is reason for Mercury's own elliptical orbital precession.

At Mercury's perihelion, the sun experiences "highest solar tidal wave" or largest bulge. The solar tsunami wave in turn increases the speed of Mercury and the accelerated boost advances the perihelion ever so slightly, totaling to 5600 arc seconds per century. The perturbations by Venus and Earth also contributed to Mercury's orbit precession, shown in the equations below. Per Newton's equations, acceleration by gravity is represented by force = Gravity constant G times mass of the sun M_s times mass of the planet M_p divided by distance R between the centers squared. $F = G \times M_s \times M_p / R^2$. Tidal equations are represented in a similar way. Tide differential = $2 \times G \times M_s \times \Delta r / R^3 + 2 \times G \times M_p \times \Delta r / R^3 + 3 \times G \times M_s \times \Delta r^2 / R^4 + \ldots$ where Δr is influencing constant in addition to the tidal gravitational effect. The gravitational constant G is 6.6734E-11 N $(m/kg)^2$, the mass of the Sun is 1.9291E+30 kg with a radius of 695800 km, and the mass of Mercury is .055 that of Earth or somewhere between 3.2847E+23 to 3.3022E+23 kg with a radius of 2440 km. Mercury's perihelion is at 46,001,200 km and its Aphelion is at 69,816,900 km, with an average distance of 57,909,050 km and average orbital speed of 47,362 m/sec. At its perihelion, sunlight takes 153.443487 seconds to reach the surface of Mercury. Gravity acceleration at the surface of Mercury is $F_g = G \times M_s \times M_p / R^2$ which comes out to about

---

[198] The key to solving Mercury's orbital anomaly is found within the relationship between the earth and its moon.

CHAPTER 6: THEORY AND PROOF OF BALANCE

3.4608 m/sec², and the resulting bulge on Mercury's liquefied surface is approximately 40.7426686 km. The gravity acceleration at the surface of Mercury number represents all perturbations from outside objects to include the sun and influences from the other planets orbiting around the sun.

F_g = 6.6734E-11 N (m/kg)² * 1.9291E+30 kg * 3.29718E+23 kg / (2440 km + 40.74km*2)²

= 3.4608 m/sec²

Bulge on Mercury = 3.4608 m/sec² * (153.443487 sec)² /2 / 1000

= 40.74267 km

Arcsec due to acceleration at Mercury's surface = degrees(arctan(bulge/perihelion)) * 3600

= (Degrees(ArcTan (40.74267 m * 2)/ 46,001,200 km)) * 3600

= 0.365372149 arcsec

Delta_r constant is calculated by getting the product of Mercury's mass and average distance divided by Sun's mass, and then add to it the acceleration on Mercury's surface.

Δr = 3.29718E+23 kg *57,909,050 km / 1.9291E+30 kg + 3.460848202 m/sec²

= 9.964519192 m/sec²

Tidal bulge on the sun due to Mercury at perihelion is sun bulge:

Sun Bulge = 153.443 arc seconds * (269.833 m/s)² /2/1000

= 5586.095 km

Acceleration on sun's surface due to Mercury's proximity is:

Gravity Constant * Mass of sun / ((Radius of sun + tidal bulge)*1000)²

= 6.6734E-11*1.9891E+30/((695800km+5586.1km)*1000)²

= 269.8329 m/sec²

At perihelion planet Mercury's tidal raising force on sun is:

Df/Dr = (2 * G * M_s * Δr / R³) + (2 * G * M_p * Δr / R³) + (3 * G * M_s * Δr^2 / R⁴) +(3 * G * M_p * Δr^2 / R⁴) + (4 * G * M_s * Δr^3 / R⁵) + (4 * G * M_p * Δr^3 / R⁵) ...

= 0.027156239

where R = 46001200 km + 2x40.74266859 km + 2x5586.09525 km =46012453.68 km

Therefore, the total Arcseconds caused by gravitational effect of tidal bulge accelerating past planet Mercury:

= Degrees(ArcTan(0.027156239)) * 3600

= 5600 arcsec per century

The equations above work perfectly when Mercury's mass is exactly 3.29718E+23 kg, meaning that it is .055208952 times that of Earth. If it is 3.2847E+23 kg, the arc seconds obtained would be just short of 5600 arcsec. If it is 3.3022E+23 kg, then we would get a number greater than 5600 arc seconds. Not only does the above equations solve the 5600 arc seconds astrophysicists know is Mercury's orbital precession, it defines Mercury's mass to be exactly =3.2971812495E+23 kg. The above solution is not at all dependent on the supposed "curvature" of space-time as proposed by Einstein. It simply shows that the tidal wave on the sun's liquid surface has a direct impact on the planet Mercury's acceleration particularly at its perihelion. The tidal action of the sun is the missing piece Newtonian experts were looking for. Einstein called this effect "frame dragging," but rather than attribute it to the curvature of space-time, we resolved that it simply and solely results from the solar tidal wave gravitational force acting on Mercury's slight bulge that nudges Mercury the additional 43 arc seconds per century. We also predict the same tidal action occurs with orbiting binary pulsars.

Einstein was a great physicist with a vivid imagination. He convinced the world with his successful fine-tuning of Newton's gravity equation with the invention of the curvature of space and time and the usage of fictitious numbers, but Newton's gravity equations and Principia did not need any adjustments, as Newton's method works just fine solving Mercury's orbital anomaly as demonstrated above. Sir Isaac Newton triumphs again! By the way, to the best of my knowledge there has not been recent confirmation of the measurements taken to show Einstein gravity bending of light around the sun (i.e. his prediction of the lensing effect) since 1919. Who knows what we may find today? We should therefore highly recommend that today's astronomers to take another picture or two and recheck the data. We might find that gravity coupled with light refraction from solar atmospheric or curvature of space causes the bending of light around the sun. That additional layer of clear colorless inert solar atmosphere or dense space (spherical dense layer of space-matter atoms) above the corona could also explain why the corona is extremely hot. Incidentally densely packed space-matter atoms surrounding the massive object in space is synonymous to curvature of space, but not time[199].

In addition to Einstein's claim of resolving Mercury's anomaly with his imaginary mental experiments, he also claimed to predict, through the usage of his relativity equations, the bending of light or the lensing effect of the sun mass as the result of the curvature of space and time together. In the next section, we will learn that that claim is also not quite accurate, even questionable. Although his equations provided acceptable predictions, it turns out that they were considered unnecessary and normally overlooked in most situations. The bending of light around the sun is not exactly the result of the curvature of space and time, but a more common effect seen on earth and the application of the natural physical law demonstrated in part 1 chapter 1, light refraction. Dense layers of space-matter surrounding the massive object in space refract and bend the light passing through it like a solid lens (without changing its color and frequency); Einstein coined this effect as the curvature of space and gravity lensing, attributed by optical density possessed of concentrated space-matter.

---

[199] In solving Mercury's orbital anomaly, we have not abandoned Einstein's work on showing how starlight bends around the sun, which of course is the same as densely packed space-matter's unique light refraction capability.

6.11 Representation of Gravity Waves

We have seen gravity waves represented in various forms from lengthening arrows to moving waves. We believe that concentric inward moving spherical waves are a better representative of gravity and outward moving concentric antigravity waves, analogous to nested Euclidean spheres like Russian dolls each with a different r radius value[200]. From our perspective, these spherical waves move toward, in the case of gravity, or away, in the case of antigravity, from objects. In the visible universe, gravity waves flow inward toward objects, the same direction space-matter density layers increase, and due to repulsion of ordinary matter antigravity waves flow outward from those objects. Therefore, antigravity waves encountering gravity waves transfers energy, intensifies, and speeds up the gravity wave moving inward, causing object to move in the opposite direction. In the invisible, Yin side of the universe gravity waves flow outward and antigravity waves flow inward. There, the gravity waves intensify antigravity waves and cause the object to move in the opposite direction.

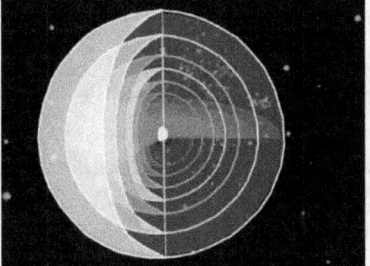
Sun's Normal Gravity Field
Planets Orbit Consistent

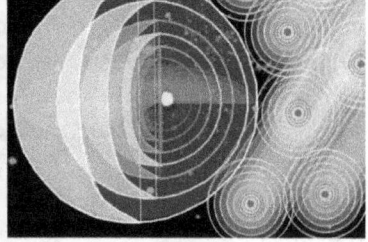
Wormholes Nudge Sun
Perturbs Mercury's Orbit?

This illustration represents gravity in a three-dimensional image as gravity affects object from any direction of the sphere 360 by 360 rather than the simplified Einstein's fabric of space and time, a flat three dimensional sheet picture implying the event horizon around a black hole is a disc and not a sphere. This could be a fatal error, which will cause one approaching too close to the event horizon to meet their end. Einstein also intended to show curvature of space-time as he calls it in layers as depicted above, but academia choose to simplify his pictures so the public can understand them. The Cosmological Balance Theory equivalent of increasing strength of gravity waves proportional to distance from the surface of the massive object also means an increasing density of space-matter layers that overlap the outer edges or upper thin atmosphere of the massive object, which Einstein calls the curvature of space-time. We, however, call it the curvature of space, not time. The more massive the object is in space, the greater the density of the layers of space-matter surrounding it, and the greater its gravitation force, and the greater the curvature of space. The contact between space-matter and normal matter is the transfer force of gravity. Part 3 defines the mechanism.

For those who prefer Einstein's General Relativity Theory flat sheet representation, they believe that Einstein's General Relativity graphical display of gravity applies to everything in the visible universe where space and time are factors in the continuum. We agree with equations Einstein used to show the bending of light by the curvature of space and gravity motion, but not gravity time dilation. In the Cosmological Balance, we predict that Einstein's Theory of General Relativity graphical display of gravity also applies to everything in the invisible part of the universe, except from an upside down from our

---

[200] This physical perception of space-matter aligns with Einstein's imaginary curvature of space.

point of view, as depicted by fabric of space, not time dilation. For antigravity objects, the fabric curvature of space is viewed from the bottom or upside down. From our point of view, a super massive white hole will bend space, not time, to form an extremely sharp peaked mountain. While intelligent beings, if they existed, in the antigravity universe, the invisible part, see their super massive white hole as a deep pointed depression in the fabric of their space, not time. The depiction below is a simplified fabric sheet three-dimensional view of a four-dimensional space without time dilation; a more accurate depiction of Einstein's fabric of space-time is quite similar to the concentric spherical bubble intensity layers we used to describe the solar system above without time dilation.

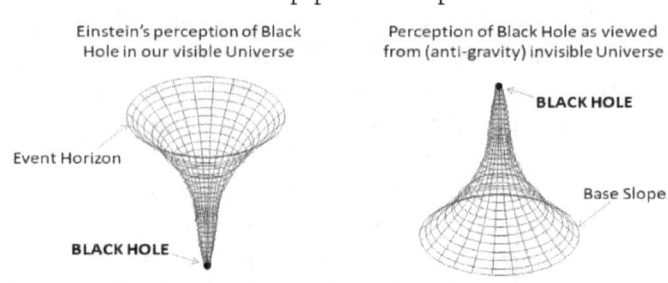

Einstein pictured one massive black hole is connected via one wormhole directly to a white hole the spits out the matter that the black hole pulled in. Therefore, Einstein's General Relativity practically does not apply to multiple "white hole orb" or "black hole orb" ejections because these orbs travel through many wormholes at the speed of light squared, a speed Einstein said was not possible. According to Einstein, time is not a factor as time might stop or run backwards for the object travelling within the wormhole tunnel. Euclidean geometry is sufficient to depict the paths of these wormholes through space as they move around stars and leave the galaxy. Moreover, Newton's gravitation equation is good enough to show antigravity force by simply setting the gravitational constant G to its negative value.

This depiction shows the tremendous energy consumed during the black holes' and white holes' excretion process.

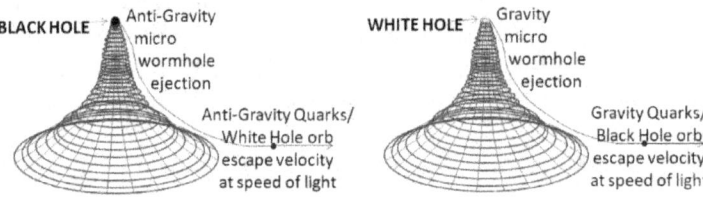

6.12 Gravitational Force Analogy

The primary reason we used Einstein's simplified depiction of space-time as a fabric is that we can all relate to motion and speed going down a hill or to falling into a hole and having a hard time getting out of it because it is too deep. These visualizations are much simpler than trying to depict the attractive or repulsive force of a magnet at varying distances from it, let alone gravity, which is similar to magnetic waves or layers, but is not limited to the speed of light. As much as we used Einstein's simplified fabric of space-time to visualize gravity and antigravity effects, it has its own flaws. Einstein visualized that massive objects like the sun warped the fabric of space-time, like a large iron ball sitting on a thin rubber sheet. The iron ball stretches the sheet downward at the point where it sits. Further explaining that the earth pushed into its orbit by space being warped by the sun's mass, specifically, the sheet is concaved inward causes the planet earth to follow that curve around the sun. This analogy is flawed. First, how can a downward force act on the mass of the sun in fabric of space-time if there is no force underneath it to pull it downward or no force above it to push it down? This is like trying to explain gravity with another space-time gravitational force beneath the sun causing the curvature. Einstein then uses three-dimensional pictures of space-time curvatures to explain a force that does not operate solely on one fabric or sheet but surrounds the object in all directions. Second, the curvature of space-time supposedly pushing the planet earth in an orbit around the sun would mechanically force the earth to rotate in a retrograde motion. However, in fact the earth and most of the other planets rotate in a prograde motion east west, that is they rotate from the west towards the east, except Venus and Uranus, which rotates retrograde. Uranus is even odder with one axis primarily pointing toward the sun. Astrophysicists conjecture that Venus and Uranus rotation were both because of collisions they encountered early in the formation of the solar system. It turns out that the simplified fabric of space-time most books used to show Einstein's work is not a true an accurate depiction Einstein himself intended. The concentric spherical intensity layered concept is more like what he had in mind[201].

Brian Greene similarly explained the curvature of the fabric of space was like a thin rubber sheet with a bowling ball placed on it representing the sun and ball bearing for the earth (Greene, The Elegant Universe: Superstrings, Hidden Dimensions, and the Quest for the Ultimate Theory, 2003, pp. 70-71). He too pointed out some of the same fallacies or shortfalls of using this two-dimensional analogy mainly that space curvature is three-dimensional, and then leaves it to the reader to imagine the true curvature picture. In synopsis, he stated that according to Einstein, the presence of mass curves space and the curvature of space create gravity, concluding that this solves the mechanism Newton searched for that caused gravity. In short, mass begets gravity, which travels at the maximum speed of light per Einstein special relativity theory; gravity is not instantaneous. This gravity speed limit is contrary to Newton's equations and Galilean relativity. It is because of these conflicts that Greene and many other mainstream scientists therefore conclude that Einstein's relativity theories replaced Newton gravity equation and Galilean

---

[201] Again, the Cosmological Balance Theory's depiction of curvature of space (space-matter wrapped around massive objects to contain them) matches exactly what Einstein's imaginary picture tells us.

relativity. The Cosmological Balance Theory challenges their conclusion and provides an alternate explanation.

Einstein believed that somehow, space was curved by the simple presence of ordinary matter; he envisioned that the more massive the object is in space the greater the curvature of space itself. Yet, he saw space as void and empty and aether as nonexistent during his time. How could nothingness have curvature? He leaned heavily on mathematics to show that curvature existed regardless of the composition of space. Einstein, like Newton, did not have enough data to present space as having substance or a type of matter, which curves or surrounds the massive objects in its realm. According to Einstein, this curvature is gravity and bends light passing at a tangent through it. However, Newton was looking for an outside source of gravity force from space, the mechanical aspect to his gravity equation, to justify his definition of instantaneous and infinite action at a distance. Space-matter satisfies both Newton and Einstein; it surrounds the massive object in the curvature of space, bends light through gravity lensing, and provides the force behind gravity, as we know it. The neutral graviton within space-matter is the Higgs Boson particle, which gives ordinary matter its mass. Again, the Cosmological Balance Theory agrees with Einstein's curvature of space, not his Special or General Relativity time dilation, and goes one step further by defining space with its own type of matter, space-matter.

If you noticed earlier, we also used another approach of representing the gravitational force when explaining wormhole tunnel effects on accelerating the sun and local star group; the picture we used is more like "bubble" layers of varying gravitational intensities, i.e. color-coded. Why relate to bubbles? A single soap bubble floating in the air is always a sphere within earth's atmosphere due to air pressure surrounding it. It is the smallest surface area and therefore the simplest shape. Naturally, a significantly large single orb in space is therefore always a spherical shape, due to the surrounding space-matter's pressure on it. When two soap bubbles floating in the air come together, they form a flat contact surface and share that one wall until it collapses and both merge into one larger bubble. The same applies to two orbs, planetoids, or stars merging in space, but at a much more devastating and rapid timeline.

The partial image here is the most accurate way of showing the curvature of space. It is space-matter wrapped in infinite layers around massive objects. Space and its space-matter respond to hold massive objects in its environment or realm, which affects other objects within in gravitational range. It conforms to the shape of the massive objects and is not necessarily perfectly spherical. This is the best depiction we can come up with showing the true gravity force, without drawing an infinite number of bubble layers, or an infinite number of arrows pointing inward with ever-increasing lengths the closer you get to the object.

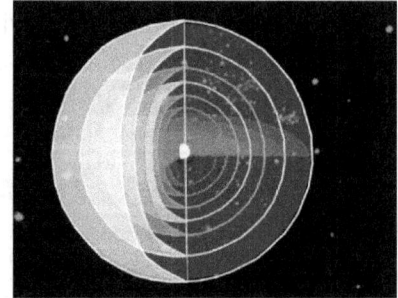

Sun's Normal Gravity Field

Planets Orbit Consistent

Nested color-coding intensity spheres were much simpler than arrows of different lengths. The spheres are like Matryoshka dolls, one inside the other all the way to the core, the massive object's surface. Gravity force occurs in an infinite number of layers surrounding an object like the sun, shown in the above picture as the center yellow orb.

Each layer shows the force intensity represented by different colors in this case the surface of the sun is the highest force intensity or strongest gravitational pull to the sun. The bubble represented by brown or dark red, the first layer up from the sun is the next highest gravity intensity. The highest concentration of space-matter overlaps the least dense part of the solar atmosphere, depicted as the orange-colored layer above; it is the thickest part of the curvature of space. The outer layer, light blue, is the lowest gravity force depicted on this chart. This light blue layer is also the least dense part of space-matter with the least curvature of space just before transitioning to flat open space. There are infinite other layers much further out from that with less gravitational attraction to the sun, which we chose not to place on the picture for simplicity sake.

Now let us look at the planet earth's orbit and rotational direction. Let us say that earth's orbit rides on the outside of the orange-colored "bubble" third from the sun's surface. The earth's velocity around the sun keeps the earth on this bubble distance or layer. Therefore, the gravitational force pushes the planet forward against the bubble while centrifugal force caused by the planet's tangent velocity pushes the planet away, also referred to as conservation of angular momentum. These opposing push actions causes the planets to rotate naturally in a prograde motion toward the east like the earth. Venus' retrograde rotation period is therefore slowing down due to these natural forces; some scientists call the force slowing down of Venus rotation an anomaly. Clearly, it is the laws of nature trying to correct Venus rotation into a prograde motion. This friction may also contribute to Venus' excessive heat buildup as the decreasing motion of the crust goes against the flow of the inner molten core. If you fast-forward time, Venus rotation would eventually slow down to a stop and then reverse direction to prograde. This stopping action will cause that side of Venus away from the sun to freeze. Uranus on the other hand, orbits around the sun on its "bubble layer" like a spinning top standing on a surface with one axis pole permanently facing the sun. The dwarf planet Pluto also rotates retrograde due to collision impact early in the creation of the solar system. Weak gravitational forces and centrifugal forces, however, have truly little impact on slowing down its rotation due to its great distance.

Mathematically, we can represent this invisible torque action or conservation of angular momentum on planet earth with a few simple calculations. Gravity acceleration near the surface of the planet is solved by multiplying the gravity constant G times the mass of the earth divided by the square of its radius to get 9.818649 meters per second squared. The Sun's gravitational acceleration at earth's orbital radius equals G times the mass of the sun divided by earth's orbital radius squared or 6.6734E-11 N $(m/kg)^2$ × 1.989E+30 kg / $(1.495979E+11 \text{ m})^2$ = 5.931044E-3 m/s^2. Earth's average velocity around the sun equals the circumference divided by the number of seconds in a year or 2*π*1.495979E+11 m / 31,557,600 seconds = 2.978525E+4 m/s. Earth experiences a slight increase of gravitational field at surface of the axis of direction tangent to the orbital path. This is calculated by adding earth's gravitational acceleration plus the sun's gravitational acceleration divided by the earth's gravity acceleration or 9.818649 m/s^2 + 5.931044E-3 m/s^2 / 9.818649 m/s^2 = 1.000604059 G's, while the trail end opposite surface of the planet experiences a decrease in gravitational field of 0.9993963 G's obtained from 9.818649 m/s^2 - 5.931044E-3 m/s^2 / 9.818649 m/s^2.

What does this gravity difference mean in layperson terms? Well, if a 100 kg person or object were standing on a weight scale on the leading axis point of the earth that person or object would weigh 100.06 kg, while the same person or object standing on the opposite point on the other side of the world would weigh 99.94 kg. On second thought, due to bodily functions replace the person completely with something more defined and consistent like a 100 kg dumbbell or free weights. You will find that anywhere within that small leading spherical cone surface of the earth the weight scale result will deviate as mentioned above give or take a few grams. Again, the solar gravitational acceleration of 5.931044E-3 m/s² is just enough to cause the Earth's atmosphere to accelerate faster than the planet. When compared to the average rotational speed of the earth is 460 meters per second, we can see that the additional solar gravitational acceleration at the earth's distance from the sun, or solar torque, has a tendency to add more speed to the earth's rotation, at the very least, sustaining its angular momentum.

How much torque force does this opposing enhanced gravity to decreased gravity create on the planet earth? This difference may not seem like much but if that enhanced gravity is multiplied times the gravitational attraction between the planet and the sun, we can calculate that torque force power. In this instance, the lead edge has additional gravitational force of

F_T = (1.000604059 - 1)*6.6734E-11 N (m/kg)² * 1.989E+30 kg* 5.972E+24 kg

/ (1.495979E+11 m)²

= 2.139589E+19 N

Trail edge has decreased gravitational force of

F_P = (1 - 0.999395941)*6.6734E-11 N (m/kg)² * 1.989E+30 kg* 5.972E+24 kg

/ (1.495979E+11 m)²

= -2.1389589E+19 N

Where 1 Newton Force = 1 kg * m/s², and the positive force F_T pushes the lead edge faster toward the sun, and the negative push force F_P can be seen as the results of centrifugal force on the planet giving it a counterclockwise rotation, east to west. Delicate digital weight scales may be used to detect gravity differences.

The above solar torque analysis clearly shows that gravitational acceleration causes the planet earth to sustain a counterclockwise rotational speed, east to west, while it moves in a prograde direction, while the orbit of the moon causes that rotation of the earth to slow down just a tiny bit every year. The earth's torque on the lead edge of the moon equals 3.26846E+17 N and the trail edge of -3.26306E+17 N bringing about a lunar rotation speed matching its orbital time, and because they are in mutual orbit about each other, the moon's torque on the earth equals to 6.65609E+14 N and trail edge of -6.65607E+14 N. When the moon is on the same side as the sun, both torques combine to accelerate earth's rotational speed. However, when the moon is on any other location, its miniscule torque on the earth decreases the sun's torque force and therefore minutely slowing down the earth's rotational speed. Adding tidal influences into the picture makes the moon a slow and steady braking force, at the same time we are in essences losing the moon as it incrementally drifts away every year about four centimeters (or 4 cm / year) from this interaction. In contrast, the earth's day will lengthen ever so slightly over a period of one hundred years amounting to only two milliseconds more time than today. Here is another picture of the sun and the inner planets as seen from the North Star viewpoint.

For a planet to move to a higher "bubble layer" farther from the sun, it would have to accelerate at a tangent perpendicular to the sun or speed up in orbital velocity. For a planet to move closer to the sun, it would have to slow down in orbital velocity. This change in speed to change orbital distance is exactly what we do to move satellites higher or lower in orbit. At the planetary level, these changes occur naturally when they have an elliptical orbit or when they are perturbed by other planets' gravitational pull. Specifically, Jupiter and Saturn are massive enough to change the orbits of each other and of Mars, Earth, and the Asteroids, as well as Uranus, and Neptune.

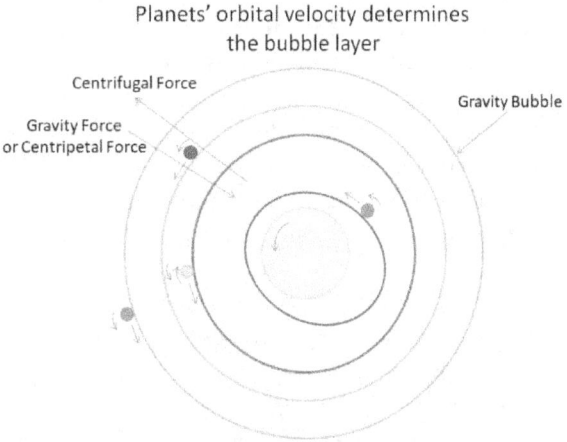

Gravitational (centripetal) Force counterbalanced by centrifugal force causes prograde rotation.

This is similar at the atomic level. To move an electron to the next higher level, it must gain energy; specifically, the electron absorbs a photon to move upward. To drop to a lower level, the electron loses a photon to shed energy. Unlike the atomic analogy with only specific wave energy levels, the gravity layers are of infinite nested levels and these spherical bubble layers do not have to be circular with the sun at the center, they are also elliptical. Earth's orbit is slightly elliptical not exactly circular. This slight difference in distance of a few million kilometers from the sun combined with solar maximum cycles coupled with increase or decrease in solar energy can cause global warming or cooling on the earth[202]. Even a slight increase in solar activity and energy output can cause all planets in the solar system to experience a warming effect, without human intervention. The historical evidence is there. All the planets' orbital path also does not have to be on one solar plane or flat disc. Gravity is Omni-directional and spherical. Planets can therefore orbit on different planes at different angles. The spherical bubble layer representation of gravity allows for the tilting of orbital planes in any direction; comets fall into that category as they come loose from the Oort cloud.

We should all do our part to improve the living conditions for our future generations. However, our effort alone is not going to change the solar heating and increasing climate temperatures. Humankind's environment carelessness and the pumping of large amounts of carbon dioxide and carbon monoxide into the atmosphere alone is not by primary cause of global warming on earth; it is the earth's position and tilt toward the sun during its

---

[202] Since most of the world is unaware that the earth's orbit is not exactly circular, our scientists tend to ignore or leave out this fact when they declare global warming is the sole doing of humankind. If they disregard the earth's distance from the sun, then they can claim anything as the cause of global warming. They could blame volcanoes if they wanted to, or forest and grassland devastation due to lightning strikes.

elliptical orbit precession. The earth's orbit is slightly elliptical, and that this ellipse moves in a perihelion precession or apsidal precession with a period of about 19,000 to 23,000 years at an angle of approximately 43 minutes per century. This slight difference of about a few million miles closer to the sun is crucial to temperatures and climate changes on earth particularly when the northern hemisphere is in its summer tilting position toward the sun. Specifically, in the summer seasons, where larger concentrations of most landmasses north of the equator absorb the excess heat, dry up, and heat up more. Similar events occur when the closest approach is during the northern winter season, where sunlight that is more concentrated brings about milder winters. However, in periods where the southern hemisphere is toward the sun during their summer perihelion, the southern oceans heat up, evaporate more, and cause more storms and more rainfall in warm regions and may also cause more severe snowstorms in the northern hemisphere winters. While the summer winter tilt during closest perihelion brings about mild and warmer than average winters. Large amounts of rainclouds dispersed in the sky tend to block out sunlight more, and cool down the heated earth, may even pushing the earth into an ice age. What happens when the earth northern hemisphere spring and autumn seasonal tilt is in the perihelion? In these cases, the earth experiences an earlier spring arrival and a longer autumn respectfully.

From Wikipedia data as an example, the Earth's distance is about 147.1 million kilometers or about 91.4 million miles from the Sun at its perihelion, which is currently in early January 2016, in contrast to about 152.1 million kilometers or about 94.5 million miles at aphelion (farthest point) in early July 2016 (Wikipedia, Perihelion and Aphelion, 2016). This amounts to a difference of 5 million kilometers between the closest approach to the sun and the furthest away. This increased distance at aphelion provides only about 93.55% of the solar radiation from the Sun when compared to the closest approach at the perihelion, a difference of 6.45%. As explained about, this difference, when the perihelion in January, tends to bring about milder winters within the northern hemisphere, which in turn allows for early spring, potentially hotter summers, and a short autumn season. In contrast the southern hemisphere, due to increased percentage of sunshine, experiences warmer oceans greater ocean current flow, that produces more rain clouds and potentially more global storms. As the perihelion precession moves around the sun, it influences the season passing through it. This results to increasing global temperatures. When the perihelion reaches around and occurs during the July month, we get an overall decrease in temperature due to the aphelion in January bringing severe drops in temperature and extended winters.

PART 3: PROOF AND PRINCIPLES

Chapter 7: Inspired Theories Analyzed

"All you really need to know for the moment is that the universe is a lot more complicated than you might think, even if you start from a position of thinking it's pretty damn complicated in the first place."

— Douglas Adams

Cas there a Big Bang? The investigation continues. The Big Bang Theory as proposed by Stephen Hawking is supported by three findings: first, the expansion of the universe; second, the presence of cosmic microwave background signal which appears to be coming from beyond the galaxy, and third, the abundance of light elements such as hydrogen, helium, lithium, and trace amounts of other elements like oxygen, carbon, nitrogen, etc.

Cosmologists argue that the Big Bang Theory as described by Stephen Hawking is not likely to have happened. It is practically impossible to scientifically believe in a theory that does not even talk about how and why the bang occurred; it says nothing at all about time zero itself or before that and appearing to lean on simple trust of the event. Yet, it was able to gain and hold the limelight for far too long. The Big Bang was able to survive due to help of Einstein's cosmological constant prediction, which Guth and Tye confirmed, that gravity under the right environment is repulsive and given such conditions the bang morphed under the umbrella known as inflationary cosmology[203]; this temporary repulsion resembles antigravity forces in action. Alan Guth and Henry Tye, American Theoretical Physicists, discovered that a super-cooled Higgs field saturates space with energy and negative pressure in exactly the same proportions as a cosmological constant, thereby exerting a repulsive gravity force expanding space itself, but at a monumental amount estimated at 10^{100} more force than Einstein's constant in a short quick burst. What Guth found was the bang for Hawking's standard big bang. However, this discovery does not support one aspect of the big bang theory; it cannot create the universe from nothing because the negative pressure the Higgs field produces can only come from a pre-existing universe, space, and time. In addition, the slightest random jump in the Higgs field can change this repulsive force to positive pressure and normal gravity (Greene, The Fabric of the Cosmos: Space, Time, and the Texture of Reality, 2004, p. 286). Such repulsive force contributes to a steady black hole excretion process different from Hawking radiation, which we already discussed in Part 2 Chapter 6. Inflationary theorists work inadvertently confirms the power of gravity-antigravity repulsion and the rapid acceleration speed of microscopic dust particles or orbs all tunneling through wormholes away from the massive white holes at the center of antigravity galaxies, known as white hole excretion.

---

[203] Inflationary cosmology attempts to explain what the big bang theory could not. Why is the universe uniform? The inflationary cosmology states that the universe started out with a slow outward movement after the bang, which gave it time to achieve uniform temperatures. Then the universe received an intense burst of rapid expansion to get to its current widely disperse regions.

The visible universe as we know it today could not have possibly spread out to its current configuration from one singularity within a matter of infinitesimal short period. Why is this? Mainstream cosmologists tell us that no matter where we look in the universe, galaxies seem to be moving away from each other but not away from one specific central point. If we roll back the clock of time, their origination will not be one single point; it is more likely to be millions if not billions of hubs. Think about that for a moment... This many originating hubs are clues, which show us that these galaxies were created from something within those hubs or points. We also have not witnessed any super massive black holes ripping apart from an extremely super powerful anti-matter explosion they internally created. If they did, we should have seen at least one in the current universe of billions of galaxies within all the time we have been looking at the sky.

We also know that if a black hole ever exploded it would send chunks of debris of various sizes in all directions at speeds significantly greater than the speed of light, anything less than that would fall right back onto itself, and the black hole survives. The scattered debris will more than likely be billions of earth-sized or smaller black holes of very condensed particles of matter and or very heavy elements that would destroy everything in its path, and certainly be an unsuitable source of hydrogen and helium to generate stars. Remember, heavy elements like iron are not the ideal fuel for stars to fuse. On the other hand, if the black hole somehow expelled only individual sub-atomic particles that fell apart, reconsolidated, and produced mostly hydrogen atoms. They may or may not coalesce and rotate around miniature black holes and merge into a large black hole. Free-floating hydrogen atoms in this case would form stellar nursery clouds, stars, and begin to make the galaxies shine. This instance would mean that the feeder universe died in billions of "big crunches" to produce billions of "multi-bangs." Repeated crunch and bang are not likely for two reasons. One, we have yet to see supersized extremely massive black hole explode, and two, the current universe expansion makes the "multi-crunches" scenario very unlikely as there is insufficient gravity to pull all the galaxies in a region together, let alone the entire universe for a big crunch. Therefore, what other source could have been at the center of each of these billions of hubs[204]? In our discussion of the Cosmological Balance, we already explained the reason for the abundance of light elements is the byproduct of super massive black holes' and super massive white holes' excretion. The other byproduct is the building blocks of space-matter. In addition, we discussed that the source of the cosmic microwave background is an emission from wormholes collapsing. Finally, we showed how antigravity galaxies interaction with normal galaxies could appear to us as an expansion of the universe.

If there was no Big Bang, then there can be no Big Crunch. For a Big Bang to occur as Professor Hawking theorized, the entire universe mass should have been compressed into one point from which it came forth. However, massive Black Hole's steady excretion prohibits an entire galaxy to be squeezed into one black hole, let alone the entire universe into one singularity. The Big Bang age is assumed at 13.8 to around 14 Billion years of age. But our analysis proves the universe is beyond 14-Billion-years casting doubt on Big Bang, likewise other tenants show they were all misinterpreted data.

---

[204] The realization that there is not one central expansion point should have caused cosmologists to abandon the big bang concept a long time ago.

7.1 *Expansion of the Universe?*

In Part 2 Chapter 6.6 Depiction of the Cosmological Balance, we presented a rough view of the struggle between the galaxies with gravity and the galaxies with antigravity and the red shift analysis of light from distant galaxies beyond 10 billion light years. Although, there are cases of galaxies colliding throughout the visible universe, most of the distant galaxies appear, according to mainstream cosmologists, to be moving away in all directions at an expanding and accelerating rate. Why is this? Our analysis of the red shift within the Hubble Deep Field image shows us that it is a result of the hydrogen and helium atoms and particles scattered within deep open space between the distant galaxies and us that cause the red shift we see. The distant galaxies are as balanced there as they are in the local galaxy clusters. Neither the galaxies with gravity, nor the galaxies with antigravity, have the necessary forces and ability to reverse or cause excessive expansion. These two types of opposing galaxies will move apart like yeast expanding bread as it grows and then deflate as counterbalance galaxies grow, two-way "breathing" effect, in and out, and essentially a static universe. In 1917, Einstein believed that universe should be static, so he modified his field equations with a cosmological constant to achieve that state (Weinberg, 2008, p. 44). As opposed to Stephen Hawking's theory, "the universe could not be static: it had to be either expanding or contracting (Hawking, The Theory of Everything: The Origin and Fate of the Universe, 2002)." Well, it is doing both as it *breathes* in an inward and outward movement[205]. Nevertheless, in the Cosmological Balance Theory, Einstein was more right about a static universe. All these galaxies, visible or invisible, will continue to move through deep space accumulating matter along its path and continue to grow gradually. They will also continue to produce material for their counterpart galaxies just as long as their massive black holes remain sufficiently dense.

A continuous unchecked expansion is one of several possible demises of the universe, endlessly increasing open space between galaxies, visible with gravity and invisible with antigravity. Again, the red shift analysis we did in the previous chapter shows that there is probably no accelerated expansion of the universe. Although our scientists believe that, the universe seems to move in this direction right now, this scenario is highly unlikely as the galaxies move through space they will eventually come into gravitational or anti-gravitational range of another galaxy of their type and begin a collision course. This collision and the merger of their black holes or white holes will seed more material for the counterpart galaxies. No matter how we look at it, at no time will all gravity galaxies occupy one part of deep space and antigravity galaxies occupy another part of space. Why is this so? Any merger of two massive black holes, between two galaxies coming together, will pour out an increase amount of antigravity white holes micro orbs, the material from which antigravity stars and galaxies are formed. Likewise, any merger of two massive white holes will release an increase number of gravity black hole micro orbs, the matter from which visible gravity stars and galaxies are born.

The Cosmological Balance Theory suggests that super massive antigravity galaxies and their massive white holes were the source that gave birth to numerous smaller normal gravity galaxies around them which eventual grew with time as they spread out away from

---

[205] Observation has shown us that some parts of the universe is moving apart, and some parts are coming closer, but the overall process remains generally static, homogenous, and isotropic.

those hubs. Temporary expansion can be the result of antigravity galaxy repelling gravity galaxies. The Yin Yang Balanced Universe scenario is the best explanation to the appearance of expansion and yet allows gravitationally bounded galaxies to collide occasionally with each other.

Depicted here is a two-dimensional view of the visible and the invisible universe, where the "voids" are the locations of the galaxies existing in the other part. The left image above taken from www.atlasoftheuniverse.com is 14 billion light years distance from the sun[206]. As a sidetrack, the age of the universe can also be represented in another form. According to Professor Fred Adams in "The Five Ages of the Universe" book (Adams, 2000); the age of the universe can be expressed in term of "cosmological decades" using scientific notation 10^n where n represents the cosmological decade. Per Adams, we are currently in the 10th cosmological decade, where he assumes an age of 14 billion years was correct. In this terminology, the first cosmological decade spans ten years. The second cosmological decade spans 100 years. The third cosmological decade spans 1000 years or 10^3. In addition, the 11th cosmological decade spans 100 billion years or 10^{11}. This is an interesting way of expressing the age of the universe. Googolplex then depicts a universe within its googol cosmological decade.

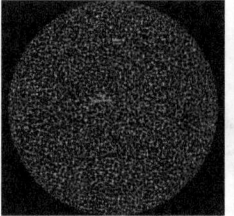 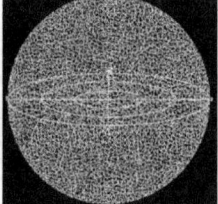

Yang "Sunny" Visible Universe   Yin "Shady" Invisible Universe

Peck flips the Theory of Relativity on its head. Michael Scott Peck, in a paper written in 2013 titled "The Theory of Everything: Foundations, Applications and Corrections to General Relativity" presents an invariant to Einstein's Theory and proves that the universe is not expanding. The two theories included in his paper (Peck M. S., 2013) are referred to as vacuum field theory and the continuous model of the universe. These deeply interrelated theories are necessary for complete consistency between general relativity and cosmology. These cosmological aspects are further applied to rule out various theories of General Relativity while also showing the universe is asymptotically flat and that red-shifted light of distant galaxies are from the relative motion and gravitational potential of the universe, he termed "tired light." He uses the Cosmic Microwave Background Radiation (CMBR) images to disprove the big bang and the expanding model. He showed that an asymptotically flat universe containing a central core would emit the observed spectrum, explain the hot ring surrounding the central cold spot in the cleaned CMBR image, the mechanics behind a steady state, and origin of the dark flow. Peck's theory inverts the perceived expansion into a contraction into the central core. His YouTube video (Peck M. , 2013) further explains that the inferred red-shift expansion is an illusion that arises from gravity alone and rules out an expanding universe. Peck overlooked the presence of dark matter, its potential photon energy absorption, and its structural support to a static universe, one of dynamic equilibrium. Strike one.

---

[206] The 14 billion light year distance is the limitation of our instruments ability to pick up information to develop pictures for us to visualize what is out there. This distance is not a reflection of the true age of the universe.

7.2 Lighter Elements and Background Radiation

In the scenario discussed in the Cosmological Balance, the concept of Big Bang Theory is very unlikely. All the normal matter we have in the galaxy in essence was produced from the decay of space-matter or from massive "white hole" excretions in antigravity galaxies as described in the previous chapter. The more massive a "white hole" becomes as it grows, the more "quarks, leptons and force carriers with normal gravity" it creates and the more micro black hole orbs it expels into wormholes and the faster it "excretes." The Cosmological Balance excretion process will never allow the white hole to get to the point where the entire universe's matter is contained in one single point as described in Big Bang Theory, prior to his theorized super colossal explosion seeding the entire universe[207]. Hawking's big bang theory is riddled with holes and at the verge of losing support. Constant and continuous emission of normal quarks, leptons, and force carriers in billions and billions of micro-blackhole orbs ejected from every massive white hole in the universe prevents such "big bang" from occurring.

Billions and billions of subatomic black hole orbs, ejected from massive white holes at the center of antigravity galaxies, traverse through wormhole tunnels far enough from their galaxy to drop below the speed of light. Once released a quick burst of microwave signal pulses outward and micro black hole emerges. The contents of the subatomic black hole reassemble into space-matter and gravity atoms, typically hydrogen and other light elements. The more massive the white hole at the center of the antigravity galaxy, the more space-matter and gravity atoms are created. Hence, billions upon billions of these new ordinary hydrogen and helium atoms are scattered throughout space. The same process applies to Super Massive Black Holes. Black hole excretion process in the Cosmological Balance prevents the Super Massive Black Hole from getting large enough to consume all the matter in a visible galaxy. The faster the Black Hole eats and grows, the more antigravity quarks, leptons, force carriers and other sub-atomic particles it creates and ejects into wormholes outward, leading to faster black hole excretion rates. The increase excretion rate limits the mass of the black hole and limits gravitational influence of stars around it. Therefore, when two black holes merge the excretion increases exponentially until enough mass is released to reach equilibrium. Likewise, billions and billions of subatomic white hole orbs, ejected from super massive black holes at the center of normal gravity galaxies, traverse through wormhole tunnels far enough from their galaxy to drop below the speed of light. Once they hit that threshold speed, they release a quick burst of microwave signal pulses outward and micro white hole emerges. The contents of the subatomic white hole fall apart and reassemble into mostly space-matter and some antigravity atoms, typically hydrogen and other light elements with antigravity properties. The more massive the black hole at the center of the normal gravity galaxy, the more

---

[207] Einstein's $E=mc^2$ equation alludes that energy can exist as a stand-alone. From this and the discovery of the black hole, Hawking envisioned all the matter of the universe coming forth from pure energy within one singularity in space and somehow spread evenly throughout the universe. If we study explosions, you will understand that this is highly unlikely, as space would resist the expansion, the distribution would be uneven, and slow to a boundary of galaxies like the Oort cloud. This concept opposes Einstein's relativity equation in which space tells matter what to do and matter tells space how to move. In addition, propulsion force of surviving particles is limited to the speed of light in two opposite directions ($c * c$) making it impossible for big bang to push all matter as Hawking declared to spread out throughout the universe.

space-matter, and antigravity atoms they create. Hence, these billions upon billions of these new antigravity hydrogen and helium atoms are scattered throughout space ready for antigravity galaxies to use, the invisible dark matter. Whether massive white or black holes, we have identified the source of the abundance of light elements spread throughout the universe, in the form of matter with gravity and matter with antigravity properties[208]. Most of this newly formed matter is hydrogen, and helium, with a few traces of other light elements. Strike two, professor.

Remnant Background Radiation?

The Hubble Telescope is capable of viewing into deep space galaxies as far away as 13 Billion light years away over a thirteen-day or more "observation" period and collection process. In those pictures, we see what appears to be black "empty voids" behind those galaxies[209] with no indicator of older light. Scientists and cosmologists tell us the "Big Bang" happened about 13.8 billion years ago. Using logic, there should be visible "light" or EM radiation from the supposed "Big Bang" lurking only 300 to 800 million light years distance behind the farthest galaxies in the "Hubble Ultra Deep Field" picture should still be glowing, from the super intense "hot" early universe. This evidence is not there. Even Hawking said he may have made a mistake in proposing the Big Bang theory. Believers of the Big Bang Theory would respond and say that the background microwave signal seen throughout the universe is proof enough. If that is the case, then show me by what means or proof you have of the microwave signal's age. Radio receivers can only detect the signal and the direction from which the signal was generated and its strength, not its age. Scientist also claimed the microwave signal is 160.2 GHz range because it has been losing energy over the last 13.8 Billion years. If so, why are we only receiving that microwave frequency? Their explanation and logic are not practical and in error, while light emitted from the farthest visible galaxies 13 billion light years away have not lost significant energy to drop its frequency below visible light to microwave as it travelled in space. We also know that photon energy of electromagnetic waves emanating from one point (the bang) cannot be lost in such perfect uniformity and simultaneously be detected everywhere in space; a spread of frequencies should exist over the spectrum to coincide with difference in distances and photon light energy loss.

According to Einstein, light photons experience no passage of time as they travel billions of years in our time. I do not necessarily agree with Einstein's perception of time in relation to speed. If cosmic radiation degrades in deep space, then where is that proof, and specifically why 160.2 GHz remains today? Let me reiterate this once more, electromagnetic radiation generated by whatever source does not lose significant energy while travelling in open space, it only dissipates or weakens as it travels great distances, unless of course it bumps into something capable of absorbing some or all of that energy. EM radiation emitted at cosmic frequencies will remain at same cosmic frequencies until it bumps into solid objects much like light remains light until it bumps into solid object like the pupils of our eyes. So, what is the source of the cosmic background radiation?

---

[208] Ordinary matter in our universe is recycled back into space and into rudimentary elements of hydrogen and some of helium.

[209] Thanks to NASA and the repairers of the Hubble Telescope, humanity is enlightened with the pristine images of the cosmos.

7.3 Microwave Radio Static Source

The second finding that supposedly supports the Big Bang Theory is the microwave background radiation found in space beyond the galaxy. The Big Bang theory suggests that the cosmic microwave background that fills all the observable space is the result of remaining electromagnetic radiation. Before the formation of stars and planets, scientists tell us that the universe was smaller, much hotter, and filled with a uniform glow from its white-hot fog of hydrogen plasma. As the universe expanded, both the hydrogen plasma and the radiation filling it grew cooler and weaker. When the universe cooled and formed stable atoms, it eventually absorbed less, and less thermal radiation and the universe became transparent instead of an opaque fog. According to mainstream cosmologists, the photons from that time, 13.8 billion or more years ago, have been propagating ever since, growing fainter and less energetic progressively moving from highest frequency to below visible wavelengths to its current microwave length[210]. The Cosmological Balance Theory disagrees with this explanation. How can electromagnetic radiation wave transition from the highest frequency to below visible light to microwave without something in space to absorb that energy? It just does not make sense. The coldness of space itself does not cause light or any other electromagnetic radiation to lose energy. Below is the depiction of the Electromagnetic Spectrum, as we know it.

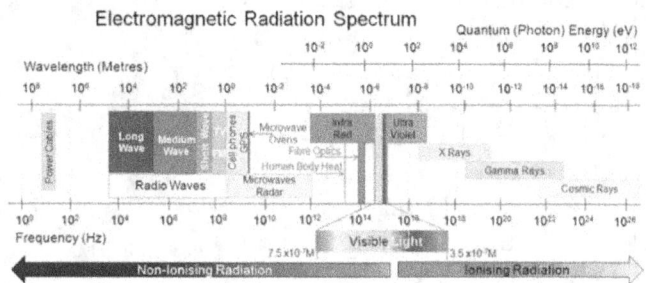

Some cosmologists will tell you that they see the Big Bang as the only explanation of the microwave existence and relate it to the expansion of the universe. We disagree. If their definition of photons losing energy as they age is correct, then all galaxies beyond 12 billion light year distance would all appear with a significant red shift of their light, in the infrared, or much lower frequencies, with or without expansion. However, that is not the case; they still appear within visible light frequencies. Age does not make light lose energy, transition through mediums does. Astrophysicists also tell us the light does not experience time and does not lose energy as it traverses empty space. So why can't the Hubble Telescope in space see the "faint glow of opaque" or the "bright hot plasma" from just after the Big Bang which should be just beyond the "Hubble Deep Field" image, an image of a time 13 billion or more years ago which took over 10 days to receive enough data to form a quality picture?

If the Big Bang happened, then that bright light should still be lighting up the night sky today, emanating from just beyond the Hubble deep field and in its background. How do we know if these microwave signals were not emitted today, a decade ago, maybe one hundred years ago, or a thousand, or less than a billion years ago? Moreover, how do we know how long they have been travelling through space? Radio receivers cannot tell the

---

[210] This is obviously a faulty explanation for the microwave background radiation of 160.2 GHz (why specifically this section of frequency). If it were true, then all the images of distant galaxies beyond 13 billion light years distance would all be radiating frequencies lower than visible light and closer to the microwave range, but there are not.

point in time of their emission. They can only provide the direction from which the signal was emitted and its strength. Did scientist have to listen for ten days or more in one spot of the sky to obtain enough data to produce one portion of the microwave signal map below, as Hubble had to do to form the Deep Field picture? No, they just gathered the data from the signal and placed it on their map. Apparently, the microwave signal was strong enough to create the map without waiting ten days to ensure data was sufficiently gathered to produce and accurate picture. The decision to use this newly found signal as justification for the Big Bang was made in haste. Therefore, we propose that these cosmic microwave signals are not the result of the Big Bang, but the result of relatively recent particles emerging from collapsing wormholes as they dropped below the speed of light velocity, thereby releasing a quick burst of photons at a peak of 160.2 GHz in the microwave range of frequency. This cosmic background radiation measures at a temperature of about 2.7 Kelvin degrees. If we are correct, this frequency shows us the exit points of wormholes and invalidates one of the tenants of the Big Bang Theory. Pictured here on the left side, a 3D image of cosmic microwave background signals stretched into a sphere to compare with the locations of antigravity galaxies in the Yin, invisible side of the universe.

In the Cosmological Balance, the radio static that scientist perceived as remnants from the creation during the Big Bang is not what they thought. The "static" they saw and "heard" from televisions and radios tuned onto an unused channel is the microwave signal that is being released when quarks, leptons, electrons, and force carriers in micro black holes orbs emerge out of collapsing wormhole tunnels from antigravity galaxies.

 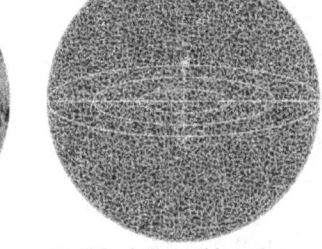

Background Microwave Signal   Yin "Shady" Invisible Universe

These quarks and other particles release a short burst of photon radiation at microwave frequency due to its low electromagnetic interaction with the collapse of the wormhole carrying them away from the antigravity galaxy. As billions and billions of these wormhole tunnels collapse, they create the static noise at the microwave frequency of 160.2 GHz[211]. The temperature of this microwave radiation registers at about 2.7 Kelvin degrees in all directions. Microwave signals released every single moment of every single day over hundreds of billions of years, if not more, has come to fill the entire sky with a uniformed pattern captured in the picture above. This microwave picture portrays the locations of antigravity galaxies scattered in the invisible part of the universe.

To explain how black hole or white hole orbs emerging out of wormholes produce these microwave signals, we need to recall how a microwave oven produced them in Part 1, Chapter 1.5. In short, the cathode in the center of the magnetron when heated up and stimulated to produce electrons travel straight toward the anode with a photon. However, the magnetic field at the base of the magnetron causes the electrons accompanied by

---

[211] This section of microwave frequency is not by chance a remnant of the big bang, but a result of trapped electromagnetic radiation released from wormhole tunnels "mouths" as they collapse, and particles exit.

photon to travel in a circular path encircling the cathode. As the electrons and photons enter the cavity, they resonant to produce the desired microwave frequency.

In comparison, the black hole or white hole orb traveling within a wormhole tunnel is of course super-energized and stimulated and is accompanied by free electrons and photons encircling it in a spiraling motion through the tube. Depicted is a wormhole crosscut. However, because of its great speed (the speed of light squared) the electrons appear to be traveling in the same pattern as though there was a magnetic field at the base of the wormhole tunnel. As a result, the electrons circulating pattern creates that magnetic field. The magnetic field and extreme speeds of the orbs traveling within the wormhole tunnel generate eddies of electrons and photons which are 1.873 millimeters in diameter and resonant the microwave radiation frequency of 160.2 GHz. The distance from the black hole or white hole orb to the wall of the wormhole tunnel it is traveling within is predicted to be slightly less than 1.8731 mm or exactly the right distance to produce the background microwave frequency of 160.2 GHz found throughout space. As these continuous flows of billions and billions of orbs in wormholes traverse far enough away from the galaxy, they slow down and drop below the speed of light. That drop in speed, causes the wormhole tunnel to collapse and release the trapped resonating 160.2 GHz microwave radiation it was carrying in a short quick outward burst. Whether these microwave frequencies were generated today or 13 billion years ago, they will always remain at that frequency level, unless they bump into something with enough mass to absorb its energy. As a result, a continuous static noise is produced at the microwave frequency around 160.2 GHz. These microwave frequency burst releases occur every moment of every single day for several hundreds of billions of years, if not more, filling the entire sky with static noise we can hear on the unused channels of the old tube television sets and am radios.

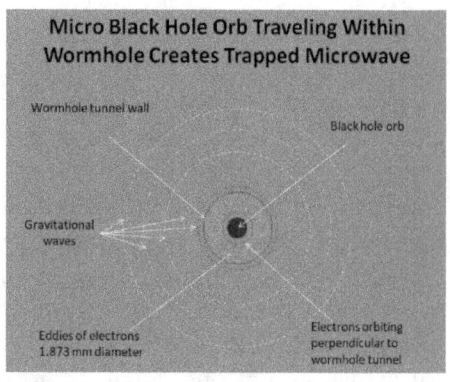

The microwave signal found throughout space did not come from the bang and is therefore not one of the justifications for the "Big Bang" Theory! Cosmologists say the universe is expanding in all directions, and that from any given point, everything seems to be expanding outward, indicating there is no one center point of origin, a contradiction to "Big Bang" concept. Even the universe map shows connection of sheets and filaments spread evenly across the vast expanse of the universe with no evidence for Big Bang origin. The speed of expansion, if such thing existed, which appears to be the same everywhere, does not support the "Big Bang Theory[212]." There is no expansion of the universe, just red-shifted light due to photon absorption of widely dispersed gases and particles over great distances; the greater the distance of the galaxies the more red-shifted its light due to "dark matter." Strike three of your tenants. Cheers Father Georges Lemaitre and Professor Hawking; rest in peace, our life goes on.

---

[212] Evidence continues to build against the Big Bang Theory to topple it eventually.

7.4 Possibility of Big Crunch?

In the Cosmological Balance, there is no Big Crunch scenario possible, which could have led to the Big Bang singularity as imagined by Stephen Hawking. Why is this so? We learned that the more massive a black hole becomes the more "antigravity" quarks and other sub-atomic particles it ejects into wormholes and thereby "excretes" faster. For argument sake, if one galaxy completely collapses on itself into its massive black hole. As it collapses, the current black hole in its center will progressively eject more and more antigravity quarks and other sub-atomic particles it pumps into wormholes as the black hole feeds. Any anti-matter created within the massive Black Hole only intensifies the creation of more antigravity sub-atomic particles. The more mass the black hole eats the faster and more antigravity mass is excreted out of it[213]. Hence, an eating and growing massive black hole can never get to the point where the entire mass of the normal stable galaxy will fit into one singularity to satisfy a subset of the "Big Crunch" as described by Stephen Hawking, let alone all the visible galaxies fitting into one singularity. Constant and continuous emission of antigravity quarks and other sub-atomic particles into wormholes prevents one "big crunch" from ever occurring.

There is an exception to not fitting the entire mass of a galaxy into one black hole, its center orb. This situation can occur if and only if a smaller galaxy is caught right in between the direct line of two merging larger antigravity galaxies. Here the smaller galaxy's mass is subject to compression anti-gravitational forces exerted by the larger antigravity galaxies, like density waves enabling the smaller galaxy's black hole to consume all of its surrounding stars and grow to its full capacity before being squeezed out of the path of the two merging antigravity galaxies way. This newly enlarging massive black hole increasingly excretes greater and greater numbers of antigravity orbs into the surrounding space, feeding the two inbound antigravity galaxies with basic building material, antigravity hydrogen and some helium, and replenishing space-matter atoms.

Excretion by massive ejection of antigravity particles and black hole orbs combined with the "evaporation" process as described by the Hawking Radiation (ejecting out one half of the matter before it passes the event horizon), controls the growth rate of Black Holes and limit its size. Exactly what is the maximum size is up for deliberations. Suffice to say that it will never get to the size imagined in the one "Big Crunch." Why is this? As the black hole grows, it will excrete more and more antigravity orbs into space, thereby controlling the maximum level of its mass.

Lastly, we have yet to see a "super massive black hole" explode with the over production of "anti-matter" within it. If we can create anti-matter within the Hadron Collider, then it is perfectly justifiable to conjecture that within Black Holes anti-matter is also created and immediately destroyed when they encounter their counterparts, or so we assume. The additional energy produced by these anti-matter matter annihilations contribute to the elevation of quarks, leptons, and force carriers to their higher energy levels and to the production of neutral and then antigravity particles as a release mechanism within black hole to prevent it from destroying itself.

---

[213] Fortunate for the universe, the big crunch is not likely due to the inherent release valve within black and white holes through the miniscule ejection and steady evaporation process.

7.5 String Theory Loophole

As mentioned in Part 1 Chapter 2, Witten gave string theory hope when he created a dictionary, which tied the loose ends of the five separate string theories together into one common baseline theory. It turns out the five were not at all independent theories, but five different interpretations or mathematical manipulations of the same theory, he called the m-theory. This triumph gave string theory hope in become the unified theory sought after by physicists, but m-theory introduced a need for another spatial dimension, making it eleven with time. All this of course depends on the assumption that string theory is right.

In part 1 chapter 2, strings were defined as ultramicroscopic loops of energy. How can this be so? Contrary to Einstein's energy-mass-momentum equation $E=mc^2$, pure energy does not exist as a standalone entity without the presence of some type of matter or particle and its motion or potential thereof. String theorists also tell us that strings are one-dimensional and possess miniscule mass. If this were true, then how can a string with miniscule mass create a mass-less subatomic particle? Evidence suggests that there must exists at least one if not two, or three, point particles entwined in mutual perpetual dance creating subatomic particles. The motions of these point particles are what scientists can be perceived as "strings" vibrating in a loop or open-ended configuration. Exact compositions of the point particles creating this perpetual dance are currently unknown. We do however know that electron-photon mutual dance keeps the electron in a certain distance away from the nuclei it orbits and allows light to travel tens of billions of light years distance. The existence of every new subatomic particle is determined by all standing subatomic particles near it, and likewise it in turn affects all other particles around it, creating sustainability and a strive for stable perpetual relationship.

Let us take this from another angle. Science also tells us that symmetry governs the laws of physics. As explained, strings in string theory are not symmetrical, they are sharp, random, and sporadic figures, and the patterns they form take on various unknown shapes. String theorists are constantly debating as to which shapes are the most accurate. String theorists predict that the vibration of strings gives subatomic particles their mass, charge, spin, and other unique characteristics, to include spherical symmetry. How can a non-symmetrical string moving in various patterns produce symmetrical orbs or spherical subatomic particles? How can a string with mass vibrate to produce a mass-less particle? If string theorists were right, then why have we not seen any string strands ripped apart by atom smashers? Instead, all we have seen are sub-atomic particles or point particles being thrown every which conceivably path possible leaving streaks and trails on detection instruments. There must be another answer more plausible than vibrating strings and membranes with mass creating subatomic particles ranging from the mass-less to the heaviest. The Cosmological Balance offers an alternate more plausible interpretation for the ideas within string theory. Did string theorists get it right? What if all matter in the universe consisted of universal energetic point particles that jiggled and vibrated in specific patterns to generate mass, charge, spin, and other characteristics within all known subatomic particles? In addition, the patterns created by these vibrating particles just happened to look like strings dancing around or wiggling as if trying to make a musical note. Whether forming loops or open-ended strands, the patterns are analogous to the trails some jet or propeller airplanes or helicopters leave behind as they twist, turn, loop, and traverse the clear sky overhead. Open-ended trails can be like movement of a pendulum swinging back and forth, as the point particle is pulled back and forth from a center spot or line, like an electron-photon interaction. These smoke trails do not create the jet or airplane; hence, these imaginary strings do not create point particles or any subatomic particles to form matter as the strings itself has no mass. It is the point particles themselves that create the "string or strand" images these string theorists have envisioned.

Let us look at this from another view; we know that the outer electron's random movement around the atom's nucleus gives the atom its radius or size and appears to be in all places at the same time as if surrounded by a cloud yet there is no "string or strand," real or imaginarily wrapped around the entire atom. The energetic pattern of this one point-particle, the electron, determines the radius and shape of say the hydrogen atom for example. Its highly energetic combined movement pattern with its partner photon gives the electron its charge, spin, and distance from the nucleus, while doing the same for the companion photon, creating a "shell" or "membrane" around the atom. This dual point particle relationship enables it to sustain perpetual motion. Similarly, one, or more, point particle's high-energy pattern and combined movement determines the shape, size, charge, spin, and other characteristics found within a particular quark. Another single or multiple point particle's pattern determines yet another type of quark, and so on and so forth, creating from various combinations of point particles every possible subatomic matter possessing gravity (embedded positive graviton), and antigravity (negative graviton), and neutral matter (zero graviton). The boundary formed by any point particle's movement is its membrane and is not necessarily spherical, just as shown by string theory.

As with the sharing of electrons by atoms bonded together, the two up and one down quark within a proton also share a wandering point particle to create a membrane around the entire proton. Likewise, the gluon's point particles enable it to hold all the protons and neutrons within the nucleus and simultaneously generate the membrane around the entire nucleus. Although some point particles, like the neutrino, with miniscule jiggles or vibration patterns possess infinitesimal mass, considered mass-less and neutral in charge, they are still highly energetic. That is not to say just because they vibrate slowly or low energy jiggle does not mean they have no energy, as they are capable of energetically zipping across the universe unaffectedly by space itself and most matter. They are so small that one of every two particles can even penetrate through one-lightyear of the element lead unaffected or so scientists speculate. In other words, the total mass of the neutrino is spread over its entire world line (path), with every point along that line having nearly zero mass. Light and other paired photon-electron electromagnetic wave of perpetual motion traveling through space remains unhindered until they collide with matter unable of re-emitting it. That light particle's total energy and mass is also spread across its entire world line, where any point along that long line yields a mass-less particle. The same applies to electron-photon orbiting around the nucleus of an atom; its combined mass is spread over the entire shell membrane the electron-photon creates. Hence, a snapshot of the electron or the photon each reveals a mass-less particle. The same holds true for the gluon particles encircling the protons, neutrons, and the nucleus of the atom.

As for the hidden dimensions revealed by Brian Greene in his book, he pointed out that each of the axis within the three known dimensions, length, width, and height, have a Planck tight turn around them in a clockwise and counterclockwise direction, which he counted as two and deduced nine dimensions. If we really think about this for a moment, we find that the width is a left-right movement, height is an up-down direction, length is a forward-backward motion, so the way we see it the turn of clockwise-counterclockwise is only one additional dimension not two. Hence, the revealing of the hidden clockwise-counterclockwise direction around the axis only reveals six spatial dimensions, not nine, unless of course our visual perceptions are not aligned with those of brilliant physicists. We can however edit string theorists' concepts by stating that one clockwise counterclockwise turn is perpendicular to the axis and one is at a 45-degree angle to the axis, this gives them two hidden dimensions, thereby producing nine spatial dimensions. There is a problem with this adjusted version; what about all the other possible angles giving many more potential 3^n dimensions. String theory calculations tell us that there must be nine spatial dimensions for their theory to work, or more specifically to create the correct number of vibration patterns to form every type of subatomic particle known by our scientists. Because of this fact, string theorists have come up with numerous different p-brane (membrane shapes) configurations that have become unmanageable and create problems for string theorists when trying to present their theory as a unified theory.

The Cosmological Balance Theory also views nine potential spatial dimensions like string theory. However, unlike string theory it is based on simple math rather than intense or complex calculations. In our theory of balance, there are three types of matter, matter with gravity, matter with antigravity, and matter with zero gravity. Each of these three types of matter move within three spatial dimensions, length, width, and height. The interactions between the types of matter cause spatial deviations to give us three times three or nine spatial dimensions and with time, if we consider it as a dimension, gives us the tenth dimension. From a non-scientific point of view, the Amazon Shamans also believed in nine dimensions, and the Viking's saw nine realms of reality in their myths, just to name a few among many other mentions of dimensions of reality in cultures around the world. Hence, for the purpose of the remainder of this book, any further discussion of string theory and its possibility to multi-verse will be kept to a minimal. Why is this? It is because "strings" are undetectable, and therefore can never be verified, while the existence of point particles, subatomic particles, or atomic particles and their patterns or ghostly trails, on the other hand, are verifiable.

Math at best can only attempt to capture what happens, not vice versa. Nature does not obey science. Science can only mimic its essence and channel and determine results. For 200 years, humanity thought Newton was perfect and right on target with his gravity equations. Einstein supposedly improved on Newton's equation with his new complex formulas in General Relativity. Einstein worked hard and made his mark on the world, but he too is not immune to errors, as we are all human. There are many books written on Einstein's blunders. So, learn from mistakes, adjust accordingly, and advance in knowledge. For it was the genius of Einstein that made humanity better and scientifically sharper, we all owed him that gratitude. However, the solutions of the past are not always pertinent to the problems of the present, given the immense knowledge and data we have observed and are still collecting. Let us work together to advance humanity.

7.6 Special and General Relativity Quandary

Well-known for his Special and General Theories of Relativity, Albert Einstein (1879-1955), according to a great many physicists, was certainly the greatest of physicists of his time, and one of the greatest scientists of the 20th Century, spending a significant amount of his time imagining solutions to situations and dreaming up the math to support his thoughts, concepts, and theories. Like all humans, he too was prone to making mistakes as he assembled his sources and worked on his equations. Among Einstein's many mistakes, the injection of a cosmological constant into his equations is the one mistake that most people think of, as he openly and publicly admitted to making this one. There he tried to add it to his equation to stabilize the universe' expansion and of course withdrew it when he learned of Edwin Hubble's finding of data leading to the declaration of an expanding universe. As a side note, Edwin Hubble was unaware of the presence of large amounts of "dark matter" when he declared that the red-shifted light of distant galaxies resulted from acceleration in an expanding universe. Was it really a mistake? No. Perhaps more importantly, by showing human mistakes, even those made by our greatest scientists, allows us, humans, to learn from them and correct the intellectual path we, and all of humanity, take. We must all recognize that these great men and women have significantly prepared the path we now walk and the accomplishments we have since achieved. However, it is time for us to adjust based on our modern and current understanding of physics, nature, and the universe, and stop going down the wrong path. Simplicity is better for the comprehension of the many.

Einstein said, "The secret to creativity is knowing how to hide your sources;" openly admitting possible plagiarism to advance the ideas in support of his theories and his innate ability to introduce made up numbers out of thin air. Despite his success, Einstein's papers were not error free; careful examination of his publications revealed mistakes, when followed through, alters the outcome of his theories. Countless physicists and mathematicians have come forward and enumerated the errors discovered, and yet Einstein's famous equations and work are still taught in Universities everywhere[214]. We partially side with those courageous physicists and mathematicians and decline usage of his Special Relativity based on his misuse of the abstract-mathematical method, and due to errors in his analysis and development of the theory. We however do not discredit his work for it has paved the way for the science of today. We just want to point out some flaws. Albert Einstein's 1905 paper "On the Electrodynamics of Moving Bodies," Special Relativity is based on two postulates, which contradict each other in classical mechanics:

1. The laws of physics are the same for all observers in uniform motion relative to one another (Galileo's principle of relativity),

2. The speed of light in a vacuum is the same for all observers, regardless of their relative motion or of the motion of the source of the light.

---

[214] Despite the many errors in Einstein's work, the bending of light around massive objects caused by the curvature of space still provides a very accurate calculation of how the forces of gravity or more accurately the atoms of space-matter are able to guide light through the geodesics path. This bend is exactly like an optical lens.

In his paper, he claimed that the velocity of the observer moving alongside changes how light reaches that observer and thereby changing time and the results of clock synchronization. The consequences resulting from his theory include:

I. Time dilation: Moving clocks are measured to tick more slowly than an observer's "stationary" clock.

II. Length contraction: Objects are measured to be shortened in the direction that they are moving with respect to the observer.

III. Relativity of simultaneity: Two events that are simultaneous to an observer A may not be simultaneous to an observer B if B is moving with respect to A.

IV. Mass-energy equivalence: $E = mc^2$, energy and mass are equivalent and transmutable. The defining feature of special relativity is the replacement of the Galilean transformations of classical mechanics by the Lorentz Transformations.

In fact, only the perception of the observer changed, while the clocks at the observed point functioned as expected. Time did not change from their point of view, nor did the perceived synchronization of the clocks, nor is energy and matter interchangeable completely into one or the other. Nevertheless, for the sake of argument let us look at Einstein's work in that paper. Einstein defines the synchronization of clocks as he initially defined $t_B - t_A = t'_A - t_B$, and then goes on to say $t_B - t_A = r_{AB} / (c-v)$ and $t'_A - t_B = r_{AB} / (c+v)$, which implies that $r_{AB} / (c-v) = r_{AB} / (c+v)$, when v = zero and that equation is false for any other value of velocity v. He disregards this one value, adopts a range of non-existent values for v, and moves on.

Below is depiction of Einstein's Special Relativity as he intended. In it, two spaceships travel at constant equal speeds, v, close to the speed of light. Both travelers, according to Einstein, each sees the other's clock running slower than their own (time dilation), and both observe the other ship being squeezed (spatial compression).

Where each traveler sees the other's clock running slower and the other ship's body squeezed more as both approach the speed of light, c.

Einstein's Special Relativity

5:24:09 PM
v < c

5:24:09 PM
v < c

Einstein presented and manipulated the equations within his Special Relativity paper to achieve the results, as he desired; they are time dilation, length contraction, relativity of simultaneity, and mass-energy equivalence. In one of Einstein's special relativity examples, he explains that two travelers each traveling at 2/3 the speed of light in the same direction, parallel with each other, observes that the other traveler's clock as running slower than theirs. Einstein uses the time difference between the speed of light and the speed of the spaceships in his equations to justify the amount of time dilation, 2/3 clock time. Einstein achieved the mirage he wanted by simply disregarding the fact that each traveler was observing the other's past clock at the same time difference, 1/3, the time information took to get to him or her. He forgot to add in or intentionally left off one important reality we all know. Light is the carrier of information, specifically historical data; the greater the observer's distance away from the object, the further back in time is their view of the event, essentially peering into the past. More specifically, the greater the distance, the more time has elapsed between the event and observation. Also, the faster the observers' motion through space, the more distance light must travel between the object and the observer, and again the further back in time to the event's occurrence. Einstein did not add this "past" time difference back into his equation. Doing so would give him another picture of the flow of time on the object's surface. Why did he leave it out? The additional of the light's travel time would in essence zero out his time dilation finding and nullify that portion in both special and general relativity theories.

The picture below relooks at Einstein's clever perception; he tells us that the closer to the speed of light each traveler gets the more he observes the other's clock running slower, and the more he notices the other ship squeezed shorter along its horizontal axis. His err is to rely too heavily on his second assumption in which the speed of light in a vacuum is the same for all observers, regardless of their relative motion or of the motion of the source of light. He should have considered the speed of light in a vacuum within the moving system, the Galilean relativity part of his approach.

Take a second look at the picture above. Pilots of slow-moving spaceships see each other's clock running normally, defined by the straight dotted line between the ships, and the image of other ship is as expected, normal. However, spaceships traveling at extreme speeds closer to the light speed give us Einstein's perception. Why is this? The pilot in the red spaceship further away sees the clock of the green spaceship, as it was back in time, one second ago, exactly the amount of time light took to travel along the hypotenuse line between the ships. Similarly, the pilot in the green spaceship sees the same of the red ship. The vacuum or space between the spaceships moving at speed v_3 affects the light traveling through it. Where $v_3 < v_1$, the light emitted from the red spaceship traveling as constant speed v_1 transitions its speed from (v_1+c) to (v_3+c) and compresses briefly toward blue shifted. The same Doppler Effect occurs with light emitted from the green spaceship. If Einstein's assumption were correct about the constancy of light speed, then there would be no compression or expansion of the light wave, and therefore no Doppler Effect.

Light is the carrier of information. To give us a true picture of reality, we must add the light travel time back into the equation; doing so shows us that both clocks run at perfect normal durations; there is no time dilation. We also know that as light travels greater and greater distances, the object emanating the light appears smaller and smaller, exactly as depicted above where the closer ship is drawn larger than the ship further away, but because both ships are traveling in parallel paths, the angled shrinking perception is more prominent along the horizontal axis not only 360 degree. To emphasize this point, let us look another example. A distant airplane flying across the sky or a ship floating on the ocean at the horizon appears tiny and slow paced to a stationary observer on the ground, where its passengers experience no slowing of time or actual spatial compression. This tells us that the perception of the observer does not change the reality of the distant object's clock or spatial characteristics. Now let us look at light escaping from significantly massive objects in space. Light appears to slow down while climbing outward (Einstein called this gravity time dilation due to red shift of light), but the same slow-moving light simultaneously generates a greater past view, which completely cancels out the perceived time dilation difference Einstein claim was in general relativity. Time dilation is false, it is only a mirage cleverly presented by special or general relativity.

Let us look at this from another angle. What is the Doppler Effect? It is the effect when waves compress at the leading edge of a moving object or expand at the trailing end of the same object in motion, assuming of course, that the observer is not in the moving object. An observer within the constant moving or stationary object does not experience the Doppler Effect; in this situation, the Galilean Relativity applies. For example, an occupant in a constant moving vehicle or car hears normal sound from his or her car speakers as if parked in a stationary position. On acceleration, he or she notices a slight difference, known as Doppler Effect, in the sound as the entire system's contents, including the air molecules and the vacuum space, attempts to catch up with the movement of the vehicle. The motion of the air molecules sound propagates through influences the speed of the sound within the vehicle; therefore, sound speed and vehicle speed are added together within the system, otherwise voices could not be heard within an aircraft, exceeding sound barrier. Light too propagates through the medium contained within the system, the vehicle, and therefore the correct light speed is the addition of the speed of light in a vacuum added to the speed of the system. The picture below depicts this principle where the pilot of both spaceships parked inside a giant ship moving close to the speed of light sees no difference in time or spatial compression.

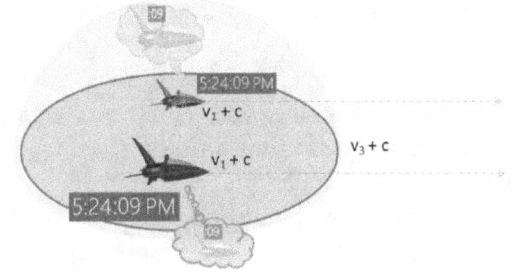

Now let us consider sound emitted from the horn at the front of a moving vehicle. As the sound leaves the horn, it starts traveling outward in a wave with the combined speed of the sound itself plus the speed of the car. When that sound enters the air molecules surrounding the leading edge of the vehicle, it recoils and slows down to assume the speed the air it is moving through, which may be zero meters per second. The compression sound wave at the front of the vehicle is then heard as a higher pitched sound, while the sound from the back end gradually increases speed, in the form of waves expanding, to be heard as a lower pitched sound. Note that the most important fact here is that the greatest compression is when the vehicle horn or siren is about to pass your position as an observer. The further the observer is away from the horn or siren moving toward them, the more normal pitch the sound seems to be. Why is this? The sound wave eventually achieves full transition to normal waves the further it travels. This fact is the opposite of what Einstein and Hubble uses; distant galaxies should then appear more normal not redder with distance despite its speed. Waves traveling within a system assume the correct total speed, by adding the speed of the medium in that system, and transitions speed when entering another system. A system accelerating is a transitioning system. Light is a wave; it too transitions in speed from one system to another and is seen as the Doppler Effect only as the object is about to pass the observer. Light emitted from the front of a moving object starts out with the total speed of the object, v, plus the speed of light in a vacuum, c, and then gradually assumes the new speed of the system into which it transitioned. Therefore, extremely distant objects should appear more as normal light wave, practically no light wave compression blue shift, or wave expansion red shift.

In our reality, it is not the clock that slows down at the source, but the frequency of light (known as Doppler Effect or what is left of it) reaching the observer. In other words, what any observer sees, regardless of their distance away or direction of motion or speed, does not affect the clock at the source location nor the spatial shape. The source clock continues to tick normally as always unaffected by what the observers within the universe around it perceive[215].

Does the Global Positioning Satellite (GPS) clock show time difference? Yes, there is a slight time difference between the satellite and the ground stations. Is it proof of time dilation? No. If this were true time dilation per Einstein, then the time difference would continue to add up or compound into a larger and larger gap. That is not what happens. This fixed time difference is well known and added back in at the receiving devices to give us accurate GPS locations. It is the result of receiving a signal from the "historical after image" of the GPS satellite back in time. Separately, the clocks on the satellite and that on the ground run normally, but the after image from the satellite clock appears as running slightly behind. As with the spaceship traveling in space, that time difference needed to be added back into equation in order to give us the proper location and time on the satellite, and thereby an accurate location on the ground determined by triangulation. What causes this time difference between surface clocks and that of the satellite? Due to the earth's movement through space, EM wave going from the orbiting object or satellite to the surface undergoes a transition of speed from $(v_1 + c)$ to $(v_2 + c)$ and on entering the earth's atmosphere becomes $(v_3 + c)$, where v_1 is orbit "acceleration" around the earth, v_2 is velocity of solar system, and v_3 is the earth's system velocity. These differences in EM speed transition causes the time delay "after image" we detect from the ground of the satellite orbiting the earth.

These perception errors and others like it propagated throughout the rest of Einstein's document on Special Theory of Relativity, invalidate certain assumptions, and significantly question the validity of the theory itself. The way we see this is that Einstein's Special Theory of Relativity equations more accurately and mathematically show the potential for Doppler Effect on emitted light and electromagnetic waves, and not the results he intended: time dilation, length contraction, relativity of simultaneity, and mass-energy equivalence. All of which are applicable to the physics of light in the quantum micro universe, and not relevant to the macro universe itself. The physics of light is not the physics of the universe and as such does not control gravity laws in the macro universe. Einstein's Special Relativity claims involving Doppler Effect mirages are nonconforming to the principle of consistency as they go against prior known laws and proofs.

For those of you that do not remember, the following equation below generalizes the Doppler Effect[216]:

$f = ((c+v_r)/(c + v_s)) f_0$,

or $f = ((c-v_r)/(c - v_s)) f_0$ for an object moving, depending on its direction,

---

[215] An observer's perception of clock duration speed at the source being observed does not actually change the speed of ticking at that source. Therefore, Einstein's imaginary experiments in special relativity are meaningless misinterpretations of mirages.

[216] Doppler Effect causes light frequencies to change not time.

where c is the speed of light, v_r is the velocity of receiver, and v_s is the velocity of the source or sender. If we replace the numerator with the light emitted from the length of Einstein's imaginary measuring bar, we have a portion of his equation, c/c-v or $1/\sqrt{(1-v^2/c^2)} = \beta$ as labeled by Einstein. Doppler Effect is great for describing visual mirages and distortions, but it does not change time and space, it only changes or shifts the frequency of the electromagnetic radiation slightly to higher or lower depending on the motion of the receiver instrument or observer's eyes. Note that the Doppler Effect formula, like Einstein's equations, ignores the direction or speed of the medium between the receiver and sender.

During his presentation of special and general relativity, Einstein failed to consider how light functions when it transitions between systems and mediums, and to add that time elapse delay back to correct the result, thereby nullifying his claim for time dilation and gravity time dilation. As we said before, Einstein's second assumption should be modified to read, "The speed of light in a vacuum is the same for all observers within the same system regardless of their relative motion or of the motion of the source of the light but is different for observers of different systems." There is no time dilation per Einstein; his first consequence is false. The physics of light does however prove that image distortions occur during extreme speeds close to the speed of light or extreme gravity; his second consequence is merely a perception seen by the observer, the observed object physically or spatially does not change. Consequence number three is valid from observers' point of views but becomes questionable when light travel time is added back into his equations. Finally, consequence four needs modifications.

Supposedly, scientist used differences of duration of muons as an example to justify the Special Theory of Relativity. They tell us that muons created from cosmic radiation colliding with the atmosphere molecules experience a lengthening of their lifespan due to time and synchronization changes attributed to Special Relativity, i.e., their internal clocks. We disagree and argue that these situations or so-called exceptions have nothing to do with Special Relativity as the observers are not traveling at excessive speed and are in a gravitational force, earth. Both are conditions for Einstein's Special Relativity theory.

Galaxies Colliding at Speeds Greater Than the Speed of Light

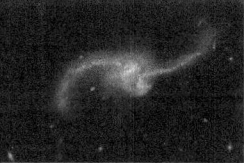

Hubble_Interacting_Galaxy_NGC_6050_(2008-04-24)

James Webb space telescope NGC2207_19Apr09

Wikipedia Atlas of Peculiar galaxies NGC2623_HLApugh

Numerous observations of any two galaxies colliding and merging show no abnormal or elongated distortions predicted by Einstein's Special Theory of Relativity. These galaxies, calculated to be traveling at the speed of light or greater, when their massive center black holes first race past each other on initial contact merger, but due to their great distances from us appears to be slow moving. These predicted speeds are based on the mutual gravitational acceleration of the Milky Way and Andromeda toward each other and their projected collision velocity. Below are photos of colliding galaxies (CDN77, 2015).

Other observations of newly discovered planets around relatively nearby stars in the galaxy also show no measurement discrepancy, or time and synchronization issues. Artists draw images of these planets, as they were, in perfectly normal elliptical orbits.

In 1915, he developed what became known as the Einstein field equations for general relativity, which relate the curvature of space-time with the mass, energy, and momentum within it. Per Einstein online (Einstein/Golm/Potsdam, 2015) 'Einstein's equivalence principle, which includes a more restricted version called the weak equivalence principle, namely that, in a gravitational field, objects which are at the same location are subject to the same gravitational acceleration - they fall at exactly the same rate ("universality of free fall").' General Relativity Theory hinges on this Equivalence Principle. In section 1.2.1 of his book, Peter Collier, (Collier, 2014) states, "find just one object that accelerates at a different rate in a gravitational field to other objects, and the equivalence principle and hence general relativity would, at the very least, be in serious trouble."

Let us look at the equivalence principle from Newton's point of view and conduct a math experiment. The force of gravity states that $F = G\, m_1 m_2/r^2$ and that gravitational acceleration is accurately written as $A = G*(m_1+m_2)/r^2$. In unique instances where the ratio of the mass of the planet is significantly greater than that of an object that is being tested on or near the surface, the acceleration equation is reduced to $A = G * M/r^2$. With these equations, let us look at the free fall of three objects (mass of Venus, one-kilogram inflated ball, and mass of Jupiter) of equal radius, 2953.728204 meters (the Schwarzschild radius of Jupiter compressed into a black hole) from the distance of 7.78716E+10 meters from the surface of the Sun. We then drop each object individually with zero speed, allowed them to fall toward the Sun, and calculate the time to impact. We find that a compressed Venus takes 12.07909722 days (1,043,634.25 seconds) to hit the sun, and the inflated ball takes 12.0791088 days (1,043,635 seconds), while the black hole Jupiter takes 12.07334491 days (1,043,137 seconds) to hit the Sun. We find only a slight difference in time of impact of but a few seconds. Now, if we release the same three orbs closer to the Sun, say 7.78716E+9 meters, we find that the compressed Venus orb hits the Sun in 9.1833 hours (33,060 seconds), the inflated ball in 9.183611 days (33,061 seconds), and the black hole Jupiter in 9.1788889 days (33,044 seconds). Again, we only see a difference of a few seconds. We find that each of these impacts is relatively close to each other but not equivalent, proving equivalence principle fails.

But if we released all three orbs at the same time from those extreme distances from the Sun, Newton gravity law tells us that the orbs would gravitationally attract and hold each other together as if they are one object. As such, they would be falling simultaneously toward the sun and impact near simultaneously only because the black hole Jupiter orb has such a significantly greater mass compared to the others. Now imaging starting out each of the three orbs with a tangent velocity of 100 m/s, we will find out that the orbital path of each orb degrades at a different rate and with each passing orbit the speed differences also become noticeable, hence each orb eventually hits the sun with a greater amount of time difference between impacts. If you do not believe this, read on.

An observable example of where equivalence principle fails is the event that occurred over twenty-five years ago, in July 1994. The Comet Shoemaker-Levy 9 (SL9), caught in Jupiter's gravitational pull, was torn apart and each piece collided with the planet in a volley one at a time. This single comet, which was made up of loosely gravitationally bound pieces, was ripped apart by Jupiter's tidal forces and pulled into a gradually degrading orbit around the planet until they finally plunged into Jupiter's upper atmosphere in spectacular splashes and left darks spots than can be seen from earth, an event the whole world witnessed. Here all the pieces of this comet started out together in this gravitational field but because of the differences in mass of each individual piece, they ended up hitting the planet at different times. Clearly, equivalence principle failed. Is General Relativity Theory still valid? According to mainstream science and academia, it is. Why is this so? Because this observed event, which of course makes General Relativity questionable, was simply overlooked, disregarded, and never used by mainstream science to show the flaws of the Einstein Equivalence Principle, and as such perpetuates the justification of his theories and his math for academia's continued usage. Equivalence principle is fine for small differences between objects compared to the mass generating the gravitational field for which the test is conducted[217]. However, the equivalence principle fails and is unacceptable for larger differences of mass falling in a gravitational field, or a gravitational field generated by extremely massive objects greater than that of the earth.

However, for the most part, any two masses tested on earth are miniscule compared to the earth's mass, and therefore their masses ignored in gravitational acceleration equations and deemed to fall at the same rate. In our imaginary experiment concerning the compressed Venus, the inflated ball, and the black hole Jupiter, the masses of the planet Venus and that of the inflated ball are both miniscule when compared to the sun, and therefore almost negligent in their gravitational acceleration toward the sun, hitting at only one second apart. However, Jupiter's mass compared to the sun, is just enough to make a difference so we can detect its more rapid acceleration toward the sun.

Some of the consequences Einstein predicted of general relativity are:

1. Time goes more slowly in higher gravitational forces. This is called gravitational time dilation. (We already discussed as invalid)

2. Orbits precess in a way unexpected in Newton's theory of gravity. This supposedly declared as observed in the orbit of Mercury and in binary pulsars, and that rays of light bend in the presence of a gravitational force. (We already disproved this earlier using Newtonian gravitational equations to calculate to the arc second and explain Mercury's anomaly and the bending of light as simply the result of gravity and solar atmospheric refraction acting on the particles within starlight passing near the sun.)

3. Frame-dragging, in which a rotating mass "drags along" the space time around it. (Newton already defined and explained this phenomenon as non-existent with Newtonian and Euclidean mathematics and his laws in Principia.)

Technically, General Theory of Relativity is a metric theory of gravitation whose defining feature is its use of Einstein's Field Equations. In his discovery of these equations, he does not identify the source of the energy to make the tensor work. He simply states that in curved space, there is an additional source of energy; Einstein calls "gravitational energy," without it the energy-momentum tensor will not give us the desired "gravity lensing[218]." Defining gravity with a gravitation force resulting in a curvature of space-time

---

[217] Equivalence principle appears to work well for proportionally tiny objects dropped at relatively short distances over massive object generating the gravitational field. This principle fails when the objects dropped are significantly increased or if the distance of fall extended many times over, as in a slow decaying orbit.

[218] The source of Einstein's "Gravitational energy" can be seen in experiments conducted by quantum physicists. Every time they try to find out information about a particle during the conduct of

is as meaningless as defining light-speed with the speed of light or a second of time with second per second.

In addition, scientists and physicists mistakenly say that the time differences between the cesium clock readings on space vessels and satellites like the Global Positioning System (GPS) than that on the cesium clock on the ground are the result of time slowing down due to speed of the orbiting object as defined in the Special and General Theory of Relativity. This is not true. It is the result of data from the earth's only based cesium clock being transmitted to the satellites and space vessels, with the error caused by frequency shift Doppler Effect, which we all know is electronically corrected by the GPS as a known clock discrepancy and has nothing to do with the time delay or shift. The manager of the world's only cesium clock, which supplies the entire world with the most efficient time based on atomic vibrations of cesium in a highly controlled and undisturbed environment. Even the cesium clock itself has good days and bad days, with ever so slightly different readings by nanoseconds depending on the vibration of the atoms that day, hour, minute, and second. It is also a known fact that separate cesium clocks run differently due to their environment and conditions. So, a cesium clock in a jet or spaceship runs at a different rate than that on the ground within a controlled environment.

Take for instance Hafele and Keating's experiment flying four cesium atomic clocks strapped to two seats within 747 jets in October 1971 traveling first eastward around the world and then westward around the world. Their results supposedly show us that the clocks traveling eastward had lost time, public released time of 59 ±10 nanoseconds, or aged more slowly and the clocks traveling westward supposedly had gained more time, public released time of 273 ±7 nanoseconds, or aged faster. Hafele and Keating concluded that the earth's rotation affected the readings slightly but supposedly still confirmed Einstein's time dilation predictions. Since the earth rotates from west to east, then an airplane traveling at an average speed of 920 km/h eastward would in essence experience a partial added acceleration due to the earth's west to east rotational speed of 1675 km/h without increasing ground speed, and its cesium clock should run slower per Einstein. While an airplane, traveling westward should experience a partial deceleration due to the earth's rotation without changing ground speed and its cesium clock should run faster per Einstein. Supposedly, Hafele and Keating confirmed both aspects by their experiment.

Let us take a careful look at Hafele and Keating experiment parameters. Were the cesium atomic clocks adequately shielded from the impact of any external magnetic field, specifically the earth's magnetic field? Yes, they were shielded from any external magnetic interference three times over. Why did they use four cesium clocks? It turns out that their accuracy was not as reliable as advertised and gave occasional unpredictable readings especially when in motion, in surges of electricity, or temperatures and humidity fluctuations, so Hafele and Keating averaged out the four clock time readings. Even the ground-based clocks they compared their averages to have drifted from the correct time. Were the cesium clocks sufficiently stabilized during flight to operate effectively? No, there were many factors such as air turbulence and contributing errors already listed above which of course invalidation of results. Were their exact time readings of all four cesium

---

an experiment, they change the speed, direction, and/or location of the particle. In other words, the cause of gravity, moving space-matter, changes the direction of the starlight photon, every time it meets, absorbs, and re-emits that starlight.

CHAPTER 7: INSPIRED THEORIES ANALYZED   249

clocks made public? Yes, they were finally published (Kelly, 1998). Did Hafele and Keating consider the time it took the Naval Observatory's cesium clock reading to propagate through circuit switches and undersea Atlantic copper cables from the east coast of the United States to London or wherever they landed while traveling around the world? Yes, however, the clock times at landing locations also had drifting discrepancies on occasion. Data collected from the four cesium clocks flown in the airplanes could not even agree with each other on which had the correct "accurate" time let alone document a correct or consistent time drift. Bottom line, the time drifts (aging slower or faster) results contained in the previous paragraph above published by Hafele and Keating supposedly confirming Einstein's time dilation prediction were shown to have been unjustifiably altered data, therefore invalid, and should be stricken from such lists used by academia to propagate Einstein's ingenious mirages.

As we said earlier, we hereby decline usage of the following: Einstein's Special Relativity and portions of General Theory of Relativity, the principle of the constancy of the velocity of light, time dilation and the speed of light as the maximum velocity an object can achieve. As discussed in Part 2 Chapter 4, "The speed of light in a vacuum 2.9979E+8 m/s is the same for all observers within the same system, regardless of their relative motion or of the motion of the source of the light but is different for observers of different systems per Galilean relativity." Einstein's second assumption as originally documented for his special relativity theory is an incorrect assumption. As such, the speed of light is not the maximum speed of any particle or object. Various circumstances throughout the universe show objects exceeding the speed of light. Hence, the principle of the constancy of the velocity of light invalidated and stricken as a principle.

The stories of the twin paradox; the motor vehicles racing past each other; and that of a rocket traveling 2/3 the speed of light fired from a plane going 2/3 the speed of light are all nonsense and unscientific imagination. Within the confines of today's technology, they are pure science fiction. The Cosmological Balance gives us many examples contradicting Einstein's theories; however, the curvature of space is only in agreement between both theories. The mergers of any two large galaxies are examples of objects (entire galaxies) potentially moving faster than speed of light as they collide, predicted based on the calculated accelerated merger of Milky Way and Andromeda Galaxies solely on mutual gravitational attraction, which is fueled by the vacuum force of space and quantum gravity.

Einstein predicted that with an increase of speed, especially those approaching light, results in relativistic increases in mass, thereby making it impossible to accelerate an object to the speed of light (Einstein/Golm/Potsdam, 2015), and claiming as a result that the accelerator at Brookhaven National Laboratory is only able to race electrons around at 99.9999 percent of light speed. According to Collier's interpretation of Einstein, he said Einstein's second postulate (the constancy of the speed of light) necessitates and destroys all assumption of absolute time and space (Collier, 2014). Really now, the way we see it, there are no mathematicians or human beings with the power to tell the universe what to do or not do, not even Einstein and his eloquent equations. As for the accelerator, these scientists failed to see that their machines are electrical and operate at rates limited by the speed of light, therefore they unable to push any sub-atomic particle faster than the electrons within their machines. The inability to reach the speed of light or exceed it has nothing to do with a relativistic increase of mass, only the mechanical limitations of their devices[219].

Let us look at this from another angle, light, and its particles. We know that the photon and electron pair traveling as light moves at the speed of light in a self-sustaining electromagnetic interaction. Even though scientists say these sub-atomic particles are "mass-less," they do in fact have some minuscule mass to be particles. If Einstein's prediction was right, then the infinitesimal mass of the photon and electron traveling at the speed of light obtains a maximum relativistic increase thereby colliding with the sunbather on the beach with a mass of a thousand times or greater, potentially punching holes into their skin. We in fact know that this is not what happens when ordinary sunlight hits the skin. The mass of particles of light remains constant as it travels at light speed, carrying with it enough energy to burn through the atmosphere and on impact with the sunbather converts to heat, as sunburn, without the presences of "holes" in their skin. Incidentally, with the aid of a magnifying glass, one can concentrate the sunlight into a highly energetic point capable of heating up and possibly burning items at the focal point.

Let us look another example. A comet passes the earth on its way toward the sun. If mass relativistic increases with speed, then with the comet's increase of speed its mass increases and it should gain additional gravitational attraction to the sun, thereby incrementally change course and plunge directly into the sun instead of following the path defined by Newton's equations since its discovery. That comet was Haley's comet, which of course does not obey Einstein's relativistic increase of mass. This comet follows the laws defined by Newton, and his $F = G\ m_1\ m_2\ /\ r^2$ equation. The same applies to the moon and satellites orbiting the earth, or to any planet or object orbiting the sun, there is no increase in mass with velocity, only an increase of energy.

Einstein also predicted that merger of two black holes would send tremendously strong gravitational waves outward, warping the fabric of space-time. Posted on the New York Times (www.nytimes.com) was an article (Overbye & Corum, 2016) stating, "A team of scientists announced on Thursday (February 11, 2016) that they had heard and recorded the sound of two black holes colliding a billion light-years away, a fleeting chirp that fulfilled the last prediction of Einstein's general theory of relativity." INFOWARS www.infowars.com (NASA, Detects Gravitational Waves, Just as Einstein Predicted, 2016) also published the same news information "Albert Einstein predicted the existence of gravitational waves in his general theory of relativity a century ago, and scientists have been attempting to detect them for 50 years. Einstein pictured these waves as ripples in the fabric of space-time produced by massive, accelerating bodies, such as black holes orbiting each other. Scientists interested in observing and characterizing these waves to learn more about the sources producing them and about gravity itself, LIGO detections represent a much-awaited first step toward opening a whole new branch of astrophysics."

Einstein predicted that gravitational waves move outward from the collision of two or more massive objects in space, namely black holes. If the gravity wave moves outward from a massive object, then how can that massive object pull other smaller objects toward

---

[219] The same dilemma applies to making spacecraft move closer to or near the speed of light; the propulsion must be at least as fast as the speed we would like to achieve and a thrust greater than the mass of the ship.

them? The resulting contact would be more of a push away than pull toward the source. The Cosmological Balance Theory states that gravity effect or space-matter vacuum force moves two objects toward each other, primarily pushing the smaller object to the massive object, not away, and augments the mutual attraction. Therefore, if the gravity wave moves inward, then how can the detection published on Feb 11, 2016 of an event "heard" on Sept 14, 2015 at 4:50:45 am Eastern Standard Time (EST) be correct. Are we detecting something that happened over 1 billion years ago or something that will happen 1 billion years from now? This is another one of Einstein's quandaries. The proximity of the Sun with its own gravity to us surely is stronger than that of the distant fading so-called gravity "ripple" caused by the merging of two black holes, a billion light years distance away.

It turns out that LIGO did not actually detect or "hear" the "ripple of gravitational waves" as they announced; it captured and recorded "space-matter wave" interferences and repulsions. What did the instruments in LIGO really detect? LIGO detected the result of the prediction set forth by the Cosmological Balance Theory. The collision and merger of two black holes in space would release and expel a quick tremendous surge of antigravity white hole sub-atomic orbs into massive amounts of wormhole tunnels in all directions, after which the new massive black hole then stabilized into its normal excretion process. This sudden mass release of antigravity orbs then travelled outward in one intense wave for few hundred light years distance or slightly less before collapsing and releasing the antigravity into space. In the part of the universe, we cannot see space-matter waves traveling outward, but to the invisible part of the universe where galaxies consist of antigravity antimatter, these antigravity waves move inward toward antigravity objects. The merger of the black holes also sent a shockwave via space-matter electromagnetic or electrostatic pulse traversing outward from one particle to next particle and space-matter atom to atom. By the time this space-matter vibration or as Einstein puts it a "ripple in the fabric of space" reached our location it intersected with normal space in this Milky Way and bent and warped ordinary matter in the solar system. The crossing of two intersecting space-matter waves coming into the solar system resulted in the interference the LIGO scientists detected. As the stable space-matter within the galaxy was hit with that of the space-matter wave of incoming particles generated by the collision of two massive black holes a billion light years away, they push the solar system in an uneven vibration pattern our instruments picked up, like ripples of intersecting waves in a lake.

Antigravity wormhole tunnels departing from the galactic center move either with or against the sun's orbital direction without crossing paths and causing interfering antigravity ripples. This typical interaction between gravity matter and antigravity antimatter is normal; it is what causes the sun's and the local star groups' rapid orbital speed around the galaxy. However, space-matter electromagnetic wave interference with antigravity and gravity coming from outside the galaxy, although minute, is detectable by the LIGO sensitive instruments. Bottom line, on September 14, 2015 at 4:50 am, we detected the quick sudden momentary disruption of the normal patterns the Sun in orbit around the galaxy and the planets get. This bending and warping caused by a wave of space-matter particles traveling near the solar system intersecting normal space to form another "ripple" wave coming into the galaxy. Our scientists then married that detected event with the confirmed observed collision of two black holes to justify Einstein's prediction and therefore justify the billions of dollars spent to construct the LIGO in Washington and Louisiana. The result shows Einstein correct only if we view the inbound wave as movement in space-matter grid. In any case, as an after effect from this event we may expect to see a few new comets and asteroids possibly knocked out of their orbit begin their plummeting descent toward the sun in the next few years to a decade, or not.

In careful and detailed examination of Einstein's papers, we can cite the contributions of Galileo Galilei, Isaac Newton Hermann Minkowski, James Clark Maxwell, Max Karl Ernst Ludwig Planck, Albert Abraham Michelson, Ludwig Eduard Boltzmann, Henri Poincaré, and others. Minkowski was known for work in combining Euclidean space with time, in a four-dimension manifold of spacetime. Maxwell was known for his work in electromagnetism and discovery that light and other electromagnetic waves in the spectrum travel at the speed of light $c = 1 / \sqrt{(\mu_o \varepsilon_o)}$, where the magnetic constant μ_o = 1.25663706E-6 and electric constant ε_o = 8.85418782E-12. Planck was famous original work in quantum theory. Michelson was known for his measurement of the speed of light. Boltzmann developed statistical mechanics to explain atoms and their properties. Poincaré worked on applied mathematics, mathematical physics, and celestial mechanics. Most of the discoveries before Einstein were in fields of mathematics, micro-atom, and quantum physics. Einstein used their work of scientists before him to develop his theories. Remember that the three forces of nature that governs the micro world, and quantum physics, does not apply to the macro and the universe where gravity rules as dominate fourth natural force. The strong nuclear force, the electromagnetic force, and the weak nuclear force reign only in the micro world of atoms and molecules, and that of quantum physics. Bottom-line, Einstein's general relativity equations are perfectly suitable for describing light and the gravity lensing of it, because light operates in the realm of quantum physics. Otherwise, general relativity is not necessary for describing gravitational effects on macro objects, like the stars, planets, asteroids, and other debris; Newton's gravity laws are fine to these objects. We will find out later that space-matter provides the curvature of space and accounts for the bending of light via a unique space-matter light refraction; it is the quantum mechanical source of Einstein's invoked additional "gravitational energy."

Einstein's work in electrodynamics of moving bodies and general relativity are based primarily on theories developed before him for micro and quantum physics, and thereby not applicable to the macro world, neither to the motion of stars and planets, nor the galaxy and the universe[220]. The extrapolation of laws governing the micro world and quantum physics into the macro world of gravity is at best problematic. Special Theory of Relativity is as we discussed a misinterpretation of the mirage seen by Einstein in his imagination caused primarily by the Doppler Effect. A mirage where he sees a sphere flattening, a bar shortening, an airplane seemingly shrinking are all examples of purely results of Doppler Effect fooling the eyes of the observer, and not achieving the results on the actual objects as he claimed. The scientists of his time are analogous to the thirsty nomadic people in the desert who see a lake at the horizon before them, only to find that when they reach the supposedly location find nothing, all deceived by the magic of Einstein's equations, concepts, and theories. General Theory of Relativity, based on the

---

[220] The work Einstein did for Special and General Relativity is based on the laws and rules governing the micro world of quantum physics, and as such is therefore only pertinent to particles within that realm. Hence, gravity Lensing accurately calculates the quantum interaction between space-matter and the particles of light, its photon and electron, and has no influence on larger objects within the macro universe, which is govern by Newtonian gravity.

work within Special Relativity, too is unnecessary and hereby will not be used as well. In the grand scheme of things, the universe does not obey Einstein or his theories and any of his sound mathematics; the universe is indifferent. Einstein believed otherwise and focused on finding that unified theory and disproving quantum mechanics during his last thirty years of his life. His dedicated efforts on this subject alone are what distanced him from mainstream physics. Scientists today have ranked the search for such a unified theory as the most important problem in modern day theoretical physics. However, obstacles like the conflicts and tensions between general relativity and quantum mechanics make this endeavor very challenging to say the least (Greene, The Fabric of the Cosmos: Space, Time, and the Texture of Reality, 2004, p. 14).

Andrew Thomas is among those scientists looking for the unified theory. Andrew Thomas, (Thomas, Hidden In Plain Sight: The simple link between relativity and quantum mechanics, 2012) notices the similarities between special and general relativity, and quantum mechanics at their fundamental principle, underlying cause, side effects, and limitation level in an effort to build an argument tying and combining the two theories together. He then lists these "glitches" in behavior common to both in a table. In doing these comparisons, he forgot to mention or observe that special relativity theory and consequently general relativity were based on the electrodynamics of moving objects and misinterpretations of the Doppler Effect of light at the quantum mechanical level, clearly the root of these similarities. Since we know that Einstein's extrapolation of electromagnetic and quantum effects does not apply to macro level of the universe, then Thomas' similarity arguments are meaningless toward finding the sought-after unified theory of quantum gravity.

Consequently, we should question the validity of all other theories and countless mathematical work based on Einstein's famous equations, concepts, and theories. Although Einstein's name is intricately linked with the physicists celebrated relation $E = mc^2$ between mass and energy, it does not make total sense. In it, he states that Energy is equal to $m * 8.9876E+16$ kg m^2/sec^2, where m is the mass and c^2 is a made-up number light speed squared, confirmed by other physicists, and not actually proven scientifically. Through this equation, Einstein predicts that the energy contained in one-kilogram mass is equal to $8.9876E+16$ Joules, an amount not only scientifically unjustified, it is impossible to have energy without the presence of matter and mass, simply because energy is defined as the motion or potential movement of matter. A critical examination of the more than half dozens of "proofs" of this relation that Einstein produced over a span of fifty years reveals that all these proofs suffer from the perpetuation of these mistakes. It is time to correct Einstein's errors in logic and math, correct his unjustified ideas and assumptions, and replace or modify his outdated concepts of low-speed and restrictive approximations. We should applaud and encourage courageous individuals such as Hans C. Ohanian, Physics Department, University of Vermont, and many other physicists like him to continue to step up to the plate and help correct our scientific endeavors and the great path of humanity. Consequently, we should also question the work of Professor Stephen Hawking as he too stood on the shoulder of Einstein to claim his place in the world of mathematics, physics, and astrophysics.

The theory of quantum gravity[221] sought after by scientists should come from the unification of the following three: Galilean relativity, Newton gravity, and Quantum mechanics theories. Qualities of Quantum mechanics contributing to unification are particle interaction, multi-valued superposition, innate connectivity, infinite range, instantaneous and simultaneous reaction, accepted macro-level absolute time defined with cesium or quark vibrations, and proposed macro-level absolute measurement defined by radius of ^{1}H hydrogen atom (protium). Qualities of Newton gravity contributing to unification are object interaction, mutual connectivity, infinite range, instantaneous and simultaneous reaction, and multi-valued motion toward and away from multiple objects, gravity equation dependent on absolute time and absolute space. Galilean relativity contributions to unification are observed motions in stationary and constant speed are identical, accelerated motion effects interactions, laws of physics apply in all states and frames of reference, and the existence of absolute time and absolute space.

The vibrations of the quarks, and consequently the motions of the protons and neutrons, within the nuclei of an atom control the speed and direction of the electrons orbiting that atom. This effect can be visualized easily if we think about David spinning the sling with the stone cradled in the pouch during his epic battle with Goliath. The faster David swings his arms the faster the stone orbits around him, until the moment he releases it and hits his target. The steadier the motions of his shoulders, and consequently his arms, the firmer and more uniformed the orbit of the stone. Hence, the time keeping vibrations of the quarks, causes the protons and neutrons to move with similar motions, and thereby causing the electrons to orbit in consistent and uniformed orbit around the nucleus. In another example, the central pendulum or rocking gear in a clock provides the steady ticking necessary for the rest of the gears within the clock to keep accurate time. Here, the center pendulum timekeeper within the atom is the quark vibrations, the gears carrying that vibration to the hands of the clock is the motion of the protons and neutrons, and the hands of the clock are the movement of the electrons. We are so obsessed with accuracy that we use the vibrations of cesium atoms to digitally count seconds for controlling worldwide electronic clocks.

Time is an illusion unchanged by observation regardless of the observer's distance, speed, or gravitation field. The actions of light, as it moves through various mediums and fields or as viewed by moving observers, slowing down and relative returning to normal speed, gives our scientists the illusion of time slowing down and speeding up. Do not let the scientific magicians fool you with their science fiction twist of the Doppler Effect as Einstein did with the professors and doctors of academia in the early 1900s. Newton on the other hand believed that absolute time is determined specifically at the source or origin clock and not elsewhere; therefore, perceptions of time or any visual distortions thereof seen by observers not at that location are exactly that perceptions of mirages, and are irrelevant in identifying the proper time at the source clock. Similarly, the source clocks' times always determine simultaneity, and any observers' timing not at that source has no impact on true result. Newton's gravity equations are based on absolute time.

To analyze Einstein's visual Doppler Effect distortions, let us take sound for instance and evaluate what happens. Hearing a higher pitch or lower pitch of a moving siren does not change when and how long the sound has been transmitting. The same applies to light.

---

[221] The Theory of Quantum Gravity it what scientists seek as the unification of quantum mechanics and general relativity (Einstein) gravity. Cosmological Balance Theory alters this slightly because general relativity gravity is actually an application of quantum mechanical physics.

Seeing a lower or higher frequency of light does not change the actual time or duration of the transmission nor the frequency emitted at the origin. What observers see and record does not change time itself at the source, just their perception of it. Einstein prediction of gravitational time dilation is inaccurate; a distant galaxy showing extreme red shift supposedly racing away at multiples faster than the speed of light cannot possibly have extremely slow or motionless clocks. Likewise, black holes cannot have little or zero in time ticking. We are but pawns guided by the time masters of the universe, absolute time all moving in unison. No matter how hard we look for loopholes to traverse into the past or jump forward to the future, there is none. We can only peer into the vast distance of space and see a glimpse of the past. Traveling in spaceships faster than the speed of light to those distant stars is not going to allow us to get to their past, only to their present.

With the questionability of Einstein's special and general relativity[222] and this new quantum defined distance and time, we can determine that a particle or object traveling specified absolute distance over a given absolute time gives us an absolute speed, and at a given direction gives us an absolute constant velocity necessary for Galilean relativity. Galilean relativity then tells us that if the whole system was moving at a relative constant speed, of say 236,000 km/h, then the absolute distance measure remains the same and the absolute time measured remains the same in the constant moving system as if both were measured from a stationary system. This gives us the same absolute constant speed or velocity relative to the whole system, whether the system is a planet with orbiting moon, or a star with orbiting planets, or a galaxy with orbiting stars, etc. Likewise, the change in speed, acceleration, or direction gives us an observed difference of motion noticeable by physical laws. Known distances between centers of objects combined with their defined total atomic mass information, which are quantum defined, gives us Newton's gravity equations; incorporating the directions and the velocity of these objects and the mutual attraction of gravity gives us the changes to their paths and velocities, and enables us to determine if these objects are in orbit or not. Similarly, Newton's gravity equation defines the heat, pressure, and internal structure of stars, to include the structure within a neutron star and a black hole; and provides insight to interactions at the quantum level. Galilean relativity tells us that that star or remnant of a star will maintain that structure even while in motion, and that quantum physics functions consistently within it, and that the planets or other objects orbiting that star follow Newton's laws. Clearly, mutual balance exists between all three.

---

[222] Reminder to readers, general relativity equations are accurately applicable to the motion of light or more specifically starlight passing through the "halo" surrounding massive objects. This halo consists of densely packed space-matter and is also known as the curvature of space.

7.7 Bending of Light by the Sun

As we saw in part 1 chapter 1.9, the earth's atmosphere refracts the sun's light allowing us to see the sun two minute before it appears above the horizon. The same refraction occurs throughout the sun's rising to vertical position and during its decent to sunset and an additional two minutes of visibility on sunset. The same refraction occurs with stars in the sky. We also saw that an astronaut, looking out the space station window, witnessing that sunrise from the surface of the moon during an earth eclipse sees the sun four minutes before it comes around the earth, about one degree from its actual position in space[223]. We can almost hear these lucky astronauts voice their acclaim that this is truly a spectacular view and awesome experience!

If we look at the bending of light by the earth's gravity alone, we find that a ray of light going from the sun or a distant star to the astronaut on the moon's surface, during an earth eclipse or dark side of the earth, calculates out to bend the light a total of 1.78E-02 meters or about 2.65E-09 degrees. However, if we consider the refraction of the same ray of light by the earth's atmosphere as it travels at least 10 km above the earth's surface on its closest approach, the bend is much greater. It will bend multiple times by different density layers totaling about one degree of refraction when seen by the astronaut within the geosynchronous space station or on the moon. Total bend observed by the astronaut with refraction and gravity comes to 6.658E+6 meters or about one degree of angle between the sun's actual location in space behind the earth and perceived location during sunrise or star-rise as seen from the moon's surface. An observer on the moon records the bending by gravity effect alone and by gravity and refraction combined.

Altitude	Bend by Earth Gravity	Gravity and Refraction
10 km	2.65E-09 degree	.992 degree
100 km	2.65E-09 degree	.496 degree
150 km	2.65E-09 degree	.397 degree
200 km	2.65E-09 degree	.298 degree
300 km	2.65E-09 degree	.1488 degree
400 km	2.65E-09 degree	.0496 degree
500 km	2.65E-09 degree	.00496 degree
1000 km	2.65E-09 degree	9.92E-05 degree

This is a depiction of the light passing near the sun and the refraction the sun's atmosphere causes.

Now let us look at observations of stars behind the sun during solar eclipses. Using the same calculations for gravity

LIGHT REFRACTION ANGLE
AFFECTED BY DENSITY

---

[223] These refraction estimates are based on atmospheric light bending not from the curvature of space around the earth. Einstein's equation can provide insight to gravity bending just above the earth's atmosphere. General relativity gravity lensing is the same as the space-matter light refractions (optical lens).

CHAPTER 7: INSPIRED THEORIES ANALYZED

alone and for gravity and refraction bending possibly caused by the clear dense colorless heated atmosphere above the corona, we get the following table:

Altitude	Bend by Sun's gravity	Gravity and Refraction
140 km	2.26E-06 degrees	2.4927 degrees
150 km	2.26E-06 degrees	2.2935 degrees
175 km	2.26E-06 degrees	1.9946 degrees
200 km	2.26E-06 degrees	1.7454 degrees
300 km	2.26E-06 degrees	1.0973 degrees
400 km	2.26E-06 degrees	0.7981 degrees
500 km	2.26E-06 degrees	0.6983 degrees
2100 km	2.26E-06 degrees	0.1995 degrees
100000 km	2.26E-06 degrees	2.26E-06 degrees

Surely, if you compare possible refraction and gravity results from the measurements taken, they are close. The sun's atmospheric refraction index is greater due to the density levels of the upper atmosphere and its energized heated particles. We call this refraction and gravity effect, where the star appears to be 140 km or greater above the surface of the sun, star-rise over the sun. This star-rise can be seen during solar eclipses, whether manmade or natural, where the star's actual position is behind the sun but appears to be above the sun, if we knew the exact density levels of the upper solar atmosphere. As always, when looking at the sun use the aid of devices to protect your eyes.

Below is a depiction of the effect of light refraction and gravity at 150 km above the sun's surface during solar eclipse:

DISTANT STAR AT 150,000 m ABOVE SUFACE OF SUN DURING ECLIPSE

Star appears to be here

Path due to sun's gravity acceleration: 2.2618E-06 degrees

Curved path due to sun's gravity coupled with density refraction, 2.29 degrees

2.29°

Normal path starlight takes when sun is elsewhere

150 km

The results of this bending effect are but one possibility caused by the sun's gravity and atmospheric light refraction combined[224]. Such results would have nothing to do with the curvature of space and time, as claimed by Einstein, if only we can prove that the solar atmospheric density at that altitude was enough to refract light. Einstein is a great physicist with a vivid imagination. In coming up with his explanation, he probably assumed that the sun had no clear or translucent atmosphere. For now, we could agree with Einstein as we

---

[224] The Cosmological Balance Theory's definition of Newtonian gravity is the combined effect of space-matter push of 6.6734E-11 Newton and quantum mechanical pull of 1 m/kg from each object. The quantum mechanical pull by itself is insufficient to bend light significantly, but the space-matter atom impact can effectively push light particles at the quantum level and make it bend with an additional "gravitational energy" as predicted by Einstein.

have yet to discover exactly how far the sun's atmosphere out actually extends and with what density levels (NASA, Layers of the Sun, 2012). Again, the lensing effect general relativity predicts is hinged on the equivalence principle, which disregards the mass of the particles in light and bends its path all the same. In any case, he convinced the world with his fine-tuning of Newton's gravity equation with the invention of the curvature of space and time, and his use of some fictitious numbers to get his equations to work. It turns out that Newton's gravity equations within Principia is accurate enough, as Newton's method works fine solving Mercury's orbital anomaly as demonstrated in the section above. **Sir Isaac Newton triumphs again!**

By the way, recent confirmations of the initial measurements taken in 1919 to show Einstein bending of light around the sun (i.e., his prediction of the lensing effect) continue to confirm the phenomena. Who knows what the solar probes may find today? We highly recommend that today's astronomers continue to analyze the solar atmosphere to figure out the cause of the corona's excessive heat buildup. Reconfirming the data may prove that refraction and gravity combined are possible factors, inadvertently hidden within the lensing equations, which give us the results we observe. The refraction of light by the sun's energized translucent outer atmosphere may partially cause the bending of the star's light, and the curvature of space-time is a perceived result. We therefore conclude that the lensing effect may be a combination of gravity and refraction. For now, Einstein's curvature of space, with normal time, is the best starting theory we have where we can make improvements. Recommend that our scientists look for a clear atmosphere composed of inert gases, probably argon, above the flaming hot corona, which is responsible for reflecting ultraviolet light and heat back toward the sun and the bending of light passing through it. Even the presence of gaseous material spread evenly in outer space can attribute to the lensing effect, particularly accumulated around galaxy halos. If not, we have a Cosmological Balance Theory solution.

As we discussed above, light refraction may be partial contributor to the bending of light, but without additional evidence, it cannot effectively replace Einstein's energy-momentum tensor work. For now, Einstein's lensing effect may stand as a good theory to calculate this phenomenon, but the Cosmological Balance Theory has a better explanation. Einstein's lensing is hinged on the principle of equivalence. The equivalence principle states that gravity and acceleration are indistinguishable. In other words, an observer in a free-falling state would not feel the effects of gravity and an observer in constant acceleration can interpret the resulting inertial force as the effects of a gravitational field. However, since we already proved that the equivalence principle is not valid for all circumstances then it is not a principle, and it fails[225]. What is the real cause for light bending around the sun? Is it gravity lensing or light refraction or both?

EINSTEIN GRAVITY LENSING EFFECT EXPLAINED

Let us look again at Newton's gravity equations and apply the equivalence principle to it. If we disregard the mass of the object as it passes by the Sun, namely using the sun's mass to estimate the path light takes as it races nearby. Refer to the "Introduction to Gravitational Lensing, Lecture scripts" by Massimo Meneghetti, for details. The

---

[225] Equivalence principle is based on experiments using macro objects over short distances and object of miniscule proportions to the mass generating the gravitational field.

equivalence principle allows us to estimate the acceleration of a body in a gravitational field, independent of its mass, composition, or structure. All we need to know is the initial vector of it as it passes by the Sun. With that said, we estimate its acceleration toward the sun as

$d^2r_v/dt^2 = -GMr_v/r^3$,

where r_v defines the location of the light particle in the gravitational field of the object, in our case the sun, with mass M, and G is Newton's gravity constant.

The above equation provides us with conic sections describing the motions of the particle as it passes mass M. Since light travels faster than escape velocity of the sun, we get a hyperbolic path, which can be written as:

$r = R(1+e) / (1+e \cos\varphi)$,

or

$r^2 d\varphi/dt = \sqrt{[GMR(1+e)]}$,

where e is the eccentricity of the conic section or orbit, R is the radius of the particle's closest approach to mass M, G is the gravity constant, and φ is the angle of bending. In polar coordinates, r and φ together define the location of the particle of light in relation to the mass M, and allows us to draw light's path given R.

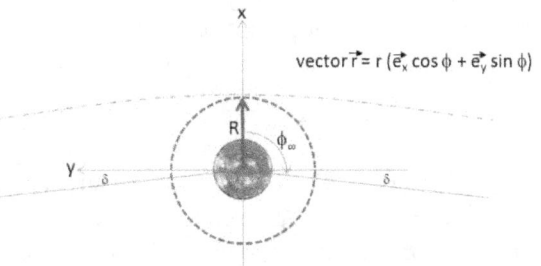

Equivalence principle & Newton gravitational bending of light.

Hence the resulting vector

$r_v = r (e_x \cos \varphi + e_y \sin \varphi)$,

where e_x and e_y are the x and y coordinates of the vector along the hyperbolic path. Therefore, the velocity v is

$v^2 = GM/R(1 + e) * (1 + 2e \cos \varphi + e^2)$.

For r →goes to infinity, the path of the particle of light achieves the angle φ in the diagram above, when $(1 + e \cos \varphi) = -1/e$, and angle φ when r reaches infinity is approximately equal to $\pi/2 + \delta$, where δ is half of the deflection angle, or $\sin \delta = 1/e$.

The equation for the eccentricity is:

$e = Rc^2 / GM + 1$,

where R is the radius of the closest approach light particles makes to the mass M, and c is the speed of light.

The mass of the sun and its radius are:

$M = 1.989 \times 10^{30}$ kg

$R = 6.963 \times 10^8$ m

This equation then resolves to

e = (6.963 x 10^8 m) * (2.998 x 10^8 m/s)^2 / [(6.6734 x 10-11 N m^2/kg^2) * (1.989 x 10^{30} kg) + 1

= 471524.9365

And the deflection angle at the surface of the sun (radius R) calculates to

2δ = 2GM/c^{2}R = 8.48313E-06 radians or 0.000486 degrees

Deflection angle at 140 km above the surface of the sun is estimated at 0.00048595 degrees, at 150 km is 0.000485943 degrees, at 175 km is 0.000485926 degrees, at 200 km is 0.000485908 degrees, at 300 km is 0.000485838 degrees, at 400 km is 0.000485769 degrees, at 500 km is 0.000485699 degrees. At 2100 km, it is estimated at 0.000484586 degrees, and at 100,000 km is 0.000425009 degrees. These estimates are greater than that calculated with just Newton's gravity acceleration equation, shown in the table at the top of this section, but is not sufficient to account for the entire bending we observed and recorded.

Einstein believed that the curvature of space-time accounted for the rest of the bending observed, he coined it as gravity lensing. To reiterate his justification, he claimed that in curved space, there is an additional source of energy associated with gravitational radiation; Einstein calls "gravitational energy[226]," which provides the necessary force for the energy-momentum tensor to complete the rest of the bending of light known as "gravity lensing." Note that he claimed the space curved around massive objects but did not explain how and what provided the additional source of energy. The Cosmological Balance Theory explains both.

In continuation of the light bending analysis, Meneghetti starts with Snell's law, which gives us the following equation sin θ_I = n sin θ_R. and then uses the index of refraction n from Fermat's principle to discuss variations of paths light takes around a massive object. Introduces the "weak lens" per Newtonian gravity (in the section above), and then goes on to solve the speed of light in a gravitational field, and how Euler equations influence its path. Meneghetti goes on to explain in various equations associated with light bending, deflection, and gravitational lensing to come up with Einstein's work and definition of the Einstein radius.

$$\theta_E \equiv \sqrt{\frac{4GM}{c^2}\frac{D_{LS}}{D_L D_S}},$$

From which we have

$$\beta = \theta - \frac{\theta_E^2}{\theta}.$$

Dividing by θ_E and setting y = β/θ_E and x = θ/θ_E, the lens equation in its dimensional form is written as

$$y = x - \frac{1}{x}$$

Multiplication with x gives us:

$$x^2 - xy - 1 = 0,$$

---

[226] Gravitational energy is just a fancy term for quantum mechanical interactions between space-matter and the photon of light. Moving space-matter in the curvature of space briefly absorbs starlight passing through it and then re-emits it at a greater refractive angle than gravity alone, Einstein's general relativity equation captures this identical effect as "gravity lensing."

CHAPTER 7: INSPIRED THEORIES ANALYZED

which has two solutions:

$$x_{\pm} = \frac{1}{2}\left[y \pm \sqrt{y^2 - 4}\right].$$

Here is an image to visualize Einstein's radius equation.

Hence, the point-mass lens has two images for any source, regardless of the distance y from the lens. If y = 0, ± x = ± 1; in other words, a source object behind the point lens has a ring-shaped image with radius θ_E. For order-of-magnitude estimates:

$$\theta_E \approx (10^{-3})'' \left(\frac{M}{M_\odot}\right)^{1/2} \left(\frac{D}{10\text{kpc}}\right)^{-1/2},$$

$$\approx 1'' \left(\frac{M}{10^{12} M_\odot}\right)^{1/2} \left(\frac{D}{\text{Gpc}}\right)^{-1/2},$$

Where

$$D \equiv \frac{D_L D_S}{D_{LS}}$$

is called the effective lensing distance.

As β → infinity, we see that $\theta_- = x_-\theta_E \to 0$, while $\theta_+ = x_+\theta_E \to \beta$; when angular separation between the lens and the source becomes large, the source is not lensed. Even $\theta_- = 0$ gives us an image.

Meneghetti goes on from there to introduce several other applications and predictions about lensing around galaxies, and other large, massive objects in space.

COSMOLOGICAL LENSING EFFECT

The Cosmological Balance does not refute Einstein's work when it comes to lensing of light passing near a massive object. We agree with his equations and the accuracy of the results it produces. We, however, have a slightly different definition of the curvature of space-time. The Cosmological Balance Theory envisions the typical grid of space-matter to form concentric spherical area surrounding extremely massive objects to contain multiple layers of space-matter arranged with increasing density approaching the massive object's outer atmosphere. The point of contact between space-matter and ordinary matter is the transfer of vacuum force into gravity on the ordinary matter. The layers upon layers of densely packed space-matter atoms are what give the sun and the planets their spherical shape. The transition of force between space-matter pressure and ordinary matter with gravity is what Einstein coined as the curvature of space-time. The Cosmological Balance Theory only refers to it as the curvature of space, and not of time.

Think of it this way. Air molecules surrounding the soap bubble, give it its spherical shape; they suspend a soap bubble floating in a room. Likewise, the space-matter surrounding the massive planet gives it its spherical shape, provides the vacuum force, and source of gravity force; space-matter also suspends it in space. Just like the air molecules, surrounding the soap bubble, the greatest contact between both is at the bubble's surface. Hence, the greatest contact between space-matter and ordinary matter with gravity is that the upper known layer of the massive object, its atmosphere, or its solid surface if it does not have an atmosphere. That contact surface area is the layer containing the densest concentration of space-matter, and the greatest light refraction or as Einstein calls it the greatest curvature of space-time[227]. We simply refer to this as the curvature of space.

Sun is depicted on both sides of picture. The right image represents the sun and its various solar atmospheres. Each solar atmospheric layer above the surface of the sun decreases in density as we move upward. The left image in the picture shows the changes in space-matter density levels from space (intra-galactic density level estimated at 1.025 atoms of space-matter per meter) to the lightest white-colored layer just before contacting the upper atmosphere of the sun, the corona.

The entire nested spherical space-matter layers' mass surrounding the sun combined contains the same mass of the object that space is trying to hold and exerts a total vacuum force of 6.6734E-11 Newton onto each consecutive level downward. Since the outside layer, far beyond the Oort cloud, is the most spread-out layer with the largest area, it is then the least dense, colored in black. Therefore, each inner nested layer becomes denser and denser packed space-matter atoms, colored in darker to lighter shades of grey, as we move toward the sun. The most densely packed space-matter layer, colored in white, is therefore the layer contacting the upper solar atmosphere. At that point, the space-matter density levels start to drop when we have contact with the outer atmosphere and continue to drop to practically nothing at the surface of the sun because that is where the densest solar atmosphere is. At that level, the combined density of the space-matter and the ordinary matter remains the same all the way to the surface of the sun, probably as at least 1/100th or less of the mass of the sun[228]. The space-matter and normal ordinary matter contact points are where the initial vacuum force transfers energy to ordinary matter as the force of gravity, through antigravity-gravity pulses. The overall spherical bubble layer around the sun and its entire solar system is also known as the solar halo, scientists may attribute this to "dark matter." Similarly, the spherical bubble layer surrounding the entire galaxy and its satellite galaxy clusters, sometimes referred to as the galactic halo, and declared by scientists as the influence of "dark matter." We see both halos as simply the edges of the start of the curvature of space around the objects.

---

[227] Einstein had to let go of preconceived assumptions to solve the tensors in general relativity. Also, recall that Einstein spent his last years of his life away from the physics community trying to solve the unification theory. We speculate that he did not return to the scientific limelight because the unified solution meant that he would have renounce his work on time dilation within special relativity and that in general relativity as the second and third greatest blunders of his life. Doppler Effect, time dilation, and gravity lensing of light are only applicable to the micro world of quantum physics and not to the macro universe of gravity. The perfect unified theory solution spelled a tremendous setback for him and his mark on the world.

[228] Exact calculations of the density of space-matter are currently undefined because the intensity of the curvature of space is dependent on the mass of the object or objects being contained within space.

Space-matter, as defined, can slow down light, absorb minuscule amounts of photon energy, and even refract light given sufficient density levels without changing its color[229]. Therefore, instead of a clear colorless upper solar atmosphere consisting of argon gas, we have multiple layers of space-matter densities that cause the unique light refraction we observed. The Cosmological Balance Theory results of gravity bending (with Newton's equations) and that of light refraction caused by densely packed space-matter combined to produce the same final numbers as Einstein's general relativity curvature of space equations, not time dilations. In other words, the presence of space-matter around objects in space is the source of the force of gravity and this densely packed space-matter's light refraction properties is what Einstein refers to as the additional "gravitational energy" required for gravity lensing energy-momentum tensor to work. However, since space-matter densities around massive objects are difficult to observe and calculate. For now, we temporarily adopt to Einstein's equations to predict with sufficient accuracy the path light takes around massive objects.

Once more, Einstein's curvature of space, without time dilation, is just another interpretation of the interaction and contact between space-matter and ordinary matter with gravity, where space is viewed and considered warped by the increased density of space-matter. These increasing density spherical layers of space-matter are nested one inside the other with the densest layer contacting the upper atmosphere of the massive object and then overlapping and decreasing in density from there to its surface. The densest spherical layer of space-matter is the source for the curvature of space and provides the additional "gravitational energy" for Einstein's energy-momentum tensor to "gravity lens" distant starlight around the sun. This additional energy is misnamed; it actually incorporates a unique type of "light refraction" caused by starlight passing through the densest of space-matter layers, like passing through perfect lens without changing the color of the light, or more specifically space-matter's occasional pulses of antigravity enhance light bending. In either case, Einstein's energy-momentum tensors effectively capture or emulate what happens to starlight passing through the curvature of space around the sun. These same types of interactions and contact also occur between space-matter and antigravity antimatter, causing the same light bending in the invisible part of the universe. The interactions, observed and partially detected as halos around galaxies, are the same density layered space-matter that occurs around the sun, and therefore bend and refract light in the same way. Therefore, the densest layer of space-matter wrapped around objects in space is the greatest curvature of space surrounding them; this occurs without any time dilation.

SUMMARY OF GRAVITY LENSING

Once again, here is the answer interweaving all points. Gravity lensing and densest space-matter light refraction work together and are caused by the same curvature of space, not time dilation. It is a balance between the Cosmological Balance Theory and General Relativity Theory. One can go as far to say that Newton's gravity equation can satisfactorily solve the gravity lensing dilemma if we incorporate Einstein's periodic pulsing increases of "mass" and "anti-mass" with each particle of light, where pulsating increase equates to "additional gravity" and antigravity, and the appropriate bend to match the light frequency.

---

[229] Space-matter, like the discovered exotic matter, possesses these features to refract light without changing its frequency or color.

General Relativity says that massive objects warp space-time; we agree with warping space, but not gravity time dilations. Nested spherical layers of space-matter are the source of the additional "gravitational energy" Einstein claimed was there due to the curvature of space. Without it his energy-momentum tensor would not work. Gravity lensing and space-matter's unique light refraction effectively yield the same result.

Cosmological Balance Theory identifies that space (space-matter) pushes on objects (and is the source of the gravity force) and at the same time gives way to accommodate the presence and motion of ordinary matter, while ordinary matter pushes back on space-matter. In the Cosmological Balance Theory, we view open space, as space-matter evenly spread in a grid of molecules and atoms at one per cubic meter, scientists coined this as "flat." In areas around ordinary matter, space-matter creates ever-increasing density bubble layers nested around it to contain it; Einstein coined this as curvature of space-time. We see it as only a curvature of space, the halo of space-matter densely packed around massive objects. This halo or curvature of space is not necessarily spherical but conforms to the shape of the massive object.

The interactions between these two different types of matter are depicted with transitioning changes in density layering from one to the other. As one leaves the planet or star, the surrounding ordinary material or gas gets thinner and thinner and goes to zero. However, from Space's point of view, when one gets closer and closer to a massive object, the layer density of space-matter increases until you encounter ordinary matter in the atmosphere. This vacuum pressure increase of space-matter is why stars and planets are spherical (the same reason why free-floating soap bubbles are spherical), while rotation provides a tendency to flatten out an object or a system.

So, the more massive the ordinary matter objects in space becomes, the denser space-matter has to get at the contact point (spherical layer) between the two types of matter in order for space to hold or support that object in space. The contact spherical layer is an overlapping of both types of matter. This gives us layers upon layers of increasing density of space-matter (which provides the "light refraction" that Einstein calls gravity lensing). Increased density and accumulation of space-matter is the same thing as warping space. Hence, both theories agree with the warping of space (or space-matter). And as such, Einstein's gravity lensing formula works with the Cosmological Balance Theory, and at the same time the light refraction (caused by densely packed space-matter with minimal photon energy absorption and without changing the color of the light) formula says the same thing. Both gravity bending and light refraction occur simultaneously. In essence, the presence of space-matter provides the "gravitational lensing." Both theories agree with warping of space. Note that the curvature of space around non-spherical objects like galaxies is not spherical. Another effect, which we will discuss later, causes the galaxy's spin and flattening.

The image here shows space-matter pushing on light with its 6.6734E-11 Newton vacuum force and simultaneously refracts the light without changing its color and frequency when light passes through it at an angle. Each greater spherical density level bends the light more. However, when sunlight escaping from the surface of the sun passes perpendicular to space-matter, the repulsions with space-matter's inward push of 6.6734E-11 Newton causes the sunlight to lose photon energy and become slightly red-shifted on its escape velocity away from the sun. This is what Einstein predicted would happen to light emitted from massive objects in space. Note: despite the red-shifted light, there is no actual time dilation at the surface of the sun, only the mirage of it.

From the physics web discussion (Exchange, 2012), the density at the band of contact between space-matter and ordinary matter of the massive object should be around n = 1×10^{-5}, this density level increases with the increase of the mass of the object. We can call the space-matter density a "halo" around the object that provides the light refraction or "gravity bending." Hence, we can expect the same type of halo around the galaxy to have enough density to refract light (or more technically provide gravity lensing). Moreover, as space-matter absorbs minimal heat from the sun, the space-matter expands and as such pushes back harder with increased pressure and force behind gravity against the source of heat. The heat energy not absorbed by space-matter is reflected to the object emitting it, resulting in corona heat buildup, like a layer of blanket keeps us warm during a cold night. This blanket keeps the cold space around the sun from overheating, and at the same time, keeps the sun toasty hot.[230] Solar mass ejection, however, punches through that barrier with escape speed sending streams of plasma and heat into deep space. The Cosmological Balance Theory still disagrees with "time" being warped per Special Relativity or gravity time dilation in general relativity.

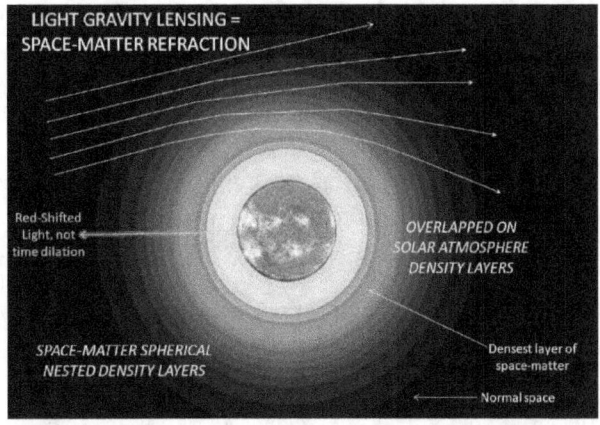

The paper entitled "Simultaneous Gravitational and Refractive Lensing" by Wojciech T. Chyla, Applied Science Enterprise, Warszawa, Poland, (Chyla, 2012) is supportive of the concept presented above in Cosmological Balance Theory. The detection of refraction in their examples is due to increase presence of ordinary matter at the contact point or spherical layer. And the paper on Gravitational Lensing Analyzed by Graded Refractive Index of Vacuum, by Xing-Hao Ye and Qiang Lin, (Ye, 2008) also points out similarities between the two. The Cosmological Balance Theory allows for both gravity lensing and light refraction justifications at the same time because they both work together to give us what we observed and recorded everywhere in the universe. The conclusion is that space-matter is the source of gravity and curvature, and that space-matter density layers bends light via a special quantum refraction assisted by the vacuum force of 6.6734E-11 Newton we know as gravity. Both effects in fact produce identical results, light travels along the geodesics, identical to rectilinearly following the path defined by the light refraction angle.

---

[230] The curvature of space keeps the sun's corona super-heated. Solar mass ejections are, however, still able to reach escape speed and punch through that curvature of dense space-matter.

7.8 Red Shift of Distant Galaxies

Earlier we quoted from Stephen Hawking's book, The Theory of Everything, where he commented, "It was quite a surprise, therefore, to find that galaxies all appeared red-shifted." He goes on to say, "Every single one was moving away from us. Even the size of the galaxy's red shift was not random but was directly proportional to the galaxy's distance from us." Concluding that, "The farther a galaxy was, the faster it was moving away (Hawking, The Theory of Everything: The Origin and Fate of the Universe, 2002, p. 23)." Hawking goes on to say, "So all we know is that the universe is expanding by between 5 percent and 10 percent every thousand million years." This claim should have alerted physicists that something must be wrong with the readings or there must be another explanation as to why they were getting this data. Edwin Hubble gathered the data and declared that the universe was expanding. The further out the galaxy was the more red-shifted was its spectrum. He concluded that this result was due to the Doppler Effect and therefore did not look for any other explanation as to why ALL the galaxies had red-shifted spectrums, such consistent results are impossible in a universe influenced with gravity, unless something else caused the spectrum shift. Could this be one of Edwin Hubble's greatest blunders? Yes.

Newton and Einstein were right to believe that the universe is more static than we think. It has some areas expanding and some areas contracting. How else can we witness the merging of numerous galaxies throughout the universe? Even Einstein adjusted his Theory of Relativity to show a static universe, he built repulsive gravity or "antigravity" into the fabric of space and time. Hawking goes on to say that, "However, the expansion of the universe meant that this light (from early universe after the Big Bang) should be so greatly red-shifted that it would appear to us now as microwave radiation (Hawking, The Theory of Everything: The Origin and Fate of the Universe, 2002, p. 28)." This is not practical as these microwave signals are not emanating from and object moving away from us. There is no object. These signals are also not as faint as starlight from galaxies beyond 14 billion light years distance, scientists did not have to wait 15 or more days to collect enough signal to create a picture. What were our scientists thinking? We suspect they were making results fit their own agenda, regardless of whether it made sense. For instance, just because a black hole results from a super massive star going super nova, and Edwin Hubble declared the universe to be expanding, does not imply that the universe emerged from a "Big Bang." Such idea assumes that there was a "Big Crunch" prior to that "Big Bang" explosion and assumes that Edwin Hubble was not wrong in declaring an expanding universe. Make your own conclusions.

With that said, let us analyze how scientists have measured the red shifted light from distant galaxies to obtain the rate of expansion. They have determined from the amount of red shift and the distance analysis that it is expanding at an accelerating rate. To measure that rate of expansion they would have to determine with great accuracy the distance to these galaxies. Astrophysicists use a method called luminosity distance to obtain it. What exactly is luminosity distance? It is a method that looks for Type 1A supernovae called "standard candles" to occur within the galaxy being measured. Scientists believe that these explosions are known to have the same absolute measurement in luminosity L = energy emitter per second. The flux F= energy per second per receiving area, observed from the supernova explosion can then be used to infer its distance via luminosity or measurement of brightness (Baumann, 2014). The relationship between the observed flux and the absolute luminosity is $F = L / 4\pi X^2$. Where F is also called the apparent luminosity, X is the luminosity distance, and L is the absolute luminosity.

"The connection between an accelerating expansion and a reduced apparent luminosity can be understood based on the naïve Newtonian cosmological model. In this model, the redshift from a distant galaxy depends on the speed the galaxy had when the light was emitted, but the apparent luminosity is inversely proportional to the square of the distance of this galaxy *now*, because the galaxy's light is now spread over an area equal to 4π times this squared distance (Weinberg, 2008, p. 50)." Weinberg goes on to say that, "Of course, it is also possible that the reduction in apparent luminosity is due to absorption or scattering of light by intervening material rather than an accelerated expansion." Such distinction is complicated business. The presence of space-matter and its effect on such light also influences the results.

Do we have a cosmic expansion or simply weakened light due to photon energy absorption? "In comparing observations of red shifts and luminosity distances with theory, we rely on the general understanding of red shifts and luminosity. One thing that might invalidate this understanding is absorption or scattering, which reduces the number of photons reaching us from distant sources. This possibility is usually taken into account by measuring the color of the source, which would be affected by absorption or scattering by some *grey* matter… Ever since the discovery of the cosmological red shift, there has been a nagging doubt about its interpretation as evidence of an expanding universe. It is possible that the universe is really static, and that photons simply suffer a loss of energy and hence the red shift naturally increasing with the distance that the photons have to travel (Weinberg, 2008, p. 57)?[231]"

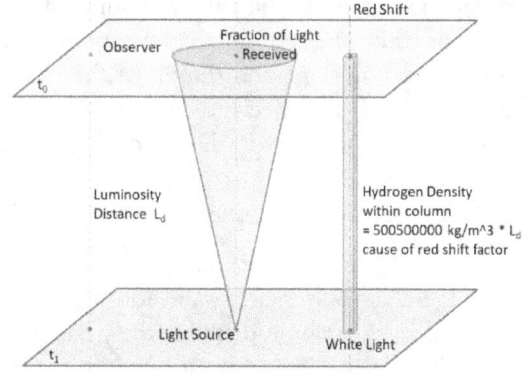

Depicted is the geometry associated with the definition of luminosity distance. To obtain a more accurate measurement of distance the above flux equation needs to be modified for the following three reasons:

1. At the time t_0 that the light reaches the Earth, the proper area of a sphere drawn around the supernova and passing through the Earth is $4\pi d^2_m$. The fraction of the light received in a telescope of aperture A is therefore $A/4\pi d^2_m$.

2. The rate of arrival of photons is lower than the rate at which they are emitted by the red shift factor $1/(1 + z)$.

3. The energy E_0 of the photons when they are received is less than the energy E_1 with which they were emitted by the same redshift factor $1/(1 + z)$.

Hence, the correct formula for the observed flux of a source with luminosity L at coordinate distance χ and red shift z is $F = L / 4\pi d^2_m (1 + z)^2 \equiv L / 4\pi d^2_L$, where we have

---

[231] Weinberg's analysis is right on target, the universe is in fact in a static and steady state.

defined the luminosity distance, d L , so that the relation between luminosity, flux and luminosity distance is the same. Hence, we find $d_L = d_m (1 + z)$.

What exactly is the red shift factor z? In physics, red shift happens when light or other electromagnetic radiation from an object is increased in wavelength or shifted to the red end of the spectrum. The red shift factor z can be calculated based on wavelength or based on frequency.

Using wavelength $z = (\lambda_{obsv} - \lambda_{emit}) / \lambda_{emit}$ and $1 + z = \lambda_{obsv} / \lambda_{emit}$

Using frequency $z = (\lambda_{emit} - \lambda_{obsv}) / \lambda_{obsv}$ and $1 + z = \lambda_{emit} / \lambda_{obsv}$

where λ_{obsv} is the observed light, and λ_{emit} is the emitted light.

Red shift is where $z > 0$ and blue shift is where $z < 0$.

Sometimes annotated as $1 + z = a(t_0)/a(t_1)$. For a(t) increasing we have a red shift by factor of $a(t_0)/a(t_1)$ equivalent to increase in wavelength by a factor of $1 + z$ (Weinberg, 2008, p. 11). Steven Weinberg, in *Cosmology* stated, "The interpretation of the cosmological redshift as a Doppler shift can only take us so far. In particular, the increase of wavelength from emission to absorption of light does not depend on the rate of change of a(t) at times of emission or absorption." He goes on to introduce several primary distance indicators and secondary distance indicators, each with inherent margins of error. For instance, the Trigonometric Parallax measures infinitesimal parallax angle to determine distance to nearby stars but ignores four factors that affect calculations. Physicists ignore the sun's orbital speed around the galaxy, assumes the star being measured orbits galaxy at the same speed as the sun, the earth's orbital diameter is infinitesimally tiny compared to the distances to the stars, and the tilt of the earth's axis at time of measurement influences the outcome. Each distance indicator developed gets better and more accurate. Still, various circumstances and factors limit accuracy. For example, "Neither galaxies nor supernovas have well-defined edges, so angular diameter distances are much less useful in studying the cosmological expansion than are luminosity distances," per Weinberg.

Based on the distance measurements and red shifts of the galaxies within the Hubble Deep Field astrophysicists have conjectured that the universe is expanding at an increasing rate. The further out in distance the galaxies are the greater the red shift and therefore they conclude that the faster these galaxies are speeding away from each other. We disagree with their deduction of an expanding universe. Why is this? Because the red shift they see, and measure is not due to Doppler Effect of the acceleration of galaxies moving away from us or from each other. It is due to the low density of matter, primarily a spread of hydrogen atoms, or a type of, and their photon energy absorption ability in the deep open space between them and us.

If you recall in Part 1 Chapter 1, we discussed how the earth's atmosphere absorption of photon energy, results in a redder light to pass through. Diagram depicts the effect:

Now, let us consider this photon-energy absorption effect over deep open space, and the resulting red shift it causes. One might think this is impossible because there is no atmosphere in space. Not exactly, there are atoms scattered everywhere, even in deep open space. We discussed this fact earlier; deep open space has a minimum density of one hydrogen atom per cubic meter to as much as 400 atoms of hydrogen per cubic meter, at estimate of 9:1 low to high-density ratio. See chapter 8 in this book for an exact explanation of this density spread. Recall how normal atoms and molecules within a star absorbed photon energy to change gamma photons to light photons and other frequencies before ejecting it from the star's surface. The same thing applies to hydrogen or helium floating in deep space; it absorbs energy from the photons hitting it. When white light from a star or a galaxy hits a significantly large amount of these hydrogen atoms and molecules, they absorb photon energy, lower the light frequency, and leave a slightly redder light to pass through to our instruments and to our eyes.

How can we demonstrate this? On earth, the density of the atmosphere at sea level is 1.2 kg/m^3, and the thickness of the atmosphere is about 480 km at vertical with most of the density below 16 km. This gives the density of a column of air 20 km straight up in the atmosphere a density of about 12,000 kg/m^2 * atmosphere cylinder 20 km long. At sunset we are looking at about two and half times that density 30,000 kg/m^2 * 2.5-atmospheres, which of course gives us the red shift effect we know so well. Although the atmosphere consists of other heavier gases than just hydrogen atoms, it takes more hydrogen atoms and molecules to equal the same density, which absorbs the same amount of photon energy to make white light have a red shift. Thus, equal density of hydrogen, even if stretched over billions of light years distance, has the same ability to absorb photon energy as the air on earth or the molecules and atoms within the sun or a star[232].

In view of this evidence, the red shift of distant galaxies could be primarily due to this very low-density of atoms of hydrogen, helium, and other particles dispersed throughout deep open space (not the perceived Doppler Effect thought to be caused by distant galaxies speeding away from us, or an expanding universe). One earth atmosphere of normal air is about equal to one atmosphere of hydrogen, helium, and other light element atoms, where white light is mostly unaffected by photon energy absorption; it remains primarily white light with a unnoticeable miniscule amount of red shift. The amount of hydrogen and other light gaseous matter in a column in space stretching out 3.5 billion light years is estimated to equate to .64 atmospheres and its photon energy absorption rate equal to no noticeable red shift effect, same as less than one earth atmosphere: none. Here white light remains mostly white light in Hubble images. The density of a column six billion light years distance is about 1.1 atmospheres of hydrogen atoms, with photon energy absorption equal to one earth atmosphere in which a tiny red shift occurs, causing some yellowish white light. For 7 billion light years distance, we have 1.28 atmospheres of hydrogen equal to red shift effect and photon energy absorption we get from 1.28 earth atmospheres. It is around this amount of hydrogen molecules where we notice the slight red shift caused by photon energy absorption of hydrogen and other atoms dispersed in space. Light reaches us with a slight yellow hue tendency.

At about eight billion light-years distance, the density of hydrogen and other atoms or molecules in space and its photon energy absorption, estimated to be equivalent to 1.46 atmospheres similar to red shift effect caused by same density of earth atmospheres, causes a minor red shift to observed galaxies, similar to yellow-orange shift at the beginning of sunset. At around 11 to 13 billion light-years (2.02 to 2.39 atmospheres), a distance that

---

[232] The physics of photon absorption in earth's atmosphere applies to the gases suspended in the great expanses of open space.

took Hubble over ten days, due to the cone fraction dispersion, to gather enough light to recognize distant galaxies, we see an ever-increasing red shift effect the further out we went. We get the same amount of red shift and photon energy absorption we get from 2.02 to 2.39 earth atmospheres. This red shift effect is like the last 30 minutes of sunset in red orange to red. For images beyond 13.6 billion light-years distance, the cone dispersed fraction of light that reaches us is so dim, and the hydrogen density and photon energy absorption greater than 2.5 atmospheres, makes these images very hazy and significantly red shifted equal to about same earth atmospheres red shift, like looking at a dust or sandstorm on the horizon during sunset. From this distance outward, the amount of density of hydrogen-type space-matter and its photon energy absorption blocks out all extremely faint cone fraction dispersed light from distant galaxies from reaching us. Therefore, we will not be able to see any light even from type 1A supernovae at 14 billion light years distance or more. This is the point at which the hydrogen gas in space would be equivalent to over 3 atmospheres equal to red shift effect of the same earth atmospheres, coupled with the decreasing fraction of cone of light gives us a pitch-black background in space. If we were somehow able to compensate for the red shift of distant galaxies caused by the photon energy absorption of hydrogen and helium atoms, and particles scattered throughout great distances in space, we would be able to simulate a more accurate picture of the universe. It would be that of a more stable and balanced universe, not excessively expanding or contracting, something very much like the galaxies within the local cluster group, somewhat gravitational bound together. Friedmann had it right the universe is homogeneous and isotropic. His second assumption about the universe states, "The universe looks identical in whichever direction we look and that this would also be true if we were observing the universe from anywhere else (Hawking, The Theory of Everything: The Origin and Fate of the Universe, 2002, p. 29)." In other words, the universe is the same in every direction from any point within the universe. The next chapter will go into detailed discussion on this topic using examples from the Hubble Deep Field Image.

Bottom line, the red shift analysis we performed clearly showed why distant galaxies appear redder the further out we look and explained why we do not see anything beyond that point due to all faint light being blocked by the low density of hydrogen and helium scattered uniformly throughout deep open space. Therefore, we can conclude that the universe is not expanding at an accelerated rate and that the Doppler Effect reasoning was faulty and made in haste. Second, the cosmic microwave background radiation is from an alternate source, the elusive white holes, and third, that source provides the abundance of light elements scattered throughout the universe. Strike three Professor Hawking, may you rest in peace!

7.9 Age of the Universe

What is the true age of the universe? Since there was no Big Bang and there is no chance of a Big Crunch in the Cosmological Balance, then how can we determine the true age of the universe? Everything is recycled in the Cosmological Balance. If you really think about it, so it is in the part of the galaxy. The Sun is a recycled, second or third generation, star. The solar system is estimated to be about 4.5 billion years old also assembled from recycled remnants of an exploded star. This implies that the first or later generation star that provided the material for the sun and the planets must have formed quickly and must have been very massive to burn through its fuel and explode. The estimate of the age the solar system is also derived from the planet surface materials, which also introduces errors. You see, the planet's surface is also being recycled as new land emerges from within the earth and old land gets pushed downward below the crust as tectonic plates move. Weinberg said, "Of course, galaxies form at various times in the history of the universe, so the age of any one galaxy does not allow us to infer the age of the universe at the time light we now see left that galaxy. However, the homogeneity of the universe implies that the distribution of cosmic times of formation of any one variety of galaxy is the same anywhere in the universe (Weinberg, 2008, p. 65)." Below is a sampling of Galaxies from the Hubble Deep Field image released by NASA on July 2004 (NASA, Hubble Site, 2004).

NASA shows us an area with smudges within the Hubble deep field image and declares those to be the oldest galaxies known, forming only 700 million years after Big Bang. Hubble deep field pictures show us that the furthest light we have collected thus far is approximately 13 billion light years away, implying that the galaxies seen in the image are at least 700 million to at most 13 billion years old. Think about that for a moment. The age of the Milky Way Galaxy, per conjecture by our scientists, is about 13 billion years old. More specifically, the same cosmologists tell us that the galaxy took about 13.2 billion years to form and become what it is today. Therefore, if the galaxies seen in the Hubble Deep Field image appear to be at least halfway or at most the same as the galaxy's development, then those distant galaxies must be at least 20 billion years old if not more, the estimated age (snapshot of the past) plus the time its light takes to get to us. Mainstream Cosmologists also tell us that the entire universe is about 13.8 billion years old and supposedly with the Big Bang occurring between 12 to 14 billion years ago. However, finding and seeing fully formed galaxies within the Hubble Deep Field shows us that the universe could be much older. How could Hubble Deep Field show us galaxies that look like ours exist in a universe that started 12 to 14 billion years ago? This argument nullifies the Big Bang Theory. Hence, we predict that the universe is certainly much, much, older than 14 billion years old. Clearly, the argument we presented in part 1 chapter 3 has merit. For all we know, it could be googolplex years old at least, constantly recycling itself per the Cosmological Balance Theory[233].

Look closely at the images posted by NASA of the galaxies within the deep field picture posted on their website. You can see that the smudges are in fact completely formed galaxies themselves. These images from the Hubble Deep Field imply that light has a limit in the distance it can travel in space, due to scattered matter occupying space or the light or glare from galaxies closer to us block them out. As light approaches distances greater than 13 billion light years, it spreads out so thinly that our instruments and telescope lens are unable to collect enough to produce a picture with enough resolution for us to see. In other words, light at 13 billion light year distance of travel encounters just enough matter, uniformly scattered through space, or glare to block that light from reaching us. We predict that any intelligent aliens, if they existed, in those distant galaxies 13 billion light years away, if they were advanced enough, would see the same view of the universe that we do of images of distant galaxies 13 billion light years away from them in all directions. This concept would then stretch the universe outward another 13 billion light years more and so forth. Imagine that, another strike against age proposed by the Big Bang Theory.

We know that this concept sounds unbelievable, but the images speak for themselves. The universe is much older than originally thought by Stephen Hawking. Let us take the upper right corner of the six pictures above and zoom into one of the faint images in the corner. It looks similar to a barred spiral galaxy, NGC 1300. A fully formed galaxy like NGC 1300, which is about 2/3 the size of the galaxy, surely took more than nine billion years to grow to its size. Now add the numbers together. Below is a magnification of the upper right corner of the Hubble Deep Field images. An unnamed barred spiral galaxy showed in the Hubble Deep Field picture below took at least 9 billion years to form and appear in that shape. Add that to fact that we are looking back in time to about 12 to 13 billion years and we get at least 21 billion years prior to that unnamed barred spiral galaxy's creation. Wait there are other fully formed galaxies in the same picture that are much further back in time; even the smallest specks or smudges are galaxies. Surely, the universe must be at least 25 billion years old if not much older than that. Again, we speculate that the exact beginning of the universe is ridiculously hard to determine and therefore its true age even harder to predict with the data we currently have. Per the Cosmological Balance Theory, it has been endlessly recycling itself. For all we know, the universe could be anywhere in between twenty billion to googolplex years old. We could be living in the googol cosmological decade, using the timetable defined by Professor Fred Adams, University of Michigan.

Hubble Deep Field image

NGC 1300, Barred Spiral Galaxy

Magnified Galaxy Resembles Barred Spiral Galaxy

---

[233] Since the Cosmological Balance is a static or steady state universe, its true age cannot be determined simply based on what information we obtained and observed.

With the *Age of Hubble* (Lucas T. , 2015), our astronomers have discovered that the visible universe expands more than some 46 billion light years in all directions. Identifying their location and plotting the representation each galaxy in a three-dimensional map of the universe, they have provided us a better view of reality. With the use of super-computers, they can analyze and simulate their movements. NASA's Hubble discoveries also show us that intergalactic space contains dust and debris, some of which they speculate came from massive black holes, mostly at the centers of galaxies that magnetically eject them at the poles. The development and usage of advance technology in the exploration of universe and space will continue to show that the age presented in the Cosmological Balance is correct; for all we know, the universe is could be infinite in age.

The Cosmological Balance will not enter the last three of *the Five Ages of the Universe* identified by Fred Adams: There will be no "Degenerate Era" where stellar evolution ends, nor the "Black Hole Era" where the only objects remaining are black holes, nor the "Dark Era" where the black hole eventually evaporate per Hawking Radiation. These last three ages speak of gloom, destruction, disintegration, and a complete end to all other possible of forms of existence, a topic less likely to encourage people to strive for improvements. Likewise, there was no first age of the "Primordial Era," or the Big Bang, or the cooling period while the universe expanded to its present size. This first age speaks of a very turbulent creation period, one awfully hard to imagine by most nonscientific people. However, the Cosmological Balance recycling of normal matter with gravity, as we know it, into matter with antigravity and subatomic material to replenish space-matter also allows for the recycling of energy between the visible and the invisible universe, and space itself. This type of future is much more encouraging to our descendants should they become space-bound or a planet hopping race. The Cosmological Balance on the largest of scales is completely homogenous, exhibiting a remarkable regularity and uniform simplicity, a balance of gravity forces acting on anti-gravitational forces. We just happen to exist at the point of the universe's history where the exact properties came together to spawn life on the earth. There is no need for multi-verse concept and no need for multiple variations of physical law concept to create the specific conditions for life, as we know it today.

7.10 Age of the Milky Way Galaxy

Now let us suppose that the universe filled with gaseous clouds of various densities scattered uniformly throughout it, as we did in Part 1 Chapter 2, this time the gaseous stellar clouds are in the Cosmological Balance. Some of these particles and atoms within the cloud groups have gravitation properties, and some clouds have anti-gravitational properties. The ones with gravitational properties tend to pull mutually together, while the ones with anti-gravitational properties pull each other together. In addition, a cloud of anti-gravitational particles and atoms exerts a repulsion force on the cloud with gravitational properties. Consolidation of all gravity particles and atoms together, and the consolidation of all antigravity particles and atoms together, are imminent. From the visible, Yang side, universe, these groupings will eventually look like filament and sheets of gaseous clouds and become the beginnings of the birthplace of galaxies.

Simulations of gaseous clouds motion in deep space or the early stages of the formations of stars conducted by university students always show that some push or density wave, from the explosion of a nearby star, must come along to compress the cloud to begin or speed up the formation of proto stars. But what happens if there are no nearby stars to explode and provide that push? Not shown in these simulations is that without some push, the cloud just sits there for eons, billions, and billions of years, unable to consolidate itself together with its own gravity. Why is this so? Simply because the atoms and particles in the cloud fight or bounce of each other if they come into contact especially as energy and temperature builds up. The cloud by itself simply has a difficult time achieving the critical mass and density to form a proto star, let alone a globular cluster galaxy. The Cosmological Balance adds the push from the existence of antigravity particles and atoms that enabled the Milky Way Galaxy to assemble and grow at a faster rate. A rate comparable to what cosmologists have conjectured is the current age of the galaxy, about 13 billion years old or as old as twenty-seven billion years or more. The age of the solar system does not determine the true age of the Milky Way. The age of the part of the galaxy could be and is most likely much younger than the central part of the galaxy, because a galaxy's core forms before it starts to accumulate more and more matter to create the stars necessary to make the galaxy "grow." Likewise, the age of the Milky Way Galaxy has nothing to do with the age of the universe[234]. The galaxy could be one of the younger galaxies of the entire universe. For all we know the galaxy could be 13 billion years younger than the Andromeda Galaxy. Then Again, the core black hole of the galaxy could be over 100 billion years old, and the outer edge where we live is the newest suburb of the galactic city. "How can we think we are the first intelligent life in the universe?" To justify this thinking, the Big Bang Theory was developed, assuming that everything was born at the same time, deviating from the norm of everything around in the universe itself, regenerated over time. Each star in the galaxy and each galaxy in the night sky is at a different stage of their development, and therefore at a different age. Not only does the galaxy's true age elude scientists, but the reason the galaxy rotates at its speeds is a mystery to astrophysicists and cosmologists today.

---

[234] The ages of each galaxy within the local region of the universe varies. Galaxies also grow at different rates, where proximity to other galaxies affects their rate of growth. A globular cluster of stars would remain exactly that if it formed in wide-open space, neither growing nor shrinking for eons.

7.11 Galactic Orbital Equation

When it comes to calculating galactic orbital speeds, we find that the orbital period analysis provided in part 1 chapter 2 section 10 is still missing one important piece, an additional force provided by the mass in motion. Some have conjectured that there is some "dark energy" lurking just beyond the galaxies edge pushing the outer stars to orbit faster than with just gravity alone. Some have said there are additional mass surrounding the galaxies that contribute to the same rapid rotation of the stars, the "dark matter." We have another explanation for this phenomenon. Remember in the solar system we summed up the mass of the sun and the inner planets to determine the orbital period for the next planet away from the sun. For Mars' orbital period, we had to sum up the mass of the Sun, Mercury, Venus, Earth, and the Moon to get the proper mass M for Kepler's third law equation. In doing the same for the galaxy's stars orbital period calculations, we discover that double summing the inner accumulated mass M_A, to account for motion of the mass (i.e., its energy), increases the orbital speed and cuts the orbital period by about half that of gravity alone (Kepler's third law). The chart shows us this relationship between motion and the mass of the inner portion of the galaxy by simply double counting each concentric circles' interior mass (highlighted column labeled inner kg)[235]:

	Schwartzschild	sectional kg	inner kg	inner	section	has	mass fraction	Radius = m	orbital period	x 31556952	years
2	1.38811E+10	9.3474E+36	9.34736E+36	Mass BH	2.45E-11	has	1.5579E-06	1.38811E+10	411.43277	1.304E-05	years
3	7.26863E+14	4.89450E+41	4.89459E+41	Inner	0.040	has	0.081575	2.27053E+19	1.68211E+14	5.330E+06	years
4	1.44669E+15	4.84725E+41	9.74184E+41	Inner	0.060	has	0.0807875	3.40579E+19	2.19042E+14	6.941E+06	years
5	2.25212E+15	5.42363E+41	1.51655E+42	Inner	0.080	has	0.09039375	4.54106E+19	2.70289E+14	8.565E+06	years
6	3.18944E+15	6.31181E+41	2.14773E+42	Inner	0.100	has	0.10519688	5.67632E+19	3.17418E+14	1.006E+07	years
7	1.05625E+16	1.20097E+42	6.97793E+42	Inner	0.200	has	0.2001624	1.13526E+20	4.98085E+14	1.578E+07	years
12	2.19471E+16	1.80003E+42	1.47789E+43	Inner	0.300	has	0.30000508	1.70290E+20	6.28757E+14	1.992E+07	years
17	3.79855E+16	2.40000E+42	2.55789E+43	Inner	0.400	has	0.40000016	2.27053E+20	7.35819E+14	2.332E+07	years
22	4.17278E+16	2.52000E+42	2.80989E+43	Inner	0.420	has	0.42000008	2.38406E+20	7.55355E+14	2.394E+07	years
23	4.56483E+16	2.64000E+42	3.07389E+43	Inner	0.440	has	0.44000004	2.49758E+20	7.74384E+14	2.454E+07	years
24	4.97469E+16	2.76000E+42	3.34989E+43	Inner	0.460	has	0.46000002	2.61111E+20	7.92945E+14	2.513E+07	years
25	5.40238E+16	2.88000E+42	3.63789E+43	Inner	0.480	has	0.48000001	2.72463E+20	8.11071E+14	2.570E+07	years
26	5.84789E+16	3.00000E+42	3.93789E+43	Inner	0.500	has	0.5	2.83816E+20	8.28791E+14	2.626E+07	years
27	8.34275E+16	3.60000E+42	5.61789E+43	Inner	0.600	has	0.6	3.40579E+20	9.12141E+14	2.890E+07	years
32	1.12831E+17	4.20000E+42	7.59789E+43	Inner	0.700	has	0.7	3.97343E+20	9.88376E+14	3.132E+07	years
37	1.46690E+17	4.80000E+42	9.87789E+43	Inner	0.800	has	0.8	4.54106E+20	1.05907E+15	3.356E+07	years
42	1.85004E+17	5.40000E+42	1.24579E+44	Inner	0.900	has	0.9	5.10869E+20	1.12529E+15	3.566E+07	years
47	8.91019E+15	6E+42	6.00000E+42	MilkyWay	1	has	1	5.67632E+20	6.0055E+15	1.903E+08	years

This double counting of mass somewhat accounts for the slingshot effect of stars has when tugging on each other and the acceleration provided by the galactic wind emanating from the black hole but is not practical. This simple relationship between motion and mass requires us to modify Kepler's third law to create a new universal law found throughout the universe. The chart above comes close, 10^{14} power, to giving us the results to match actual observations: sun's distance to center of galaxy 2.36513E+20 to 2.46731E+20 meters, and an orbital period around the galaxy of 7.57367E+15 seconds (2.40E+08 years). This speed estimates to about 720,000 km/h (450,000 mph) or about 200,000 m/s. The new close approximation adjusted formula of

$$T^2 = 2 \cdot 4\pi^2/(GM_G) \cdot a_g^3 = 8\pi^2/(GM_G) \cdot a_g^3,$$

where M_G represents gravity matter in motion (its momentum), G is the gravitational constant, a_g is the length of the semi-major axis of the gravity matter system being analyzed, and T is the stars orbital period in the galaxy or the galaxies in the universe.

---

[235] This approach to produce orbital acceleration of stars around the galaxy is not new; it has been tried before.

This gives us two other simplified equations quite like Einstein's equations:

$T^2 = 8\pi^2/(GM_G) * a^3$, for matter with gravity and

$T^2 = 8\pi^2/(GM_A) * a^3$, for matter with antigravity.

However, the above equations are close but not good enough. In rethinking the motion of the galaxy, we realize that the galactic wind (generated by antigravity wormhole tunnels) provides the addition boost to accelerate the stars throughout the entire galaxy and discover that Euler's constant $e = 2.718281828459$ play an important part in solving gravity and antigravity interaction within the galaxy. If we accurately represented this galactic orbital period formula in this book, we could then apply it to other galaxies, and in turn for the entire universe.

In order to tackle this dilemma properly, we need to divide the galaxy into at least 100 concentric circles from the center to the edge, and further distribute the mass of galaxy appropriately with higher concentrations of density toward the center and low density at the edges of the galaxy. Let us then combine the circumference of each concentric formula $2\pi R$ (in meters) with the mass formula M_S from the Schwarzschild Radius $R_S = 2GM_S/c^2$ (in meters) where the product of Mass $M_S = c^2 R_S/(2G)$ and acceleration of that mass ($\sqrt{2GM_S/R_S^2}$) on that circumference gets us the force that propels outer stars

Force = Mass * Acceleration

$= c^2 R_S/(2G) * \sqrt{2GM_S/R_S^2}$

$= c^2/(2G) * \sqrt{2GM_S}$

$= c^2 * \sqrt{2GM_S/(2G)^2}$

$= c^2 * \sqrt{2GM_S/(4G^2)}$

The constant G cancels out and the 2/4 reduced to ½, giving us

$= c^2 * \sqrt{M_S/(2G)}$

where c is the speed of light, M_S is the mass within the Schwarzschild Radius, and G is Newton's gravitational constant.

The equation above is great progress, but it still does not account for the force behind the antigravity wormhole nudges we discussed earlier in part 2 chapter 6. It turns out that this nudge is simply represented by Euler's number $e=2.718281828459$. Inserting this clarification into the galactic rotational force, we now have:

Galactic_Rotational_Force = Mass * Acceleration * e

$= c^2 * \sqrt{M_S/(2G)} * e$

How can we then calculate the galactic wind force[236]? We simply take the galactic rotational force and divide it by the sectional mass within the Schwarzschild radius.

We get: $G_{WF} = G_{RF} / M_S$

---

[236] The galactic wind is the force exerted by antigravity antimatter moving through the galaxy.

CHAPTER 7: INSPIRED THEORIES ANALYZED

$$= c^2 * \sqrt{(M_S/(2G))} * e/M_S$$

$$= c^2 * \sqrt{(M_S/(2GM_S^2))} * e$$

The mass M_S cancels out to give us

$$= c^2 * \sqrt{(1/(2GM_S))} * e$$

To solve for the Cosmological Balance orbital period of stars within the galaxy, we need to revisit Kepler's Third Law $T^2 = 4\pi^2/(GM)*a^3$, the orbital period to travel the circumference of a circle or an ellipse is 2π or the radians in the circle, and the Schwarzschild radius formula $R_S = 2GM_S/c^2$. We start with Kepler's law

$$T = \sqrt{(4\pi^2/(GM)*a^3)}$$

then multiply by the galactic wind force acting on the circumference for that portion of the galaxy (i.e. rotational force from that concentric black hole) divided the radians of a circle or ellipse 2π divided by 2. Note: $2\pi/2 = \pi$. We get the formula for Cosmological Balance galactic orbital period T_c as

CBU_Orbital_Period = Kepler_Orbital_Period x Galactic_Wind_Force / π

$$T_c = [\sqrt{(4\pi^2/(GM_S)*a^3)}] * [c^2 * \sqrt{(1/(2GM_S))} * e] / \pi \quad (7.8.1)$$

$$= c^2 e \sqrt{(4\pi^2/2(GM_S)^2)*a^3)} / (2\pi/2)$$

The $\sqrt{(GM)^2}$ reduces to $\pm GM$ to account for gravity and antigravity instances

$$= c^2 e /(GM_S) * \sqrt{(4*4\pi^2/(2*4\pi^2)*a^3)}$$

The $4\pi^2$ cancels out and the 4/2 reduces to 2 to give us

$$= c^2 e /(GM_S) * \sqrt{(2a^3)}$$

Therefore, the galactic orbital period is shortened to

$$T_c^2 = 2c^4 e^2/(GM_S)^2 * a_c^3 \quad \text{or} \quad 1 = 2c^4 e^2/(GM_S)^2 * a_c^3 / T_c^2$$

where T_c is the Cosmological Balance orbital period in seconds of stars going around a specific galaxy given the distance a_c from its center in meters, M_S is the accumulated mass of all the matter within the concentric circle of radius a_c, c is the speed of light, the G is Newton's gravity constant, and e is Euler's number.

To summarize the above discussion, Kepler's third law is applicable to solar systems and the Cosmological Balance equation or law is applicable to rotation of galaxies and the universe. Why is this? Since the mass at the center of a solar system is not a black hole sufficiently large enough to produce white holes with antigravity antimatter and associated wormhole tunnels, then Kepler's third law $\sqrt{(4\pi^2/(GM) * a^3)}$ is applicable to that system. Large systems like the Milky Way Galaxy possess massive black holes at its center capable of ejecting antigravity white hole orbs into wormhole tunnels. The resulting push is the product of Kepler's third law $\sqrt{(4\pi^2/(GM) * a^3)}$ and the galactic wind $c^2*\sqrt{(1/(2GM))}*e/\pi$ thereby accelerating stars orbiting around the galaxy to speeds calculated by the abbreviated equation of the galactic orbital period equation (7.8.1)

$$T_c = \sqrt{(c^4/(GM)^2*2a_c^3 e^2)}, \quad \text{or} \quad T_c = c^2 e/GM*\sqrt{(2a_c^3)}.$$

Again, this Cosmological Balance equation is applicable to the motion within a galaxy as well as the motion of all the visible galaxies (matter with gravity) within the entire universe. Consequently, this equation is applicable to the universe itself as the motions of all visible gravity galaxies influence each other in the same way as motion of the stars within the galaxies influence each other. Below is an excel spreadsheet showing the Cosmological Balance galactic equations and their results:

	C	D	E	F	H	J	K	L	M	N	O
1	$R_s=2GM/c^2$	6E+42 * fract	$\sqrt{(M/2G)} *c^2*e$	$c^2*\sqrt{(1/2GM)}*e$		Mass Distribution	60K ly * sect	$\sqrt{(4\pi^2/(GM))*a^3}$	$c^2/GM*\sqrt{(2a^3)}*e$	sec-->yr	
2	Schwartzschild	sectional kg	mass in motion	galactic wind	section	mass fraction	Radius = m	Kepler orb period	CBU orb per'd (s)	x 31556952=	years
3	1.38811E+10	9.34736E+36	6.46535E+40	6.91676E+03	2.45E-11	1.55789E-06	1.38811E+10	411.4327697	9.05841E+05	2.870E-02	years
4	6.72125E+14	4.52600E+41	1.42267E+43	3.14333E+01	0.020	0.075433	1.13526E+19	4.37315E+13	4.37557E+14	1.387E+07	years
5	7.26849E+14	4.89450E+41	1.47945E+43	3.02269E+01	0.040	0.081575	2.27053E+19	1.18944E+14	1.14442E+15	3.627E+07	years
6	7.81572E+14	5.26300E+41	1.53414E+43	2.91495E+01	0.060	0.087716677	3.40579E+19	2.10725E+14	1.95523E+15	6.196E+07	years
7	8.77456E+14	5.90867E+41	1.62552E+43	2.75108E+01	0.080	0.098477788	4.54106E+19	3.06194E+14	2.68133E+15	8.497E+07	years
8	1.00078E+15	6.73911E+41	1.73600E+43	2.57600E+01	0.100	0.112318529	5.67632E+19	4.00687E+14	3.28550E+15	1.041E+08	years
13	1.79649E+15	1.20973E+42	2.32591E+43	1.92266E+01	0.200	0.201622202	1.13526E+20	8.45877E+14	5.17678E+15	1.640E+08	years
18	2.67496E+15	1.80128E+42	2.83817E+43	1.57564E+01	0.300	0.300213632	1.70290E+20	1.2735E+15	6.38711E+15	2.024E+08	years
23	3.56433E+15	2.40017E+42	3.27618E+43	1.36498E+01	0.400	0.400028141	2.27053E+20	1.69854E+15	7.37993E+15	2.339E+08	years
24	3.74245E+15	2.52011E+42	3.35705E+43	1.33210E+01	0.420	0.420018764	2.38406E+20	1.78349E+15	7.56238E+15	2.396E+08	years
25	3.92060E+15	2.64008E+42	3.43602E+43	1.30149E+01	0.440	0.440012153	2.49758E+20	1.86843E+15	7.74046E+15	2.453E+08	years
26	4.09876E+15	2.76005E+42	3.51322E+43	1.27288E+01	0.460	0.460008345	2.61111E+20	1.95337E+15	7.91451E+15	2.508E+08	years
27	4.27694E+15	2.88003E+42	3.58877E+43	1.24609E+01	0.480	0.480005567	2.72463E+20	2.03831E+15	8.08479E+15	2.562E+08	years
28	4.45513E+15	3.00002E+42	3.66277E+43	1.22091E+01	0.500	0.500003715	2.83816E+20	2.12324E+15	8.25154E+15	2.615E+08	years
33	5.34612E+15	3.60000E+42	4.01235E+43	1.11454E+01	0.600	0.600000498	3.40579E+20	2.5479E+15	9.03916E+15	2.864E+08	years
38	6.23713E+15	4.20000E+42	4.33383E+43	1.03186E+01	0.700	0.700000074	3.97343E+20	2.97255E+15	9.76342E+15	3.094E+08	years
43	7.12815E+15	4.80000E+42	4.63306E+43	9.65221E+00	0.800	0.800000015	4.54106E+20	3.3972E+15	1.04375E+16	3.308E+08	years
48	8.01917E+15	5.40000E+42	4.91410E+43	9.10019E+00	0.900	0.900000011	5.10869E+20	3.82185E+15	1.10707E+16	3.508E+08	years
54	8.91019E+15	6.00000E+42	5.17992E+43	8.63320E+00	1	1	5.67632E+20	4.2465E+15	1.16695E+16	3.698E+08	years
55			Euler's # = 2.71828E+00				5.67632E+20 m		$\sqrt{(c^4/(GM)^2*2a^3*e^2)}$		

Note that in the highlighted area in green shade in column C, D, K, and L, we have Kepler's third law application, which does not reflect the sun's true orbital period given its distance from the center of the galaxy. The columns E, F and M provide the additional mass in motion, galactic wind, and results showing the adjusted Cosmological Balance galactic orbital period T_c. The highlighted orange shade areas in rows 23 and 24 intersecting with columns K, M and N indicate that the sun's estimated distance and observed orbital speed falls within this band. But the velocity of outer stars is excessive.

This is the final galactic equation (7.8.2)

$$T_c = [\sqrt{(4\pi^2/(GM_S)*a^3)}] * [c^2 * \sqrt{(1/(2GM_S))} * e/\pi]$$

In the next two chapters, we will use this galactic equation to develop a Cosmological Balance Equation to replace Einstein's relationship equation for the entire universe[237]. Note that this equation shows only half of the forces acting on the galaxy. The missing piece is that when the galactic wind crosses half the galaxy's mass, it starts to encounter resistance from the space surrounding the galaxy. Hence, the force exerted by the galactic wind (Euler's constant e) is progressively reduced by increments obtained from e-1 divided by the product of two times total number of segments. The resistance ϱ expressed as e/ϱ. So, the corrected galactic equation reduces to $T_c = c^2/GM * \sqrt{(2a^3)} * e/\varrho$.

---

[237] The Cosmological Balance Equation solves for the movement of macro objects in space, and the physics of the universe, gravity. It does not solve the motion of subatomic particles and electromagnetic waves since these are of the micro universe and affected by quantum physics and mechanics. In these instances, Einstein's relativity equation is applicable, as it deals with the motion of light, subatomic particles, and electromagnetic radiation by calculating and identifying their geodesic path around massive objects.

Chapter 8: Space Balances Gravity Antigravity

"The universe is all about balance. The forces of light and darkness are meant to keep a check on one another. If one becomes too powerful and starts overrunning the other, that balance will be upset. For the tyranny of virtue is as unbearable as the stranglehold of vice."

— Shatrujeet Nath, The Guardians of the Halahala

Henry More, the Cambridge Platonist, believed that space was not empty; the divine spirit constituted it and therefore was also known as divine absolute space. To modern physicists, space may be filled with the ocean of the Higgs field, composed of Higgs particles. Other scientists also see space filled with some type of "dark matter" and "dark energy[238]." Wikipedia explains, "Dark matter neither emits nor absorbs light or any other electromagnetic radiation at any significant level. According to the Planck mission team, and based on the standard model of cosmology, the total mass–energy of the known universe contains 4.9% ordinary matter, 26.8% dark matter and 68.3% dark energy. Thus, dark matter is estimated to constitute 84.5% of the total matter in the universe, while dark energy plus dark matter constitutes 95.1% of the total mass–energy content of the universe." Stephen Weinberg further discussed dark matter, he said, "For instance at the end of the era of inflation there was a time of so-called reheating, during which the energy of the vacuum was transferred to ordinary matter and radiation, but we have no idea what particles were first created during reheating or how the energy transfer took place. Later, there was presumably a time when some particles effectively stopped interacting with the rest of matter and radiation and became what is now observed as cold dark matter, but we can only guess when this was and how it happened (Weinberg, 2008, p. 245)."

Andrew Thomas (Thomas, Hidden in Plain Sight 2: The equation of the universe, 2013) suggested that the universe is spatially flat and had a total of zero energy, and therefore must follow the Schwarzschild radius equation, similar to that of a black hole. In making this claim, he concluded that the radius of the universe is $R_u = 2 G M_u / c^2$, where M_u is the total mass of the known universe, G is the gravity constant, and c is the speed of light. Particles or objects trying to escape from the universe' outer radius limit (event horizon or edge) will be pulled back into the universe, as light particle is pulled back into the black hole. This given radius, determined by a given mass, implies that the universe must therefore inevitably assume a steady state or one possibly fluctuating between slightly greater and slightly less than that radius, like a soap bubble expanding and shrinking due to pressure changes in the atmosphere. The unknown factor here is that exact mass of the total universe and therefore the exact radius is unknown. Thomas' mistake, however, is his assumption that the big bang theory was still valid despite his equations, and that he was looking for a hybrid theory between it and the old "steady state" theory. Coincidentally, Thomas' analysis reached the same final state conclusions[239] as the theory we presented in

---

[238] The terms "dark matter" and "dark energy" describe matter and energy unknown to astrophysicists and cosmologists. It does not describe the elements color or that they are not radiating light, only that they are still undefined.

[239] Thomas resolved that the universe was homogenous and isotropic in a steady static state.

part 2 chapter 5, a Cosmological Balance. Let us take a closer look at space and its structure.

In the introduction and in the first chapter in part 1, we stated up front that there are three types of matter in the universe: those with gravity, those with antigravity, and those that make up space. Brian Greene alluded to another type of matter in space when he posed the question physicists are working to resolve, "What is the framework of space? What constitutes the 'molecules' and 'atoms' of space (Greene, The Fabric of the Cosmos: Space, Time, and the Texture of Reality, 2004, p. 335)?" This third type of matter, space-matter, is the source of the forces of gravity and antigravity, it makes space what it is. In review, gravitational force governs all matter in the visible universe, pulling objects together in space and on earth. Anti-gravitational force governs matter in the invisible part of the universe, making antigravity objects come together in space. Space-matter only exists in a gaseous or more specifically a super cold liquid-gas form with practically no viscosity; it occupies deep open space to maintain the void, keeping space what it is. It is the buffer between the other two types of matter and contributes to the cosmological balance. Space-matter is neutral in the sense that it no gravitational attraction to itself or to matter with gravity or with antigravity. Space-matter fills space, like the Higgs field and its ocean, it does not affect the overall speed of light in open space, but instead of tiny Higgs particles, we have unique energized space-matter atoms. The space-matter atom itself contains the Higgs particle, which is the neutral graviton particle that provides space a minuscule amount of viscosity, giving the characteristic of mass to ordinary and antigravity particles or objects[240]. Hence, the space-matter grid and its ocean also consist within it the Higgs field and its ocean. So, when we talk about space-matter grid think of the Higgs field overshadowing it. Space-matter is the one piece of what scientists refer to as dark matter, antigravity antimatter is the other piece. Space-matter is very exotic; it pulses with gravity-antigravity alignment bursts at a rate equal to one-five-hundredth that of the speed of light (599,119.05 hertz), essentially exhibiting a near zero balance between gravity and antigravity, with just enough strength to mutually repel each other to create the vacuum of space, a force powerful enough to cause galactic rotations. The vacuum force of space suspends ordinary matter and antigravity antimatter within itself, through increased density layers, known as the curvature of space. The gravity-antigravity frequency results from periodic alignment of neutral quark (*hygratium*) internal motions, which is roughly two times ninety-nine factorial divided by product of ninety-six factorial times π, or $(2 * 99!) / [(99-3)! * \pi]$, where 99 is the 100 to 1 ratio of strong force to electro force, 2 is the inward and outward motion, 3 is the needed number of hyperbolics to form a tetrahedron, and pi for the spherical shape.

The occasional decomposition of space-matter yields both gravity matter and antigravity antimatter simultaneously per the energy, mass and momentum relationship discussed in Part 2 Chapter 5 section 5.6. Therefore, if ordinary matter is estimated correctly to be about 4.9% of all matter and energy in the universe, then it is logical that antigravity antimatter should be closer to 4.9% of all matter and energy as well, leaving 24.35% to be space-matter. Space-matter is the unknown mass our scientists refer to as

---

[240] The Cosmological Balance Theory goes so far as to predict that the space-matter atom contains the Higgs particle, which is also the sought-after neutral graviton particle, it gives ordinary matter and antigravity antimatter its mass.

"dark matter" and the "dark energy" to be 65.85%, which they reference is the force emanating from the "dark matter," what we shall call space-matter and its energy. We will explain this energy force later. For now, the distribution of matter is as follows, matter with gravity constitutes 14.35% and matter with antigravity constitutes 14.35%, and space-matter therefore constitutes approximately 71.30% of the total matter in the universe dispersed throughout deep open space and its adjusted energy to be about 65.85%. Thus, the space-matter and its energy force constitute about 90.20% of the total mass-energy of the universe. Note that all mass of matter with gravity, matter with antigravity, and space-matter (neutral) combined only occupies about .0001% of the entire universe we live in, leaving the 99.9999% to be completely void of all particles of matter. This estimate applies to an infinite universe, lots, and lots of nothingness in between actual particles of matter.

In an article originally written by Steve Carlip and consecutively updated by him with Matthew Wiener and Geoffrey Landis in 1998 and again by himself in 2011, *Does Gravity Travel at the Speed of Light?* In it (Carlip, 2011), he begins by saying that the speed of gravity has not been directly measured because it is beyond the capabilities of today's technology and cannot be duplicated in a laboratory because of its weakness. Carlip's paper compares Newton and Einstein views. He describes gravity in the Newtonian model as an instantaneous force, acting over unlimited distance, at the positions of the object's currently locations, and not in a previous retardation position. Then he talks about gravity in Einstein's General Theory of Relativity, which propagates at the speed of light, through space-time curvature distortions, thereby influencing the motion of objects in space. Carlip's paper debates, which is right and how can gravity act if it was traveling at the speed of light and goes on explaining the motion of objects A and B in terms of damping, retardation, field propagation, electrodynamics, radiation, time delay, and angular momentum all creating workarounds to the light speed limitation. If general relativity is correct in calculating the physics of the universe, then the force of gravity traveling at the speed of light causes complications to Newton's equations. If Newton's equations are correct in defining gravity as instantaneous, then Einstein's general relativity is in error for macro objects, but accurate for calculating paths of particles and electromagnetic waves such as light in the micro universe of the quantum world. In attempt to answer his question, his paper implied that speed of gravity is not infinite based on damping measurements, and therefore must be closer or equal to the speed of light. In closing his article, Carlip proposed detecting and measuring the speed of gravitation waves from a supernova, in comparison with neutrino bursts, and visual identification, to back up his answer and add another proof to Einstein's GR. Carlip did not prove Newton wrong[241].

Simplified solutions to capturing reality with equations are usually the right ones and the most useful in the real world, especially when repeatedly proven valid. Hence, the Cosmological Balance sides with Newtonian gravity equation as an instantaneous force, and with Galilean's relativity where the laws of motion are the same in all inertial frames or in constant speed; and decline usage of Einstein's General and Special Theories of Relativity, its space-time curvature with the speed of light maximum, and the speed of gravity, light. Gravity is not only a "pull" force which radiates outward and travels at infinite speed from object A to object B to pull object B to A and vice versa. Trying to

---

[241] It is interesting that Carlip points out the difficulties in using Einstein's equation on gravity, but does not use this to prove Einstein wrong instead, he confirms Einstein's work as the way ahead.

measure the speed of this so-called gravitational radiation or wave is irrelevant. Gravity is also the result of a "push" force caused by the vacuum force of space throughout the universe, which acts instantaneously upon object A and object B at their present locations to push them toward each other and readjusts instantaneously as each object moves in time. Scientists have called it dark matter and dark energy because they have not identified it, but they know it is there. We call it space with its space-matter and vacuum pressure, energy, and force. This is the force Newton predicted was there but was unable to resolve its existence during his lifetime; therefore, he wrote Principia based on the accepted scientific and mathematical knowledge of his time as a philosophical proof rather than mechanical, and merely mentioned this unknown mechanical force, some called aether or ether, in his closing statement to Principia Mathematica. The end result of the transfer of space vacuum force from the essence of space, a unique type of space-matter, to visible objects of normal matter possessing "gravity," the ability to clump together, simplifies into Newton's gravity equation $F = G(m_1 m_2)/r^2$. This gives us the perception that the mass of two objects are instantaneously attracting each other with a force inversely proportional to the distances between their centers and react accordingly. The internal graviton mechanical effect provides the pull aspect part of gravity, which bonds atoms and holds mass together. How can we visualize this effect?

Imagine for a moment that you were riding in a free-floating hot air balloon drifting over the ocean, and a tropical storm began to develop just 5 miles away. As the storm strengthens you find your hot air balloon changes direction moving ever closer toward the storm as if the storm were pulling your airship toward it. We know that the earth's total atmospheric air pressure provides the force to push your hot air balloon toward the low-pressure system, the tropical storm, and the closer the airship moved toward the center of the storm, the stronger the winds pushed. Incidentally, the stronger the low pressure in the atmosphere, the larger the storm becomes. Now replace the tropical storm with the sun, the hot air balloon with the planet earth, and the atmospheric pressure with that of galactic space. In this analogy, the total vacuum pressure of space simultaneously pushes the earth toward the sun and the sun toward the earth with the greatest effective movement on the earth, exactly as predicted by Newton's gravity equation, $F = G(m_1 m_2)/r^2$. The more massive the star, the stronger its inertial resistance and less likely to move toward the smaller object, effectively appearing to pull the smaller object toward it. The vacuum force acts equally on both objects A and B, but since the larger more massive object A resists movement more, the smaller object B moves toward the massive object B, exactly as calculated by Newton. This source of vacuum pressure, energy, and its force is attributed to the presence of space-matter uniformly spread throughout space. Again, space-matter has practically no gravitational properties or characteristics on itself. Space-matter atoms contain a neutral graviton; the space-matter grid is the Higgs Ocean. Suppose Newton concept of an infinite universe is correct. If we assume that all matter in the universe was governed by gravity, and we ignore the tremendous distances, then the universe would have gathered itself to its center, where the density is expected to be extremely high. But, since only 14.35% of all matter in the universe is with gravity, that mass lacks sufficient attractive force to consolidate itself due to the great distances between objects, let alone that that matter only occupies .0001% of its total volume.

8.1 What Makes Outer Space Appear Void?

Scientists have convinced humanity that space is completely empty and void. It is far from it. Space-matter occupies space, it is a unique type of matter, unaffected by gravity or antigravity, which makes outer space appear void. Sir Isaac Newton, even after successfully defining gravity and its properties, hated his own theories about gravity being an "action at a distance" force. He believed so strongly that the gravitational force must be the result of some type of material interconnecting objects in space. He was one of the first scientists to seriously suggest there was a mysterious substance in space some called the aether (sometimes spelled ether) that connected all objects in the universe." He was right in his reasoning, but had insufficient data to justify his believes, so he developed his equations based on philosophy rather than mechanics. Rather than using the term aether, we call Newton's mysterious substance space-matter[242].

What exactly is space-matter? Space-matter is vaguely like the old concept of "aether." It attracts neither matter that has gravity nor matter that has antigravity; and is not neutral in mass. This space-matter exists in a super cold gas-spacious form primarily of unique substantial high-energy hydrogen and in rare instances, helium, uniformly scattered throughout deep open empty space and does not draw itself together, like ordinary matter or antigravity antimatter; it repels away from itself and spreads out. Normally, extreme cold air or gas in the atmosphere on earth is very dense and has higher pressures due to gravity. However, because space-matter does not mutually attract each other gravitationally or anti-gravitationally, their super electrostatic charges and balanced pulsating gravity-antigravity make them spread out much more than gravity matter or antigravity antimatter. The particles themselves although large are hard to see, because of their fast speeds and constant rapid jiggling, phasing in and out. The spacing within and in between space-matter molecules is what we see and perceive as void empty space, like that in between galaxies. These countless atoms are larger in radius and dispersed in space, and its massive super dense electrons and nuclei elusive and undetectable by our eyes or instruments for the same reason antigravity antimatter are undetectable to us. Their presence is associated with the Higgs field and its ocean. They are there with gravity and antigravity gaseous matter and cause the spectrum red shift of extremely distant galaxies. Light from visible galaxies and invisible galaxies with antigravity passes through them and loses a miniscule amount of photon energy as it collides with tauon (very high-energy dense electron) particles. The highest and last form of matter, space-matter, sustains deep space, and somewhat neutralizes the attractive effects of the other two types of matter by creating great distance in the expanse of space. There is a fourth energy force associated with space-matter itself; its internal potential energy, pressure, and force is the source of the gravitational force ordinary matter experiences and the source of the anti-gravitational force affecting antigravity antimatter. This force is quite like air pressure pushing on lightweight objects within a room or measured as atmospheric pressure. This space-matter energy force keeps space practically empty and appears almost void of atoms and molecules. Space-matter's graviton is a balanced gravity-antigravity, or neutral force, almost

---

[242] Unable to justify a mechanical push force of gravity with aether as the source, Newton decides to make his gravity equation a pull force from a distance and in order to produce the right results he inserts a factitious additional term into the equation $(m/kg)^2$ to cancel out appropriate pieces. He did not reveal from where this extra term came.

as if switched off, yet it is clearly the primary attribute for the sustainment of the vacuum force of space.

If you recall, the galaxy consists of about .0001% matter with mass and 99.9999% void empty space, thanks to the balance of gravity and centrifugal forces, and the other three natural forces. Similarly, the atoms on earth and throughout the visible universe, is primarily void empty space 99.999999999% and only a fraction of it is particles with mass and sustained through a balance of the interaction of three of the four natural forces. The same result occurs in antigravity galaxies. Antigravity antimatter with mass occupies only but a miniscule fraction of the total space where the antigravity antimatter resides in space, just as in the galaxies we know.

The relationship between normal matter with its fourth energy force gravity on itself is represented by the formula $F = G * m_1 * m_2 / r^2$, results in mutual attraction, or more specifically written $F = F_v * 1m/kg * m_1 * 1m/kg * m_2 / r^2$, where vacuum force F_v = 6.6734E-11 Newton[243]. The relationship between antigravity antimatter with its fourth natural energy force on itself is similarly represented by the formula $F = G * -m_1 * -m_2 / r^2$, results in antigravity mutual attraction, or more specifically written $F = F_v * -1m/kg * m_1 * -1m/kg * m_2 / r^2$, where vacuum force F_v is 6.6734E-11 Newton. In addition, the relationship between antigravity objects and gravity objects is represented by the formula $F = G * m_1 * -m_2 / r^2$ or $F = G * -m_1 * m_2 / r^2$, results in mutual repulsion, or more specifically written $F = F_v * -1m/kg * m_1 * 1m/kg * m_2 / r^2$, where vacuum force F_v is 6.6734E-11 Newton. The energy force on matter with gravity and/or matter with antigravity is either mutual attraction or mutual repulsion. Incidentally, Brian Greene described the Casimir's force as a quantum vacuum force with a ghostly grip that can pull together two very thin low-mass plates that are in close proximity to each other in which mutual gravity alone could not per calculations (Greene, The Fabric of the Cosmos: Space, Time, and the Texture of Reality, 2004, p. 332). In the Cosmological Balance Theory, the Casimir's quantum pulls force or quantum fluctuation when combined with space-matter vacuum force of 6.6734E-11 Newton hitting the outside of the plates pushes the two sheets together. Vacuum force will be explained in detail later.

If you recall in Part 2 Chapter 6, we discussed how super massive black holes and white holes both generate high-energy quarks, leptons, and force carriers, which were ejected outward with antigravity and gravity forces respectively into wormhole tunnels away from their galaxies. When these wormhole tunnels collapse, the majority, about 71.30% of the high-energy quarks, leptons, and force carriers' gravitons flip back to neutral and form dense-particle space-matter such as super-cooled substantial hydrogen or high-energy helium is over 1700 times more massive, and size-wise larger than normal hydrogen, yet illusive in time. The space-matter neither has only gravity or antigravity characteristics, yet both, neutral. Moreover, some remaining high-energy quarks, leptons, and force carriers, no more than about 14.35%, remain as antigravity or 14.35% remain as gravity atoms and continue their journey away from the galaxy. In the frigid temperatures of deep space and minimum gravity or antigravity influence, these unique space-matter hygratium atoms quickly form chemical bonds with each other to become stable neutral

---

[243] The Cosmological Balance Theory redefines Newton's $(m/kg)^2$ as the result of quantum entangled attraction of all normal matter with gravity of 1 m/kg for each object in space. The same type of quantum entangle attraction occurs for antigravity antimatter at -1 m/kg for each object in space.

molecules of matter. Their intense dense particles and great size causes them to spread out into a uniformed grid primarily consisting of empty space like air molecules spreading out, or more specifically, evenly occupying deep open space we know as the void between galaxies, 99.999999999999% empty space! Exactly like the void within lighter atoms found in the atmosphere on earth. Finding space-matter in space is like trying to see air molecules in a pitch-black dark room. All we see is the empty space between atomic particles, unable to find or detect the space-matter.

Once again, it has been over a century since Einstein first published his renowned papers laying out his greatest intellectual achievement, the crown of his career, the theory of special and general relativity. In his theories, Einstein showed that outer space was malleable and could twist, distort, and curve under the influence of matter. Since the "shape" of space was affected by the presence and distribution of matter and energy, and matter moved around it, this further means that the "shape" of space is dynamic; it twists and bends and changes with time (Lincoln, 2015). This idea was truly revolutionary and catapulted Einstein to fame. As Einstein wrote his first papers describing a basic theory of gravity relativity in November of 1915, he and others soon applied this behavior to the whole universe, and discovered one implication, that is all matter attracts all other matter, and the static universe concept would no longer remain static. The gravitational attraction would cause all matter to collapse into itself in a single point, assuming there is not enough rotation to counter this. And even if one did not start with a static universe, the mass distribution of the universe should evolve. Hubble's discovery in 1929 indicated that the universe was not static. It was expanding. This observation led Belgian priest Georges Lemaître to propose in 1931 that the universe originated from a small and compact state, what he called a "Cosmic Egg" and what is now called the Big Bang, after Einstein's comment to Lemaître's work.

According to mainstream science, dark energy almost certainly exists and was anticipated by Einstein around 80 years before its discovery. The tale of why Albert Einstein inserted the cosmological constant and then removed it from his theory as his greatest blunder is often told, the bottom line is that Einstein regretted its temporary inclusion. Today, something like it must exist, so the removal of the cosmological constant was the real error. The point is that the malleability of outer space is the result of the attributes of space itself. Space is thereby able to reconfigure and mold itself into a structure or shape capable of suspending matter within itself. The method for which this is done will be discussed in greater details in this chapter. Yes, space has structure that keeps it perfect dynamic equilibrium and static. Einstein and others called this structure, the fabric of space.

8.2 Structure of Space

The space-matter atoms in deep space are a unique massive form of hydrogen. These atoms consist of high-energy Tau lepton or Tauon, a dense proton (with 5-top and 7-bottom high-energy quarks), a dense neutron (with 4-top and 8-bottom high-energy quarks), a high-energy gluon, and a dense high-energy photon to hold the high-energy tauon in its orbital level. This space-matter naturally becomes more stable when they form a unique hydrogen molecule in the cold temperatures of deep space. It contains the Higgs particle characteristic, which we predict is also a neutral graviton. Its super dense particles and high-energy structure gives this atom and the molecule its great volume and radius and creates more distance and empty space between molecules. Its characteristics are similar to super cooled liquid gas rather than a high energetic gaseous state, but much more spread out due to lack of gravitational or anti-gravitational attraction. Instead of being fast moving like gaseous matter, they are slower deliberate molecules holding steadfast their distance but still able to give way to other objects passing through its area. Again, it acts more like super cold liquefied gas, very densely packed almost solid around massive objects, in the calm still sea of open space that gradually gives way as a galaxy pushes through it, like a slow-moving boat splits through the super salty dense waters. Its overall total mass and energy suspends galaxies in space by surrounding with a curvature sometimes seen as a halo[244], and gives the homogenous universe the sheet and filament characteristic it has, like newly spun cotton candy.

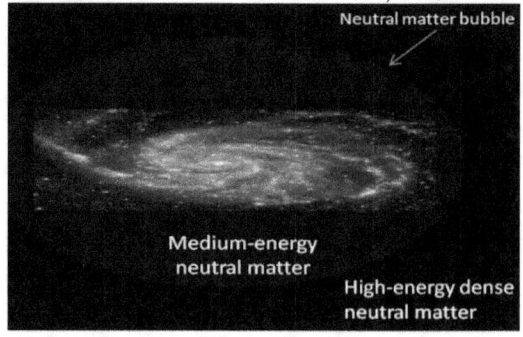

Above is a picture of a galaxy moving through deep space. When a galaxy moves through open space, the dense-particle space-matter entering the bubble around the galaxy destabilize when they enter the heated regions at the edge of the galaxy, which excites the components of the space-matter, first into medium energy space-matter and then into more denser spherical layers. The space-matter within the galactic halo contains same amount of mass as the galaxy it contains; Einstein coined this as the curvature of space. Vera Rubin was on to something when she speculated that galaxies reside inside a sphere causing the stars to move faster than anticipated by gravity alone. This perceived sphere or boundary is the resulting interaction between matter with gravity and that of the gravity-antigravity pulses of space-matter.

As a result, each galaxy is surrounded by medium energy space-matter bubble filled with escaping wormhole tunnels. Medium energy in terms that these space-matter atoms and molecules are more energized, they are in a gaseous state, not a super cold liquid-gas. Each star is also surrounded by a bubble, curvature of space, where all of the medium energy space-matter resists decays and holds the magnetic froth of sub-atomic particles and lighter, more normal sized atoms, which solidify to form the gaseous cloud or chunks of ice, known as the Oort cloud at the edge of the solar system. On direct contact with a

---

[244] Space-matter is a perfect gaseous liquid substance because it exerts equal pressure, equal energy, and equal force on objects in space from all directions. This property enables it to contain those objects.

star's corona, space matter strengthens significantly, as it absorbs the high energy and reflects heat back onto the star, resisting decays into smaller basic sub-atomic particles and lighter atoms. In rare instances, space matter becomes the source of intergalactic sub-atomic particles. Weinberg calls the surrounding bubble around the galaxy, spherical halos. "The cold dark matter particles could not lose their energy through radiation cooling, so they remained in large more-or-less spherical halos around these galaxies (Weinberg, 2008, p. 403)." Another way to explain this concept is to imagine that the space-matter dense-particle atoms and molecules mutually push each other away, due to natural electrical charge they possess, and maintain enough distance to nullify most of the attraction or repulsion of the three natural forces it contains, similar to air molecules maintaining gaseous state in the atmosphere. Let us imagine for a moment that we have a glass box 10 km in width by 10 km in depth by 10 km in height filled with air from the atmosphere. In a warm day, the molecules of air are excited and bouncing off each other, and in doing so, spread out throughout the box, even to the point of moving to fill the top part of our box, fighting the effects of gravity. However, in a very cold day, the air molecules slow down, sink, and accumulate along the bottom of the box, as gravity pulls them downward, becoming very dense air there.

Now, let us imagine the same box of air transported into outer space between the gravitational influences of the moon and the earth, zero gravity. As the sunlight hits the air molecules, they get excited, bounce rapidly off each other, and fill the entire box evenly. As the earth's shadow eclipses the box, the air molecules shed energy, slow down, and even become almost motionless, but are still able to maintain that uniformed spread throughout the box, as there is no gravity to pull them to any one particular side of the box. Air pressure still exists within the box, but evenly spread to all sides of the box, as opposed to one side on earth due to the pull of gravity. This cold stationary gaseous state is similar to how space-matter sustains deep open space, evenly spread, and yet uniformly unique, resisting any effect from either force of gravity matter or antigravity antimatter. Weinberg calls this force and its energy "vacuum energy." What do we see when we look between galaxies is empty space? We do not see the three-dimensional grid of molecules evenly spaced like that of a perfectly formed diamond, we only notice the great expanse and void created. Below is a depiction of space-matter in the form of high-mass super dense-particle hydrogen molecules spread evenly over super cooled liquid-gas space at a density of one atom per cubic meter in open flat space.

The basic space-matter is called *hygratium*; it is the unique neutral quark that generates gravity-antigravity pulses. A grid of hygratium quarks makes deep open space appear void and flat. Depicted above is space-matter in the form of high-mass super dense-particle *hygratium* hydrogen molecules spread evenly over super-cooled liquid-gas space at a density of one atom per cubic meter in open flat space. Let us analyze the density of space-matter, which coexists with normal matter and antigravity antimatter, to be about 71.30% of all matter in the universe. If we take an area of deep space, 14 billion light years cubed, it gives us a volume of 2.32378E+78 cubic meters, and at 71.30% we have an adjusted volume of 2.32377E+78 cubic meters of space-matter in space. We get a distribution of one space-matter atom per cubic meter throughout all 14 billion light years cubed except where normal matter with gravity and matter with antigravity exists. Open space between galaxies is occupied by space-matter and normal matter or by space-matter and matter with antigravity, at which point the density of those locations in open space could be as much as 400 atoms per cubic meter. Of which 399 atoms are normal matter with gravity matter or antigravity antimatter, which absorb photon energy, lower spectrum frequency, and cause the red shift. In the deep open space between galaxies, the density of space-matter by itself amounts to one atom per cubic meter at the low end, and in the presence of normal matter or antigravity antimatter, we get a higher density of about 400 atoms per cubic meter at the higher end.

How does this affect objects possessing gravity or antigravity? Each atom in the grid above pushes outward on all directions on the unique atoms next to it and they in turn push outward on the ones next to them creating the void in space. This force emanating from the space-matter is the source of what we call gravity or the "pulling force" of normal matter. When space-matter absorbs photon energy intensity from light hitting it, the atom and molecule does not heat up, it becomes more stable and expands its radius. Its photon energy absorption slightly slows down the speed of the photon but does not change its color, like light experiment conducted by Lene Hau, a Harvard University physicist. "An entirely new state of matter first observed four years ago, has made this possible. When atoms become packed super-closely together at super-low temperatures and super-high vacuum, they lose their identity as individual particles and act like a single super-atom with characteristics like a laser. Such an exotic medium can be engineered to slow a light beam 20 million-fold from 186,282 miles a second to a pokey 38 miles an hour[245]. "In this odd state of matter, light takes on a more human dimension; you can almost touch it," says Lene Hau, a Harvard University physicist (Cromie, 1999)." In direct opposition, increasing the speed of light, John G. Cramer conducted an experiment where he shined a laser beam through a chamber filled with super excited energized cesium atoms. As the laser enters the chamber, the cesium atoms absorb the light and release it with additional speed and energy, so by the time the light reached the other end of the chamber it escapes with 310 times the speed of the original laser beam, at faster than light-speed. (Cramer, 2002) Both of these experiments together prove that the speed of the medium within a system or the added motion of the system itself influences the speed of light within it. We will explore this in detail with other examples later in this chapter.

Weinberg calls the matter that has this property, *grey* matter; we call it space-matter, as it also has no gravity or antigravity characteristics on itself. It spreads out with a pushing force generated by super cold gaseous, almost liquid, space-matter suspended in deep open space. Weinberg also calls the gas radius spreading force that creates empty space, *vacuum* energy. Objects with gravity and objects with antigravity easily move past space-matter because it pushes them along without any additional changes to their gravity or antigravity forces, respectively. Space-matter provides the constant energy force (pulsating antigravity) that pushes one gravity object toward another gravity object, while also providing the push

---

[245] If exotic matter can be manufactured on earth using simulated vacuum and extreme cold temperatures naturally found in space, then it is reasonable to predict that such exotic type matter naturally occurs in the open expanses of space. We call it space-matter.

on the other gravity object. Space-matter affects antigravity objects similarly by mutually pushing them toward each other, through pulsing gravity forces.

Imagine again for a moment that you somehow have a collection of some space-matter atoms enclosed in an airtight box. That box would appear to us void of normal matter and empty like outer space, a natural vacuum, but not black due to the background behind the box. This experiment would be difficult to do on earth because the space-matter may destabilize and convert or decay to a more stable normal matter in the environment, into normal ordinary hydrogen and helium, or slip through back into space. Similarly, space-matter dispersed in space is also like air molecules in the atmosphere on a clear black night on earth. The air molecules push each other apart to sustain its gaseous form while fighting gravity. Two stars in space can be compared to two hot air balloons floating in the air each tied to a long thin line holding them together, as gravity holds two stars bound together. Space suspends the stars and galaxies and yet allows them to bind.

How then should we represent the relationship between space-matter and its energy force? We defined space-matter's energy force as being neither attractive on itself or with normal matter or with antigravity antimatter. For this energy force to fit all three of these conditions, it must be repulsive to at least itself, and exert a force on normal gravity matter as well as antigravity antimatter, pressure and energy. That force on itself is primarily electrical in nature and comes from trillions and trillions of extremely dense Tauon each orbiting their space-matter nuclei dispersed throughout deep open space between galaxies, or space-matter intergalactic dense-particle hydrogen, and enhanced by rapid pulsating gravity-antigravity forces. Space-matter is 71.30% of all matter, so its pressure or energy force acts on the galaxies, the solar systems, and the planet as the push gravity force in the visible universe and push antigravity force in the invisible part of the universe. Space-matter vacuum pressure forces converts into gravity forces in the visible universe and into antigravity forces in the invisible part of the universe. The visible matter is only 14.35% and the invisible antigravity antimatter is about 14.35% of all mirrored antimatter. Although space-matter consists of very dense particles, its overall effect on mass is zero when it comes to weight or more specifically attraction toward each other is zero, as it has both gravity or antigravity pulsating characteristics, overall neutral and balanced in nature. The space-matter antigravity pulse is the pressure applied to matter with gravity, and its gravity pulse pushes on antigravity objects. Space-matter around objects is its curvature.

The Cosmological Balance space-matter, primarily consisting of unique type of dense "hydrogen" atoms, is different from the neutral intergalactic hydrogen atoms described in "Lyman α forests." Lyman states that clouds of intergalactic neutral hydrogen atoms (baryons, proton nuclei without orbiting electron) would completely block any light with a frequency about the red shifted Lyman α line from sources beyond $z=5$. "In 2001, the spectrum of the quasar SDSSp J103027.10+052455.0 with red shift $z = 6.28$ discovered by the Sloan Digital Sky Survey was found to show clear signs of a complete suppression of light in the wavelength range from just below the red shifted Lyman α wavelength (Weinberg, 2008, p. 78)." How is this different from space-matter? In the Cosmological Balance, space-matter does not form clouds; space-matter or dense-particle hydrogen atoms spread out and form the void of space, and the energy it contains. It is not the normal ionized hydrogen or baryon (Lyman's version of neutral intergalactic hydrogen) with gravity or antigravity that forms these clouds in and around the cosmological space-matter atoms and molecules that the Sloan Digital Sky Survey detected and those supplied by the spectrum of intense gamma ray sources known as *gamma ray bursters.*

What could be the structure of the space-matter "hydrogen" atom? Again, space-matter has practically no gravitational or anti-gravitational characteristics; their graviton is neutral, the Higgs particle. We predict that the space-matter atom consists of a high-energy proton, a high-energy neutron, and a tauon orbiting it held in place by a high-energy photon. The nucleus has one proton consisting of five top quarks and seven bottom quark, one neutron consisting of four top quarks and eight bottom quarks, and a high-energy dense gluon. The following mass estimates are gathered from Wikipedia.com. The mass of the top quark is estimated to be about 3.1000E-25 kilograms. The mass of the bottom quark is estimated to be approximately 7.44889E-30 kilograms at low end. The proton's mass is then approximately 1.55006E-24 kilograms. The neutron's mass is then approximately 1.2401E-24 kilograms, and the gluon is approximately 6.04445E-31 kilograms. The mass of the Tauon orbiting the nucleus is estimated at 2.1478E-26 kilograms and the mass of the high-energy photon holding the Tau lepton in place is 3.8786E-35 kilograms. The combined total mass of the super dense-particle space-matter hygratium atom is 2.8116E-24 kilograms[246].

"Space-matter" Atom = High-Energy Photon + Tauon + High-Energy Proton + High-Energy Neutron + High-Energy Gluon

= Photon + Tauon + (5TopQuarks + 7BottomQuark) + (4TopQuark + 8BottomQuark) + Gluon

= 3.8786E-35 kg + 2.1478E-26 kg + (5*3.1000E-25 kg + 7*7.44889E-30 kg) + (4*3.1001E-25 kg + 8*7.44889E-30 kg) + 6.04445E-31 kg = 2.8116E-24 kg

The radius of a normal hydrogen atom on earth is approximately 5.3×10^{-11} meters. However, due to the extremely low temperatures of space, 2.7 Kelvin Degrees, the super dense-particle space-matter "hydrogen" atom will have a condensed radius for its mass. Therefore, the radius of the space-matter atom estimated at 3.2358E-11 meter is larger and more dense-particle than a normal hydrogen atom of 2.6349E-13 meter. In the coldness of space, these dense-particle atoms are about 100 times larger than normal hydrogen atoms on earth but are expected to move much more slowly, while still maintaining distance between other space-matter atoms and molecules, an average density of one atom per cubic meter. Dense-particle space-matter atom acceleration due to electrostatic repulsion is estimated at 2.3735E+13 meters per second squared. The vacuum energy causing this acceleration is approximate 8.83919E+15 Newton meters. This vacuum energy is not gravity; it is cause by pressure in space produced by electrostatic repulsion between slow moving super dense-particle high-energy space-matter atoms and molecules. The same energy causes gas atoms and molecules on earth to spread out and bounce off each other. However, in this case, the vacuum energy, pressure, and force make space what it is, appear void and empty. The Force or more specifically Vacuum Force exerted by these dense-particle space-matter atoms is obtained by the product of its mass times its

---

[246] This is the Cosmological Balance Theory's best guestimate of the content of each atom of space-matter. Detailed scientific investigation needs to be performed to accurately show the exact structure, characteristics, and properties.

acceleration or more specifically 2.8116E-24 kg x 2.3735E+13 m/sec² = 6.6734E-11 kilograms meters per second squared or 6.6734E-11 Newton.

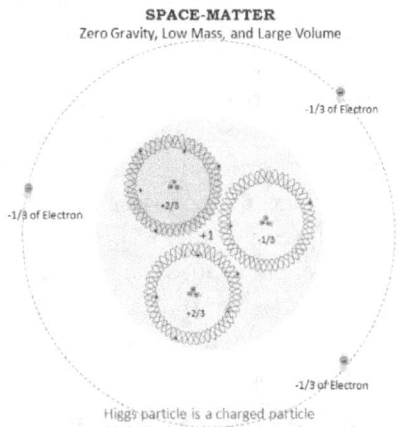

Space matter produces practically no gravity nor antigravity on itself yet aligns in pulses gravity and antigravity at a rate equal to about one-five-hundredths (1.9984E-03) of the speed of light (599,119.05 cycles per second or hertz). This happens because the quark core consists of three negative point particles attached to both neutrinos and antineutrinos to form a total charge of minus one. Orbiting the core are five sets of positive point particles bonded to a neutrino and an antineutrino with each group possessing a +1/3 charge. The five sets total charge amounts to plus 5/3 creates the space matter up quark. This simplified space matter nuclei, consists of two up quarks and one down quark, is orbited by three pairs of negative point particles each attached to two neutral point particles (a neutrino and an antineutrino) with each having a negative 1/3 charge. The three negative pairs together make up the electron orbiting this space matter atom. Total volume of the space matter atom is just under one-meter diameter, and practically massless. Space-matter (*hygratium*) in intergalactic open space will consist of neutral space-matter quarks and no electrostatic shell around it. However, space-matter near antigravity objects will consist of two antimatter down-quarks and one up-quark with a positron orbiting about it. Depicted below is an image of space matter near normal gravity objects.

Each outward motion of the trio of positive point particle generates gravity. However, the inward motion of the same positive point particle group generates the same amount of antigravity to counterbalances it. The dashed spherical outline represents the spherical layer where the generated gravity cancels the generated antigravity. Total net gain of space matter is zero gravity. Space matter's miniscule point particle motion also equates to minimal mass generation, and therefore practically registers as a massless atom with no gravitational or antigravity attraction to other atoms. The other three natural forces, however, remain active within and among space matter to create the grid and ocean of space. As balanced simple atoms with absolutely zero charge and zero gravity their tendency is to spread out in great distances and form a grid of at least one-meter intervals. It makes space appear void and empty. This neutral space matter atom can and does absorb photon energy from normal light passing through it, and in doing so eventually red shifts the light traveling extremely far distances numbering in the hundreds of millions of light years. As the photon enters the space matter atom, it causes the outer electron to gain energy move to a higher level and then re-emit the photon with slightly less energy. This tiny energy absorption maintains the 2.7-Kelvin degree temperature in open space. Extremely turbulent space can force two space matter atoms to collide, merge nuclei, and decay down to gravity matter and antigravity antimatter, while releasing neutrinos and antineutrinos. These loose neutral point particles collide and transition to positron and electrons, and back to neutral point particles or charged particle pairs, elementary pairs for either antigravity antimatter or gravity matter.

8.3 Pressure and Energy of Space

Let us look at this space-matter pressure to gravity energy transfer from another perspective. Suppose we inflate a balloon to about 20 centimeters and examined it. Its size seems to stay relatively consistent throughout the hour of observation. Take that same balloon and place it inside an enclosed glass box with one opening, and then seal that opening. Now vacuum out half the air within the box and observe the balloon. It begins to expand and grow, in size, as the air is removed from the box. However, if air is pumped into the box, the balloon begins to shrink in size as more and more air goes into the box. The air pressure in the box causes the balloon to shrink smaller; likewise, space-matter vacuum energy causes the bubble around an object to move inward toward the center of the more massive object assisting gravity pulling smaller objects toward the larger object. In this example, the balloon represents the invisible gravitational barrier around a star, say the Oort cloud.

The invisible gravitational barrier transfers energy from one bubble layer inward to an interior nested bubble layer with ever-increasing gravitational intensity as energy force gets closer to the star's surface[247]. This increase in gravity intensity equates to an increase in space-matter density layers or the increase of curvature of space. The strong invisible gravitational bubble layer or curvature of space above the corona is what causes the super heating of the star's corona, and beyond that star's visible corona is yet another invisible layer of clear or translucent space-matter. It stops 90% or more of the heated particles from escaping or radiating heat outward. The average pressure exerted by space-matter occupying all of deep space amounts to 3.80379E-63 Newton per meter squared, which exerts a vacuum energy of 8.83919E+15 Newton meters and the vacuum force resulting from space-matter itself is 6.6734E-11 Newton. This vacuum force and the energy throughout space provide the gravitational and anti-gravitation constant force. Weinberg said that, "The individual cold dark matter particles are assumed to move too slowly for them to produce an appreciable pressure or anisotropic inertia (Weinberg, 2008, p. 259)." The Cosmological Balance proves otherwise, the space-matter pressure is there and effective only because the high-energy and density of the space-matter itself, we know of its effects as the force of gravity, through gravity-antigravity pulses. However, gravity or antigravity fluctuations do not affect space-matter they just pass right through. "That is, the coordinate mesh is tied to the "dark" (space-matter) matter particles in such a way that they remain at rest (unaffected) despite fluctuations in the gravitational force in which they move (Weinberg, 2008, p. 259)."

Pressure and Energy equations are applied to gaseous and liquid matter on earth as well as to dense-particle space-matter hygratium atoms in space. Vacuum Pressure equals Force divided by Area or Force times distance divided by Area times distance, which is the same as Energy divided by Volume. Where Force equals mass of dense-particle atom times acceleration; 2.8116E-24 kg * 2.3735E+13 m/sec^2 = 6.6734E-11 kg m/sec^2. Volume of space 14 billion light years cubed is 2.32378E+78 meters cubed[248]. Therefore, Vacuum

---

[247] The invisible bubble layer of space-matter around objects is also referred to as the curvature of space. They are the same.

[248] We used 14-billion light year distance to calculate space pressure, energy, and vacuum force because that seems to be the observation limit of our current technology.

Pressure is Force times distance divided by volume or 6.6734E-11 kg m/sec² * 1.32454E+26 / 2.32378E+78 m³ = 3.80379E-63 N/m². Vacuum Energy equals Vacuum Pressure times Volume of space or 3.80379E-63 N/m² * 2.32378E+78 m³ = 8.83819E+15 kg* m²/sec² = 8.83819E+15 Joules. The same equations are written out below in several lines for readability:

Vacuum Force = mass * acceleration

= 2.8116E-24 kg * 2.3735E+13 m/sec²

= 6.6734E-11 kg m/sec²

Vacuum Pressure = Force / Area = Force * distance / (Area * distance)

= Force * distance / Volume

= 6.6734E-11 kg m/ sec² * 1.32454E+26 m / 2.32378E+78 m³

= 3.80379E-63 N/m²

Vacuum Energy = Pressure * Volume

= 3.80379E-63 N/m² * 2.32378E+78 m³

= 8.83919E+15 kg*m²/sec²

We expect the atoms of space-matter to be extremely spread out gaseous high-energy dense-particle matter with firm almost liquid characteristics. Since the atoms and molecules of space-matter are dispersed uniformly spread throughout space at a ratio of one atom per cubic meter, its vacuum force, basically, the same anywhere in flat space and more densely packed especially in curved space around massive objects. By the same reasoning, the vacuum pressure and associated vacuum energy is normally the same for any position or coordinates in flat Euclidean space. The exceptions to this norm are considered systems of high-pressure and low-pressure build-ups. In all instances, this vacuum energy in space is transferred from space-matter substantial atoms and molecules to other atoms and molecules simply by atomic and molecular electrostatic repulsion, enhanced by pulsating gravity-antigravity force. It provides the force, which allows matter with gravitational characteristics to "pull" mutually together two objects with mass per Newton; technically, they are also pushed, as he originally intended, together as kinetic vacuum energy is transferred between space-matter atoms to matter with gravity or antimatter with antigravity. Gravity in normal visible matter and antigravity in the invisible part of the universe provides the natural tendency for these molecules to bind mutually together, chemically, and molecularly, a quality proven by quantum mechanical entanglement based on shared origin.

8.4 Vacuum Force of Space

In the previous two sections, we discussed the structure of space, as well as the pressure and energy inherent to space-matter. What we did not explain is how that resulting Vacuum Force becomes the Gravitational constant we see in Newton's equation? Yes, Newton, discovered that unseen force between bodies and found that it was a constant number followed by Newton square meters per square kilograms. He assigned the square meters per square kilograms to cancel out when multiplying the masses of two objects and dividing by the square of the distance between the centers of the two objects in meters.

The Cosmological Balance takes a slightly different approach to Newton gravitational equations. The Cosmological Balance theory takes the resulting vacuum force of space-matter and defines how such force becomes the gravity we see in every day[249]. First, let us consider a lone object in space; say one rogue planet, floating in deep open space between galaxies. Such an object would be receiving equal force from all directions, and thereby assume a spherical shape, just like a soap bubble floating in the air as it leaves the blower's launching loop. This rogue planet's bubble shell is its atmosphere, is the point of transfer of forces and energy toward the center of the planet as gravity and suspends the planet in space.

Now let us consider two objects in space not orbiting each other. The vacuum forces act upon each object from all possible angles with equal force except for the cones between the two objects. Above is a depiction of the relationship between the two objects and their cones:

Using the distance between m_1 and m_2 as the length of adjacent side of the right triangle, and the radius of m_1 as the opposite side, we can solve for the angle intersecting the center of m_2 from the formula $\tan^{-1}(opp/adj)$. The angle obtained φ is ½ of the total angle of the m_2 spherical cone. Then by using the radius of m_2 and the angle φ, we can solve for the radius of the cone from $r_2 = m_2 radius \times \tan(\varphi)$. From there we solve for the volume of the spherical cone m_2 from equation $V_{m2} = 2/3 \pi \times m_2 radius^2 \times h_2$, where $h_2 = r_2 - $ square root $(r_2^2 - r_{2c}^2)$. Then we obtain the mass of m_2 cone from equation $m_2 cone = $ (mass of m_2) $\times$ (volume $m_2$cone)/(volume m_2 sphere) = $m_2 \times V_{m2}/V_{m2s}$, where volume of sphere m_2 is solved with equation $4/3 \pi \times m_2 radius^3$. We recall that space-matter exerts a push force of 6.6734E-11 N in all directions. We therefore use this force to solve for the push force on m_2 by multiplying space-matter force times mass of m_2 cone times mass of m_1 shielding divided by the distance squared. This gives us force$_2$ = $F_{sm} \times m_{2cone} \times m_1 / d_r^2$. Then, we determine the amount of resistance to movement or inertia force with equation:

---

[249] Vacuum force of 6.6734E-11 Newton exerted in all directions equally by space-matter around an object in space simply suspends it in space. But if we consider two objects within a sufficient angle of view, they experience a proportionate force difference based on distance enough to push them together at a rate exactly equal to Newton's gravity equation.

CHAPTER 8: SPACE BALANCES GRAVITY AND ANTIGRAVITY

$m_2$resist = (one square meter per square kilogram) x (spherical volume m_2) / (cone volume m_2). We then solve for the total push effect by equation (push force on m_2) times (resistance of m_{2cone}) = force$_2$ x $m_{2resist}$ = G_{f2}.

For earth as m_1 at a mass of 5.972E+24 kg, the moon as m_2 with a mass of 7.34767E+22 kg, the distance between the two at 3.844 x 10^8 m, the radius of earth as 6.378E+6 m, and the radius of the moon as 1.738E+6 m, we get the resulting push force of 1.98175E+20 N.

Right Triangle m1 to m2			Right Triangle m2 to m1		
r1=opposite	6.3781E+06		r2=opposite	1.7381E+06	
distance r=adjacent	3.8440E+08		dist r=adjacent	3.8440E+08	
hypotnuse	3.8445E+08		hypo h2	3.8440E+08	
1/2 angle=tan^{-1} (opp/adj)	0.016590829	radians	1/2angle=atan(opp/adj)	0.004521561	radians
full angle from m2 to dia m1	0.033181659	radians	angle from m1 to m2	0.009043123	radians
r$_{2c}$=radius m2 sphere cap	28835.19754	m	radius m1 sphere cap	28838.87172	m
cone volume m$_2$ sphere	1.5135E+15	m^3	cone volume m$_1$ sphere	5.5549E+15	m^3
surface area m2 cone	1.60064E+11	m^2	surface area m1 cone	5.80469E+11	m^2
m$_2$ cone mass kg	5.0561E+18	kg	m$_1$ cone mass	3.0524E+19	kg
SM Push Force = F$_{sm}$	6.6734E-11	N	SM push force=F$_{sm}$	6.6734E-11	N
F$_{sm}$*m$_{2cone}$*m$_1$/d$_r^2$ = force$_2$	1.36369E+16	N kg^2/m^2	m*m$_{1cone}$*m$_2$/d$_r^2$ = force$_1$	6.49253E+00	N kg^2/m^2
force$_2$ * m$_{2resist}$ = G$_{f2}$	1.98175E+20	N	force$_1$ * m$_{1resist}$ = G$_{f1}$	1.98175E+20	N
h$_2$ = r$_2$ - sqrt(r$_2^2$-r$_{2c}^2$)	239.2054072	m	h$_1$ = r$_1$ - sqrt(r$_1^2$-r$_{1c}^2$)	65.19847391	m

We solve for the angle φ intersecting the center of m_2 from the formula $\tan^{-1}$(opp/adj).

φ = tan-1(6.3781E+6 / 3.8440E+8) = 0.01659 radians

This angle is ½ of the total angle of the m_1 spherical cone.

2 * φ = 0.0331817 radians

Then by using the radius of m_2 and the angle φ, we can solve for the radius of the cone from r_{2c} = $m_2$radius x sin(φ).

r_{2c} = 1738100 m * sin(0.01659 radians)

= 28835.19754 m

We then solve for the volume of the spherical cone m2 from equation V_{m2} = 2/3 π x $m_2$radius^2 x h$_2$. Where h$_2$ = r$_2$ – sqrt(r$_2^2$ - r$_{2c}^2$).

h$_2$ = 6.378E+6 m - sqrt((6.378E+6 m)^2 - (28835.19754 m)^2)

= 239.2054072 m

V_{m2} = 2/3 π * (1738100m)^2 * 478.4437394 m

= 1.5135E+15 m^3

Then we obtain the mass of m$_2$cone from equation m$_2$cone = (mass of m$_2$) x (volume m$_2$cone)/(volume m$_2$ sphere) = m_2 x V_{m2}/V_{m2s}, where volume of sphere m$_2$ is solved with equation 4/3 π x m$_2$radius.

Volume of m$_2$ sphere = M_{2sv} = 4/3 π * (1738100m)^3

$$= 2.19944\text{E}+19 \text{ m}^3$$

$$M_{2\text{cone}} = 7.34767\text{E}+22 \text{ kg} * 1.5135\text{E}+15 \text{ m}^3 / 2.19944\text{E}+19 \text{ m}^3$$

$$= 5.0561\text{E}+18 \text{ kg}$$

We then incorporate space-matter force 6.6734E-11 N to solve for the push force on m2; by multiplying space-matter force times mass of m2 cone times mass of m_1 shield divided by the distance squared: $\text{force}_2 = F_{sm} \times m_{2\text{cone}} \times m_1/d_r^2$.

$$\text{force}_2 = 6.6734\text{E}-11 \text{ N} * 5.0561\text{E}+18 \text{ kg} * 5.972\text{E}+24 \text{ kg} / (3.844\text{E}+8 \text{ m})^2$$

$$= 1.36369\text{E}+16 \text{ N kg}^2/\text{m}^2$$

Then we determine the amount of resistance to movement or inertia force with equation one square meter per square kilogram times (spherical volume m2) / (cone volume m2).

$$M_2\text{resist} = 1 \text{ m}^2/\text{kg}^2 * 2.19944\text{E}+19 \text{ m}^3 / 3.0272\text{E}+15 \text{ m}^3$$

$$= 1.4532\text{E}+04 \text{ m}^2/\text{kg}^2$$

We then solve for the total push effect by equation (push force on m_2) times (resistance of $m_{2\text{cone}}$ m^2/kg^2) = $\text{force}_2 \times m_{2\text{resist}} = G_{f2}$.

$$G_{f2} = 1.36369\text{E}+16 \text{ N kg}^2/\text{m}^2 * 1.4532\text{E}+04 \text{ m}^2/\text{kg}^2$$

$$= 1.98175\text{E}+20 \text{ N}$$

This solution coincides with and confirms Newton's gravity force source is the medium permeating space itself, space-matter.

For the Sun as m_1 and the Earth as m_2, we get:

Right Triangle m1 to m2		Right Triangle m2 to m1		
r1=opposite	6.9600E+08	r2=opposite	6.3781E+06	
distance r=adjacent	1.4960E+11	dist r=adjacent	1.4960E+11	
hypotnuse	1.4960E+11	hypo h2	1.4960E+11	
1/2 angle=tan⁻¹(opp/adj)	0.004652373 radians	1/2angle=atan(opp/adj)	4.26344E-05	radians
full angle from m2 to dia m1	0.009304746 radians	angle from m1 to m2	8.52687E-05	radians
r_{2c}=radius m2 sphere cap	29673.19223 m	radius m1 sphere cap	29673.51334	m
cone volume m_2 sphere	5.8810E+15 m³	cone volume m_1 sphere	6.4176E+17	m³
surface area m2 cone	5.9734E+11 m²	surface area m1 cone	6.48853E+13	m²
m_2 cone mass kg	3.2315E+19 kg	m_1 cone mass	9.0385E+20	kg
SM Push Force = F_{sm}	6.6734E-11 N	SM push force=F_{sm}	6.6734E-11	N
$F_{sm}*m_{2\text{cone}}*m_1/d_r^2$ = force$_2$	1.91658E+17 N kg²/m²	$m*m_{1\text{cone}}*m_2/d_r^2$ = force$_1$	3.91872E+01	N kg²/m²
force$_2$ * $m_{2\text{resist}}$ = G_{f2}	3.54192E+22 N	force$_1$ * $m_{1\text{resist}}$ = G_{f1}	3.54192E+22	N
$h_2 = r_2 - \sqrt{r_2^2-r_{2c}^2}$	69.02550147 m	$h_1 = r_1 - \sqrt{r_1^2-r_{1c}^2}$	0.632555604	m

The above spreadsheet shows us that this solution also coincides with Newton's equation and is equal to his gravitational force between earth and the sun. However, note that the amount of space-matter vacuum force moving the earth is far greater than that on the sun, clearly showing that the earth is pushed toward the sun or falling faster to the sun and by Newton's detailed calculations, orbits the sun while barely wobbling the sun toward

CHAPTER 8: SPACE BALANCES GRAVITY AND ANTIGRAVITY

the earth. This force $G_{f2} = G_{f1} = Gm_1m_2/r^2 = G * 1.989E+30 * 5.972E+24 / (1.496E+11)^2 = 3.54192E+22$ N, is pushing the earth and the sun toward each other.

For the Earth as m_1 and a 100 kg Cannon Ball as m_2 with radius of 1.25 m, we get:

Right Triangle m1 to m2		Right Triangle m2 to m1	
r1=opposite	6.3781E+06	r2=opposite	1.2500E+00
distance r=adjacent	6.3782E+06	dist r=adjacent	6.3782E+06
hypotnuse	9.0201E+06	hypo h2	6.3782E+06
1/2 angle=tan^{-1} (opp/adj)	0.785390226 radians	1/2angle=atan(opp/adj)	1.9598E-07 radians
full angle from m2 to dia m1	1.570780452 radians	angle from m1 to m2	3.9196E-07 radians
r_{2c}=radius m2 sphere cap	0.883876461 m	radius m1 sphere cap	1.249980157 m
cone volume m_2 sphere	1.1981E+00 m^3	cone volume m_1 sphere	1.0474E+07 m^3
surface area m2 cone	6.346392096 m^2	surface area m1 cone	25046347.45 m^2
m_2 cone mass kg	1.4644E+01 kg	m_1 cone mass	5.7554E+10 kg
SM Push Force = F_{sm}	6.6734E-11 N	SM push force=F_{sm}	6.6734E-11 N
$F_{sm}*m_{2cone}*m_1/d_r^2$ = force$_2$	1.43463E+02 N kg^2/m^2	$F_{sm}*m_{1cone}*m_2/d_r^2$ = force$_1$	1.70215E-08 N kg^2/m^2
force$_2$ * $m_{2resist}$ = G_{f2}	9.79649E+02 N	force$_1$ * $m_{1resist}$ = G_{f1}	9.79649E+02 N
$h_2 = r_2 - \sqrt{r_2^2 - r_{2c}^2}$	0.366109508 m	$h_1 = r_1 - \sqrt{r_1^2 - r_{1c}^2}$	1.22935E-07 m

Here the earth and cannonball solution also coincide with Newton's equation. The amount of force transferred from space-matter onto the atmosphere and then to the cannonball causes it to fall to the earth, while practically no effect on the earth movement to the cannonball. This force $G_{f2} = G_{f1} = Gm_1m_2/r^2 = G * 5.972E+24 * 100 / (6.3782E+6)^2 = 9.7965E+02$ N, is pushing the earth and the cannonball toward each other. For the Sun as m_1 and Mercury as m_2, we get:

Right Triangle m1 to m2		Right Triangle m2 to m1	
r1=opposite	6.9600E+08	r2=opposite	2.4400E+06
distance r=adjacent	5.7910E+07	dist r=adjacent	5.7910E+07
hypotnuse	6.9841E+08	hypo h2	5.7961E+07
1/2 angle=tan^{-1} (opp/adj)	1.487783515 radians	1/2angle=atan(opp/adj)	0.042109439 radians
full angle from m2 to dia m1	2.97556703 radians	angle from m1 to m2	0.084218878 radians
r_{2c}=radius m2 sphere cap	2431597.652 m	radius m1 sphere cap	29299508.87 m
cone volume m_2 sphere	2.7902E+19 m^3	cone volume m_1 sphere	6.2597E+23 m^3
surface area m2 cone	5.29452E+13 m^2	surface area m1 cone	6.67629E+16 m^2
m_2 cone mass kg	1.5119E+23 kg	m_1 cone mass	8.8160E+26 kg
SM Push Force = F_{sm}	6.6734E-11 N	SM push force=F_{sm}	6.6734E-11 N
$F_{sm}*m_{2cone}*m_1/d_r^2$ = force$_2$	5.98406E+27 N kg^2/m^2	$m*m_{1cone}*m_2/d_r^2$ = force$_1$	1.48029E+01 N kg^2/m^2
force$_2$ * $m_{2resist}$ = G_{f2}	1.30502E+28 N	force$_1$ * $m_{1resist}$ = G_{f1}	1.30502E+28 N
$h_2 = r_2 - \sqrt{r_2^2 - r_{2c}^2}$	2237681.293 m	$h_1 = r_1 - \sqrt{r_1^2 - r_{1c}^2}$	616984.116 m

Here the Sun and Mercury solution also coincides with Newton's equation. The amount of force transferred from space-matter onto the planet causes it to orbit the Sun, while practically no effect on the Sun's movement to the planet Mercury. In the spreadsheet above, this force $G_{f2} = G_{f1} = Gm_1m_2/r^2 = G * 1.989E+30 * 3.29718E+23 / (5.791E+7)^2 = 1.305E+28$ N, is pushing the Sun and the Mercury toward each other.

The vacuum pressure, energy, and force exerted by space-matter's gravity-antigravity pulses fulfill Sir Isaac Newton's original intent for a medium between bodies. The tables inserted above, extracted from excel spreadsheets, shows us that the vacuum force exerted by space-matter on normal matter provides the same gravitational effect captured by Newton's equation $G \times m_1 \times m_2 / r^2$. Sir Isaac Newton great work through his definitions, axioms, propositions, problems, and theorems in Principia solves the universal motion and

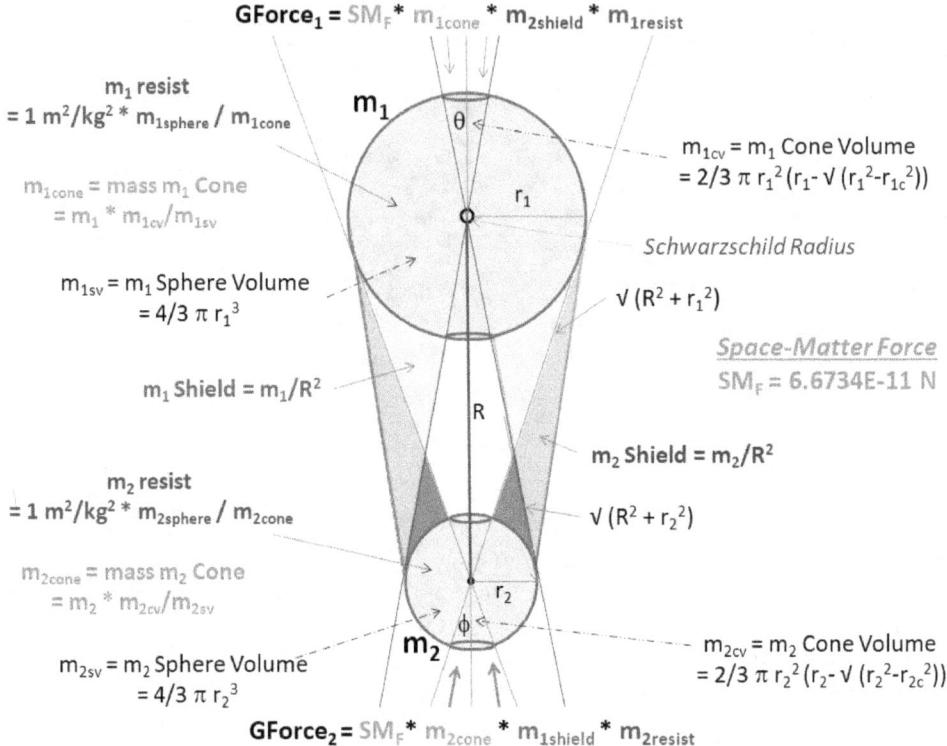

interaction of bodies. He deduced the gravitational constant but avoided showing the transfer of mechanical force and energy in his equations, and thereby simplifying it for humanity to use, $F = G * m_1 * m_2 / r^2$. Below is a diagram showing the relationship of the forces and their equations:

Since the vacuum force's push on bodies and objects produces the same results as gravity, it applies to planets orbiting stars and rectilinearly correcting the planet's path into a curved elliptical or circular motion. Space-matter force causes the effect of gravity, pushing objects toward the center of its Schwarzschild radius as depicted above. It also applies to whole galaxies and the movement of galaxies in the universe. All of Newton's work written in Principia remains pertinent and valid and its contribution to modern day mathematics immense and space travel immeasurable. Let me reiterate that the Cosmological Balance prediction of space-matter and antigravity does in no way refute Newton's great work; it confirms his initial expectation of a medium and mechanical source of gravity. Without proof of this source, Newton had to use a philosophical approach to present definitions, laws, axioms, and gravity equations in Principia[250].

Newton's gravity law, action from a distance, is valid even when viewed from a push aspect on two point masses. Another way to show that space-matter vacuum force's push is also inversely proportional to the square of the distance between the centers of the

---

[250] For simplicity sake, Newton's gravity equation is preferred over the vacuum force transfer equation.

objects affected; we need to look at it from the perspective of surface area of the imaginary sphere drawn from the center of one object to the center of the other object. From a geometric perspective, a sphere's surface area is calculated by $A = 4\pi r^2$ and the area of a circle is similarly calculated by $A = \pi r^2$. In both instances, the total area is proportional to the square of the radius, and in this case, the radius is the distance between the centers of both objects in space. As the smaller object's center orbits along the surface of the imaginary sphere or bubble gravity layer, the force of its gravity exerted on it by the larger object is precisely the same as the vacuum force generated by space-matter pushing on that smaller object. Therefore, space-matter push force is inversely proportional to the square of the distances between the centers of both objects; in short, this is exactly as Newton predicted by his gravity equations.

The vacuum force's push on bodies and objects toward each other within the antigravity universe also produces the same results as "antigravity attraction" there, except for the repulsion between gravity and antigravity objects we saw in black hole and white hole excretions. To reiterate, the repelling force, amid only gravity matter and antigravity antimatter, originates at the atomic level, ejecting the two inside cones and adjoining halves apart are twice as strong as the vacuum push force and is similar to like magnetic poles separating. However, when it comes to an antigravity object orbiting another, the space-matter's vacuum force rectilinearly adjusts the antigravity planet's path around its antigravity star. When seen from the "north star" position, the rotation of their planets would be in the opposite direction, retrograde, and its orbital path, opposite to the visible universe, clockwise. The antigravity galaxy's stars orbit its center in the clockwise direction, when seen from their "north star" position. Turning an antigravity galaxy upside-down does not make it a normal galaxy with gravity; the primary difference is at the atomic level and the ability for opposites to repel each other, as well as tendency for like matter objects to clump and bond together, and action we call gravity. The point here is that all of Newton's laws and equations remain pertinent and therefore applies universally to the antigravity part of the universe, even when invisible to us, as well as it does in the visible universe of gravity.

Below is a simplified relationship between space-matter, matter with gravity, and matter with antigravity.

Let me reiterate this important result. The Cosmological Balance Theory, for simplicity sake, therefore, defines the interaction of all three types of matter in the universe as such. The gravitational constant, known in Newton's gravity equation, is the product of vacuum force and natural gravitational attraction of like objects, where the vacuum force is 6.6734E-11 Newton and the natural gravitational of two normal objects with gravity in the visible universe is one meter per kilogram. The revised equation should now read 6.6734E-11 N x (1 m/kg x m_1) x (1 m/kg x m_2) / r^2 = 6.6734E-11 N m^2/kg^2 x m_1 x m_2/r^2 = Gm_1m_2/r^2. The same applies within the invisible part of the universe, antigravity antimatter, and their galaxies, where vacuum force is 6.6734E-11 Newton and the natural anti-gravitational attraction of negative one meter per kilogram. The revised antigravity equation reads 6.6734E-11 N x (-1 m/kg x m_1) x (-1 m/kg x m_2) / r^2 = 6.6734E-11 N m^2/kg^2 x m_1 x m_2/r^2 = Gm_1m_2/r^2.

However, for space-matter on itself, the gravitational constant as defined per Newton's gravity equation has no impact on it because the natural attraction of two space-matter particles or atoms is zero meters per kilogram, or no natural attraction. The revised equation reads 6.6734E-11 N x (0 m/kg x m_1) x (0 m/kg x m_2) / r^2 = zero. The interaction equation between space-matter and normal matter with gravity is 6.6734E-11 N x (1 m/kg x m_1) x (0 m/kg x m_2) / r^2 = zero; and between space-matter and antigravity it reads 6.6734E-11 N x (-1 m/kg x m_1) x (0 m/kg x m_2) / r^2 = zero. Finally, the equation between matter with gravity and matter with antigravity is written as 6.6734E-11 N x (1 m/kg x m_1) x (-1 m/kg x m_2) / r^2 = -6.6734E-11 N m^2/kg^2 x m_1 x m_2/r^2 = $-Gm_1m_2/r^2$, a repulsion force opposing gravity or antigravity, depending on which part of the universe we perceive this[251]. How can space matter have no gravity field? Simple answer, its quark point particles pattern does not generate gravity or antigravity, outer positive neutral point pairs of the quark orbit in spherical path without oscillating in and out. See graviton mechanism discussed in previous chapters in part two.

One idea that is common between the Cosmological Balance and other theories is that space itself with its unique matter (space-matter) tells matter with gravity and matter with antigravity what to do and where to move, while matter with gravity and matter with antigravity moves or flows through space itself, somewhat telling space to give way and allow passage. This is very similar to the relationship we get between ships out in the water and the ocean as a whole, where the ships are the galaxies, and the ocean is space itself. Although ships can move themselves through the ocean, the ocean and its currents still influence their direction and speed. The directions galaxies move is determined by their relationship with their surroundings, whether force is applied to obtain attractive or repulsive interaction between space itself, matter with gravity, and matter with antigravity. Space itself suspends and moves galaxies, while galaxies travel through space.

In addition, the space-matter surrounding massive systems also provides a unique effect, space powerfully enhances the rotation of the galactic system, and a halo also known as the curvature of space. It is likened to channeling an alternating current into electromagnets within an electric motor to spin the rotor housing of permanent magnets. More specifically, a galaxy of ordinary matter is like permanent "gravity" poles, while space matter surrounding that galaxy alternately vibrates "gravity-antigravity" pulses at a rate of one-five-hundreds times (1.9984E-03) the speed of light (or 599,119.05 hertz) to rotate and flatten the galaxy. This frequency rate creates the gravity constant force, 6.6734E-11 N, at the same instance generates the antigravity constant, and simultaneously causes space-matter to maximize separation between them, producing vacuum force and temperature 2.7 Kelvin. Details of how this works will be discussed below.

---

[251] Again, the Cosmological Balance Theory uses a modified Newton gravity equation.

8.5 Importance of Space

Let us reiterate this important point; the presence of the space-matter in intergalactic space is the reason why space sustains its temperature of 2.7 Kelvin Degrees. Space-matter has practically no gravitational or anti-gravitational attraction; it is space. Due to its extreme cold and dense-particle matter, space-matter absorbs minuscule photon energy from light passing through it, thereby temporarily slowing down the photon's speed to a crawl but not changing its color, while the dense-particle space-matter hydrogen atom or "*hygratium*" increases its electrostatic charge slightly, expands space around itself through pulses of gravity-antigravity, and maintains the vacuum of space. However, an extremely high concentration of space-matter around massive objects, known as the curvature of space, does eventually red shifts the light passing through it. The vibrations or jiggles of the dense particles within the space-matter emit the energy to raise and maintain the temperature of space to 2.7 Kelvin Degrees[252]. This temperature is the result of the vacuum force of space. Anything higher than that may begin to destabilize the space-matter atom's structure and reduce it to medium energy space-matter, or even normal particles of matter. Without the presence of space-matter and its "vacuum energy" or ability to spread out and create the "void" of deep space, gaseous normal matter with gravity would simply spread out evenly throughout space. It would not be able to attract each other gravitationally to achieve the Jeans critical mass, let alone form the stars and galaxies in the current visible universe. We owe our existence to the presence of space-matter spread throughout flat open space and curved around massive objects and the antigravity antimatter in the invisible part of the universe. In deep open space, space-matter is the norm at such high percentage, while matter with gravity and matter with antigravity is the rare exception. Count your blessings.

An age-old question comes to mind. Which came first the chicken or the egg? Did space-matter come after space or did space come about because of space-matter? Neither. they are one, and the same. Is the universe, as we know it, in a bubble whose boundaries are held outward by vacuum energy? If so, what is on the other side of those boundaries? Did matter with gravity or antigravity come after or before space-matter? We can go on and on with these types of questions and get nowhere fast. Edwin Hubble, unaware of the existence of "dark matter," interpreted the red-shifted light he saw from distant galaxies in his telescopes and declared the universe to be expanding exponentially, coupled with Einstein's Special and General Relativity equations, gave way to Hawking's "Big Bang Theory," a highly unlikely unimaginable explosion. Hawking's Big Bang claim and Einstein's inflationary theory led to countless mathematicians and physicists working out extraordinary and elaborate formulas for every possible scenario along the way to show that this universe came out of nothingness and has gone through eons of developmental stages. Solely for the purpose to justify our own existence for us, and after we are long gone, according to Hawking, the universe returns to cold empty nothingness or ends in a "Big Crunch." No one in academia really knows for sure how the universe came about and what it has in store for us. Since space-matter defines space, then it is logical that matter with gravity and matter with antigravity are byproducts of small amounts of space-matter decaying or breaking down to its lower energy sub-atomic particles.

---

[252] Although the average temperature of open space is 2.7 Kelvin, there exists within it pockets where the temperature can get as low as 0 Kelvin if there are no particles moving through that pocket.

Remember the 10 cubic kilometer box filled with air floating between the earth and the moon in zero gravity. The air within the box is evenly distributed and invisible for all practical purposes to anything residing within the box. We are like a speck of dust on a gnat hovering with what we and it believes is a group of tiny young fireflies within the box, unable to recognize objects beyond 5 kilometers, we squint through our telescope to get a better glimpse. One of the lightning bugs blows up, or so we thought, and scatter its guts everywhere in all directions. This event is equivalent to a time traveler sent back in time to observe a star in a small galactic cluster going super nova, the first in a region with no other stars. The gnat is her spaceship, and the fireflies are the galactic cluster of stars. Shortly after the explosion, we see debris slow down exponentially as it moves outward from the blast. Why is this? Because the gaseous invisible matter in the vicinity resists the blast outward spread. Perhaps we are nothing and everything, or simply too small, to grasp the universe we live in. This scenario demonstrates another example of why the Big Bang Theory is not plausible. Even if the super nova we just witnessed as a time traveler were from a super colossal explosion of a black hole destroyed from within by anti-matter matter annihilations, its outward blast would still be slowed down by the atoms and molecules already occupying space. Remember, space and space-matter exist as one, and are inseparable. Therefore, the pressure of the surrounding matter will resist any explosion within a region of space. The Big Bang did not spread matter throughout the universe. Time, space, and physical laws, we know today, do not allow it. So why should we believe that it had allowed it in the past[253].

How can we, as a human race, continue to be so egotistical and believe such destructive dismal circumstances? In any case, the true explanation of how the universe came to be is certainly beyond our grasp. We can continue to spend countless hours, day after day, dreaming a whole library or trillions of terabytes of possibilities until the day we all die, or accept the universe as it is, a Cosmological Balance. The universe has been in existence much longer than cosmologists and astrophysicists' conjecture and will continue to remain infinite and timeless. With that said, here is one possible Cosmological Balance creation scenario.

---

[253] The laws of physics today is assumed to apply to the past as well as the future; for if it did not, all the universe would all be in total chaos of sub-particles if they are somehow still able to sustain themselves.

8.6 Cosmological Balance Creation

In the beginning, assuming there was one, there was darkness, but not complete emptiness. There was no visible light or matter as we know it to be seen anywhere in the vast universe. Space appeared completely empty and void, but it was not. Space-matter evenly spread throughout the cosmos gave space and the universe volume and expanse; it came into existence with time and space. There is no such thing as pure energy, as energy is defined by matter in motion or the potential for motion. The energy (matter in motion of uniformly scattered sub-atomic particles) that caused this creation became space-matter and space simultaneously during the first era in the beginning of the universe, day zero, according and adhering to the laws of physics then. Space-matter itself does not have gravitational or anti-gravitational characteristics or properties and therefore does not clump together but spreads out to fill all of space. Infinite energy (infinite particles in motion without extreme heat) became high-energy quarks and dense sub-atomic particles, which became space-matter and space, creating the great expanse of the infinite universe, as we know it. Space-matter and space stabilized and became a cold liquid-gaseous spacious fluid. The majority of universe and the space-matter remain stable and slow-moving fluid like the seas of a great ocean of air, with relatively even pressure maintained throughout flat open space. But from time-to-time, pressure and movement caused space-matter to come together in eddies and develops the beginnings of a "storm."

Turbulence in the flow of space-matter develops low-pressure and high-pressure regions in space. Much like the atmosphere on earth, low-pressure regions in space develops into instability of the space-matter atoms and molecules at the center of these rotations. These high-energy high-density atoms and molecules collided more frequently, and small fractions eventually start to decay with each collision with other high-density atoms and molecules. The collisions break down the high-energy high-density space-matter atoms into low energy subatomic particles, which reassemble into the hydrogen atoms and on occasion the helium atoms we know on earth, while simultaneously creating antigravity counterparts, which are repelled away to form two clouds, gravity and antigravity. During the break down process from high-energy high-density space-matter to the hydrogen atoms, that we have on earth and in the sun, the gravity switch activated; the first creation of matter with gravity, and the natural tendency to bind together. This gravity activation is counterbalanced by the first antigravity particles of matter[254]. The pressure and energy provided by space-matter became the source we know as gravity force on earth and the source of the antigravity force in the other part of the universe, the invisible side.

This turbulence continued to grow as the low pressure increased with each newly created hydrogen atom, matter with gravity. More and more hydrogen atoms and molecules begin to consolidate in these eddies caused by the low-pressure region in space, and under mutual attraction of gravity as defined by Newton. Simultaneously, the mutual

---

[254] The decomposition of the neutral graviton (Higgs particle) and the grand unification force in space-matter gives way to positive graviton in ordinary matter, and a negative graviton particle in antigravity antimatter.

attraction of antigravity material also assembles an opposite rotating group, which repels the gravity matter eddies. A storm begins to brew, and a stellar nursery starts to take shape into a gaseous cloud in space. Cosmologists see these new invisible stellar clouds as gravitational lensing areas filled with "dark matter," which somehow can bend the lights emitted from distant galaxies in the opposite side of the universe from us. We learned that surrounding high concentrations of space-matter, the curvature of space, around the massive stellar clouds bend the distant light. As the low-pressure region continues to grow, the storm gets stronger, and the stellar nursery clouds compress into new stars, lighting up one at a time, until a new spherical cluster galaxy begins to shine. A new galaxy destabilizes space-matter in the region as heat and energy hit this already weakened space-matter. The storm strengthens and the galaxy grows, attracting more hydrogen atoms. For every galaxy, there exists an antigravity galaxy.

The perceived nothingness of space and its space-matter creates what we know as stars and galaxies in the visible universe, all because of turbulence within pressure and energy eddies. Simultaneously, antigravity antimatter is also created based on the opposite spin of the low-pressure system in space. Like storms on earth, northern hemisphere storms spin in the opposite direction from southern hemisphere storms. The solar systems and galaxies created by antigravity atoms spin in the opposite direction as those created by matter with gravity. The planets orbiting antigravity solar systems also tend to move in retrograde or opposite direction that the planets travel. In addition, the atoms themselves are left-handed atoms, opposite when compared to those within the gravity-based matter world. The natural stable space-matter gives way to create both matter with gravity and matter with antigravity. So, to balance things out in the Cosmological Balance, matter with gravity and matter with antigravity eventually returns into space the building blocks of space-matter. We learned of this return as the excretion process of super massive black holes and super massive white holes, ejecting high-energy high-density top and bottom quarks with neutral characteristics that is having no gravity or antigravity properties. These space-matter building block particles are provided through the wormhole tunnels leaving the galaxies collapse and release them into space. The newly provided high-energy high-density sub-atomic particles, through the laws of physics, combine quickly and for survival reasons, to become a stable dense-particle hydrogen space-matter (*hygratium*), and take their place in the grid of space in the great expanse of the universe. This newly formed space-matter possesses neutral gravitons also known as the Higgs particle, which gives the characteristic of mass to ordinary matter and antigravity antimatter. Space-matter transitions from the turbulent pressure in space to matter with gravity or antigravity in galaxies, and then returns back to space-matter from the central hub of the galaxies, their massive black or white holes, like the eye of the cyclone, hurricane, or typhoon, appear as calm normal atmosphere, a gateway to provide space-matter back into space. As pressures in the sea or ocean of space change, the storms brewing can fall apart and return to normal calm space and the newly formed matter with gravity or antigravity that began to consolidate dissipates and scatters. This matter with gravity or antigravity simply floats in space until their respective galaxies, rogue stars, or planets attract them. The pools of matter red shift distant galaxies. On occasion, galaxies in space merge to become larger more massive and gravitationally stronger galaxies than its feeder systems. Not only are we, remnants of stars and of space-matter space itself and its energy. We are nothing and everything, created from seemingly nothing and yet possess the energy within everything.

8.7 Cosmological Balance Equation

Earlier in this chapter, we reevaluated the composition of known and conjectured matter in the universe. Space-matter's mass constitutes 71.30% of the total matter in the universe, matter with gravity is 14.35%, and matter with antigravity is 14.35%. In chapter 7 in this book, we also solved for the galactic orbital period with the equation $1 = 2c^4e^2/(GM_S)^2 * a_c^3/T_c^2$ and then derive the equation

$$G = 2c^4e^2/(M_S)^2 * a_c^3/T_c^2 = 2c^4e^2 * a_c^3/(T_c^2 * M_c^2),$$

which is shorthand for

$$T_c = [\sqrt{(4\pi^2/(GM_S)*a^3)}] * [c^2 * \sqrt{(1/(2GM_S))} * e/\pi].$$

By substituting C_m as the representation of the motion of normal matter M_G in the visible gravity part of the Cosmological Balance as it interacts with antigravity antimatter, where $C_m = (a_c^3/(M_G * T^2) * kg/m)$; its energy relationship to mass and momentum is kg^2/m^2, the sub-atomic particle attraction at the quantum mechanical level. Since we have learned that the motion of space-matter S_m produces the force 6.6734E-11 N from which the gravitational constant G is realized, we represent all space-matter and its motion with S_m vacuum force and get:

$$G = 6.6734\text{E-}11 \text{ N m}^2/\text{kg}^2$$

$$= S_m * m^2/kg^2 \qquad (8.7.1)$$

Now substitute S_m m²/kg² for the G in the equation $G = 2c^4e^2 * (a_c^3/(T_c^2 * M_C))$, we get roughly these equations for the galactic orbital period

$$S_m \text{ m}^2/\text{kg}^2 = 2c^4e^2 * (a_c^3/(T_c^2 * M_C)),$$

$$S_m = 2c^4e^2 * (a_c^3/(T_c^2 * M_C) * kg^2/m^2)$$

$$S_m = 2c^4e^2 * C_m, = 1.19372\text{E+}35 * C_m, \qquad (8.7.2)$$

Similarly, by substituting A_m to represent all antigravity antimatter and its motion in the invisible part of the universe when it interacts with matter with gravity, where $A_m = \sqrt{(a_a^3/(M_A * T_A^2))} * kg/m)$. We get:

$$S_m = 2c^4e^2 * A_m, = 1.19372\text{E+}35 * A_m, \qquad (8.7.3)$$

However, each of the three types of matter in the Cosmological Balance and their motions affect the motions of the other two types of matter. More specifically, the equations above when combined read:

$$S_m = 2c^4e^2 * A_m = 2c^4e^2 * C_m, (8.7.4)$$

$$2c^4e^2 * A_m = 2c^4e^2 * C_m,$$

$$A_m = C_m,$$

$$0 = C_m - A_m, \qquad \text{or} \qquad 1 = C_m / A_m,$$

and since space-matter has no attraction with itself, its motion can be represented with something like the Kinetic Energy formula for predicting speed of gaseous molecules. However, instead of the constant 2/3 used for air molecules on earth, we have the constant 2.3723E+37, which is derived largely from the volume of space 14 billion light years by 14 billion light years by 1 meter.

$$KE_{avg} = 2.3723E+37 \, kT \qquad (8.7.5)$$

where k is Boltzmann's constant, 1.38×10^{-23} joules per Kelvin (J/K), and T is the temperature in Kelvin of space. Scientists speculate that the average temperature of the deep space between galaxies is 2.7 Kelvin, intergalactic temperature. Plugging in these numbers in the Kinetic Energy equation $KE_{avg} = 2.3723E+37 \, kT$ above we get

$$KE_{avg} = 2.3723E+37 * 1.38 \times 10^{-23} \, kg*m^2/sec^2/k * 2.7 \, k$$
$$= 8.83919E+15 \, kg*m^2/sec^2$$

In section 2.3 above we discussed that Vacuum Energy of deep space was equal to $8.83919E+15 \, kg*m^2/sec^2$.

We can therefore calculate the repulsion vibration velocity of space-matter using the formula: Velocity of space-matter equals the square root of 2 times its Kinetic Energy divided by the mass of its atoms.

$$V_{SM} = \sqrt{(2*KE_{avg}/m_{sm})}$$
$$= \sqrt{(2*8.83919E+15 \, kg*m^2/sec^2 / 2.8116E-24 \, kg)}$$
$$= 7.9295E+19 \, m/s$$

As we discussed above, S_m represents the motion of all space-matter S_{sm} in the universe, the resulting vacuum force from such movement is 6.6734E-11 N. We can show this relationship for all three types of matter in the universe in the following equation.

$$S_m * 1 = S_m * 1$$
$$S_m * A_m = C_m,$$
$$S_m = C_m / A_m.$$

COSMOLOGICAL BALANCE UNIVERSAL EQUATION:

However, the above relationship equation is still incomplete. It lacks separating out the interactions between gravity matter and antigravity antimatter. In order to show this portion, we need to go back to the Cosmological Balance equation above (2.7.1):

$$T_c = [\sqrt{(4\pi^2/(GM_C)*a^3)}] * [c^2 * \sqrt{(1/(2GM_C))} * e/\pi],$$

Redefining the equation into two parts gives us the clarification we need to effectively define space-matter, gravity matter, and antigravity antimatter interactions.

$$T_c^2 = [(4\pi^2/(GM_C)* a_c^3)] * [c^4 * 1/(2GM_C) * e^2/\pi^2], \qquad (8.7.6)$$

where the first part of the galactic orbital period equation after the equal sign is basically a slightly modified Kepler's third law. The second part of the equation is the acceleration resulting from the interaction between gravity matter and antigravity antimatter as the objects move about in a circular or rectilinear path once around the circumference of the galaxy divided by two, $2\pi/2$, times Euler's number e. We have defined and explained this interaction between gravity matter and antigravity antimatter within the galaxy as "galactic wind" in part 2 chapter 6. As for the interaction between gravity matter galaxies and antigravity antimatter galaxies, we shall call it the flow of the "cosmic currents," as it pushes, pulls, and redirects the movement and path of both types of galaxies, gravity matter and antigravity antimatter, within the universe as a whole. By consolidating the a_c^3 and M_C in the above equation (8.7.2), gives

$$T_c^2 = [4\pi^2/G] * [(c^4 e^2/2GM_C)/\pi^2] * a_c^3/M_C,$$

and then divide by T_C^2 to both sides, we have:

$$1 = [4\pi^2/G] * [c^4 e^2/2GM_C\pi^2] * a_c^3/(T_c^2 * M_C), \qquad (8.7.7)$$

Here we have the three parts we are looking for: Kepler's third law, the interaction of antigravity and gravity matter known as the galactic wind, and its motion.

If you recall, the representation of all gravity matter and its movement is:

$$C_m = a_c^3/(T_c^2 * M_C), \qquad (8.7.8)$$

and space-matter and its motion produce the vacuum force 6.6734E-11 N, so

$$S_m * m^2/kg^2 = G$$

$$S_m = G * kg^2/m^2 = 6.6734\text{E-}11 \text{ N}$$

So, by substituting C_m in equation (8.7.7) above, we get:

$$1 = [4\pi^2/G] * [c^4 e^2/2GM_C\pi^2] * C_m, \qquad (8.7.9)$$

$$S_m = [4\pi^2 * kg^2/m^2] * [c^4 e^2/(2GM_C \pi^2)] * C_m,$$

where the movement of space-matter affects or tells gravity matter how to move.

Similarly, antigravity antimatter and its motions resolve to the following:

$$1 = [4\pi^2/G] * [c^4 e^2/2GM_A\pi^2] * A_m, \qquad (8.7.10)$$

$$S_m = [4\pi^2 * kg^2/m^2] * [c^4 e^2/(2GM_A\pi^2)] * A_m,$$

where the movement of space-matter tells antigravity antimatter how to move.

MODIFICATIONS TO MIMIC EINSTEIN'S FORMULA:

The presentation in this section below is just for demonstration only.

Note: equation (8.7.6) can also be rewritten in terms Einstein used in his relationship equation between space and matter with gravity ($G_{\mu\nu} = 8\pi G/c^4 T_{\mu\nu}$).

Start with equation (8.7.6)

$$1 = [4\pi^2/G] * [c^4 e^2/2GM_C\pi^2] * C_m,$$

then multiply by $1 = (2G/c^4\pi) / (2G/c^4\pi)$ to get

$$1 = [4\pi^2/G * 2G/c^4\pi] * [c^4 e^2/2GM_C\pi^2 * c^4\pi/2G] * C_m,$$

$$1 = [8\pi G/c^4] * [c^8 e^2/4G^3 M_C\pi] * C_m,$$

and factor out the extra G constant to the right side of the equation to get:

$$(1 \text{ m}^2/kg^2) S_m = [8\pi G/c^4] * [c^8 e^2/4G^2 M_C\pi] * C_m,$$

$$(1 \text{ m}^2/kg^2)^2 S_m^2 = [8\pi G/c^4] * [c^8 e^2/4GM_C\pi] * C_m,$$

where the 1 on the left-hand side represents space $G_{\mu\nu}$ and on the right-hand side we can let $T_{\mu\nu} = [c^8e^2/4G^2M_C\pi] * C_m$, to ineffectively mimic Einstein's famous equation $G_{\mu\nu} = 8\pi G/c^4\, T_{\mu\nu}$. Similarly, antigravity equation resolves to

$$1 = [8\pi G/c^4] * [c^8e^2/4G^3M_A\pi] * A_m,$$

$$(1\ m^2/kg^2)\, S_m = [8\pi G/c^4] * [c^8e^2/4G^2M_A\pi] * A_m,$$

$$(1\ m^2/kg^2)^2\, S_m^2 = [8\pi G/c^4] * [c^8e^2/4GM_A\pi] * A_m,$$

Therefore, we are able to build an equation similar to that of Einstein:

$$[8\pi G/c^4] * [c^8e^2/4GM_A\pi] * A_m = [8\pi G/c^4] * [c^8e^2/4GM_C\pi] * C_m,$$

Multiply both sides by $(2/c^2\pi)$ to restore the second part of each side to the proper gravity antigravity interaction.

$$[8\pi G/c^4] * [c^8e^2/4GM_A\pi * (2/c^2\pi)] * A_m = [8\pi G/c^4] * [c^8e^2/4GM_C\pi * (2/c^2\pi)] * C_m,$$

$$[8\pi G/c^4] * [c^4e^2/2GM_A\pi^2] * A_m = [8\pi G/c^4] * [c^4e^2/2GM_C\pi^2] * C_m, \quad (8.7.11)$$

Now include the presence and motion of space-matter S_m on the left side to get

$$S_m * [8\pi G/c^4] * [c^4e^2/2GM_A\pi^2] * A_m = [8\pi G/c^4] * [c^4e^2/2M_C\pi^2] * C_m, \quad (8.7.12)$$

Similarly, if we transpose the equations (7.8.7) above, we get

$$S_m * [8\pi G/c^4] * [c^4e^2/2GM_C\pi^2] * C_m = [8\pi G/c^4] * [c^4e^2/2GM_A\pi^2] * A_m. \quad (8.7.13)$$

Equations (8.7.12) and (8.7.13) above tells us that motion of space-matter and antigravity antimatter within a system of consisting of gravity matter, i.e. the galaxy or the entire universe, provides a force, $[c^4e^2/2GM_C\pi^2]$. This force is a repulsion force that accelerates objects within that gravity system, thereby making the system rotate or move faster than with just gravity alone as calculated with Newton's or Kepler's equations. The same applies when looking at this from the antigravity antimatter view represents the push of gravity, a repulsion force that accelerates objects, thereby making it rotate or move faster within that system.

How should we then represent gravitational systems with no antigravity particles moving through it or for antigravity systems with no gravitational particles moving within or through it? In the absence of antigravity particles within a gravity matter solar system, the influence or interaction goes to one; specifically, we divide both sides by $[c^4e^2/2M_C\pi^2]$ in equation (8.7.12) to get:

$$S_m * ([8\pi G/c^4] * A_m) = ([8\pi G/c^4] * C_m),$$

where the left-hand side of the equation constitutes the movement of all of space (without consideration of the presence of antigravity antimatter or its effects).

We get the same view from the antigravity perspective:

$$S_m * ([8\pi G/c^4] * C_m) = ([8\pi G/c^4] * A_m),$$

where the left-hand side of the equation constitutes the movement of all of space (without consideration of the existence of gravity matter) as seen from an alien, if they existed, in the antigravity part of the universe perspective. More specifically Einstein's notation for space equates to

$G_{\mu\nu} = S_m * ([8\pi G/c^4] * A_m)$, from gravity viewpoint, or $G_{\mu\nu} = S_m * ([8\pi G/c^4] * C_m)$,

as seen from antigravity point of view. As you can see from the "Einstein-like" equations above, math allows us to manipulate data into whatever form we want to fit the imagination, even if it is a mirage. Specifically, the Einstein general relativity equation $\mathbf{G_{\mu\nu} = 8\pi G/c^4\ T_{\mu\nu}}$ does effectively calculate orbits around massive objects but is not necessary to effectively show the movement of the solar system and hardly the galaxy, or any other galaxy. For now, we understand why academia and the mainstream scientific community continue to say it is valid for the universe despite the contradicting knowledge we have gained and learned about physics and cosmology over the last 100 years. Nevertheless, Einstein's general relativity equation is the best we have for calculating the gravity lensing of light around massive objects[255]. We agree with his equation without the gravity time dilation. In order for the Cosmological Balance Theory to replace his equation completely, it would have to define accurately the exact density of each layer of space-matter in the curvature of space around the massive object and calculate to the same resolution how it refracts light. This data manipulation shows almost any configuration desired.

COSMOLOGICAL EQUATION FOR ALL MOTION UNIVERSE:

The Cosmological Balance Equation will not retain the modified equations within the section above to accommodate Einstein. Let us redo our analysis the right way. The equation defined in (8.7.6) and (8.7.7) stands as written:

$1 = [4\pi^2/G] * [c^4 e^2/2GM_C\pi^2] * C_m$,

where $C_m = a_C^3/(T_C^2 * M_C)$ is gravity matter in motion and G is the gravitational constant. T_C is the orbital period of gravity matter, a_C is the length of the semi-major axis, and M_C is the inner accumulated mass of the system.

$1 = [4\pi^2/G] * [c^4 e^2/2GM_A\pi^2] * A_m$,

where $A_m = a_A^3/(T_A^2 * M_A)$ is antigravity antimatter in motion and G is the gravitational constant. T_A is the orbital period of antigravity antimatter, a_A is the length of the semi-major axis, and M_A is the inner accumulated mass of the system.

$[4\pi^2/G] * [c^4 e^2/2GM_A\pi^2] * A_m = [4\pi^2/G] * [c^4 e^2/2GM_C\pi^2] * C_m$,

$S_m = G * kg^2/m^2 = 6.6734E-11\ N$

By including the presence of space-matter $S_m = 1$ on the left side, gives

$S_m * [4\pi^2/G] * [c^4 e^2/2GM_A\pi^2] * A_m = [4\pi^2/G] * [c^4 e^2/2GM_C\pi^2] * C_m$,

By transposing or switching perspectives to antigravity antimatter, gives

$S_m * [4\pi^2/G] * [c^4 e^2/2GM_C\pi^2] * C_m = [4\pi^2/G] * [c^4 e^2/2GM_A\pi^2] * A_m$,

where space-matter's vacuum force provides the push of gravity and antigravity. Therefore, the motion of antigravity and space-matter tells gravity matter what to do. And vice versa, the movement of gravity and space-matter tells antigravity antimatter how to

---

[255] General relativity is a quantum equation and is best approach to solve the gravity lensing of starlight traveling through the curvature of space around massive objects in space.

move[256]. To reiterate, the first part of the right-hand side of the equal sign is Kepler's third law and the second part is the acceleration caused by the galactic wind. The second part goes to "1" when there is no interaction between antigravity and gravity matter, when planets are shielded by the mass of a star.

COSMOLOGICAL BALANCE EQUATION SUMMARIZED:

In review: To produce the complete and accurate relationship between all three types of matter throughout the entire universe, the Cosmological Balance combined Kepler's laws, Newton's laws, Schwarzschild, and geometry principles to create an equation applicable to the whole universe. We did this by starting with the fact that the presence and movement of any antigravity particles directly affects the movement of normal gravity particles within its path, a 1 = 1 action equals reaction. Replace the "1" on the left side with antigravity antimatter and its motion A_m and the "1" on the right side with gravity matter and its motion C_m to get:

$1 = A_m = C_m = 1$,

Therefore, the motions of space-matter and normal gravity matter tells antigravity antimatter what to do, and vice-versa, the motions of space-matter and antigravity antimatter tells gravity matter what to do. Below is a simplistic representation of this relationship. In this case, we have vast distances between floating clouds of antigravity particles and clouds of gravity particles without interaction between the clouds, independent of each other.

$S_m * A_m = C_m,$ $S_m * C_m = A_m$ $S_m = 1$

Remember space-matter, as defined, does not attract each other, it makes space what it is, appear empty and void, a vacuum. We can now include Kepler's third law orbital period for gravity and antigravity antimatter only (multiplied to both sides of the equation), assuming no overlap or direct interaction between gravity-antigravity, to get:

$S_m * [4\pi^2/G] * A_m = [4\pi^2/G] * C_m,$ and (8.7.14)

$S_m * [4\pi^2/G] * C_m = [4\pi^2/G] * A_m,$ (8.7.15)

where the antigravity system or the gravity system is independent due to extreme distances and has no interaction or sufficient effect on the other, specifically that portion showing antigravity-gravity interaction becomes one. Examples of regions of space where there are no interactions between gravity matter and antigravity antimatter include the solar system where the sun or star's gravitation shields the planets from any influences from antigravity particles. In this instance, the $[c^4 e^2/2GM\pi^2]$ was divided out to both sides of the (7.8.2) equation to give us only Kepler's third law.

The interaction linking gravity and antigravity antimatter, $[c^4 e^2/2GM\pi^2]$, in large galaxies gives us:

$S_m * [4\pi^2/G] * [c^4 e^2/2GM_A\pi^2] * A_m = [4\pi^2/G] * [c^4 e^2/2GM_C\pi^2] * C_m,$ (8.7.16)

and

---

[256] These Cosmological Balance Equations are the capstone of this book. It covers all motions of all types of matter at the macro universe level. Einstein's general relativity equation covers the micro or quantum level effect on light in the curvature of space.

$$S_m * [4\pi^2/G] * [c^4 e^2/2GM_C\pi^2] * C_m = [4\pi^2/G] * [c^4 e^2/2GM_A\pi^2] * A_m, \quad (8.7.17)$$

where antigravity particles traverse through the gravitation system and vice versa thereby significantly accelerating the rotation of the system under observation. Stars move faster and orbit the galaxies at an accelerated speed than calculated with gravity or antigravity alone. These last two equations (8.7.16) and (8.7.17) also apply to interactions between antigravity galaxies and gravity galaxies as they move through space.

The presence and motion of space-matter produces the vacuum-force of 6.6734E-11 N, which when combined with the inherent quantum mechanical attraction of 1 m/kg of gravity matter, provides the force for gravity to function as defined by Kepler and Newton laws. The presence of antigravity antimatter moving in wormhole tunnels through the galactic system at the speed of c^2 provides the acceleration boost for the second part of the equation below. The resulting orbital period equation for galactic systems with gravity and the universe is:

$$T_c = [\sqrt{(4\pi^2/(GM_C)*a_c^3)}] * [(c^2 * \sqrt{(1/(2GM_C))} * e/\pi], \quad (8.7.18)$$

where the second part represents antigravity particles or objects that move through it. The equation is combination of Kepler's law, Schwarzschild radius, quantum mechanics, and the principles of geometry. Moreover, the absence of antigravity antimatter in a system is represented by setting the second part to one, "1," by dividing itself out. Examples of systems without antigravity particles or objects moving through them include shielded solar systems, or a gaseous cloud floating in deep open space, not influenced by opposite gravity or antigravity particles moving through them.

$$T_C = [\sqrt{(4\pi^2/(GM_C)*a_c^3)}] * \{[(c^2 * \sqrt{(1/(2GM_C))} * e/\pi] / [(c^2 * \sqrt{(1/(2GM_C))} * e/\pi]\},$$

$$T_C = [\sqrt{(4\pi^2/(GM_C)*a_c^3)}], \quad (8.7.19)$$

where T_c is the orbital period of an object in the system with semi-major distance a_c, M_C is the accumulated mass of the inner part of the system, G is the gravitational constant 6.6734E-11 N m²/kg², c is the speed of light, e is Euler's number, π is the constant pi. Hence, resulting orbital period for galactic and universe systems with antigravity is

$$T_A = [\sqrt{(4\pi^2/(GM_A)*a_A^3)}] * [(c^2 * \sqrt{(1/(2GM_A))} * e/\pi], \quad (8.7.20)$$

and its shielded solar system orbital period is:

$$T_A = [\sqrt{(4\pi^2/(GM_A)*a_A^3)}], \quad (8.7.21)$$

where T_A is the orbital period of an object in the system with semi-major distance a_a, M_A is the accumulated mass of the inner part of the system and not just the center massive object, G is the gravitational constant 6.6734E-11 N m²/kg², c is the speed of light, e is Euler's number, π is the constant pi. So, there you have it, a simplified unification equation for the universe, which brings together Newton and Kepler Laws, quantum mechanics, Schwarzschild radius and mass, Galilean relativity, and the principles of geometry and polar coordinate systems, and others[257]. These equations are incomplete and need to merge both halves of the forces acting on each side of the universe.

---

[257] The Cosmological Balance Theory is a simplified unified theory with a simple cosmological equation for all macro matter within the physics of the universe.

8.8 Space-Matter Affects Time & Curvature of Space

Physics and natural laws are assumed to work well within the realm of each of the three types of matter, except for quantum level mutual attraction of space-matter atoms and molecules. Space-matter is the original common and unique atom naturally assembled solely for survivability in the frigid vacuum environment of space. Once space-matter particles form a "dense-particle highly energized atom" (for survivability), it has no attraction for other atoms; they repel and bounce of each other like gases in the air on earth but with greater distances between atoms, until they stabilize into a gaseous almost semi-liquid grid we perceive as space. Its structure, based on the conjectured amount of detectable "dark matter" in the visible universe and experiments conducted on stable vacuum frigid exotic matter, gives stability to space and is the source of the other two types of matter. The Cosmological Balance Theory developed a predicted estimate of what exotic space-matter atoms would look like and defined some of its characteristics.

The composition of space-matter is nothing like ordinary matter or antigravity antimatter, its proton and neutrons would be comprised of at least nine each super high energy level, above or at tau and muon tau quarks, a state not yet found, where the graviton is neutral. The presence of the neutral graviton, known as the Higgs Boson particle, gives normal matter and antigravity antimatter its mass. Space-matter falls under the category of an exotic form of matter. Space-matter's "electrons" are at least at the level of tauon or higher energy. These dense-particle, super high-energy, atoms electrostatic charges repel each other, highly resisting situations pushing them together. In areas of space where there is no ordinary matter or antigravity antimatter present, space-matter forms a uniformed grid of atoms spread out at one each per cubic meter, known as flat open space. Each atom's interior has 99.999% empty space, practically a vacuum. This thin spread of atoms appears to us as "a void" almost totally and completely empty "space," hence our academia's perception of open space. The high-energy space-matter's exotic structure is what enables it to maintain itself as space and to sustain the temperature of intergalactic space at an average of 2.7 Kelvin. Under rare excessive pressures and collisions with other space-matter, they become unstable and on rarer occasions decay into both ordinary gravity and antigravity antimatter types simultaneously, which repel away from each other immediately. These new types of matter, created from the same reaction, activated their graviton, are, hence, interconnected at the quantum level and mirrored the movement of their anti-type matter. Large quantities ordinary matter within massive objects is surrounded by nested spherical layers of increasing denser space-matter, also known in science as the curvature of space. At the opposite end of the lifecycle of matter, black holes and white hole eject the materials necessary to reassemble as space-matter, high-energy quarks, leptons, and force carriers with neutral gravitons, replenishing the stability of space.

When normal ordinary light collides with, absorbed by, and passes through space-matter atoms in open space, it briefly slows to a crawl and then is re-emitted in the same line or path it was originally traveling without changing color at least to a degree not detectable. The overall speed of light through open space is maintained at light speed since it is mostly a vacuum. So large quantities of space-matter spherically wrapped around massive objects like the sun form something like a dense blanket, which bends and guides light to follow the geodesies around the sun, or any massive object, as predicted by Einstein; gravity alone as defined by Newton was insufficient to cause this result on starlight.

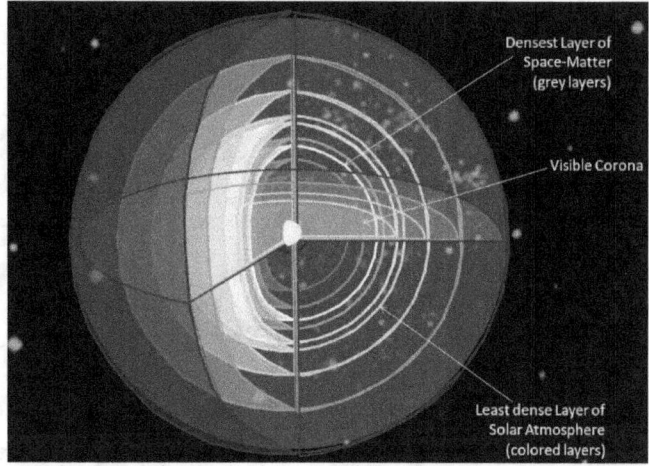

The image above depicts decreasing density layers of ordinary matter and increasing layers of space-matter; Einstein called the curvature of space: This blanket of space-matter provides the vacuum force known as gravity that holds the sun in its spherical shape, while the natural quantum attraction of molecules and atoms retains that hold for ordinary matter to stay together. Since space-matter is a completely different type of matter than ordinary matter, its light bending properties cause light to change its direction gradually without changing the color of light or its wavelength, resulting in what looks like a simple curved light path along the geodesies or straightest route around the sun, where multiple paths resulting in multiple images are normal. This light bending ability is known as refraction or lensing. The website (Cromie, 1999) provides some ideas used in the discussion above, where Lene Hau tested the type of matter we might find in space-matter's structure. As for energy density, energy cannot exist without the presence of some type of matter. Therefore, energy density must equate to some type of matter density; "dark energy" or vacuum energy is the motion of space-matter or "dark matter." Energy is nonexistent by itself.

TIME CHANGES EVERYTHING

As for time dilation in Special Relativity Theory, we saw Einstein's work in Special Relativity as a misinterpretation of visual mirages created by the Doppler Effect that he saw in his mental experiments, where objects were stretched, shorten, etc... and with his faulty concept and math errors, Einstein concluded that observers saw time ticking differently, and developed time dilation. We looked at the original Special Relativity paper (in English translation), and found out that he used equations remarkably similar, almost identical, to that showing the Doppler Effect of light and other electromagnetic waves. We also found unacceptable errors in his math. We have therefore decided not to trust the definition Einstein set forth of time dilation in Special Relativity Theory.[258] Time dilation supposed proofs include cesium clocks, global positioning system satellites, and clocks in space station and time paradox stories in imaginary spaceships to name a few, which may

---

[258] An observer's time perception of a source clock does not change how fast or slow the source clock runs. It runs exactly the same regardless of who is observing it, just as much as the moon is there in orbit around the earth, regardless, if anyone is looking at it. In other words, absolute time is determined only at their source or origin clock, so perceptions of time or any distortions thereof seen by observers not at that location are exactly that perceptions and are irrelevant in identifying proper time at the source clock. Similarly, the source clocks' times always determine simultaneity, and any observers' timing not at that source has no impact on true result.

not be conclusive due to discrepancies and inaccuracies internal to the clocks used. As for experiments involving moving cesium clocks or other atomic clocks, their time readings become highly unstable in any such moving environments. We saw and believed that testers and evaluators have edited or adjusted their data output to affix their name among others that have "proven" Einstein right. They sacrificed their integrity just to hop on the bandwagon of Special Relativity and General Relativity Theories. Kelly, A. G. (1998) uncovered the alterations performed during the Hafele & Keating Tests, and wrote an article exposing them "Did They Prove Anything?" Celbridge, Ireland: HDS Energy Ltd (Kelly, 1998).

We also are aware of the Global Positioning System (GPS) time difference and other examples that mainstream science have announced as proof of Einstein's time dilation, but we are not convinced of such time paradoxes stories. Space-matter density itself in the curvature of space around massive objects has everything to do with it. We will elaborate more on this later. For now, let us investigate the credit given of "gravitational waves" detection to Einstein's general relativity light speed constant, particularly the announcement made by the two Laser Interferometer Gravitational-Wave Observatory (LIGO) scientists. Did LIGO have solid data? Did they rule out earthquake in their regions? Will they willingly release all their data over the month prior to discovery and months after it for a qualified third party to look at? How many other "false" alarms did they go through before the two stations agreed to release this "confirmed" reading and discovery after tying it to the merger of black holes?

If LIGO data is true, then it truly detected the warping of space, which is not the same as a gravitational wave. We view normal undisturbed space-matter per the Cosmological Balance Theory as constantly exerting a vacuum force on an object in space, from which is the force of gravity. However, the ripple of space is not necessary a gravitational wave. It is like comparing the atmospheric pressure to the sound wave we hear. The force of gravity is a constant vacuum force exerted by space-matter, while the ripple of space is more as an electrostatic current passed from atom to atom at the speed of light, akin to a lightning bolt moving through air or electricity running through a wire. The space-matter grid itself passed a dissipating wave of electrostatic pulse generated by the merger of two black holes from 1 billion light years distance away to us here on earth. The electrostatic pulse or wave traveling through a medium, regardless of the medium, is strictly limited to the speed of light; like electrons traveling through copper wire or light traveling through fiber optics. After travelling over 1 billion light years distance from the source, these waves are but ripples within the space-matter grid. The ripple moves the Higgs field within the space-matter grid, which in turn warps the atoms on earth with a momentary twist and bend our devices picked up. In other words, the intersection of this space-matter wave pulse with ordinary matter on earth briefly warped the planet at the atomic level, something analogous to a space induced subatomic quake. LIGO picked up that momentary space ripple.

TIME DILATION

Time paradox stories used to demonstrate Einstein's thought experiments are all examples of science fiction, not science fact. The day we have machines or spacecraft to move us close to the speed of light is the day we can perform these Einstein mental experiments in real life. The Cosmological Balance Theory predicts that some situations will confirm Einstein, and some will not. All it takes is one example to set the

CHAPTER 8: SPACE BALANCES GRAVITY AND ANTIGRAVITY

questionability of all of Einstein's theories; Comet Shoemaker-Levy 9 is the example, where his Principle of Equivalence fails, and therefore General Relativity is at risk. All the comet pieces started out together as one entity within Jupiter's gravitational field, but due to minor differences in mass and the number of orbits around the planet, each piece fell to its doom at different rates, one by one, they created new scars on the planet. We already confirmed and demonstrated this mathematically with Newton's equations earlier with the mass of Venus, a cannonball, and the mass of Jupiter falling toward the sun.

Here is one for you, the reader, who is a hardcore follower that sides with Einstein when it comes to time dilation and the speed of light as a constant. If time slows down to a crawl or stops within objects at extremely high speeds close to or at the speed of light, or within an intense gravitational field, then why do black holes furiously continue to consume stars and other ordinary matter around it. As attracted matter races around the event-horizon, it accelerates to speeds near or exceeding the speed of light before plunging into the abyss. Therefore, if Einstein were right with time dilation, then all the matter that the black hole attracts would simply endlessly accumulate at the edge of the event horizon and never pass or enter into the black hole itself, since the attracted matter's clocks have theoretically halted per Einstein and some say even reversed in time. Nevertheless, observations tell us otherwise and black holes consume matter and continue to grow in mass.

Similarly, the merger of the two black holes with each other, announced by NASA on February 11, 2016, that we just observed, which occurred a billion years ago, should not have occurred per Einstein, as their clocks would have to stopped as they approached each other at orbital speeds close to if not exceeding the speed of light. Their intense gravitational field should also have stopped their clocks and suspended their approach on each other. The merger of black holes is contrary to Einstein time dilation theory, and so is everything else in the universe. His math errors in the development of Special Relativity is the reason I use, supported by actual observations contradictory to his claim of time dilation, are apparent contributions toward its demise. There is no time dilation, only the mirage perception of it created by the Doppler Effect twist[259]. Bottom line, absolute time is determined only at the source clock, so distorted time perceptions seen by observers not at that location are exactly that misperceptions and are irrelevant in identifying proper time at the origin's clock. Incidentally, source clocks' times always determine simultaneity, and any observers' timing of two events not at those sources have no impact on the true result, regardless of how convinced they are. General Relativity inherited errors from Special Relativity theory. If time functions normally and the speed of light constant is disregarded, then the GR equation is fine in showing behavior of light traveling near massive objects.

---

[259] There is no time dilation, only absolute time. Einstein's mental experiments work to show that Doppler Effect is apparent at great speeds and in intense gravitational fields. These perceptions do not affect time itself at any of the sources he investigated. Absolute time duration or ticking of it occurs at its origin, and so any other observer's perception of that time other than at the source is irrelevant and unable to change that source clock's speed physically. Simultaneity is always determined at the source clocks, perception of observers' not at the source have no effect on whether the event occurred simultaneously or not.

8.9 Space-Matter Affects Acceleration and Time

In the Cosmological Balance Theory, time is universal throughout space and flows at the same rate. It speeds neither up nor slows down with an increase or decrease in gravity or antigravity forces, or with increase or decrease in velocity and acceleration. Time is absolute. In Cosmological Balance Theory, ordinary gravity matter approaching a massive object with intense gravity encounters greater density layers of space-matter as it gets closer and closer to the object's upper atmosphere or surface if it has no atmosphere, which cause it to barely lag or slow down its descent through the denser space-matter. Starlight penetrating these same dense layers of space-matter around earth encounters a slight loss of photon energy and fractional red shift, sometimes seen on earth as starlight flickering. In addition, an object orbiting or one in a decaying orbit within that dense layer of space-matter surrounding a massive object will encounter resistance due to repulsions experience a loss or gain of speed, thereby slightly altering position locations in orbit, and this delay misinterpreted as time-dilation proof. In terms of Einstein might have used in mathematical proof, a falling body in a decaying orbit encounters the curvature of space around the massive object, and due to increasing speeds closer to light speed c, its ticking of time slows down. That same falling object encounters the resistance of space-matter blanketed around the massive object, and since the density of space-matter equates to curvature of space, it thereby experiences a slight slowing in its orbit or descent. The observer within the object perceives slowing in speed not time. In both instances, the falling accelerating object cannot and does not have time dilation nor stop in mid tracks or in time because gravity does not fail. The falling object plunges onto the surface of the massive object attracting it, regardless, if it was able to accelerate closer to the speed of light. This is the closest to acknowledging minor time distortion, which is only due to resistance of higher concentration density of space-matter in the curvature of space. Again, only its speed has changed not time.[260]

Acceleration and Time Within Cosmological Balance

Let us start by comparing an American civil war smooth bore long gun to a modern-day rifle. When we fire a smooth bored long gun, the round projectile goes out about a hundred yards, loses a great deal of momentum (speed) due to air friction, and hits the ground short of its target. However, a streamlined bullet with a properly pointed tip, fired from a modern rifle (twisting the bullet), pierces the air much more efficiently and goes a lot further before gravity and air resistance overcome it. Now, consider this idea to spaceships traveling in space.

If we build a spaceship with a round or blunt tip and expect it to achieve speeds near the speed of light, we will fail, as this ship will encounter or hit the maximum number of atoms, whether space-matter, or ordinary matter, along its path. The larger its width, the more space-matter it will encounter in its path. Designing a spaceship with a sharper tip and more streamlined hull to move and split through matter (whether ordinary, antigravity, or space-matter) like a sharp bullet or jet plane would be more likely to achieve maximum speed commensurate with its propulsion. Rotating the ship on its axis of travel also provides dual purpose, artificial gravity, and rifling effect. This simple analogy provides the

---

[260] Resistance when passing through large amounts of space-matter can affect speed and be misinterpreted as time delay.

reason why a spaceship regardless of how perfectly streamlined traveling through space has a difficult time reaching speeds near the speed of light. The faster the ship tries to move through space, the slower it seems to move through it, due to the increase frequency of repulsions with space-matter, and the slower time appears to be running along the journey, for example the pioneer anomaly leaving the solar system. Remember that the space-matter grid contains within it the Higgs Ocean, which slows down any objects with mass moving through it. The more and more fuel we burn to propel us faster and faster, the more time seems to be slowing down. In fact, time within the spacecraft has not changed rate or ticking duration, it just feels that way because space affects the ship's speed. In this situation, we are visualizing what a round bullet fired from the civil war rifle might see, rapid acceleration, gradual slowing in speed where time seems to slow down, and finally stop, but time does not slow down nor stop, only the round bullet's motion. It does not matter how streamlined we make the spaceship; we will encounter more and more space-matter the faster it moves through space. For every meter, the ship of 100 kg at 1 meter in diameter width travels, it will encounter an average one space-matter atom, due to its density of one per cubic meter. A two meter by one-meter diameter ship will encounter an average two atoms per meter travel, etc... In addition, the spaceship from time to time will also encounter ordinary gaseous matter and slow down even further. So, a spaceship traveling 2 million meters per second will encounter 2 million atoms per second of space-matter along the way, with each space-matter atom's mass of $2.8116E{-}24$ kg moving at $2.3735E{+}13$ m/s^2 pushing back with a vacuum force of $6.6734E{-}11$ N /atom.

Assume for the moment that if intergalactic space were completely void and empty then a spaceship set in motion in it would glide continuously through it without losing speed, and likewise easily accelerate through it with minimal energy; however, we know otherwise, space is occupied sparsely with space-matter atoms, which are negligent at slow speeds. However, at high speeds say one quarter that of the speed of light, we can easily see that a spaceship would have to burn a lot of fuel just to maintain its speed, and a lot more to accelerate. The closer the spaceship got to the speed of light, the more and more fuel it needed to accelerate it faster. This analogy is partially agreeable with Einstein's perception that mass "increases" with velocity, especially at speed closer to the speed of light, specifically need more, if not infinite, energy in needed to propel that mass to achieve the speed of light, but without gaining mass.

The force necessary to maintain a 2 million m/s speed for a 100 kg spaceship = (100kg * 1 * 2.0E+6 m/s) * (2.0E+6 atom/sec / 6.6734E-11 kg m/s^2 /atom) = 5.99431E+24 kg thrust, where 6.6734E-11 N is the vacuum force of space and the source for the gravity constant. The same spaceship at 2 meters width by 2 meters height needs (2x2) x 5.99431E+24 kg thrust to maintain that same speed.

The force to maintain 299.8E+8 m/s speed for the same 1-meter diameter 100 kg spaceship = 1.34692E+31 kg thrust.

The force to push a 1 kg spaceship to keep its 299.8E+8 m/s speed through space-matter, we need 1.34692E+29 kg thrust. To reach the speed of light surely requires significant if not "infinite" energy, surely a lot more than the ship can transport. This exponential increase of thrust force is the "practical" analogy or Cosmological Balance Theory reason science says that the speed of light is constant. This concept situation agrees with Einstein theory; however, it does not make light speed a "constant."

As we already stated, it would then take almost infinite energy (fuel) to enable the ship, traveling through space-matter, to reach the speed of light. This is essentially Einstein's concept of the speed of light constant from the Cosmological Balance Theory perspective[261]. In the Cosmological Balance Theory, the speed constant is valid if we were fighting against space-matter, analogous to a speeding round bullet encountering air resistance, and is less applicable when the wind is accelerating the bullet. However, if space was completely void of any type of matter, then the spaceship only needs a small finite amount of fuel, due to no resistance, to reach the speed of light. Ion propulsion should be sufficient. The constant c or light speed is also invalid where space-matter pushes objects together as gravity or when sub-atomic objects travel in wormhole tunnels, where its EM repulsive shield protects contents from colliding with space-matter atoms, slipping right in between them. Speed constant of c is also not true when two galaxies or two black holes merge; they will accelerate toward each other and in most cases exceed speeds beyond the speed of light since gravity law does not fail when it is accelerating these objects. The common basis of this concept with Einstein's speed of light constant demonstrates and confirms that space has some type of exotic matter scattered uniformly throughout it. The Cosmological Balance Theory calls it space-matter. Open space is not void of matter; it contains some form of "dark matter." Again, if intergalactic space was completely empty and void of any type of matter, then there will be no resistance to speed and an object going 100 million m/s will continue to move at the same speed (Law of Inertia), unless some force changes it. How does this relate to time? Einstein tells us that the closer we move toward the speed of light in a spaceship the slower our clock runs, going to zero as we reach light speed. Let us conduct our own analysis of possible scenario and situations.

Scenario: Let us use Brian Greene's (Greene, The Fabric of the Cosmos: Space, Time, and the Texture of Reality, 2004) explanation of time dilation with Bart going 2/3 the speed of light on his super scooter and Lisa watching him speed off. For every 2 hours Bart rides, Lisa experiences 3 hours of time passing. We will use the same speed relationship in the situations below.

Situation 1: To put this into perspective, let us first look at a routine situation on earth involving travel by airplane, where we are comparing airspeed versus ground speed (disregarding all other accelerations – planet, sun, galaxy, etc.). The airplane air speed readout says it is going 600 mph, but its ground (relevant) speed tells us something different depending on the direction of the wind in relation to the direction of the plane is heading. If you had a tail wind of 50 mph, then the plane experiences a ground speed ("relevant or absolute speed") of 650 mph. If the plane was encountering a head wind of 60 mph, then the plane's ground speed is 540 mph, give or take a few miles per hour. The same physics apply in the space and affects the "absolute" speed of the spaceship.

Let us assume for the moment that spaceship captain Bart believes that intergalactic space is completely empty and void of all matter. Now extrapolate the airplane travel situation into intergalactic space aboard a sleek 100 kg spaceship 1 meter in diameter going 199.8667 Million m/s or 2/3 speed of light for 3 hours (in an imaginary experiment). At

---

[261] The only way Einstein could have effectively perceived the speed of light as a limit was to imagine that space resisted the movement of mass through it. As that mass approached light speed, it would require infinite energy to push it there. This idea is analogous to the Higgs Ocean, which resists the motion of mass through it. This is the same as the Cosmological Balance Theory.

that rate of speed and time within the ship, Bart calculates he should have gone 2.15856E+12 meters (or 2.15856 trillion meters) at the end of 3 hours (ship's time). Let us assume for a moment that time passes normally for those in the ship, per Einstein. The spaceship captain Bart re-verifies his spatial coordinates and finds that he actually traveled 3.23784E+12 meters over the last 3 hours (ship's time) and concludes that his speedometer must be wrong because in order to travel that distance in 3 hours' time he would have to be moving at the speed of light. He somehow manages to speak to Lisa for a time check, and she tells him that he has been travelling about 4.5 hours, where Bart's completed travel distance agreed for a ship at 2/3 c in a 4.5-hour time. Captain Bart is dumbfounded swearing that he only traveled for 3 hours and determines that either Lisa's clock was running fast, or his clock was running 2/3 slower. Then he remembered what Einstein said about time dilation and ecstatically acknowledged that Einstein was right, my ship's clock ticks slower; Bart's two hours at high speeds equals Lisa's three hours in the fixed space station. Hence, Bart's three hours equals Lisa's 4.5 hours. This gives Bart a headache, as he feels something is not right here but just cannot place his finger on it. He needs help to understand the dilemma.[262]

Situation 1a: Let us conduct the same experiment in the Cosmological Balance with Ethan "Makani Blazer" and Veronica "Bugs Hora": Now let us assume we have been traveling in our sleek 1-meter diameter 100 kg spaceship for 10,800 seconds (3 hours) at 2/3 speed of light. In Ethan's initial calculations without resistance, a 3-hour journey at 2/3 speed of light would have covered 2.1585 trillion meters, enough to reach his destination (Veronica's space station at a fixed spatial coordinate, their rendezvous point). However, in the Cosmological Balance, a spaceship travelling in intergalactic space (between galaxies) will encounter a "head wind" of space-matter resistance at 1 atom per meter cubed each with vacuum force of 6.6734E-11 N. Since academia told Ethan that space was completely void of matter (especially intergalactic space), he did not apply additional thrust to compensate for any resistance.

Hence, after 3 hours of traveling at 2/3 speed of light or so he thought, Ethan checked his spatial coordinates and discovered that he only went 1.8650 trillion meters instead of 2.15856E+12 meters (absolute distance to destination) so he kept going until he travels just under 4.25 hours when his alarm indicated he reach his destination. If Ethan's ship went through additional pockets of scattered ordinary matter (invisible intergalactic gaseous clouds), these added repulsions would have slowed his ship down closer to 4.5 hours. Using the time of travel and the distance, he determined that either he was going 1.33241E+08 m/s (slightly slower than 2/3 c velocity) or his ship's clock ticks was off (slower). Ethan finally remembered what Einstein said in the Special Relativity theory something about time dilation. Ethan's clock read 4.25 to 4.5 hours. The trip, which should have taken him 3 hours in his ship to reach his destination, took him 4.25 hours (4.5 hours if we accounted for additional scattered intergalactic gaseous matter with gravity). According to Einstein's prediction, Ethan's 4.25 hours (4.5 hours with addition collisions with gaseous matter) inside his struggling ship speeding at 2/3 c speed should have been 6.375 hours (6.75 hours due to other collisions) on a clock outside his ship at Veronica's station. Instead of gaining time as Einstein predicted, Ethan lost time according

---

[262] All the situations investigated here point to no time dilation. Time continues to function as expected.

to the ship's clock. Ethan spoke to Veronica at the destination and she agreed with the 4.25-hour (or 4.5-hour) travel time reading and his findings. If Einstein were correct with time dilation, then the spaceship's collisions with space-matter and ordinary matter in space would somehow alter time within the ship but it did not. Hence, we did not confirm Einstein's prediction of time dilation and consider it false.

Nevertheless, let us assume for a moment that Einstein was right about time dilation, then 2 hours within a ship traveling at 2/3 the speed of light would be equal to 3 hours outside the ship in a stationary object (as in Situation 1 above). For this to be true, we must assume that the resistance the traveling spaceship encountered with space-matter was the source of the time slowing effect. Then the faster the ship goes the more space-matter collisions it will encounter and thereby somehow altering and decreasing the rate of time the ship occupant experiences. However, time functioned normally within the spaceship to its occupants, per Einstein. Applying math, we find that 4.5 hours times at 2/3 the speed of light should have caused a 2/3 slowing of time. This equals to about 3 hours of passage of time to observers outside the ship. The occupants within the traveling spaceship did not sense time slowing down until they finally reached their destination. The 3-hour trip seemingly took them 4.5-hours of struggle to complete, due to collisions with space-matter somehow slowed their clocks to about 2 hours. This situation is highly unlikely and impractical for a solid object like a spaceship but conceivable for a particle of light when we consider it encountering exotic matter, which slows it down while absorbing its photon energy. This confusing tale of science fiction hurts our heads, fabricated to demonstrate how people were gullible to Einstein's genius, the mirage of the physics of light. That is right the physics of light is of the quantum world, which is not applicable to the macro universe. In the Cosmological Balance Theory, time dilation if any is the result of time delay caused by the ship's ordinary matter colliding with space-matter during travel. Time dilation as defined by Einstein also not confirmed in this instance. Only the photons within light, on the other hand, does slow down slightly (losing time) when passing through space-matter. As an exotic matter, space-matter can and does slow a particle of light down to a crawl for a few nanoseconds and then it resumes light speed when re-emitted in the same direction of travel. This time lose in photons is most notable within dense concentrations of space-matter around massive objects. Here, Einstein time dilation is right when describing the movement of light itself in open space and passing tightly around massive objects. The physics of the quantum (micro world) do not apply to the macro universe of gravity objects. We can now see that the Einstein thought experiments were just an outright honest misinterpretation of the Doppler Effect as applied to the visual wavelengths of light played out in the mental mirages conceived within his head. This is science fiction at its best passed off as science fact; do you not agree Forest Gump (main character played by Tom Hanks from movie of the same name)?

Situation 2: In careful examination of his spaceship's systems, Ethan found that his spatial coordinate system did not work properly when he was traveling at high speeds. As a result, he recalibrated his spatial coordinate system, and accelerated back up to 2/3 speed of light. This time, his cruise control worked more efficiently with the recalibrated spatial coordinate system directing the thrust controller to apply the necessary accelerations to counter that "dark energy in space" in order for the spaceship to maintain 2/3 speed of light. After travelling for 10,800 seconds (3 hours), he revalidated his spatial coordinates, found out that he moved exactly $2.1585E+08$ meters distance, and reached his destination.

However, for some reason when he compared his clock with the rendezvous spaces station's clock, he found that his clock was in perfect synchronization with Veronica's clock. The rendezvous station Captain Veronica told Ethan that he had been traveling for 10,800 seconds (3 hours). She asked Ethan, "How did you do it?" He replied, "I had to apply additional thrusts to maintain 2/3 c speed in order to fight the unknown resistance." He told Veronica that Einstein time dilation could not be correct because both our clocks are synchronized. However, the more thrust energy Ethan's ship expelled the more resistance (from some "dark energy") he encountered; after 10,800 seconds at 1.99862E+08 m/s, he encountered a total force equal to about 5.601E+07 m/s of resistance. He practically exhausted his fuel reserves to get to the destination in a timely manner. Here in situation 2, the accelerating spaceship encountered a tremendous amount space-matter resistance, but this time Ethan expelled more energy to maintain its "absolute" 2/3 light speed with the assistance of properly functioning spatial coordinate system checks. Again, a spaceship's collisions with space-matter does not cause time to slow down within the moving object; Upon arrival, Ethan's clock within his spaceship read the same as Veronica's clock at the space station. Time dilation not confirmed; the presence or collisions with space-matter did not slow time within a high-speed spaceship traveling near the speed of light.

Situation 3: Ethan's spaceship is again in deep open intergalactic space. Let us now suppose that he had a genius on board who installed a modification to his spaceship with the capability to make a shield, by theoretically pushing energy into every molecule within the hull of the ship (creating an electromagnetic-gravity or gravity-electrostatic charge at the nose of the ship). Combining this modification with the normal ion or photonic (laser) propulsion system, Ethan accelerated his ship to 2/3 c speed, and turned on his cruise control. The target is Veronica's space station again. The cruise controller turned on and off the thrusters as needed, and his ship maintained 2/3 light speed. He counted 10,800 seconds (3 hours) and checked his position. He discovered that he traveled 2.1585057E+12 meters as expected. In terms of Cosmological Balance Theory, his spaceship pushed space-matter out of the way while the ship tunneled a path through space, significantly minimizing collisions with space-matter. After 10,800 seconds of minimal collisions with space-matter along the way, he confirmed that his ship traveled 2.185 billion meters, concluding that he maintained an "absolute" velocity of about 199.862 million m/s (2/3 light speed) along the way, and as such, reached his destination exactly as calculated, in 3 hours. His clock showed 10,800 second had passed, the stationary destination clock showed 10,800 seconds passed. Absolute speed matched absolute distance, zero "wind" resistance. Both clocks agreed. Effectively no time dilation detected in this situation; Einstein time dilation not confirmed. Ethan also concluded that in this situation, if he had been actively engaging his thrusters and accelerating, he could have easily exceeded the speed of light barrier. He took a note to try it on the next journey.

Situation 4: Ethan's spaceship is again in deep open intergalactic space, slightly over 1,000 light years from the nearest galaxy's edge, with that galaxy's radius of 50,000 Light Years (LY) and a mass of 6E+42 kg. Suppose he built another modification to his spaceship, the capability to multiply effectively its actual mass (100-fold), by theoretically pushing additional energy into every molecule within the hull of the ship to create electrostatic charge, increase its mass, and focus its direction. In addition, Ethan designed and installed a repulsive cone of laser beams projecting to a point one kilometer in front of the ship capable of defecting matter in the ship's path. The point where lasers beams intersect is where they cancel each other out to form the tip of the cone shield. Combining these modifications with the normal ion or photonic (laser) propulsion system, Ethan aimed his ship toward the nearest galaxy, accelerated it to 2/3 light speed, turned on the focused mass multiplier, and activated on the cruise control. The target galaxy's gravitationally pull then grabbed hold of the ship's enhanced mass, and started pulling Ethan toward it, with acceleration at 1.71999E-07 m/s^2. The cruise controller eventually minimized and, in some instances, turned off the thrusters; his ship continued to accelerate toward the galaxy. He counted 10,800 seconds, turned off his mass multiplier, and stopped to check his position. Ethan discovered that he traveled 2.1588031E+12 meters and reached speeds close to the speed of light (1.99898E+08 m/s) unknowingly. In terms of Cosmological Balance Theory, space-matter pushed the enhanced mass of his spaceship toward the galaxy with acceleration equivalent to the galaxy's gravity force (at 1000 LY distance from the galaxy edge). This is analogous to the "wind" pushing the sailboat through water or the "tail wind" on an aircraft on earth, while the electrostatic charge of the ship's hull split a path through space-matter ahead of it minimizing collisions with space-matter, and the laser cone deflected away any normal matter within its path. After 10,800 seconds of minimal collisions with space-matter and normal matter along the way, he confirmed that his ship traveled 2.1588031E+12 meters, concluding that he somehow accelerated to about 1.99898 million m/s along the way, and as such, gained time, reaching his destination early and completely past it at 3 hours' time. His clock showed 10,800 second had passed where he overtook the destination, while the stationary destination clock showed 10,798 seconds (2.99 hours) passed when Ethan zoomed by Veronica. The closer Ethan moved to the galaxy the faster his spaceship accelerated. Nevertheless, the traveling ship's clock matched that of the space station's clock. If Ethan had been applying thrusters, his acceleration would have been much greater and would have reached Veronica in less than two hours easily, at speeds beyond the speed of light. Effectively time dilation not detected; Einstein equation not confirmed again (with minimal space-matter collisions).

The spaceship in situation 4 above is the type of ship conceived to "fold time and space" without traveling in a wormhole (in essence propelling itself through space in shorten time). It agrees with the Cosmological Balance Theory definition of space-matter and duplicating how the exotic matter in space repels each other, using its "energy," and gravity force to accelerate the spaceship exponentially. Bottom line: Since space contains space-matter, which some scientists called it, "dark matter" and "dark energy," then Einstein's time dilation is not confirmed and not universally applicable. It is valid only for light itself when light goes around massive objects in space, it slows down slightly as it curves, or when light travels distances beyond 14 billion light years.

8.10 Antigravity and Antimatter: It's About Time

If you thought those space and time examples above were confusing, then this section will be a bit harder to comprehend. Here we will introduce a completely outlandish concept, anti-time, one that will obviously be awfully hard to prove or disprove with our current instruments and current technology. Somewhat like antigravity and antimatter was when it was originally conceived, but unlike them anti-time is expected to be rejected as impossible. With the advent of sharper and better observations and techniques, we will begin to gather more evidence to confirm or deny its existence and function. For now, all we can do is investigate this through mental experimentation and imagination. For the effects of antigravity and antimatter are completely hidden right before us, invisible to our sophisticated instruments and advanced technology, a mindboggling mirage, only detectable through the motion of visible matter and visible galaxies in space, our final frontier.

Notwithstanding, let us start by talking about the symmetrical structural differences between normal ordinary gravity matter and that of antigravity antimatter at the atomic and subatomic particle level. When we defined antigravity antimatter atoms in the previous chapters and sections, we defined it as possessing a mutual pull that operates with a gravitational constant that is invoked on a negative mass. When we investigated and predicted the antigravity quark within an antigravity atom, we designed them by a reversal of positive and negative point particle motions, and in doing so, they were structurally stable anti-matter. If an electron collides with an anti-electron (a positron) because of their mutual, electrical attraction, per mainstream science, would instantly annihilate each other instantly. Matter annihilates anti-matter on contact, releases energy and transitions. However, as we build upon the structural level or symmetrical system, there is a tendency to develop systems leaning more and more towards a preservation of existence through the reversal in the function of time. In other words, the atoms and quarks within normal matter gravitationally repel an antigravity atoms and quarks within antimatter, and as such avoid collision, and sustain mutual coexistence. Mutual gravity attraction of normal matter with each other requires time to move in the direction we are very familiar with, forward, while in mutual attraction of antigravity antimatter appears backwards in motion, as seen from our perspective, but functions normally in time. In essence, the forward passage of time, unique, predefined, and predetermined by the motions of the gravity particles within normal matter, and likewise the passage of time, within the antimatter particles that constitute antigravity antimatter, is defined and determined normal for such system, as well as it is unique to that system, but in reverse motion. One can theorize another possibility of obtaining effective space travel. However, for now, let us now look at difference between light emitted from a gravity bound star and anti-light "emitted" from an antigravity star.

What enables us to detect visible galaxies, and all the matter in the visible universe? We detect them through their light and electromagnetic waves. The development of telescopes has enabled us to learn about the sun, the planets, their moons, comets, and the debris in the solar system. Advanced telescopes incorporating electromagnetic (EM) or "light" detection over the entire spectrum of waves, have enabled us to observe distant galaxies, find exoplanets in the galaxy, and speculate the amount of dark matter and dark energy out there. Dark matter and dark energy, estimated primarily on the motions of galaxies and the forces that may have caused these variations, amount to about ninety-five percent of the total composition of the universe, as we know it. Dark matter and dark energy were responsible for motions that went against that expected with the application of Newton's gravity equation and Einstein's general relativity; we see this as antigravity force. For instance, the spiral arms of the galaxies rotated faster that gravity alone could move them. Therefore, if "light" reaches our instruments, we were able to "see" and observe the source from which it came. In addition, through the merger of source frequencies, we can computer generate spectacular-colored images of every target detected and now visible to us. However, we are still unable to detect the invisible dark matter and dark energy, now seen as the effects of antigravity antimatter and space-matter itself, essentially because everything "emitting anti-light" is undetectable and invisible to our instruments and us. We do not read data carried by positrons. It is that simple. Not so fast, how is it possible?

Defining material assembled from antimatter particles into antigravity antimatter is considerably a long stretch of our imagination. Visualizing how antimatter particles form is another level in and of itself. On the other hand, it is relatively easy to understand how antigravity antimatter and its particles would attract each other. Antigravity antimatter mutually attracts each other and functions the same as gravity matter attracts other ordinary matter as we know it, but in reverse motion. Hence, the densest formation of antigravity antimatter is what we would call an antigravity "black hole" we called a white hole, or a mirror opposite of a gravity black hole. Nevertheless, our definition of a white hole requires that it ejects or spews out inordinate amounts of material, where nothing enters its event horizon, not even light. As we learned from our analysis in previous chapters above, gravity matter repels antigravity antimatter, and vice versa, antigravity antimatter repels ordinary gravity matter. There is only one literal way that the laws of physics will operate perfectly and allow such conundrum to exist; it is through the motions of particles contained within normal matter and antimatter. More specifically, the motion from the perspective of ordinary matter and its subatomic particle composition, and functionality, versus the motion from antigravity antimatter, and its subatomic anti-particle composition, and functionality, are mirrored reflections of each other.

From our naïve perspective, time always functions in one single flow direction, forward. For our simple minds, it is unimaginable that time can run backwards, or at least it exists only within science fiction. However, we as human beings are able to remember precious past experiences, reaching backward within our memory banks to bring forward the circumstances of the past, in order to elaborate that experience to others around us, and share ideas. In a sense, we can "see" backwards in time, relive experiences, at least as far back as we can remember. From our naïve perspective, we cannot change our past, nor the passage of time in the world, and we think that of the universe as well. In order to get a "feel" of the invisible part of the universe, we need to step out of our comfort zone, and learn to realize that motion, not time, can run backwards, and to those operating within that parameter see it as normal. In fact, numerous movies and films portray that realm, and the repercussions it presents to the current time, at least from our perspective. Billiard balls bounce off each other on the billiard table the exact same way whether the film is played forward or backwards in time. The motion of gravity matter system is the forward playing of the film in normal time, while the motion of antigravity antimatter system would be the reverse playing of the film in the same normal time.

What exactly is anti-time? Anti-time is defined as time running backwards as compared to what we experience physically. If we were able to "observe" physically an anti-time system, we would see that its time within that system appear to run backwards or clocks "anti-time," time in reverse. However, if viewed from within that anti-time system, time flows forward to them, and vice versa. Einstein would have jumped on this concept in a heartbeat, claiming that the antigravity part of the universe operates in anti-time, insisting that antigravity "black holes" are the long sought after "white holes." He then would have imagined a massive wormhole tunnel connecting the gravity "black hole" to the anti-time anti-gravity "black hole," which would be spewing out antimatter like a "white hole." However, that is far from the truth. There is no anti-time system or part of the universe. Antigravity antimatter is just a mirrored system when compared to ordinary gravity matter. Antigravity antimatter operates in normal forward time. An antigravity system is invisible to our instruments and us simply because it is not synchronized or in phase with the functions of our eyes and detection devices, which look for and measures electrons not positrons. Specifically, our devices are not designed nor calibrated to look for and read positrons within antimatter inverted-electromagnetic waves. Coining an antigravity antimatter black hole with the name "white hole" is not a mistake, because it literally spews out normal gravity matter in all directions during its true Cosmological Balance excretion process, and the cosmic background signals are the visible wormhole tunnels exit points mapped. For the same reason, a black hole within visible galaxies is antigravity "white holes" since it ejects antigravity antimatter into space. Clearly, time must flow forward throughout the universe, whether in the gravity and antigravity portions of it. Anti-time is science fiction.

From our understanding of normal gravity and time in the laws of physics, a star emits light outward into space. The antigravity star would then shine "anti-light" outward in normal time as expected within that antimatter system, and from our perspective, this anti-light travels as expected outward and away from the antigravity star's center, as time flows normally. However, its anti-light is invisible to our eyes and instruments, because the unexpected information carrier is positrons. This EM cannot be detected as it moves towards us, in the normal flow of time. To reiterate, the flow of time in a gravity system as see from the antigravity's perspective is the same, normal, but the gravity system itself is undetectable. Therefore, from an antigravity and normal flow of time, an ordinary matter system operates in mirrored physics, the light emitted from a gravity star shines outward and in normal time. However, that ordinary light is undetectable or "invisible" to antigravity instruments as their instruments and "eyes" are looking for the information carried by positrons, and in doing so ignore the presence of electrons. Both parts of the universe view the unknown on the other side as consisting of large amounts of "dark matter and dark energy."

How can time flow normally in an antigravity antimatter system and in a gravity system yet the systems are mirrored? Time is just one other orderly control aspect in the laws of physics. As we discussed numerous times, physics operates the same forwards or backwards in time, depending on the perspective of the observer and the system in question, gravity, or antigravity antimatter. For simplicity, we will say a gravity system operates forward motion as expected and antigravity system functions in reverse or in a mirrored motion. Let us put it another way, from within the antigravity system where time runs forward, the "anti-light" functions as expected, emitted outward and away from antigravity star and flows according to forward time into space. Similarly, from within the gravity system where time runs forwards, light functions as expected, emitted outward from the normal gravity sun. The coexistence of the visible and invisible parts of one universe, where the instantaneous and infinite forces of gravity and antigravity, the motions of matter and antimatter in time, energy, and space balances into a homogeneous and isotropic universe. Gravity and antigravity galaxies move like two ships passing each other in the ocean of space, invisible to each other, one moving forward motion and the other moving in a mirrored backward motion in normal time, where each perceives the nudge of an "ocean" current of overwhelming "dark energy and dark matter" around them. The anti-light and ordinary light too pass each other in space, as out-of-phase particles. Never shall the two encounter each other, gravity and antigravity, matter and anti-matter, in time and space.

Just in case, this is still confusing to you. The structural configuration and particle motions within the up or down quarks create gravity or antigravity, while the neutral balanced particle motions of the quarks within space-matter create near zero gravity effect by oscillating gravity-antigravity pulses. The creation of antigravity in antimatter up and down quarks makes its motion appear to run backward, or mirrored to gravity, from our perspective. Space-matter, being the neutral medium between both enables gravity matter or antigravity antimatter to pass through or traverse within it, practically unhindered by its presence, and synchronizes to the ticks of time, phasing in and out as "dark matter" and then "dark energy." The three forms of matter inherently operate within the following parameters: Ordinary gravity matter functions in the positive flow of time. Antigravity antimatter operates in the same flow of time, but mirrored motions. Moreover, zero gravity space-matter is the timekeeper, the active pendulum fluid which gravity and antigravity antimatter are suspended, flow through, and exist simultaneously. The alternating pendulum time swings of space-matter also have one other side effect. Since the basis of gravity and antigravity is the result electromagnetic-weak-strong force interactions in the quark, the alternating pulses of gravity-antigravity from space matter, generates rotational motion on larger gravity or antigravity systems, such as galaxies, in as much as an alternating current on electromagnet drives the permanent magnetic rotor inside an AC motor. Thus, the gravity-antigravity pulses cause the stars in the galactic arms to move faster than expected by gravity or antigravity alone. This same spin induction can push a stellar nursery into solar systems, flatten large spherical globular clusters into a disc, and boost the rotating spiral arms of galaxies. Gravity is forward motion in positive time, antigravity appears to us as mirrored motion, which is actually normal from the antigravity perspective, and the alternating gravity-antigravity pulses of space-matter is the force that drives the spin or rotation of matter (whether permanent gravity or antigravity) on either part of the universe, the visible or the invisible. Thus, the gravity-antigravity pulses make gravity matter turn one way, and antigravity antimatter turn in the opposite direction mirroring each other.

Whew! I am glad that is settled. Gravity matter moves forward in time. Antigravity antimatter appears to have mirrored motion, in normal time clock. How does this affect the absolute void of space? Empty space moves neither forward or backwards; to be more exact, it pulses gravity and then antigravity like a pendulum clocking motion repeatedly, it exists essentially outside the effects of gravity or antigravity for all eternity. This pendulum ticking enables space-matter to exist in both realms and allows gravity matter and antigravity antimatter universe to flow right through it, and yet suspends all matter, and as such makes detection practically impossible. Space-matter's rate of vibration, at the speed of light, is practically neutral except for periodic alignment of gravity and antigravity pulses; at one-five-hundredth times the speed of light, at a rate of about 599,119.05 cycles per second (hertz), provides the gravity and antigravity constant of 6.6734E-11 Newton. At 299,560 cycles per second of antigravity, and 299,560 hertz of gravity. The formation of space-matter, the composition of space, likewise has no gravity or antigravity mutual pull on itself because of its delicate subatomic structural balance and basically zero gravity-antigravity. This quality enables space-matter to allow gravity matter to traverse through it, and simultaneously, enables antigravity antimatter to pass its vastness.

In addition to allowing gravity matter and antigravity antimatter to slip though the ocean of space-matter, we get one interesting side effect and inadvertently reveal the extreme forces and the energy hidden within the vastness of the universe. Space-matter has two swings on the pendulum of time, gravity and then antigravity and back. As space-matter pulses forward gravity, it allows the gravity galaxy to pass through normal time in the area or space it occupies. However, as space-matter pulses antigravity in what appears to be backwards motion, it applies a spherical antigravity force or bubble of energy pushing onto the gravity galaxy itself, amounting to an equivalent to Newton's gravitational constant G. It is just enough to suspend the galaxy in space and sufficiently strong to influence the acceleration of stars particularly at the outer edges of a rotating galaxy, to speeds anywhere from 200 to 275 kilometers per second. The same applies as antigravity galaxies traverse through space. The space-matter's antigravity pulse backward motion allows the antigravity galaxy to pass easily in time, while the space-matter's forward gravity pulse applies gravitational pressures to keep the antigravity galaxy suspended in space and adds the addition forces to speed up the stars at the rotating antigravity galaxy's outer edge or spiral arms at rate of 200 to 275 km/s. Scientists have been theorizing about this "dark energy" for a while. It is concealed right before their eyes, and practically undetectable to instruments. In other words, the "push" factor on gravity matter and antigravity antimatter caused by the vacuum force of space applies everywhere throughout the universe. Said another way, space-matter's randomized gravity antigravity omnidirectional spiked pulses enable these powerful atoms to maximize mutual repulsion and separation, while providing the alternating gravity antigravity forces to apply pressure on both types of matter to suspend each in space and in doing so also cause sufficiently large masses to rotate according to their time motion synchronization. This action likened to running an electric motor where an alternating current applied to the coils of electromagnets provides the forces to turn the permanent magnets on the center rotor in either direction forward or backward. This is dark energy scientist are searching for.

Now that we have a complete run down of all aspects of the Cosmological Balance, let us formalize the information into working definitions, conjectures, corollaries, postulates, lemmas, axioms, laws, and finally principles.

Chapter 9: Cosmological Balance Principles

"The magic is only in what books say, how they stitched the patches of the universe together into one garment for us."

— Ray Bradbury, *Fahrenheit 451*

We are all part of the universe we live in, and the universe in us is trying to teach us its secrets. It is time we open our minds to this Cosmological Balance Theory and its stupendous possibilities to raise our human intellectual level. It is one of our history's examples of ordinary people offering what academia at first considered an unorthodox idea only to find out the theory presented was in fact accurate and correct. We should applaud thinkers for they give humanity the necessary boost it needs to move onto the next level of consciousness, a step that benefits the progression of science. The idea set forth in this book is one of these concepts that obviously go against accepted agreements of mainstream scientists. Historical world changing theories like the planet is not flat, the planet is not the center of the solar system, the solar system is not the center of the galaxy, and the galaxy is not one of two galaxies, it is among billions if not trillions of other galaxies scattered throughout the visible universe. This chapter summarizes the highlights of the principles discussed.

The universe is much more simplified than we are led to believe, a theory based on the existence of three types of matter. There are elements made up of matter with gravity with which we are all quite familiar, those with antigravity with which we learned about, and those that make space what it is, this third type of matter, occupies the firmament of the heavens. As gravitational force governs the movement of all matter, we know and see it as the visible universe, antigravity force likewise governs all matter in the invisible part of the universe. Space-matter exists in an extremely cold gaseous form, floating and occupying deep open space, making space what it is. Interactions of these three types of matter have always existed and will continue to exist indefinitely[263].

The Cosmological Balance Theory then explained the relationship between the three types of matter by redefining the most ferocious orb in the galaxy, the massive black hole at the center of it. In our discussion, we learned that the black holes are not cold and desolate, but homogenous and a "soup" of subatomic particles, leptons, and force carriers elevated to their highest energy levels, with an occasional transition to antigravity, which are released from the surface of the black hole. These released white hole orbs travel

---

[263] Each new discovery about space is revealing that space is not as empty and void as originally thought. First, scientists believed that space filled with ether that provided the means for light to traverse from the sun and a star to earth, only to find out that light was a self-perpetuating electromagnetic wave that needed no medium to move through a vacuum, unlike sound waves. Einstein then believed and predicted that gravity was limited and travelled at the speed of light, implying that a gravity wave was similar to some type of electromagnetic wave. Then Professor Higgs believed that space filled with the Higgs field or Higgs boson particle gives ordinary particles of matter its mass, by resisting its motion. Later, cosmologists discover significant amounts of some unknown "dark matter" and "dark energy" permeating the universe. This new additional mass and energy does not make galaxies come together and the collapse the universe.

through wormhole tunnels beyond the edge of the galaxy, to primarily replenish space-matter, and secondarily create lighter gases elements for the antigravity part of the universe. This antimatter ejection release combined with the ferocious consumption of black holes dubs it as dual mirrored white holes overlaid in the same space. We also learned that antigravity galaxies did the same thing in their massive antigravity black holes at the center of their galaxies, replenish space-matter, and secondarily create gravity matter building blocks, lighter gases like hydrogen and helium. The antigravity black hole release of gravity matter while simultaneously eating of antigravity material makes the antigravity black hole also a white hole. Therefore, from our perspective the antigravity black hole spews out gravity material just as if we expect a white hole to do. In both cases, a "black hole" is also a "white hole" occupying the same space, as it eats one type of matter and ejects the opposite type of matter.

The theory then declares that there is only the curvature of space, not of time, and that the universe itself was not ever expanding. We did this by first learning that light bends around the sun by two factors, first gravity itself gradually pulls light as it races past the sun at its tremendous speed, and second by refraction caused by clear heated gases just above and space-matter surrounding the corona of the sun, known as the curvature of space. Gravity by itself causes only a slight deviation. The primary bending of starlight passing near the sun occurs as that starlight enters the gases at the outer edge of its atmosphere, above the corona, continues to bend as it passes each atom within different density layers of space-matter, and then makes its final bend as it enters the "vacuum" of space once again on its way to earth. In addition to bending light, these gases absorb photon energy and shift the color of the light slightly to the red part of the spectrum. Since space is not empty, billions and billions of light year distance of space-matter and the gravity and antigravity gases it has causes any light traveling through it to lose photon energy and appear redder to our eyes and our instruments and telescopes. Distances beyond 13.5 billion years make the distant starlight appear very dark red or almost black. Extreme distances beyond 14 billion light years completely block all starlight from reaching us.

The Cosmological Balance then explains how gravity and antigravity interacted with each other, and with space-matter. The presence of antigravity antimatter leaving galaxies causes the stars in it to accelerate and the entire galaxy to rotate in unison at speeds greater than predicted with gravity alone. The same applies to gravity matter leaving antigravity galaxies; it rotates quicker than with just antigravity motion only. We go on to explain gravity as a layer intensity force, like the skin of bubbles, layers upon infinite layers of spheres from the strongest closest to the massive object to the weakest furthest away.

The theory then uses the power of Newton's equation to solve the orbital dilemma of Mercury's anomaly. The solution to this planetary anomaly is rooted within a phenomenon that occurs on earth, tides. Tidal effects on earth cause the moon to accelerate and move higher in its obit, gaining distance from the earth and an orbital precession. At the same time, the moon is causing the earth to slow down in rotation, thereby gradually lengthening the period of one rotation. The plasma surface of the sun similarly forms a rising of the tides as Mercury makes its closes approach, while the surface of the planet Mercury too creates a bulge on its molten surface. Both additional masses align and gravitationally attract each other, thereby causing the rest of Mercury's orbital precession to occur.

The Cosmological Balance Theory then defines space-matter and how it makes space what it is, by presenting a brief discussion on the origins of gravity matter and antigravity antimatter, and how the latter two replenish space-matter. This trio-matter balance is fixed. Decay of space-matter is limited to a fixed percentage, and likewise gravity matter and antigravity antimatter existence are limited to a certain fixed percentage. Excessive gravity matter or antigravity antimatter means more massive black holes or more massive white holes respectfully, and greater production and release of antigravity white hole orbs or gravity black holes respectfully, thereby greater replenishing of space-matter. The merger of two massive black holes would therefore send out a quick and tremendous burst of antigravity white hole orbs in all direction within massive amounts of wormhole tunnels. Picture such an ejection as a colossal wave initially moving at the speed of light squared, and then gradually reducing to the speed of light before collapsing two hundred light years from the galaxy's edge. Space-matter on the other hand will continue to carry the ripple of the shock outward from atom to atom via electrostatic transfer of energy or collisions at the speed of light primarily in a waveform. The same type of instruments designed to record the Einstein predicted "gravitational wave" should easily detect the movement of this space-matter wave and its interference with the galaxy's normal space-matter gravity waves. To be more accurate, we could properly name these instruments as "space-matter and gravitational wave interference detectors[264]."

Two black holes merging emit practically no Hawking radiation. For the maximum Hawking radiation, we also learned that an entire star falling or rapidly wandering too close into the event horizon of a black hole could be ripped apart and dragged in quickly, sending almost half of the star's mass flung outward in a massive ejection from the event horizon, but not from the black hole itself. This massive amount of Hawking radiation evaporation would appear as though streams of escaping matter were reconsolidating above the black hole and racing away as one large object, and possibly reigniting into a small star. In any case, we expect large amounts of gamma and x-ray radiation to accompany such event-horizon solar mass ejections or major flare (Whitney, 2016). After all, the first level of electromagnetic wave released during fusion within a star is gamma and most of the radiation is at x-rays when the star war ripped apart. In addition, x-rays can penetrate the fog of matter between the black hole and our telescopes.

The theory then takes a leap and redefines planetary motions and the force of gravity. We did this by first listing Kepler's and Newton's laws. We then discuss how space-matter transfers vacuum force into gravity or antigravity forces. The theory then explains how gravity and antigravity forces function as instantaneous push forces of infinite range. The force of gravity begins with vacuum force 6.6734E-11 Newton of space-matter repulsive interaction with normal matter at the upper spherical layer of the object, known as the curvature of space, and then transfers this force from atom to atom toward the center of the object in space as the force behind gravity. Normal matter and its quantum entanglement attraction enable this force transfer to move more rapidly toward the center

---

[264] Einstein envisioned "gravitational wave" moving outward from the merger of black holes. The Cosmological Balance Theory views this outward wave as that of space-matter transferring energy in waveform via atom to atom at the speed of light. This is the same speed limit of electrostatic or electrical movement through a medium. The end result detection by LIGO instruments is the same.

at a rate of 1 meter per kilogram. The same force transfer results occur in the antigravity environment or realm, the invisible part of the universe.

The Cosmological Balance Theory then presented a planetary and galactic orbital period equation applicable to both situations, and to the universe. The first part of the equation is pertinent to all situations. The second part is valid only if antigravity antimatter is traveling in wormhole tunnels through and around gravity objects or vice versa if gravity matter is traveling in wormhole tunnels through and around antigravity objects. The orbs traveling within the wormhole tunnels transfers repulsive force and accelerate the objects in its path. The second part of the equation goes to "one" in systems where there is no interaction between antigravity and gravity, like a star or solar system, or significantly decreases in effect, approaching "one" for extremely distant galaxies. The theory then presents the cosmological balance equation defining the relationship between space, its space-matter, and its motions, and that with antigravity antimatter and its motions, and with gravity matter and its motions. This relationship stems from the galactic and planetary orbital period equation. It takes us to the next level beyond Einstein's space and ordinary matter equation.

The Cosmological Balance Theory stands firm on the following claims. The big bang did not happen. The universe is not expanding nor collapsing on itself. The cosmic microwave and the abundance of lighter elements are replenished daily. There is only the curvature of space and not of time. Time is universal throughout space with steady and consistent durations or ticks. It speeds neither up nor slows down with an increase or decrease in gravity or antigravity forces, or with increase or decrease in velocity and acceleration. Open space is spatially flat and uniformly dispersed, and its visible matter isotropic. Space-matter surrounds and curves around ordinary massive objects to contain it. We learned that massive black holes and white holes excretion, not hawking radiation or evaporation, are essential to the cosmological balance. Moreover, this balance has always been in existence and will continue for eons upon eons, as infinite as time and space, long after we are gone.

The entirety of this book up to and until this point was designed to present its information in terms for the scientific inclined layperson to follow along and learn about another potential unified theory worth reading and pondering about. However, to be plausible and credible as a theory, the remainder of this chapter is written for justification aimed mainly at satisfying mathematicians and scientists. In it, we gathered the most pertinent portions of the whole theory into this chapter, the Principles of Cosmological Balance. Throughout the presentation of this book, we have relied heavily on Newton's work in Principia, as well as the accomplishments and groundbreaking proof of many other unification scientists and physicists, while making appropriate modifications and refuting some. With it, we presented a mathematical justification to this revolutionary and all-encompassing unified theory of the universe while closely adhering to the principle of consistency. The Cosmological Balance Principles stands as the representation of a quite simple concept that explains and presents a solution to the phenomena that baffles scientists today. Again, for ease of reference, the remainder of this chapter consolidates the highlights of the Cosmological Balance Principles in a near mathematical format.

9.1 Principal Types of Matter

The Cosmological Balance Theory brings forward all the mathematical principles listed within Sir Isaac Newton's Principia, along with the works from many other prominent scientists and physicists, and then adds to them the clarifications and modifications contained within the following:

Definition 1, Normal Gravity Matter: All matter in the visible universe are composed of an arrangement of molecules and atoms made up of particles, protons, neutrons, and electrons, which are further assembled from subatomic particles of leptons, quarks, and force carriers, and are quantum mechanically interconnected or entangled, an instantaneous property of infinite distance. Matter appears to us in three primary forms: gaseous, liquid, or solid. Secondary forms of matter include plasma, and super fluids, etc. There are three dimensions: length, width, and depth. Time is somewhat of a fourth dimension, which only moves in one direction in our realm, forward, and at the speed regulated by the vibrations of subatomic particles, the speed of light. The actual particles of an atom occupy just a microscopic portion of the atom's volume. About 99.99999% of the atom's volume is just empty space[265]. These atoms and molecules possess four natural forces, the strong nuclear force, the electromagnetic force, the weak nuclear force, and the gravity force. Based on shared common origin, quantum theory mathematically represents the interaction of all these particles and sub-particles at the subatomic level, like a telekinesis link. Most gravity matter is configured to be right-handed at the molecular level, except living or organic matter, which is composed of lefty molecules that are uniquely influenced at the subatomic level by the antigravity universe and capable of resisting gravity to live.

Definition 2, Energy: In physics, energy is the ability to move something, or do work, or make a change to the matter. The amount of energy something has refers to its capacity to cause change or something to happen. Energy has a few defined properties. It is always conserved; it cannot be created or destroyed. It can be transferred, however, between objects or group of objects by interaction of forces. Energy comes in many different forms such as heat, light, chemical energy, and electrical energy. It can be kinetic or potential energy, the ability to bring about change or to do work. Kinetic energy is that of motion. An object moving has kinetic energy as it rolls on the ground or moves through space. An object in motion tends to stay in motion. Potential energy is the energy contained in an object not in motion.

<u>Laws of Thermodynamics:</u>

First Law of Thermodynamics: Energy can be changed from one form or type into another. Energy cannot be created or destroyed. The total energy and matter in the universe remain constant, merely changing from one form into another. The First Law of Thermodynamics is also known as the Law of Conservation.

The Second Law of Thermodynamics: The second law states, "In all energy exchanges, if no energy enters or leaves the system, the potential energy of the state will

---

[265] If normal matter has 99.9999% empty space within it, then an exotic atom with neutral graviton would have wider gaps between atoms and truly appear as open space with significantly more void and emptiness.

always be less than that of the initial state." This is commonly known as entropy. Entropy is a measure of its disorder, the natural inanimate environment state of balance: living cells are structured and so have low entropy, it contains structure not typically found in nature, inanimate objects in the environment.

Matter, energy, momentum, and time are all interconnected. Energy is the motion of matter, and matter cannot exist without energy. Without the steady ticking vibrations of time embedded within all quarks shielded by a super strong subatomic force and the energy they give us; the universe would not exist. In other words, if all quarks or the smallest subatomic particles in the universe suddenly stop vibrating or jiggling, they would cease to exist and with their disappearance all atoms and matter would cease to exist, time halts or fails to flow at the speed of light, no energy produced, and the universe as we know it would completely vanish. The universe cannot exist without space, matter, and energy. So, what good would time be in such a universe? Fortunate for the universe we live in, natural laws prevent this demise from ever happening.

Definition 3, Gravity: The weakest of the four natural forces, gravity, governs the macro world of the universe. It pulls all matter on earth to the planet, holds the moon in orbit around the earth, and holds the planet earth in orbit around the sun. For that matter, it holds the entire solar system together, and keeps the sun and the millions of stars orbiting the galaxy. Its attraction is both instantaneous and infinite in range. Normally, star systems are gravitationally bound to other nearby star systems in what is known as the local star group, which travel together. Even galaxies are bound gravitationally by a very weak pull to other nearby galaxies in the local group. Newton represents its attraction or pull force as $F = G\, m_1 m_2 / r^2$, where m_1 and m_2 are the mass of two objects, G is the gravitational constant roughly estimated at $6.6734E\text{-}11$ N m^2/kg^2, and r is the distance between the center of both objects. Newton suspected the existence of an unknown source, which eluded him, that provided such a force.

Corollary 1, Quantum Mechanical Graviton: The quantum structure within quarks consists of four different point particles: neutral (neutrino and antineutrino), positive (non-paired positron), and negative (non-paired electron). These point particles make up all micro components of subatomic matter particles (point particles) and are as miniscule if not smaller than electrons orbiting atoms, therefore as elusive and undetectable. The collisions of a positive and negative point particle create a neutral (neutrino) and anti-neutral (antineutrino) point particles. However, the collision of a charged point particle and a neutral point particle results in formation of a point particle pair capable of tapping into the strong nuclear force. Neutral point particles are the messenger particle of and the source of the strong nuclear force; when paired with a charge point particle, neutral particles have the ability to absorb 2/3 fraction of the charged particle's energy in order for the pair to survive; this results in partial charges of 1/3. The neutral point particles paired with charged particles control when positive and negative repel apart by aligning themselves between the two charged particles (- n n +) applying the strong force and turn off the repulsive force by placing themselves on opposite ends to halt the motion and allow the negative and positive point particles to attract (n - + n). As a result, the neutral particle slingshots paired point particles outward with a rapid acceleration for a miniscule distance then allows the charged particles' attraction to overwhelm the outward motion and pull inward with slower "freefall" descent. At the point particles' closest approach to the center, the neutral point particles realign, and its energetic strength then overwhelms the charged attraction and again slingshots the positive particle pair outward. Each outward hyperbolic swing creates the jiggling quality of the quark surface and gravity.

Quantum physics has shown us that the more energetic point particles move the more mass they achieve, per Einstein's energy-mass equation, $E=mc^2$, and that acceleration equals gravity in his equivalence principle. Rapid acceleration outward provided by the strong force in the neutral point particles results in generated gravity, and the slow return inward motion provided by the electromagnetic force results in zero gravity, or the total combined "graviton" effect at 6.6734E-11 N $(kg/m)^2$. The total energy of the point particles within each quark determines its mass, while the generated rapid outward acceleration and slow inward motion causes its gravitational field, the "graviton." Specifically, the three natural forces working together create the fourth natural force, gravity. The "graviton" is not a single spin-2 massless particle, but the result of point particle motions within quarks, and among quarks in nucleons, and among subatomic particles within the nucleus. These three layers of "gravitational field" enhance their internal layer's graviton range. The more atoms are in an object the greater its mass and the greater its gravitational field. No strings attached in this concept. Below are the illustrated hyperbolic paths of point particle pairs within an up quark:

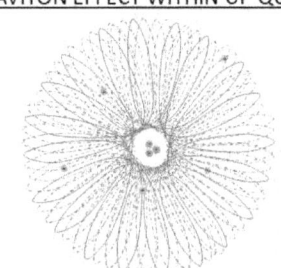

GRAVITON EFFECT WITHIN UP QUARK

The "graviton" within every atom in the visible universe results from neutral point particles, the messenger particle's ability to turn on and off the strong nuclear repulsion force to avoid collisions of positive and negative point particles. These types of motion also caused by gluons within nucleons to accelerate quarks rapidly outward away from and then cushion

Point particle's outward acceleration of 2.9222E-38 seconds exceeds cushioned inward motion of 2.9222E-37 seconds creates gravity field, the "Graviton" of 6.6734E-11 N $(m/kg)^2$

descent to the center quark; thereby increases the gravitational field and enhances the "graviton's" reach. We observed this motion as jiggling of quarks, which explains why larger atoms have both more mass and gravitational pull, and why electron levels around atoms are fuzzy layers in quantum flux. The "graviton" action envelopes the entire quark and generates a very tiny amount of gravitational field in every direction amounting to a miniscule quantum contribution to the overall 6.6734E-11 N $(kg/m)^2$ constant. This "graviton" effect combined with those of all the quarks and nucleons within every atom in the earth creates the earth's total gravitational field, one of the mechanical sources for Newton's gravity based on Einstein's insight in the equivalence principle.

Using the distance of Planck length (1.6E-35) divided by 100 of 1.6E-37, and the equation for accelerated travel time equals square root of $(2r/(A - A/r^2))$ where r is distance and A is acceleration, the outward motion of accelerating point particles at light speed is

= SQRT(2*1.0000001m/(299792458 m/s² – 299792459 m/s² / 1.0000001m²) * 1.6E-37

= 2.99222E-38 s

And the inward motion time is

= SQRT(2*1.0000001m/(2997924.58 m/s² − 2997924.59 m/s² / 1.00001m²)*1.6E-37

= 2.99222E-37 s

Since the strong nuclear force is one hundred times greater than the electromagnetic force, we obtain a ratio of outward acceleration at 99 kg/s² to inward at -1 kg/s². Using the outward and inward time above, we estimate the "graviton" effect within a quark

= [6.6734E+27 N (kg/m)² * 2.9369E-37 s * 99 kg/s² /1E+38

+ 6.6734E+25 N (kg/m)² * 2.9222E-36 s * (-1 kg/s²) /1E+36]

/ (99 kg/s² * 2.9369E-37 s − 1 kg/s² * 2.9222E-36 s)

= 6.6734E-11 N (kg/m)²

Again, the "graviton effect" occurs within every quark in every atom on earth and in the visible universe, validated within the fusion process of a star, like the sun. The most significant of the fusion process is when one hydrogen 1 atom (single proton) fuses with another. During the fusion, three positive/neutral point particle pairs are broken away from one of the up quarks colliding. This converts that up quark into a down quark making its core more loosely bound and energetic. The released point particles reassemble into one positron (a group of three positive and two neutral point particles) and one neutrino (neutral point particle).

In gravity matter, the internal core of the quark consists of antineutrino and negative point particle pairs making it an "antigravity" core to counterbalance the "gravity" field developed by the neutrino/positive pair generating the "graviton" motion effect. Therefore, reversing of neutrino and antineutrino pairing with charged particles may be one way to create antigravity accelerated anti-graviton effect for the invisible part of the universe, while equal speeds of inward and outward motion results in neutral zero gravity matter or space-matter. This is an illustration of this fusion step:

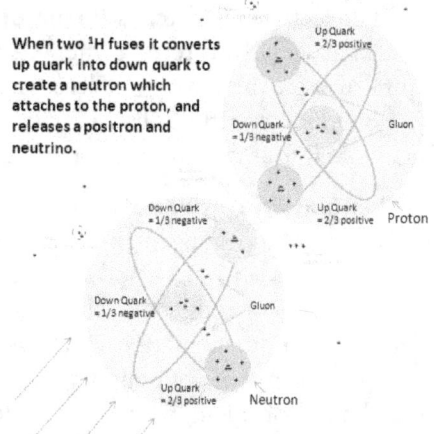

When two ¹H fuses it converts up quark into down quark to create a neutron which attaches to the proton, and releases a positron and neutrino.

Definition 4, Three Types of Matter: There exist three types of matter in the Cosmological Balance[266]. Normal matter as we know it with gravity, it exists in everything we see in the visible universe and is gravitationally attractive in nature. Antigravity antimatter, it exists in the portion of the universe we claim is invisible, primarily undetectable by our eyes and our instruments, and is anti-gravitationally attractive with each other by nature, but repulsive toward gravity matter. Likewise, gravity matter is repulsive when in contact or in proximity with antigravity antimatter. The third type of matter is space-matter. It existed long before any gravity matter or antigravity antimatter came into existence; it is their source. Space-matter's motion generates energy and

---

[266] The foundation of the Cosmological Balance Theory is built upon the existence of the three types of matter: space-matter, ordinary matter with gravity, and antigravity antimatter. This definition is derived from observations and data collected of the universe.

pressure, and the vacuum force of 6.6734E-11 Newton, which we associate with as the gravity and antigravity force. This vacuum force combined with the natural quantum entanglement of particles enables gravity in the visible universe to be instantaneous and possess infinite range, and simultaneously in the antigravity invisible part the universe it too is instantaneous and possess infinite range toward like particles. Quantum mechanical entanglement is repulsive between gravity matter and antigravity antimatter, with an inherent desire to move apart at the speed of light in opposite directions, or the speed of light squared for the micro orb ejected from a super massive object. The Cosmological Balance Theory takes Newton's gravity constant and redefines it as the combination of push force from perfect liquid-gas or ocean of space at 6.6734E-11 Newton and the quantum entangled attractive force of each ordinary matter object with common origin at 1 m/kg, each antigravity antimatter object at -1 m/kg, and space-matter at 0 m/kg. This makes Newton's gravity constant on earth read as

Gravity constant = 6.6734E-11 Newton * 1 m/kg * 1 m/kg,

and for antigravity universe it reads

Antigravity constant = 6.6734E-11 Newton * -1 m/kg * -1 m/kg

and for space-matter reads

Non attraction = 6.6734E-11 Newton * 0 m/kg * 0 m/kg.

The electroweak force is the unification of electromagnetic force and the weak nuclear force. The gravity force, or the "graviton" identified above, is the unification of three natural forces: the strong nuclear force, the electromagnetic force, and the weak nuclear force working together. The gravity-antigravity repulsion force is the "grand unification force" of all four natural forces combined.

Postulate 1, Antigravity Antimatter: All matter in the invisible part of the universe is composed of an arrangement of antigravity molecules and atoms made up of antigravity particles, amassed as anti-protons, anti-neutrons, and anti-electrons, which are further assembled from antigravity subatomic particles of leptons, quarks, and force carriers. Antigravity antimatter appears to us in three primary forms: gaseous, liquid, or solid. Secondary forms of antigravity antimatter include plasma, and super fluids, etc. There are three dimensions: length, width, and depth. Time is considered to be somewhat of a fourth dimension, which only moves in one direction, forward in time (with backward motions from our view), and coincidentally also at the speed of light, which is regulated by the vibrations of antigravity subatomic particles. The actual particles of an atom occupy just a microscopic portion of the atom's volume. About 99.99999% of the atom's volume is just empty space. These antigravity atoms and molecules possess four natural forces, the strong nuclear force, the electromagnetic force, the weak nuclear force, and the antigravity force. Quantum theory mathematically represents the interaction of all these particles and sub-particles of common origin at the subatomic level. Most antigravity antimatter is configured to be left-handed at the molecular level and when compared to their counterpart right-handed gravity matter is predicted to be identical in chemical properties and characteristics.

The norm in the antigravity universe is lefty molecules, except for their living or organic antigravity antimatter, which is composed of right-handed molecules that are uniquely influenced at the subatomic level by the gravity universe, giving it the vibrations and capability of resisting antigravity to live. Below is the gravity antigravity relationship between both parts of the universe. Antigravity galaxies and other antigravity objects and particles in space cause chance left-handed molecules within the gravity universe to vibrate and resist gravity and enable them to consolidate and have the qualities of life. At the same time, gravity galaxies and other ordinary objects and particles in space cause chance right-handed molecules within the antigravity universe to vibrate resisting antigravity and enabling them to consolidate and have the qualities of life there.

Antigravity antimatter motion seems to flow backwards, in matching ticks, or clocking, and appear in reverse motion to ordinary matter. Motion in an antigravity system operates effectively as a mirrored image to the way ordinary gravity matter functions. To simplify terminology and concepts, when we refer to time in an antigravity system, we are essentially running laws of physics backwards in direction. As a side effect, the antigravity system is invisible, undetectable by any conceivable instrument or technology, completely out-of-sync to matter and time, and repulsive to ordinary gravity matter, as we know it. Specifically, existing detection devices look for electrons to read information, but anti-light operates with inverted EM waves, where positrons are the carrier of information. Since antigravity antimatter is an opposite of that of permanent gravity matter, then space-matter gravity-antigravity pulses drives the antigravity galactic motion backwards. An antigravity system is a mirrored image to that of a typical gravity system. The arrows represent the direction of motion of the stars depicted:

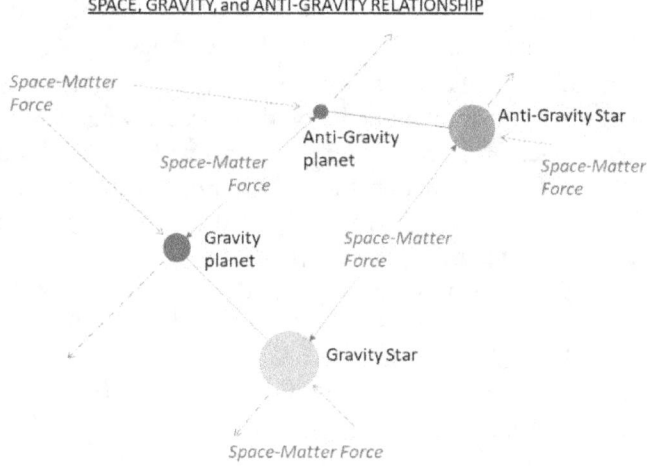

Conjecture 1, Infinite Space and Time: Space and time existed before the creation of matter with gravity or antigravity. Space has infinite volume, and deep open space maintains an average temperature of 2.7 Kelvin, due to the presence of space-matter, which generates vacuum energy and minute pressure, and the vacuum force. Space and its space-matter timing also ticks at the speed of light, forward and backwards, which is regulated by the vibrations of neutral space-matter subatomic particles and is the pace setter controller for antigravity and gravity matter jiggling speed. Thereby, forward, and backward time is universal for all space-matter and space is infinite in distance and age.

9.2 Principle of Black and White Holes

Lemma 1, Remnants of Dead Stars: A neutron star is the second to the densest orb in the galaxy, next to a black hole. A neutron star too is the remnant of a star that went super nova, typically of small radius (about 30 km) and very high density, composed predominantly of closely packed neutrons.

If we look at all the remnant orbs of dead stars, to include these neutron stars, we find that all have extreme pressures and temperatures, so why not the black hole which is more massive and denser than a neutron star and possess just enough gravity to prevent its light and any electromagnetic radiation from escaping. Black holes too are remnants of a super massive star that went supernova. However, their resulting gravity pressure breaks down even its neutrons into a "soup" of free-flowing subatomic particles[267]. In addition, matter within the Schwarzschild radius contains a repulsive nature that causes these subatomic particles within that radius to be homogenous and free flowing. The Schwarzschild radius for the black hole is just within the event horizon, and obviously encompasses the entire black hole.

Corollary 2, Black Hole Pressure and Temperature: Assume a black hole exists, as scientists have conjectured, not the cold desolate singularity some visualize, but an orb maintaining a spherical shape. How can this be? Friction generated by quantum graviton quarks fighting against gravity, with a force stronger than that of neutrons within a neutron star, elevates pressure and temperatures within a black hole and gives it its spherical shape. Let us look at estimating these pressures and temperatures. Now imagine a thin cylinder with area $A = \pi r^2$ with length R_{BH} from surface of the black hole to its center, see image below. The volume of that cylinder is $V = \pi r^2 \times R_{BH}$. Assuming the black hole has a constant homogenous density ϱ and Newton's law of gravity $F = G \times m_q \times M_{BH} / R_{BH}^2$. Where gravity constant $G = 6.6734 \times 10^{-11}$ N (m/kg)^2, the mass of the black hole $M_{BH} = 9.3474 \times 10^{36}$ kg, and the mass of a quark $m_q = 3.0$ to 5.5×10^{-30} kg (averaged at 4.25×10^{-30}). Then the density of black hole is $\varrho = 1.61 \times 10^{26}$ kg/m^3, the radius of the black hole $R_{BH} = 1.39 \times 10^{10}$ km and set $r = 1.1$m then the area $A = \pi r^2 = 3.801327111$ m^2.

Pressure p(r) at the center of the black hole can also be roughly estimated using:

$p(r) = \varrho \times G \times M_{BH} / R_{BH}$

$= 1.61 \times 10^{26}$ kg/m^3 $\times 6.673 \times 10^{-11}$ N (m/kg)^2 $\times 9.3474 \times 10^{36}$ kg $/ 1.39 \times 10^{13}$ m

$= 7.2246\text{E}+42$ N/m^2

This pressure generates extreme temperatures within the black hole. Using Boltzmann constant $k=1.4 \times 10^{-23}$ J/K, the temperature within the black hole can be approximated with the formula

$T = G * m_q * M_{BH} / (k * R_{BH})$

And m_q is the mass of an average quark (4.25×10^{-30} kg).

---

[267] The latest merger of two black holes into one demonstrates that they combine easily and confirms that they are more like a "soup" of fluid particles rather than a hard cold dark singularity. Sufficiently hard solid objects would have bounced off each other, like billiard balls.

CHAPTER 9: COSMOLOGICAL BALANCE PRINCIPLES

= 6.6734x10^{-11} N (m/kg)^2 x (4.25 x 10^{-30} kg) x (9.3474 × 10^{36} kg) / (1.4x10^{-23} J/K x 1.39x10^{10} km)

= 13,833,023,682.89 K

= 1.38 x 10^{10} Kelvin degrees

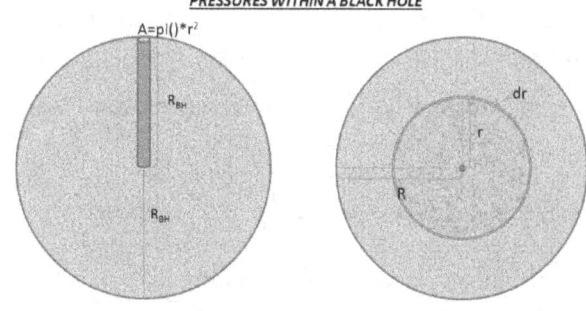

Clearly, these extreme pressures and exuberantly high temperatures do not conform to the predicted hard cold desolate objects that some mainstream cosmologists believe of black holes. Black holes are most likely a "soup" of particles constantly colliding with each other, emitting, and absorbing tremendous energy, causing elevated changes in subatomic particle energy intensity levels, forms, and properties. In short, the super strong quark's outward force against gravity enabling homogeneity within a black hole is augmented by the energy released from antimatter matter annihilations and a steady stream of micro-antigravity orb production and ejection.

Definition 5, Black Hole Existence: There exists at least one massive, or a super-massive, black hole at the center of most well-developed normal gravity matter galaxies. Some galaxies contain numerous other smaller black holes also scatter throughout the galaxy's body and arms. Similarly, there exists at least one massive, or super-massive, white hole at the center of most well-developed antigravity antimatter galaxies, some of which contain several smaller white holes. Both massive black holes and white holes are capable of breaking down their matter into its basic subatomic particles, and in doing so, produce anti-matter particles, which are annihilated instantly on contact with their counterparts, and in turn release extremely high energy levels that elevate other adjacent existing subatomic particles to their highest energy potential level. Once these subatomic particles reach their highest energy value, they are susceptible to become the opposite antigravity or gravity particle: black holes produce antigravity particles and white holes produce gravity particles, most of which float upward, eventually released and ejected outward, and take with its microscopic amounts of mass and energy away with it.

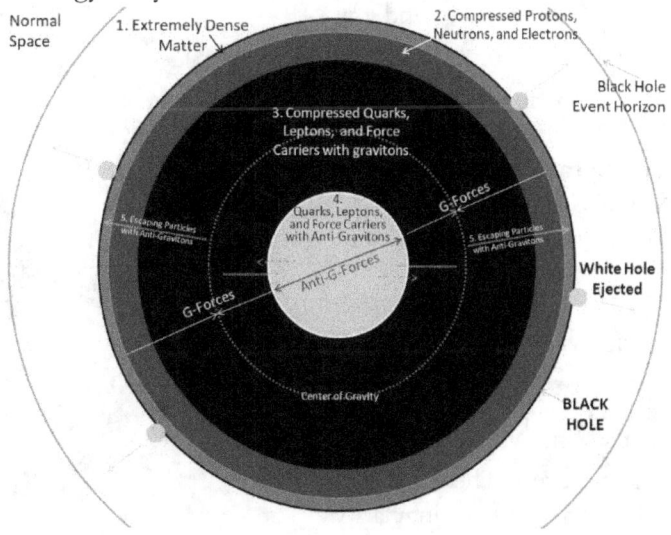

Definition 6, Black Hole Excretion: Massive black holes at the center of galaxies excrete minute amount of mass and energy by creating microscopic white hole orbs which float to the surface and are released at the speed of light squared. These white hole orbs are transported through multiple wormhole tunnels from the center of the galaxy to beyond its edge. Since about 99.99% of the galaxy is just empty space, these orbs easily weave their way around stars on their way outward. Their release pattern and rotation of the black hole are what determines the shape of the gravity galaxy, see image below. These white hole orbs replenish space-matter and are also a source of antigravity antimatter building blocks.

Axiom 1, Black Hole Excretion Rate and Path: The sun is estimated to emit 1 x 10^{45} photons per second and the black hole in the center of the galaxy is 4.7 x 10^6 times the solar mass. Using the Schwarzschild radius formula $R=(2GM)/(c^2)$, the black hole is predicted to emit white hole orbs consisting of between two billion to four billion sub-atomic particles, at a rate of 1.175E+42 to 2.35E+42 orbs per second in all directions. The mass of each set of white hole orbs is estimated between 7.37E-19 kg to 1.47E-18 kg and has a radius between 1.09449E-45 mm and 4.36881E-42 mm.

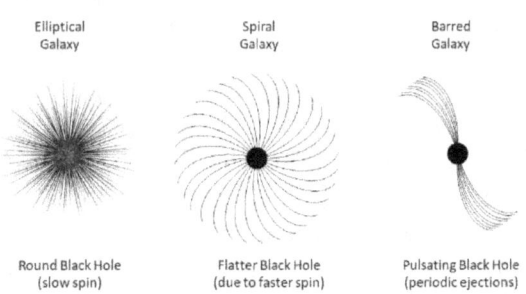

The white hole orbs ejected from the black hole surface will instantaneously reach escape speed of the speed of light squared, accelerated by the energy = $pc^2 \sqrt{2}$. These orbs **enter into wormhole tunnels, with** the energy in the $\sqrt{2}$ or square root of two, and move outward away from the center of the galaxy primarily traveling in straight lines, unless they encounter stars that cause their path to temporarily curve away and around it. The angle of departure of these orbs from the black hole will be minimally affected toward rotation of the black hole's rotational spin if it was rotating near or at the speed of light; otherwise, its angle is exactly 90 degrees perpendicularly to the surface of the black hole and directly outward. The stronger the gravity of the object in its path, the sharper the convex curve away the orb will take. The massive stars orbiting in close proximity to the super massive black hole at the center of the galaxy acts like shutters on a ship's Morse code light device, letting groups of white hole orbs pass through in segments of time toward the sun and its nearby stars, like a lighthouse beacon matching the orbiting star's cycle. Each massive group of wormholes carries with it enough antigravity "galactic wind" to nudge the sun due to its size and mass, with minimal effect on disturbing the orbit of the planets and the Oort cloud.

Definition 7, White Hole Pressures and Temperatures: The formulas used in Corollary 2 to determine pressures and temperatures also apply to massive white holes or "antigravity black holes." White Holes too have extreme pressures and exuberantly high temperatures. They are not hard cold desolate singularity

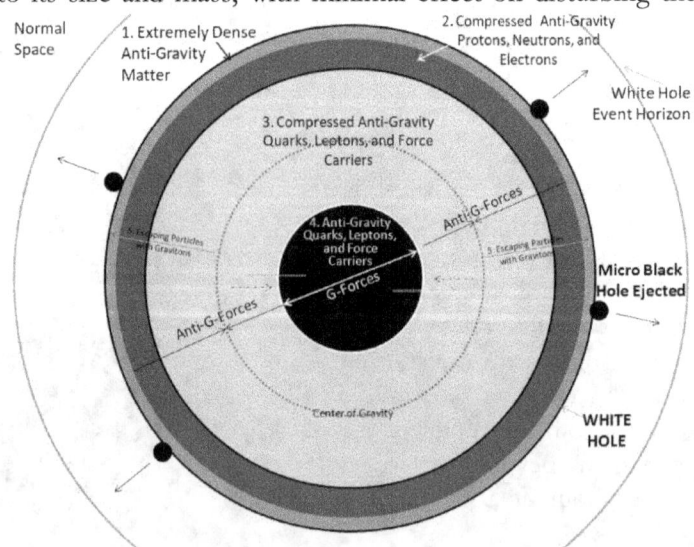

objects that some mainstream cosmologists believed of black holes or of white holes. White holes, when viewed from within the antigravity perspective and flow of time, are very much like black holes so they too are likely to be a "soup" of particles constantly colliding with each other, emitting, and absorbing tremendous energy, causing elevated changes in subatomic particle energy intensity levels, forms, and properties. Consequently, from our view of time, in forward ticking clock, the white holes eject matter with gravity, exactly as defined.

Definition 8, White Hole Excretion: Massive white holes at the center of antigravity galaxies excrete microscopic amounts of mass and energy through multiple wormhole tunnels transporting microscopic black hole orbs from the center of the galaxy to beyond its edge. Similarly, about 99.99% of the antigravity galaxy is predicted to be just empty space, thereby allowing these micro-orbs to easily weave around stars on their way outward. Their anti-time release pattern and the rotation of the white hole are what determines the shape of the antigravity galaxy; see image below[268]. These miniscule black hole orbs replenish space-matter and are also a source of gravity matter building blocks.

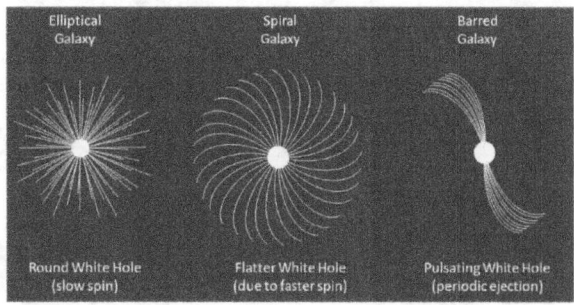

---

[268] The release pattern of wormholes leaving the galaxy creates the necessary "galactic wind" to move the stars orbiting the galactic system to conform into a matching pattern. Barred release pattern creates a barred galaxy. Spiral release creates a spiral galaxy. Elliptical release pattern defines the elliptical galaxy. Transitions between patterns cause different variations of these three types of galaxies.

9.3 Principle of Light Bending & Energy Absorption

Lemma 2, Light Bending on Earth: Refraction occurs as light transitions from one medium into another medium of a different density. People living on earth at least once if not more times in their lives witness the bending of light, or refraction. In addition to refraction, the photon energy of light is also absorbed by atom it encounters in its path, making the reemitted light redder. Regardless of whether we view it or not, it occurs with every sunrise and every sunset. Yes, whether you know it or not, as we see the top edge of the sun appearing at the horizon, we are looking at the sun, which is physically two-minutes below the horizon. At sunrise, we see it two minutes before it is at the horizon. In addition, at sunset, we view it an additional two minutes longer as it goes below the horizon[269]. The sunlight during the day, and the moon and every star's light at night we see above our head is refracted by the earth's atmosphere as it moves upward in the sky. The point of zero refraction is directly overhead; see image to the right.

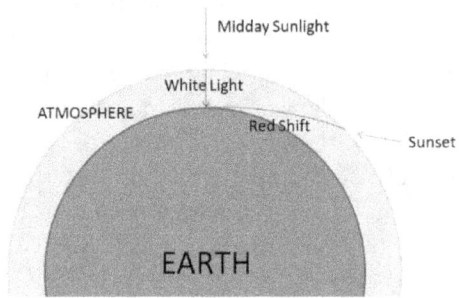

This index of refraction is defined as

$n = c/v$,

where c is the speed of light in a vacuum and v is the speed of light in the medium.

The equation showing changes in wavelengths are

$\eta = \lambda/\lambda_m$,

where λ is the wavelength of the electromagnetic radiation in a vacuum and λ_m is the wavelength same EM radiation in the medium.

Despite light's changes in speed and wavelength, the frequency of the light or EM radiation remains the same.

The relationship between speed, frequency and wavelength is

$v = f * \lambda$,

where v is the speed, f is the frequency, and λ is the wavelength.

Mirages are also caused by refraction of light by difference in air temperature, and thereby difference in density. People have witnessed mirages at least once in their lives.

Lemma 3, Light Bending Around Sun: The bending of light around the sun is caused by effects of gravity combined with light refraction or more specifically optical lensing from densely packed space-matter gases at the outer edge of the corona, the curvature of space per Einstein, without the curvature of time or time dilation.

---

[269] Atmospheric red shift seen during sunrise and sunset make for the most memorable occasions and breathtakingly beautiful pictures.

Scientists know that the Sun possesses an atmosphere with an outer edge they call the corona and from earth it takes a solar eclipse to block out the sun to see the outer edges of the corona. What is still unknown is exactly how much distance out from the sun is its atmosphere above the corona and how much of it is clear and colorless? Surely, the sun, which contains about 99.87% of the solar system's total mass, has an extensive and much more complex multilayered atmosphere than the one earth. This extra thick, partially clear space-matter layer both reflects and super heats solar atmosphere and is responsible for the unique bending of light from distant stars behind the sun to appear to be over the sun. The starlight's refraction angle is dependent on the position of the star behind the sun and the thickness of the solar atmosphere during its light's approach above the sun. Space-matter light refraction is equivalent to "gravity lensing" results as determined by Einstein's Special Relativity equations. The motion of space-matter toward the sun effectively grabs the starlight passing through it, briefly absorbs it, and reemits it at the angle equivalent to Einstein's calculated "gravity lensing" equation without changing its frequency or color. Einstein refers to this additional curvature ability as "gravitational energy" we contribute it to the optical density of space-matter light refraction. Both are identical in results.

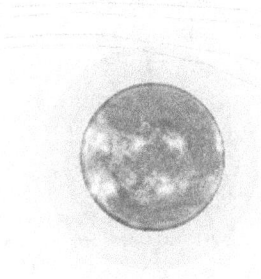

LIGHT REFRACTION ANGLE AFFECTED BY DENSITY

Now let us compare observations of stars behind the sun during solar eclipses and calculations for bending with gravity alone and for bending with gravity and refraction. We get the following table:

Altitude	Bend by Sun's gravity	Gravity and Refraction
140 km	2.26E-06 degrees	2.4927 degrees
150 km	2.26E-06 degrees	2.2935 degrees
175 km	2.26E-06 degrees	1.9946 degrees
200 km	2.26E-06 degrees	1.7454 degrees
300 km	2.26E-06 degrees	1.0973 degrees
400 km	2.26E-06 degrees	0.7981 degrees
500 km	2.26E-06 degrees	0.6983 degrees
2100 km	2.26E-06 degrees	0.1995 degrees
100000 km	2.26E-06 degrees	2.26E-06 degrees

Light bending with gravity and space-matter light refraction matches the observation readings; see image below. Refraction and gravity bending results confirm the need for "curvature of space" explanation, without the curvature of time or dilation thereof. Densely packed space-matter surrounding the massive object is the curvature of space[270].

---

[270] Since you cannot curve nothingness or emptiness, the Cosmological Balance Theory simply inserts a unique type of matter (specifically exotic space-matter) to define accurately Einstein's curvature of space.

Postulate 2, Atmospheric Red Shift: All molecules in the atmosphere absorb a portion of the photon's energy as light is passed from molecule to molecule. This gradual loss of energy is what causes the light to become redder as it encounters great amounts of gaseous atoms and molecules. Three earth-atmospheres thickness at the horizon level causes the sun to appear redder as it goes down during sunset and as it comes up during sunrise. Practically everyone on living on earth has witnessed this phenomenon at one point in his or her life.

Proposition 1, Cosmic Red-Shift: The Red Shift amount detected and observed of distant galaxies beyond 10 billion light years is correlated directly to the distance and is thereby caused by the photon-energy absorption of gaseous matter in the path line of the light emanating from the distant light source to our instruments. Space-matter itself is primarily responsible for a great portion of the photon energy absorption of light, in addition to the presence of traces of normal matter and that of antigravity antimatter. While matter with gravity, resulting from colliding and on occasion decaying of space-matter, accumulates in low-pressure areas, absorb light photon energy from distant galaxies and causes the red shifts in their light. In addition, matter with antigravity, similarly created by excessive colliding and decaying of space-matter absorbs antigravity light photon energy from distant antigravity galaxies and causes red shifts in their light; see image below.

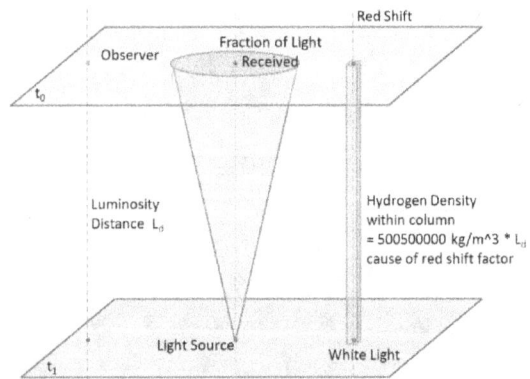

The bulk of radiated energy is absorbed by space-matter is for the maintenance of the structure of space itself and is not available as heat. In other words, light photon energy emanating from stars with gravity or antigravity becomes the energy that sustains space-matter and maintains space as it is, a vacuum. This space-matter also has the ability to absorb photon energy directly from the light passing through it and slow down the light to a crawl and in some instances without changing its color, and then reemit the light to its original speed and original direction. Same effect applies whether it is light or anti-light. Larger concentrations of space-matter, however, do alter the direction of the light passing through, known as refractive properties due to its optical density.[271]

---

[271] Distances beyond 13.8 billion light years contains sufficient densities of all three types of matter blocks out all starlight and all electromagnetic waves from reaching us. Such distances therefore appear pitch black.

CHAPTER 9: COSMOLOGICAL BALANCE PRINCIPLES

9.4 Principle of Gravity and Antigravity

Lemma 4, Quantum Gravity Equation: Newton's gravity equation $F=Gm_1m_2/r^2$ is modified into a quantum gravity formula that reads

Quantum Gravity Force = V_f x (1 m/kg x m_1) x (1 m/kg x m_2) / r^2

where the vacuum force V_f which is generated by space-matter equals 6.6734E-11 N, and the 1 m/kg is the quantum mechanical graviton and its associated entanglement with the universe. The image below depicts the complicated interactions between vacuum force push on cones and quantum mechanically entanglement on objects. This transfer of push force and quantum attraction for normal gravity matter simplifies to Sir Isaac Newton's gravity equation we all know and accept.

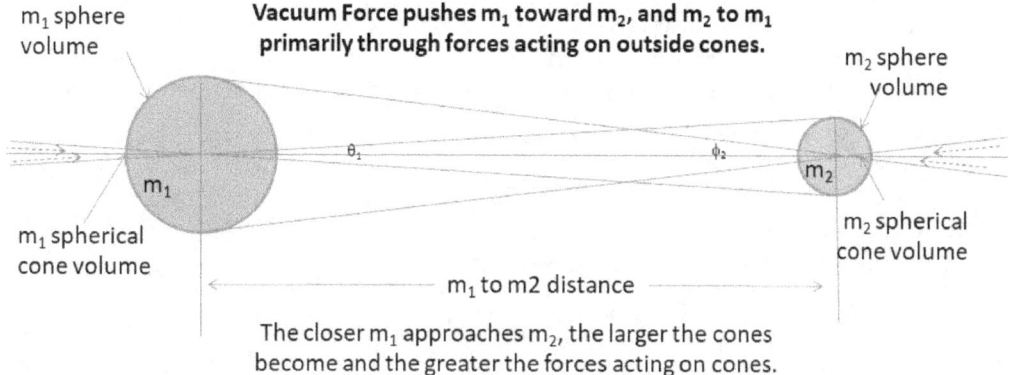

The closer m_1 approaches m_2, the larger the cones become and the greater the forces acting on cones.

Definition 9, Antigravity Equation: The relationship between antigravity antimatter with its fourth natural energy force on itself represented by the formula:

$F = G * -m_1 * -m_2 / r^2$,

results in antigravity mutual attraction, or more specifically written:

$F = F_v * -1m/kg * m_1 * -1m/kg * m_2 / r^2$,

where vacuum force F_v is 6.6734E-11 Newton. In addition, the relationship between antigravity objects and gravity objects is represented by the formula:

$F = G * m_1 * -m_2 / r^2$

or

$F = G * -m_1 * m_2 / r^2$,

results in mutual repulsion, or more specifically written:

$F = F_v * -1m/kg * m_1 * 1m/kg * m_2 / r^2$,

where vacuum force F_v is 6.6734E-11 Newton. The energy force on matter with gravity and/or matter with antigravity is either mutual attraction or mutual repulsion.

Lemma 5, Cosmological Energy Mass Equation: Einstein's energy mass equation $E=mc^2$ is modified to read

$E = \pm \sqrt{(p^2v^4 + m^2c^4)}$,

where E is the energy, p are the particles within the mass m and v is its velocity (vibration), m is the mass, and c is the speed of light. Energy is defined as the movement of matter.

ENERGY, MASS, MOMENTUM RELATIONSHIP

Energy = E (Force * Distance = Joule)
Particle* v^2 = pv^2 (Vibration=Joule)
Mass * c^2 = mc^2 (Joule)

Axiom 2, Gravitational Escape Speed: Escape speed equation for most objects to break the gravitational attraction of mass M with radius r is

$$V_e = \sqrt{(2GM/r)},$$

where G is the gravitational constant.

Axiom 3, Black Hole Escape Speed: Massive black holes eject subatomic antigravity orbs and massive white holes eject subatomic gravity orbs. Normal matter with mass m repels antigravity antimatter with mass −m from the surface of the black hole, of which all its matter was broken down into sub-atomic particles p, ie. m equals p, or m = p, and vice versa. The energy released is:

$$E = c^2 \sqrt{[(p^2 + p^2)]}$$
$$= c^2 \sqrt{(2p^2)}$$
$$= \pm pc^2 \sqrt{2},$$

where the particles p, in this case grouped into a white hole orb of antigravity antimatter is ejected at light speed squared, c^2, travels away from the black hole's surface in a wormhole tunnel[272]. Mathematically, we represented this as $E = \pm pc^2 \sqrt{2}$, where the square root of two ($\sqrt{2}$) represents the energy creating the wormhole tunnel and the ± represents repulsion from either the surface of a black hole or a white hole.

Lemma 6, Gravity Layered Intensity: Gravity field intensity is inversely proportional to the square of the distance and best represented in concentric spherical layers or bubble layers of intensity surrounding the massive object or system. A planet visualized orbiting the sun on such a bubble layer follows prograde rotation and adherence to the law of conservation of angular momentum. A planet not following the natural prograde rotation would experience increased temperatures and gradual slowing of rotation speed as nature tries to correct its direction.

This invisible solar torque action on a planet is calculated using the solar gravity acceleration at planet's orbital radius, which is equal to G times the mass of the sun divided by orbital radius squared. For the earth, we have 6.6734E-11 N $(m/kg)^2$ x 1.989E+30 kg / $(1.495979E+11\ m)^2$ = 5.931044E-3 m/s^2. The Earth's average rotational speed is about 460 meters per second and with the solar acceleration of 5.931044E-3 m/s^2 alone it can gradually increase speed and cover the distance of 460 meters in 393.84867 seconds. However, with every degree the earth turns the point of torque moves with it like a ratchet wrench resetting for the next turn. The rotation and increased acceleration combined

---

[272] Escape speed from any black hole is derived from power of the grand unification force where one object has a negative graviton and the other a positive graviton are repelled away from each other at the speed of light. Since the micro orb is infinitesimally smaller than the massive black or white hole, it receives the product of both speeds, c^2.

sustains the earth's orbital speed. How can we more accurately show the effect? Let us look at this acceleration and the orbital speed together. Earth's average velocity around the sun equals the circumference divided by the number of seconds in a year or 2*π*1.495979E+11 m / 31,557,600 seconds = 2.978525E+4 m/s. Earth's slight increase of gravitational field at the axis of direction tangent to the orbital path is 1.000604059 G's derived from earth's gravity acceleration of 9.818649 m/s² plus 5.931044E-3 m/s² divided by 9.818649 m/s², while the trail end, opposite surface of the planet, gravity is 0.999396306 G's. This difference may not seem like much, but when multiplied by the gravity acceleration between the sun and the earth we can visualize the power of that torque force. Here, the lead edge has the additional gravitational force of

F_T = [(1.000604059 - 1)*6.6734E-11 N (m/kg)² * 1.989E+30 kg* 5.972E+24 kg] / (1.495979E+11 m)²

= 2.139589E+19 N

Trail edge has decreased gravitational force of

F_P = [(1 - 0.999396306)*6.6734E-11 N (m/kg)² * 1.989E+30 kg* 5.972E+24 kg] / (1.495979E+11 m)²

= -2.138297E+19 N

Where 1 Newton Force = 1 kg * m/s², and the positive force F_T pushes the lead edge toward the sun, and the negative push force F_P can be seen as the results of centrifugal force on the planet giving it a counterclockwise rotation, east to west, as the planet travels prograde around the sun. Together this torque force and the additional solar gravitational acceleration sustain the earth's rotational and orbital speeds.

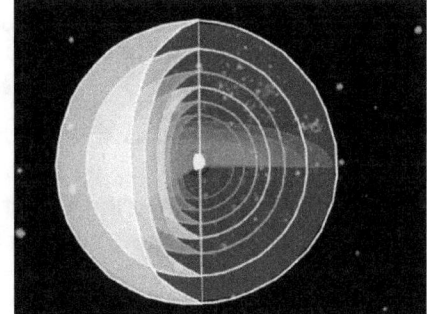

Sun's Normal Gravity Field

Planets Orbit Consistent

9.5 Principle of Tidal Forces and Mercury's Anomaly

Lemma 7, Tidal Effects on Earth-Moon System: The moon's gravitational attraction causes tides to rise and fall on Earth, and in doing so, the moon also slows down the rotational spin of Earth at a rate of a couple of milliseconds every hundred years. Similarly, the earth's pull on the moon combined with the extra mass from the rising of the tide causes the moon's orbit to increase slightly in speed and therefore gradually drift away at a rate of 3.78 centimeters per year[273].

Theorem 1, Mercury's Anomaly Solved: Cosmological Balance solves Mercury's orbital period anomaly using Newton's gravitation force equation in conjunction with the gravitational tidal effects of the Sun's plasma surface to Mercury's molten side bulges.

Mercury's extremely close elliptical orbit around the sun causes the sun's plasma fluid to form a rising tide as Mercury approaches its perihelion. Its gravity causes the sun to react with a tidal wave that builds and travels along the Sun's surface without opposition. With each passing hour, the solar tidal wave grows more and more in height and increases in speed until Mercury just reaches its perihelion. At Mercury's perihelion, the solar tsunami tidal wave achieves its greatest peak then surpasses Mercury's orbital speed, gravitationally tugs more on the planet, and accelerates it forward. Einstein calls this action "frame dragging." This gravity tugging holds the planet's perihelion distance and nudges it slightly forward, resulting in the additional orbital precession. See image below:

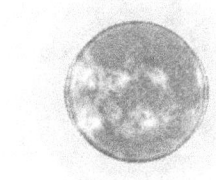

The solar tsunami wave in turn increases the speed of Mercury and the accelerated boost advances the perihelion to 5600 arc seconds. The perturbations by Venus and Earth are included in Mercury's orbit precession equation. Per Newton's equations, acceleration by gravity is represented by force = Gravity constant G times mass of the Sun M_s times mass of the planet M_p divided by distance R between the centers squared.

$$F = G \times M_s \times M_p / R^2.$$

Tidal equations are represented in a similar way.

Tide differential = $2 \times G \times M_s \times \Delta r / R^3 + 2 \times G \times M_p \times \Delta r / R^3 + 3 \times G \times M_s \times \Delta r^2 / R^4 + \ldots$

where Δr is influencing constant in addition to the tidal gravitational effect. The gravitational constant G is 6.6734E-11 N (m/kg)², the mass of the Sun is 1.9291E+30 kg with a radius of 695800 km, and the mass of Mercury is .055 that of Earth or somewhere between 3.2847E+23 to 3.3022E+23 kg with a radius of 2440 km. Mercury's perihelion is at 46,001,200 km and its Aphelion is at 69,816,900 km, with an average distance of 57,909,050 km and average orbital speed of 47,362 m/sec.

At Mercury's perihelion, sunlight takes 153.443487 seconds to reach its surface. Gravity acceleration at the surface of Mercury is:

---

[273] Our moon's elliptical orbit also has a precession that moves around the earth, similar to that of Mercury around the Sun.

CHAPTER 9: COSMOLOGICAL BALANCE PRINCIPLES

$F_g = G \times M_s \times M_p / R^2$

which comes out to about 3.4608 m/sec², and the resulting bulge on Mercury's liquefied surface is approximately 40.7426686 km. The gravity acceleration at the surface of Mercury number represents all perturbations from outside objects to include the sun and influences from the other planets orbiting around the sun.

F_g = 6.6734E-11 N (m/kg)² * 1.9291E+30 kg * 3.29718E+23 kg / (2440 km + 40.74km*2)²

= 3.4608 m/sec²

Bulge on Mercury = 3.4608 m/sec² * (153.443487 sec)² /2 / 1000

= 40.74267 km

Arcsec due to acceleration at Mercury's surface = degrees (arctan (bulge/perihelion)) * 3600

= (Degrees(ArcTan(40.74267 m * 2)/ 46,001,200 km)) * 3600

= 0.365372149 arcsec

Delta_r constant is calculated by getting the product of Mercury's mass and average distance divided by Sun's mass, and then add to it the acceleration on Mercury's surface.

Δr = 3.29718E+23 kg *57,909,050 km / 1.9291E+30 kg + 3.460848202 m/sec²

= 9.964519192 m/sec²

Tidal bulge on the sun due to Mercury at perihelion is sun bulge:

Sun Bulge = 153.443 arc seconds * (269.833 m/s)² /2/1000

= 5586.095 km

Acceleration on sun's surface due to Mercury's proximity is:

Gravity Constant * Mass of sun / ((Radius of sun + tidal bulge) * 1000)²

= 6.6734E-11*1.9891E+30/(((695800km + 5586.1km)*1000)²

= 269.8329 m/sec²

At perihelion planet Mercury's tidal raising force on sun is:

Df/Dr = (2 * G * M_s * Δr / R^3) + (2 * G * M_p * Δr / R^3) + (3 * G * M_s * Δr^2 / R^4) + (3 * G * M_p * Δr^2 / R^4) + (4 * G * M_s * Δr^3 / R^5) + (4 * G * M_p * Δr^3 / R^5) ...

= 0.027156239

where R = 46001200 km + 2 x 40.74266859 km + 2 x 5586.09525 km = 46012453.68 km

Therefore, the total Arcseconds caused by gravitational effect of tidal bulge accelerating past planet Mercury:

= Degrees(ArcTan(0.027156239)) * 3600

= 5600 arcsec per century

The equations above work perfectly when Mercury's mass is exactly 3.29718E+23 kg, meaning that it is .055208952 times the mass of Earth. In addition to solving the 5600 arc seconds for Mercury's orbital precession, it defines Mercury's mass[274] to be exactly equal to 3.2971812495E+23 kg.

Conjecture 2, Galaxy Age-Rings: Galaxies grow by consecutively building upon and merging of smaller galaxies into larger and larger ones, with each merger taking millions if not billions of years. Therefore, the Milky Way Galaxy is predicted to be older than 23 billion years in the making, therefore the universe itself is much older than 14 billion years the mainstream scientists speculate with the Big Bang Theory. Depicted by NASA above is a spiral galaxy with the oldest stars at its center and the youngest or newly added stars at its edges. Several small newly formed globular cluster galaxies surround most of these galaxies.

---

[274] The exact mass of Mercury is necessary to solving its orbital anomaly.

9.6 Principle of Balance in Types of Matter

Conjecture 3, Three Types of Matter in Balance: There are three types of matter in the universe: space-matter, normal gravity matter, and antigravity antimatter. Balance and symmetry are maintained between all three types of matter. Space-matter makes space what it is, empty and appear to be void, and exerts a vacuum force of 6.6734E-11 Newton universally. This vacuum force provides the push behind gravity and antigravity. Miniscule amounts of space-matter decays through excessive collisions with itself in low-pressure disturbances and develops into systems spawning both gravity and antigravity eddies, which may or may not produce stars and galaxies. At the center of well-developed galaxies are massive black or white holes. Massive black holes and massive white holes gradually excrete, eject microscopic orbs into wormhole tunnels, and provide the material to replenish space-matter. Excessive gravity galaxies merge to produce super massive black holes, which in turn produces more space-matter and more antigravity building blocks of lighter elements, giving way to more antigravity galaxies until balance is achieved again.

Definition 10, Space-Matter: What exactly is space-matter? Space-matter makes space what it is, appearing void and empty; it is vaguely like the old concept of "aether." It attracts neither matter that has gravity nor matter that has antigravity; and is not neutral in mass. It exists in super cold gaseous form primarily substantial space-matter hydrogen atoms (*hygratium*) uniformly scattered throughout deep open empty space, timeless. It does not draw itself together, like ordinary matter or antigravity antimatter; it repels away from itself and spreads out, like normal gases and vapor on earth but at a wider distance. Space-matter's super electrostatic charges make them spread out much more than normal matter. The particles themselves although large are hard to see, because of their fast speeds and constant rapid jiggling. The spacing within and in between atoms and molecules is what we see and perceive as void empty space. The atoms are larger in radius and more dispersed in space, and its super dense electrons and nuclei elusive and undetectable by our instruments for the same reason antigravity antimatter are undetectable to us. They cause the spectrum red shift of extremely distant galaxies; as light from visible galaxies passes through them, they lose some photon energy before reemission. The highest and last form of matter, space-matter, sustains deep space, and somewhat neutralizes the attractive effects of the other two types of matter by creating great distance in the expanse of space, and powers galactic rotations. With balanced vibrations of gravity antigravity, there is no attractive fourth natural force associated with space-matter itself; its internal potential energy, pressure, and force is the source of the gravitational force ordinary matter experiences and the source of the anti-gravitational force affecting antigravity antimatter. Space-matter could absorb and reemit photon energy from light passing through it without changing its frequency or color. Space-matter packed around massive objects has a great deal of optical density and a unique light refraction that provides same result as the "gravity lensing" Einstein captured in his general relativity equations.

Postulate 3, Space-Matter Decay: The Cosmological Balance Theory does not claim a specific age of the universe, and therefore, does not identify when the creation of normal matter, as we know it, began. It does assume, however, that space and time always existed even at the beginning of the universe, like that calm state and clear skies before a storm. In such a state, there was complete darkness of space, but not void, as it was already occupied by a high-energy unique space-matter, which contained only .00015 % of actual particles with mass, a mass neutral of gravity or antigravity, yet both, characteristics for which its sole purpose is sustaining space itself. This completely alien form of matter to us is the true normal state of space-matter that repels or bounces off each other, like hydrogen atoms and molecules creating the gaseous upper atmosphere on the planet or in an inflated balloon, but instead of atmosphere, space-matter creates the vacuum of space. In the beginning, it formed a gaseous grid of slow moving almost liquid gas evenly spaced at about one to 1.00015 atoms per cubic meter. The space-matter and space came into existence simultaneously and resulted in the vacuum perceived as empty space. Their simultaneous existence gave space its pressure, force, energy, and infinite volume. The energy of space or sometimes as it is called, vacuum energy, and its force as it transfers through collisions of space-matter with normal matter, as we know it, is the source of the phenomena we call gravity and the source of antigravity force on antigravity antimatter.

The majority of universe and the space-matter remain stable and slow-moving fluid like the seas of a great ocean of air, and pressure evenly maintained throughout flat space. However, from time-to-time, pressure and movement caused space-matter to come together in eddies and develops the beginnings of a "storm." Turbulence in the flow of space-matter develops low-pressure and high-pressure regions in space; these are areas of curvature of space. Low-pressure regions in space develop into instability of the space-matter atoms and molecules at the center of these rotations. These high-energy high-density atoms and molecules collided more frequently and eventually start to decay with collision with other high-density space matter atoms. Small amounts of collisions break down the high-energy high-density space-matter atoms into low energy subatomic particles, which reassemble into the normal hydrogen atoms and on occasion the helium atoms we know on earth. Somehow, during the break down process from high-energy high-density space-matter to the hydrogen atoms, we have on earth and in the sun, the gravity switch activated, the first creation of matter with gravity, and the natural tendency to bind together. The pressure and energy, and force provided by space-matter became the source we know as gravity force on earth.

Definition 11, Space-Matter Properties: Space-matter atoms are a super-dense massive form of hydrogen that consist of high-energy Tau lepton or Tauon, a dense proton (with 5-top and 7-bottom high-energy quarks), a dense neutron (with 4-top and 8-bottom high-energy quarks), a high-energy gluon, and a dense high-energy photon to hold the high-energy tauon in its orbital level. Timeless space-matter is naturally more stable as a dense hydrogen molecule in the cold temperatures of deep space. Its super dense particles and high-energy structure gives the atom and the molecule its great volume and radius and creates vast distance and empty space between molecules. Its characteristics are like super cooled liquid frozen gas rather than an energetic gaseous state, but much more spread out. The total mass of the unique space-matter hygratium atom is $2.8116\text{E}-24$ kilograms.

Space-matter Atom = High-Energy Photon + Tauon + High-Energy Proton + High-Energy Neutron + High-Energy Gluon

= Photon + Tauon + (5TopQuarks+7BottomQuark) + (4TopQuark+8BottomQuark) + Gluon

= $3.8786\text{E}-35$ kg + $2.1478\text{E}-26$ kg + ($5*3.1000\text{E}-25$ kg + $7*7.44889\text{E}-30$ kg) +

($4*3.1001\text{E}-25$ kg + $8*7.44889\text{E}-30$ kg) + $6.04445\text{E}-31$ kg

= 2.8116E-24 kg

The radius of a normal hydrogen atom on earth is approximately 5.3×10^{-11} meters. However, due to the extremely low temperatures of space, 2.7 Kelvin Degrees, the unique dense-particle space-matter hygratium atom will have a condensed radius for its mass. Therefore, the radius of the space-matter atom estimated at 3.2358E-11 meter is larger and denser than a normal hydrogen atom of 2.6349E-13 meter. In the coldness of space, these unique atoms are about 100 times larger than normal hydrogen atoms on earth but are expected to move much more slowly, while still maintaining distance between other space-matter atoms and molecules, an average density of one atom per cubic meter. Substantial space-matter atom acceleration due to electrostatic repulsion is estimated at 2.3735E+13 meters per second squared. The vacuum energy causing this acceleration is approximate 8.83919E+15 Newton meters. This vacuum energy is not gravity; it is cause by pressure in space produced by electrostatic and gravity-antigravity repulsion between slow moving super high-energy space-matter atoms and molecules. The same energy causes gas atoms and molecules on earth to spread out and bounce off each other. However, in this case, the vacuum energy,pressure, and force make space what it is, appear void and empty. The Force or more specifically Vacuum Force exerted by these substantial space-matter atoms is obtained by the product of its mass times its acceleration or more specifically 2.8116E-24 kg x 2.3735E+13 m/sec^2 = 6.6734E-11 kilograms meters per second squared or 6.6734E-11 Newton. Space-matter has minimal effects on itself. The gravitational constant as defined per Newton's gravity equation has no impact on space-matter because the natural attraction of two space-matter particles or atoms is zero meters per kilogram, or no natural attraction. The revised relationship equation reads 6.6734E-11 N x (0 m/kg x m$_1$) x (0 m/kg x m$_2$) / r^2 = zero. The interaction equation between space-matter and normal matter with gravity is 6.6734E-11 N x (1 m/kg x m$_1$) x (0 m/kg x m$_2$) / r^2 = zero; and between space-matter and antigravity it reads 6.6734E-11 N x (-1 m/kg x m$_1$) x (0 m/kg x m$_2$) / r^2 = zero. Space-matter can do this because it is spread evenly across space at about one substantial atom per cubic meter.

Proposition 2, Intergalactic Lighter Elements, Background Microwave Radiation: The abundance of light elements throughout the universe is the result of continuous breakdown of gravity matter and antigravity antimatter within massive white and black holes, released through excretion, and ejection into wormhole tunnels, and released beyond their galaxy. This process thereby renews the formation of light gaseous antigravity and gravity matter floating in space and rebuilds space-matter[275].

To understand this principle fully we need to think about how a magnetron operates within a microwave oven. The magnetic field at the base of the magnetron causes the electrons, accompanied by photon, to travel in a circular path around the cathode, a rod at the center. As the electrons and photons, enter cavity chambers surrounding the magnetron, they resonant in a frequency matching the size of the chamber to produce the desired microwave radiation. In comparison, the microscopic black hole or white hole orb traveling within a wormhole tunnel is of course super-heated and stimulated cathode and is accompanied by free electrons and photons encircling it in a spiraling motion through the tube, the wormhole tunnel. However, because of the orbs great speed (the speed of light

---

[275] The Cosmological Balance is a renewable system moving matter and energy from space into galaxies and back into space.

squared) the electrons encircling it appear to be traveling in the same pattern as though there was a magnetic field at the base of the wormhole tunnel. As a result, the electrons circulating pattern creates that magnetic field. The magnetic field and extreme speeds of the orbs traveling within the wormhole tunnel generate eddies of electrons and photons which are 1.873 millimeters in diameter and resonant the microwave radiation frequency of 160.2 GHz. The distance from the microscopic black hole or white hole orb to the wall of the wormhole tunnel it is traveling within is predicted to be slightly less than 1.8731 mm or exactly the right distance to produce the common background microwave frequency of 160.2 GHz found throughout space.[276]

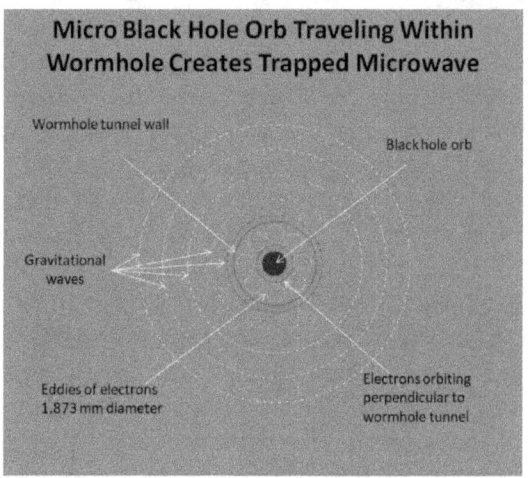

As billions and billions of these microscopic orbs travel in their individual wormhole tunnels traverse far enough away from the galaxy, they slow down and drop below the speed of light. That drop in speed, causes their wormhole tunnel to collapse and release the trapped resonating 160.2 GHz microwave radiation it was carrying in a short quick outward burst. It does not matter if these microwave frequencies were generated today or 13 billion years ago, they will always remain at that frequency level, unless they bump into something with enough mass to absorb its energy. As a result, the continuous static noise we detect throughout the universe is produced by the release of energy from collapsing wormhole tunnels and read at the microwave frequency of around 160.2 GHz. The emerging microscopic black hole from the wormhole tunnel regains its ability to expand to normal space between particles, and quickly assumes one of two options. These high-energy subatomic particles, leptons, and force carrier are most likely to transition the graviton switch from gravity matter to neutral and reassemble as exotic high-energy space-matter atoms. This process extracts energy from some of the other subatomic particles. Secondary option is that about 15% of these high-energy particles lose considerable energy and become low-energy quarks, leptons, and force carriers. These particles become normal gravity hydrogen and helium atoms, the basic building blocks of normal gravity objects, stars, and galaxies. Similarly, the emerging microscopic white hole from their wormhole tunnel regains its ability to expand to normal space between particles, and quickly assumes one of two options. These antigravity high-energy subatomic particles, quarks, leptons, and force carrier are most likely to transition the graviton switch from antigravity antimatter to neutral and reassemble as unique high-energy space-matter atoms. This process extracts energy from some of the other subatomic particles. The secondary outcome is that about 15% of these high-energy particles lose considerable energy and become low-energy quarks, leptons, and force carriers. These particles become antihydrogen and antihelium atoms, basic building blocks of antigravity stars, and galaxies. Space-matter is sustained and perpetuated repeatedly.

---

[276] The accidental discovery of the magnetron within microwave ovens and transmitters provided the insight to solving where the cosmic microwave background comes from.

9.7 Principle of Planetary Motion and Gravity

Lemma 8, Kepler's Three Laws: Also known as Kepler's laws of planetary motion.

1. The orbit of a planet is an ellipse with the Sun at one of the two foci. In other words, the planets follow the path of ellipses, and that a circular orbit is a specific type of ellipse where both foci are one in the same location.

2. A line segment joining a planet and the Sun sweeps out equal areas during equal intervals of time. Similarly, a ray from the sun to a planet sweeps out equal areas in equal times.

3. The square of the orbital period T^2 of a planet is proportional to the cube of the semi-major axis a^3 of its orbit. This constant of proportionality is independent of the individual planets, or stated in another way, each planet has the same proportional constant. Kepler's formula now reads $T^2 = (4\pi^2/(G*M)) * a^3$, where M is the mass of the sun and G is the universal gravitational constant Newton discovered[277].

Lemma 9, Newton's Three Laws: The laws of motion.

1. This law states that an object or body remains at rest or in uniformed motions in a straight line unless another force external to it acts upon it. This resistance property is known as **law of inertia** and the mass of that object is a measure of its inertia, which is also referred to as inertia mass.

2. This law states that when a net force acts on an object, it will accelerate that object or body in the direction of that force, called **force law**. The relationship between the mass m, acceleration a, and the force F is written as $F = m*a$.

3. Similarly, the relationship between mass m, the momentum p, and the velocity is written as $p = m*v$.

4. This law states that when one object or body exerts a force on another object or body, the second object exerts an equal force in the opposite direction against the first object. This law is simply the **action-reaction law**, and applies to objects in all situations, whether the objects are in motion in a uniform velocity or accelerating, or stationary.

Newton's gravitational equation: $F = G\, m_1 m_2 / r^2$, where m_1 and m_2 are objects with mass, r is the distance between the centers of the objects, G is the gravitational constant $6.6734\text{E}{-11}\ N\ m^2/kg^2$, and F is the force acting on m_1 and m_2 alike.

Corollary 3, Vacuum Force Transfer to Gravity: Space-matter collision or repulsive interaction with gravity matter transfers its vacuum force into the force for the push we know as gravity. In addition, this force when transferred with collisions between space-matter and antigravity antimatter provides the force for the anti-gravitational attraction of antigravity antimatter. Fueled by vacuum pressure and energy, the vacuum force of $6.6734\text{E}{-11}\ N$ within one kilogram of unique high-energy space-matter at one meter per kilogram colliding with normal gravity matter at one meter per kilogram transfers the total force of $6.6734\text{E}{-11}\ N\ m^2/kg^2$, which is equivalent to Newton's gravitational constant G used on normal matter equations.

---

[277] Kepler gave Newton the necessary insights to solve gravity, and simultaneously provided a baseline for solving the Cosmological Balance Equation.

Pressure and Energy equations are applied to gaseous and liquid matter on earth as well as to exotic space-matter hydrogen atoms (*hygratium*) in space. Space-matter Vacuum Force, Pressure, and Energy are

Vacuum Force = mass * acceleration

= 2.8116E-24 kg * 2.3735E+13 m/sec^2

= 6.6734E-11 kg m/sec^2

Vacuum Pressure = Force / Area = Force * distance / (Area * distance)

= Force * distance / Volume

= 6.6734E-11 kg m/ sec^2 * 1.32454E+26 m / 2.32378E+78 m^3

= 3.80379E-63 N/m^2

Vacuum Energy = Pressure * Volume

= 3.80379E-63 N/m^2 * 2.32378E+78 m^3

= 8.83919E+15 kg*m^2/sec^2

We see the atoms and molecules of space-matter as extremely spread out gaseous unique high-energy matter, with firm almost liquid characteristics. Since the atoms of space-matter are dispersed uniformly spread throughout space at a ratio of one atom per cubic meter, its vacuum force is the same anywhere in space. By the same reasoning, the vacuum pressure and associated vacuum energy is normally the same for any position or coordinates in Euclidean space. The exceptions to this norm are considered systems of high-pressure and low-pressure build-ups. In all instances, this vacuum energy in space is transferred from space-matter substantial atoms to other atoms simply by atomic and molecular repulsions. It provides the force, which allows matter with gravitational characteristics to "pull" mutually together per Newton; technically, they are also pushed, as he originally intended, together as kinetic vacuum energy is transferred between space-matter atoms and molecules to atoms with gravity or antigravity. Gravity in normal matter and antigravity in the invisible part of the universe provides the natural tendency for these molecules to bind together, chemically, and molecularly, a quality proven by quantum mechanical graviton and its entanglement, a sort of telekinesis that is both instantaneous and infinite in range, some call string theory.

Theorem 2, Vacuum Force Equals Gravity Constant: First, let us consider a single object or spherical orb in space floating in an area of space not under the influence of any other objects. Such an orb would receive equal vacuum force from all directions, and thereby achieve a spherical shape assuming it was massive enough, just like a soap bubble reconfigures itself in the air as it leaves the blower's launching loop. The orb's bubble shell, gravity layer, or in some cases the atmosphere is the point of transfer of forces and energy toward the center of it as the force we call gravity. Large objects in space are spherical because this shape has the least amount of surface area and the best contact for the transfer of force; it is the reason soap bubbles are spherical when floating by themselves. The space-matter's bubble layer or shell surrounding the massive object in space is what

Einstein refers to as the curvature of space. Thereby, this vacuum force molds floating objects and simultaneously suspends it in space. Now let us consider two objects in space not orbiting each other but sufficiently close to attract each other at the quantum entanglement level. The vacuum forces act equally on each object from all possible angles except two cones.

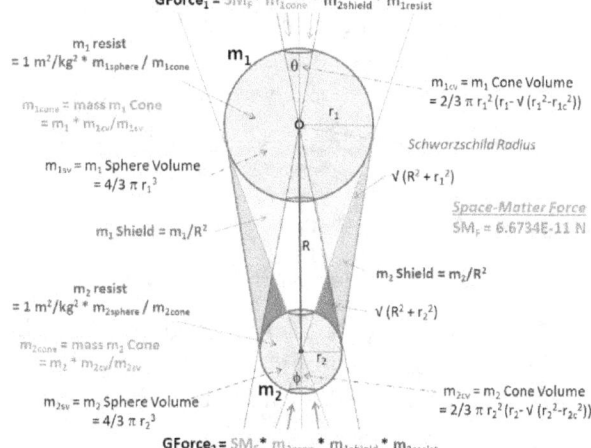

These cones calculated by drawing intersecting lines between the edges of the two objects to the centers of the other object. The cones on the opposite sides of the objects are the surfaces where the vacuum force transfers into gravity. As the objects approach each other, the cone surfaces become increasingly greater, or as Newton put it, inversely proportional to the distance.

Using the distance between m_1 and m_2 as the length of adjacent side of the right triangle, and the radius of m_1 as the opposite side, we solve for the angle intersecting the center of m_2 from the formula $\tan^{-1}$ (opp/adj). The angle obtained φ is ½ of the total angle of the m_2 spherical cone. Then by using the radius of m_2 and the angle φ, we can solve for the radius of the cone from $r_2 = m_2\text{radius} \times \tan(\varphi)$. From there we solve for the volume of the spherical cone m_2 from equation $V_{m2} = 2/3 \pi \times m_2\text{radius}^2 \times h_2$, where $h_2 = r_2 - $ square root $(r_2^2 - r_{2c}^2)$. Then we obtain the mass of $m_2$cone from equation $m_2\text{cone} = $ (mass of m_2) x (volume $m_2$cone)/ (volume m_2 sphere) $= m_2 \times V_{m2}/V_{m2s}$, where volume of sphere m_2 is solved with equation $4/3 \pi \times m_2\text{radius}^3$.

We then use space-matter's vacuum force of 6.6734E-11 N to solve for the push force on m_2 by multiplying space-matter force times mass of m_2 cone times mass of m_1 shielding divided by the distance squared. This gives us force$_2$ = $F_{sm} \times m_{2cone} \times m_1 / d_r^2$. Then, determine amount of resistance to movement or inertia force with equation:

$m_2$resist =(1 square meter per square kg)x(spherical volume m_2)/(cone volume m_2).

We then solve for the total push effect G_{f2} by equation (push force on m_2) times (resistance of m_{2cone}) or G_{f2} = force$_2$ x $m_{2resist}$.

This solution coincides with and confirms Newton's gravity force source is the medium permeating space itself, space-matter, and its vacuum force. Since the vacuum force's push on bodies and objects produces the same results as gravity constant, it applies to planets orbiting stars and the rectilinearly correcting of the planet's path into a curved elliptical or circular motion. Space-matter force causes the effect of gravity, pushing objects toward the center of its Schwarzschild radius as explained above, while quantum mechanical entanglement provides the instantaneous attractive tendency over infinite distances[278]. It also applies to whole galaxies and the movement of galaxies in the universe.

---

[278] Fortunate for us this complex vacuum force transfer and quantum entangled attraction reduces to Newton's simple gravity equation.

9.8 Principle of Planetary & Galactic Orbital Period

All of Newton's work written in Principia remains pertinent and valid and its contribution to modern day mathematics immense and space travel immeasurable. Let me reiterate that the prediction of space-matter and antigravity does in no way refute Newton's great work; it confirms his initial expectation of a medium and mechanical source of gravity. Without proof of this, Newton had to use a philosophical approach in presenting his definitions, laws, axioms, and gravity equations in Principia.

Lemma 10, Circumference of Circle Formula: The circumference of a circle is calculated by formula:

Circumference = $2\pi R$,

where R is the radius of the circle, and pi is the constant 3.14159265358979323846...

Lemma 11, Schwarzschild Radius Formula: The Schwarzschild radius (also known as the gravitational radius) is the radius of a spherical object with enough compressed mass where the escape speed from the that object is the speed of light. The formula reads:

$R = 2GM/c^2$

where M is all the mass is compressed within a sphere with radius R, resulting in escape speed of the speed of light, c.

Corollary 4, Planetary Orbital Period: Modified Kepler's third law integrating Schwarzschild radius and mass: The square of the orbital period T^2 of a planet is proportional to the cube of the semi-major axis a^3 of its orbit. Formula reads:

$T_c^2 = (4\pi^2/(G*M_S)) * a^3$

where M_S is the sum of all inner orb masses in the system contained by orbit with a^3 semi-major.

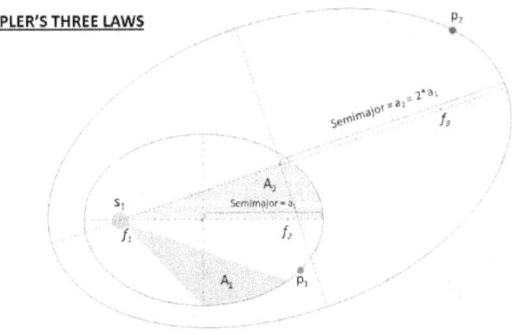

Combining Kepler's third law with Schwarzschild radius and mass gives us the following calculated orbital periods for the solar system:

kg	Mass of Orb	Accumulate kg	Schwarzschild	orbital per = s	Radius = m	orbit (days)	orbit(yrs)
1.98891E+30	Sun	1.9889146E+30	2.953601E+03	10020.57907	6.9630E+08	0.1159789	0.00032
3.29718E+23	Mercury	1.9889149E+30	2.953602E+03	7600261.545	5.7910E+10	87.96599	0.24084
4.86732E+24	Venus	1.9889198E+30	2.953609E+03	19410581.68	1.0820E+11	224.65951	0.6151
6.04709E+24	Earth and Moon	1.9889258E+30	2.953618E+03	31556904.03	1.4960E+11	365.24194	1
6.39000E+23	Mars	1.9889264E+30	2.953619E+03	59335337.47	2.2790E+11	686.75159	1.88026
1.89813E+27	Jupiter	1.9908246E+30	2.956438E+03	374436105	7.7850E+11	4333.7512	11.8654
5.68300E+26	Saturn	1.9913929E+30	2.957282E+03	934968762.4	1.4330E+12	10821.398	29.628
8.68100E+25	Uranus	1.9914797E+30	2.957411E+03	2659672376	2.8770E+12	30783.245	84.2817
1.02400E+26	Neptune	1.9915821E+30	2.957563E+03	5199198052	4.4980E+12	60175.903	164.756
1.30900E+22	Pluto	1.9915821E+30	2.957563E+03	7823083905	5.9063E+12	90544.953	247.904

where the orbital period of Mercury is 87.966 earth days; Venus has 224.66 earth days; Mars has 686.75 earth days (1.88 earth years); Jupiter has 4,335.82 earth days (11.87 earth years); and Saturn has 10,821.40 earth days (29.63 earth years). While Uranus has orbital period of 30,783.25 earth days (84.28 earth years); Neptune has 60,175.90 earth days (164.76 earth years); and Pluto has an estimated orbital period of 90,544.95 earth days (247.90 earth years).

This Schwarzschild modified Kepler orbital period equation is applicable to all solar systems within the galaxy and within all other galaxies given that we can figure out the length of each planet's semi-major axis.

Theorem 3, Planetary and Galactic Orbital Period Formula: Planetary and Galactic orbital periods are defined with the formula:

$$T_c = [\sqrt{(4\pi^2/(GM_C)*a_c^3)}] * [(c^2 * \sqrt{(1/(2GM_C))} * e/\pi], \text{ or}$$

$$T_c^2 = (4\pi^2/(GM_C)*a_c^3) * [c^4 \, e^2/\pi^2 / (2GM_C)],$$

where the first part is the modified Kepler's third law with summation of inner masses, and the second part represents the acceleration caused by gravity-antigravity particle repulsion or object interactions. This equation unifies Newton and Kepler laws, the Schwarzschild radius and mass, quantum mechanics, and the principles of geometric and polar coordinate systems.

GALAXY ORBITAL PERIOD EQUATION

$T_c = [\sqrt{(4\pi^2/(GM_C)*a_s^3)}] * [c^2 * e/\pi \, \sqrt{(1/(2GM_C))}],$ Schwarzschild Radius Adjustments

To obtain this equation, we first combine the circumference formula of each concentric circle $2\pi R$ (in meters) with the mass formula M_S from the Schwarzschild Radius $R_S = 2GM_S/c^2$ (in meters). The product of Mass $M_S = c^2 R_S/(2G)$ and acceleration of that mass ($\sqrt{2GM_S/R_S^2}$) on that circumference gives us the force that propels stars just outside that circumference edge

Force = Mass * Acceleration

$= c^2 R_S/(2G) * \sqrt{(2GM_S/R_S^2)}$

$= c^2/(2G) * \sqrt{(2GM_S)}$

$= c^2 * \sqrt{(2GM_S/(2G)^2)}$

$= c^2 * \sqrt{(2GM_S/(4G^2))}$

The gravitational constant G cancels out and the 2/4 reduced to ½, giving us:

$= c^2 * \sqrt{(M_S/(2G))}$

where c is the speed of light, M_S is the mass within the Schwarzschild Radius, and G is Newton's gravitational constant.

The equation above is great progress, but it still does not account for the force behind the antigravity wormhole tunnels nudging gravity objects. This force is simply represented by Euler's number e=2.718281828459. We now have:

Galactic_Rotational_Force = Mass * Acceleration * *e*

$$= c^2 * \sqrt{(M_S/(2G))} * e$$

To calculate the galactic wind force, we take the galactic rotational force and divide it by the sectional mass within the Schwarzschild radius. We get:

$$G_{WF} = G_{RF} / M_S$$

$$= c^2 * \sqrt{(M_S/(2G))} * e/M_S$$

$$= c^2 * \sqrt{(M_S/(2GM_S^2))} * e$$

The mass M_S cancels out to give us:

$$= c^2 * \sqrt{(1/(2GM_S))} * e$$

To solve for the Cosmological Balance galactic orbital period of stars within the galaxy we start with Kepler's Third Law, the orbital period to travel and the Schwarzschild Radius formula $R_S = 2GM_S/c^2$.

$$T = \sqrt{(4\pi^2/(GM_S)*a^3)}$$

then multiply by the galactic wind force acting on the circumference for that portion of the galaxy (i.e. rotational force from that concentric black hole) divided the radians of a circle 2π divided by 2, since most stars orbiting path around their galaxies are circular[279]. Note: dividing by $2\pi/2 = \pi$ gives us the acceleration force of the galactic wind at any one point on the circular path of the star orbiting its galaxy. We get the formula for Cosmological Balance galactic orbital period T_c as

	C	D	H	J	K	L	M	N	O	P	Q
1	$R_S=2GM/c^2$	6E+42 * fract	Mass Distribution		60K ly * sect	$\sqrt{(4\pi^2/(GM)*a^3)}$	$c^2/GM*\sqrt{(2a^3)}*e/\rho$	sec --> yr		Resist ρ	Seg
2	Schwartzschild	sectional kg	section	mass fraction	Radius = m	Kepler orb period	CBU orb per'd (s)	x 31556952= years		e/ρ	0
3	1.38811E+10	9.3474E+36	2.45E-11	1.55789E-06	1.38811E+10	1118.390221	9.05841E+05	2.870E-02	years	1.0000	1
4	6.72125E+14	4.52600E+41	0.020	0.075433	1.13526E+19	1.18875E+14	4.37557E+14	1.387E+07	years	1.0000	2
5	7.26849E+14	4.89450E+41	0.040	0.081575	2.27053E+19	3.23323E+14	1.14442E+15	3.627E+07	years	1.0000	3
6	7.81572E+14	5.26300E+41	0.060	0.087716677	3.40579E+19	5.72811E+14	1.95523E+15	6.196E+07	years	1.0000	4
7	8.77456E+14	5.90867E+41	0.080	0.098477788	4.54106E+19	8.32322E+14	2.68133E+15	8.497E+07	years	1.0000	5
8	1.00078E+15	6.73911E+41	0.100	0.112318529	5.67632E+19	1.08918E+15	3.28550E+15	1.041E+08	years	1.0000	6
13	1.79649E+15	1.20973E+42	0.200	0.201622202	1.13526E+20	2.29933E+15	5.17678E+15	1.640E+08	years	1.0000	11
18	2.67496E+15	1.80128E+42	0.300	0.300213632	1.70290E+20	3.46172E+15	6.38711E+15	2.024E+08	years	1.0000	16
19	2.85253E+15	1.92085E+42	0.320	0.320142425	1.81642E+20	3.693E+15	6.59834E+15	2.091E+08	years	1.0000	17
20	3.03031E+15	2.04057E+42	0.340	0.340094953	1.92995E+20	3.92414E+15	6.80254E+15	2.156E+08	years	1.0000	18
21	3.20823E+15	2.16038E+42	0.360	0.360063305	2.04348E+20	4.15518E+15	7.00048E+15	2.218E+08	years	1.0000	19
22	3.38625E+15	2.28025E+42	0.380	0.380042207	2.15700E+20	4.38617E+15	7.19278E+15	2.279E+08	years	1.0000	20
23	3.56433E+15	2.40017E+42	0.400	0.400028141	2.27053E+20	4.61711E+15	7.37993E+15	2.339E+08	years	1.0000	21
24	3.74245E+15	2.52011E+42	0.420	0.420018764	2.38406E+20	4.84803E+15	7.56238E+15	2.396E+08	years	1.0000	22
25	3.92060E+15	2.64008E+42	0.440	0.440012513	2.49758E+20	5.07893E+15	7.74046E+15	2.453E+08	years	1.0000	23
26	4.09876E+15	2.76005E+42	0.460	0.460008345	2.61111E+20	5.30982E+15	7.91451E+15	2.508E+08	years	1.0000	24
27	4.27694E+15	2.88003E+42	0.480	0.480005567	2.72463E+20	5.5407E+15	7.95338E+15	2.520E+08	years	1.0165	25
28	4.45513E+15	3.00002E+42	0.500	0.500003715	2.83816E+20	5.77157E+15	7.98759E+15	2.531E+08	years	1.0330	26
33	5.34612E+15	3.60000E+42	0.600	0.600000498	3.40579E+20	6.92591E+15	8.10212E+15	2.567E+08	years	1.1157	31
38	6.23713E+15	4.20000E+42	0.700	0.700000074	3.97343E+20	8.08023E+15	8.14798E+15	2.582E+08	years	1.1983	36
43	7.12815E+15	4.80000E+42	0.800	0.800000018	4.54106E+20	9.23455E+15	8.14877E+15	2.582E+08	years	1.2809	41
48	8.01917E+15	5.40000E+42	0.900	0.900000011	5.10869E+20	1.03889E+16	8.11941E+15	2.573E+08	years	1.3635	46
54	8.91019E+15	6.00000E+42	1	1	5.67632E+20	1.15432E+16	7.97854E+15	2.528E+08	years	1.4626	52

CBU_Orbital_Period = Kepler_Orbital_Period x Galactic_Wind_Force / π

$$T_c = [\sqrt{(4\pi^2/(GM_S)*a_c^3)}] * [c^2 * \sqrt{(1/(2GM_C))} * e/\pi]$$

---

[279] The rotational speed of the center black hole indirectly influences the orbital speed of all the stars orbiting the galaxy via the galactic wind it produces. No galactic wind gives us globular clusters.

CHAPTER 9: COSMOLOGICAL BALANCE PRINCIPLES

where T_c is the Cosmological Balance galactic orbital period in seconds of stars going around a specific galaxy given the distance a_c from its center in meters. M_C is the accumulated mass of all the stars within the concentric circle of radius a_c, c is the speed of light, the G is Newton's gravity constant, π is the constant pi, and e is Euler's number.

Above is the table showing the calculation results for the galaxy's orbital period for a given distance from the center of the galaxy, and resistance ϱ of galactic wind e/ϱ to edge.

where we can see that the calculations match actual observations. The sun's distance to center of galaxy is between 2.36513E+20 to 2.46731E+20 meters fall into the range above (column K, lines 24 and 25), and its orbital period around the galaxy 7.57367E+15 seconds (2.40E+08 years) falls within the range identified by (columns M and N, lines 24 and 25). So, the CBU galactic orbital equation reads as $T_c = c^2/GM * \sqrt{(2a^3)} * e/\varrho$.

For those not convinced of the antigravity-gravity "galactic wind," then think of it as representing the rate of angular momentum transfer between stellar objects, specifically it is the rate stellar gravity interacts in galaxies. Simultaneously, the "galactic wind" force is also generated by space-matter's inherent gravity-antigravity pulses from its particles motion within its quarks. This alternating gravity-antigravity hum from space rotates galaxies, likened to an alternating current electromagnet spinning a permanent magnet rotor. The force of the galactic wind e is incrementally diminished after passing one-half of the galaxy's mass. This effect smooths the rotational velocity toward galactic edge.

This galactic orbital period equation is applicable to all galaxies in the universe.

9.9 Principle of Cosmological Balance Equation

Proposition 3, Quantum Entanglement Observation: Quantum entanglement shows us that the particles and objects considered intertwined instantaneously with other particles or objects of the same origin respectfully as if they are one particle or object[280]. Quantum mechanics tells us that we should treat the universe as one entangled object.

Proposition 4, Source Disclosure of Cosmic Microwave: As described in detail in Proposition 2, the cosmic microwave background radiation is the result of black hole orbs dropping out from wormhole tunnels ejected from antigravity galaxies, and not the remnants of the Big Bang. Likewise, antigravity cosmic microwave background radiation is the result of white hole orbs dropping out from their wormhole tunnels departing normal gravity galaxies.

The black hole or white hole orb traveling within a wormhole tunnel, of course, is super-heated and accompanied by free electrons and photons encircling it in a spiraling motion through the tube, which was caused by the $\sqrt{2}$ energy. However, because of its great speed (the speed of light squared) the electrons appear to be traveling in the same pattern as though there was a magnetic field at the base of the wormhole tunnel. As a result, the electrons circulating pattern creates that magnetic field. The magnetic field and extreme speeds of the orbs traveling within the wormhole tunnel generate eddies of electrons and photons which are 1.873 millimeters in diameter and resonant the microwave radiation frequency of 160.2 GHz. The distance from the black hole or white hole orb to the wall of the wormhole tunnel it is traveling within is predicted to be slightly less than 1.8731 mm or exactly the right distance to produce the background microwave frequency of 160.2 GHz found throughout space.

As these billions and billions of orbs moving in wormholes traverse far enough away from the galaxy, they eventually slow down and drop below the speed of light. That drop in speed, causes the wormhole tunnel to collapse and release the trapped resonating 160.2 GHz microwave radiation it carries in a short quick outward burst. Whether these microwave frequencies were generated today or 13 billion years ago, they will always remain at that frequency level as they propagate through space, unless they bump into something with enough mass to absorb its energy. As a result, a continuous static noise at the microwave frequency around 160.2 GHz is produced and picked up by our instruments.

Theorem 4, Space and Matter Cosmological Balance Equation: One idea that is common between the Cosmological Balance and other theories is that space itself with its unique matter (space-matter) tells matter with gravity and matter with antigravity what to do and where to move. Moreover, as matter with gravity and matter with antigravity moves or flows through space itself tells space to give way and allow passage. The Cosmological Balance Equation defines this relationship between all three types of matter in the universe (space-matter, normal gravity matter, and antigravity antimatter) and is expressed in the following formula:

---

[280] All matter with gravity shares the same origin. Likewise, all antigravity antimatter shares similar origin of creation. They both come from space-matter. Therefore, quantum entanglement applies to both gravity and antigravity realm.

$$S_m * [4\pi^2/G] * [c^4e^2/2GM_A\pi^2] * A_m = [4\pi^2/G] * [c^4e^2/2GM_C\pi^2] * C_m, \qquad (9.9.1)$$

where e is Euler's constant, G is the gravity constant, S_m represents space-matter and its motions, A_m represents antigravity antimatter and its motions, C_m represents normal gravity matter and its motions, $C_m = a_c^3/(T_c^2 * M_C)$. As antigravity particles traverse through the gravitation system, they significantly accelerate the rotation of the system under observation; stars move faster than with gravity or antigravity alone. In these equations, space-matter and antigravity tells normal matter how to move, and likewise space-matter and gravity matter tells antigravity antimatter how to move.

In Theorem 3, we have the Cosmological Balance galactic orbital period equation:

$$T_c = [\sqrt{(4\pi^2/(GM_C)*a^3)}] * [c^2 * \sqrt{(1/(2GM_C))} * e/\pi],$$

Redefining the equation in terms of T_c^2 gives us the clarification we need to effectively define space-matter, gravity matter, and antigravity antimatter interactions.

$$T_c^2 = [(4\pi^2/(GM_C)* a_c^3)] * [c^4 * 1/(2GM_C) * e^2/\pi^2],$$

where the first part of the galactic orbital period equation is a modified Kepler's third law. The second part of the equation is the acceleration resulting from the interaction between gravity matter and antigravity antimatter, times Euler's number e, as the objects move once around the circular circumference of the galaxy divided by two to determine acceleration, $2\pi/2$, the "galactic wind." For the interaction between gravity matter galaxies and antigravity antimatter galaxies, we shall call this flow the "cosmic current," as it pushes, pulls, and redirects the movement and path of both types of galaxies within the universe as a whole.

By consolidating the a_c^3 and M_C in the above equation, we get

$$T_c^2 = [4\pi^2/G] * [(c^4e^2/2GM_C) / \pi^2] * a_c^3/M_C,$$

and then divide by T_C^2 to both sides, we have

$$1 = [4\pi^2/G] * [c^4e^2/2GM_C\pi^2] * a_c^3/(T_c^2*M_C),$$

Here we have the three parts we are looking for: Kepler's third law, the galactic wind, and matter with gravity and its motion.

We defined all Cosmological Balance gravity matter and its movement as

$$C_m = a_c^3/(T_c^2 * M_C),$$

and space-matter and its motion produce vacuum force 6.6734E-11 N,

$$S_m * m^2/kg^2 = G$$

$$S_m = G * kg^2/m^2 = 6.6734\text{E-}11 \text{ N}$$

By substituting C_m in equation above, we get

$$1 = [4\pi^2/G] * [c^4e^2/2GM_C\pi^2] * C_m,$$

$$(S_m)^2 = [4\pi^2 * kg^2/m^2] * [c^4e^2/(2(kg^2/m^2)M_C\pi^2)] * C_m,$$

where the movement of space-matter affects or tells gravity matter how to move.

Similarly, antigravity antimatter and its motions resolve to the following:

$1 = [4\pi^2/G] * [c^4e^2/2GM_A\pi^2] * A_m,$

$(S_m)^2 = [4\pi^2 * kg^2/m^2] * [c^4e^2/(2* kg^2/m^2 M_A\pi^2)] * A_m,$

where the space-matter affects or tells antigravity antimatter how to move.

The above gravity and antigravity equations are both equal to 1. This gives us:

$[4\pi^2/G] * [c^4e^2/2GM_A\pi^2] * A_m = [4\pi^2/G] * [c^4e^2/2GM_C\pi^2] * C_m,$

By including the presence of space-matter $S_m = 1$ on the left-hand side, we get:

$S_m * [4\pi^2/G] * [c^4e^2/2GM_A\pi^2] * A_m = [4\pi^2/G] * [c^4e^2/2GM_C\pi^2] * C_m,$ (9.9.2)

And by transposing or switching perspectives to antigravity antimatter, we get:

$S_m * [4\pi^2/G] * [c^4e^2/2GM_C\pi^2] * C_m = [4\pi^2/G] * [c^4e^2/2GM_A\pi^2] * A_m,$ (9.9.3)

where space-matter's vacuum force provides the push of gravity and antigravity. Therefore, the motion of antigravity and space-matter tells gravity matter what to do, and vice versa, the movement of gravity and space-matter tells antigravity antimatter how to move[281]; see image below.

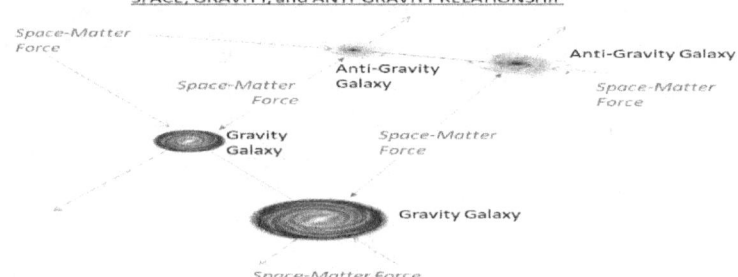

To reiterate, the first part of the right-hand side of the equal sign is Kepler's third law with an accumulated Schwarzschild mass and the second part is the acceleration caused by the galactic wind, gravity antigravity interaction. The second part of the equation goes to "1" when there is no interaction between antigravity and gravity matter, i.e. when planets are shielded by the mass of a star, or in a standalone system.

When an orbital period exists without the presence of gravity-antigravity interaction, then the system's gravity matter and its movement C_m is strictly defined as

$C_m = a_c^3/(T_c^2 * M_C),$

where the second M_C from antigravity-gravity interaction is not included within the system because its effects occur only when the circular motion of matter is enhanced with antigravity movement through the system, while space-matter and its motion S_m continues to produce the vacuum force 6.6734E-11 N, the primary source of gravity and antigravity. The above equations (9.9.1) and (9.9.3) are applicable to the entire universe, the Cosmological Balance Equation simple because it reduces to the basic acceleration and gravity equations discovered by Sir Isaac Newton.

---

[281] Unlike Einstein's general relativity equation, which primarily applies to gravity lensing of the particles and waves of light; Cosmological Balance equation applies for all macro-objects and systems in the universe.

9.10 Principle of Basic Acceleration and Gravity

Theorem 5, Acceleration and Gravity in Cosmological Balance Equation: To complete the final analysis of the Cosmological Balance Theory equation, we need to consider what happens to the equation when there is no orbital period and there is no influence due to antigravity particles passing through the area or system[282]. We already learned that the absence of the influence by antigravity particles means that the second part of the equation becomes "1." We also anticipate that if there is no orbital period and there is practically an absence of influence by antigravity particles or their systems, then the equation simply defaults to Newton's gravity force equation, $F = Gm_1m_2/r^2$, with the acceleration toward the larger mass as $\sqrt{(2GM_c/r_c^2)}$. The orbital period T_c becomes non-existent and a collision course eminent, where each passing second produces no orbit, and the measurement of circumference of the ellipse, approximated by $C = 2\pi \sqrt{((a^2+b^2)/2)}$, or of the circle $Cir = 2\pi a_c$ approaches the measurement r_c. Where the semi-major distance a_c times two becomes the distance r_c, a straight line, between the centers of mass m_1 to mass m_2.

$$T_c^2 = [(4\pi^2/(GM_C)* a_c^3)]$$

$$a_c \pi^2 / T_c^2 = GM_C/(2a_c)^2 \quad (9.10.1)$$

$$\pi^2 a_c / T_c^2 = Gm_{1C} / 4a_c^2,$$

$$m_{2C} * \pi^2 a_c / T_c^2 = Gm_{1C} m_{2C} / (2a_c)^2, \quad (9.10.2)$$

In the presence of an orbital period in a solar system, the sun is at one focal point, and the planet orbits around the sun and the other focal point in an ellipse. This ellipse is represented by $f^2 = a^2 - b^2$, where a is the length of the semi-major, and b is the length of the semi-minor, and f is the distance from the center of the ellipse to the focal point. The distance from the planet at point P to the focal point f_1 is labeled PF_1 and the distance from P to f_2 is PF_2. The eccentricity $\varepsilon = 2f/2a = f/a$, and the two lengths, focal point to the planet, added together is equal to 2a,

$$PF_1 + PF_2 = 2a,$$

To change this ellipse into a straight line, with no orbital period, the semi-minor b becomes zero. As the semi-minor b goes to zero, the eccentricity goes to "1," specifically 2f=2a, moving the focal point f_2 onto the planet P, changing its trajectory directly into f_1 and dividing the period T by two, to $T/2$ or the time to impact Q_c. This shortens the length of the semi-major a, from the center between the two objects to the sun f1, and the same length a, from the center to the planet P, thereby setting $2a = r$, where r is the distance between the center of the sun to the center of the planet m_2. Similarly, the distance around the circumference $2\pi*sqrt((a^2+b^2)/2)$ goes to the length of r, where b goes to zero, $b \rightarrow 0$, we also have $2\pi a/\sqrt{2} = \sqrt{2} * \pi a \rightarrow r$, and $\pi/\sqrt{2} \rightarrow 1$ for a flat line. Therefore, substitute $2a_c$ for r_c into the equation (9.10.2) above, we get:

$$\pi^2 a_c/(T_c/2)^2 = Gm_{1C} / r_c^2,$$

---

[282] Where there is no orbital period, the Cosmological Balance Equation simplifies to Newton's gravity equation or his acceleration equation.

$$4\pi^2 a_c / T_c^2 = Gm_{1C} / r_c^2,$$

There are two ways to interpret the above equation: first as acceleration, and second as Newton's gravity equation.

NEWTON'S ACCELERATION EQUATION

As acceleration, we have

$$\sqrt{(2r_c \pi^2 / T_c^2)} = \sqrt{(GM_C / r_c^2)},$$
$$\sqrt{(2 \cdot 2r_c (\pi/\sqrt{2})^2 / T_c^2)} = \sqrt{(GM_C / r_c^2)},$$
$$\sqrt{(2 \cdot 2r_c (1)^2 / T_c^2)} = \sqrt{(GM_C / r_c^2)},$$
$$\sqrt{(4r_c / T_c^2)} = \sqrt{(GM_C / r_c^2)},$$
$$\sqrt{(r_c / (T_c^2/4))} = \sqrt{(GM_C / r_c^2)},$$
$$\sqrt{(r_c / (T_c/2)^2)} = \sqrt{(GM_C / r_c^2)},$$
$$\sqrt{(r_c / Q_c^2)} = \sqrt{(GM_C / r_c^2)},$$
$$A_C = \sqrt{(GM_C / r_c^2)},$$

where acceleration of object m_2 going toward M_C is equal to $\sqrt{(r_c / Q_c^2)}$.

NEWTON'S GRAVITY EQUATION

As Newton's gravity equation, we multiply both sides of the equation by m_2, to get:

$$(2r_c \pi^2 / T_c^2) \cdot m_2 = (G\, m_{1C}\, m_{2C} / r_c^2),$$

Multiply left side by $2/(\sqrt{2} \cdot \sqrt{2})$, to get

$$(2 \cdot 2r_c (\pi/\sqrt{2})^2 / T_c^2) \cdot m_2 = (G\, m_{1C}\, m_{2C} / r_c^2),$$

As semi-minor b goes to zero, then $(\pi/\sqrt{2}) \to 1$, giving us:

$$(4r_c (1)^2 / T_c^2) \cdot m_2 = (G\, m_{1C}\, m_{2C} / r_c^2),$$
$$(4r_c / T_c^2) \cdot m_2 = (G\, m_{1C}\, m_{2C} / r_c^2),$$
$$r_c / (T_c/2)^2 \cdot m_2 = (G\, m_{1C}\, m_{2C} / r_c^2),$$
$$r_c / Q_c^2 \cdot m_2 = G\, m_{1C}\, m_{2C} / r_c^2,$$
$$F_c = G\, m_{1C}\, m_{2C} / r_c^2,$$

where the force of gravity is F_c.

GRAVITY AND ACCELERATION

To reiterate, in the absence of a periodic orbit, we have collision course where m_1 and m_2 are moving directly toward each other in a straight line with distance r between the centers of each object. The relationship between space-matter in conjunction with antigravity antimatter affects normal gravity matter in the following equation,

$S_m * A_m = C_m,$

where S_m represents space-matter and its motions, A_m represents antigravity antimatter and its motions, and C_m represents normal gravity matter and its motions.

Converting motion of periodic orbit into direct collision course

$S_m * [4\pi^2/G * a_a^3/(T_a^2 * M_A)] = [4\pi^2/G * a_c^3/(T_c^2 * M_C)],$

reverts to both the acceleration and the gravity equations:

$S_m * G\, m_{1C}\, m_{2C}/r_c^2 * F_a = G\, m_{1A}\, m_{2A}/r_a^2 * F_c,$

$S_m * \sqrt{(GM_C/r_c^2)} * A_a = \sqrt{(GM_A/r_a^2)} * A_C,$

where the top formula uses Newton's gravity equation with F as the force of gravity, and the bottom uses the acceleration equation with A as acceleration.

ANTIGRAVITY, GRAVITY, & REPULSION ACCELERATION

Since $(G\, m_{1C}\, m_{2C}/r_c^2) / F_c = 1$ and $\sqrt{(GM_C/r_c^2)} / A_C = 1$, we can reposition them to either side of the equation, to get

$S_m * A_C * A_a = \sqrt{(GM_A/r_a^2)} * \sqrt{(GM_C/r_c^2)},$

$S_m * A_C * A_a = G\sqrt{M_A}\sqrt{M_C}/r_{ac}^2,$

and

$S_m * F_c * F_a = G\, m_{1A}\, m_{2A}/r_a^2 * G\, m_{1C}\, m_{2C}/r_c^2,$ (9.10.3)

$S_m * F_c * F_a = G\,(-m_{1A})(-m_{2A})/r_a^2 * G\, m_{1C}\, m_{2C}/r_c^2,$

where both right-hand interactions are attractive.

By definition of antigravity antimatter, their interaction with gravity matter is represented with a negative number specifically the part associated with G becomes −1 m/kg and is repulsive. We defined space-matter, as the source of force on gravity matter combined with the natural attraction in quantum entanglement m/kg is F. Hence, by commutative law the above formula (9.10.3) can be changed to show gravity antigravity repulsion, and becomes:

$S_m * F_{ac} * F_{ca} = G\,(-m_{1A})\, m_{2C}/r_{ac}^2 * G\, m_{1C}\,(-m_{2A})/r_{ca}^2,$

$S_m * F_{ac} * F_{ca} = (-G\, m_{1A}\, m_{2C}/r_{ac}^2) * (-G\, m_{1C}\, m_{2A}/r_{ca}^2),$ (9.10.4)

where equation (9.10.3) shows antigravity galaxies attracting each other while gravity galaxies attract each other simultaneously, and in equation (9.10.4) we also have at the same time the antigravity galaxy repels the gravity galaxy away, and the distance between the center of mass of both galaxies is represented by r_{ac}. This four-way interaction enables the whole universe to remain homogenous, maintaining the Cosmological Balance.

BASIC ATTRACTIVE EQUATION

The basic equation of attraction between two gravity matter galaxies is

$S_m * F_c = G\, m_{1C}\, m_{2C}/r_c^2,$

where m_{1C} and m_{2C} represents two gravity galaxies. The distance r_c between gravity galaxies is inversely proportional to the attraction of the masses of m_1 and m_2.

This formula can also be written as,

$S_m * F_c = 6.6734E\text{-}11 \text{ N} * \text{m/kg}*m_{1C} \text{ m/kg}*m_{2C} /r_c^2$,

where 6.6734E-11 N is the force provided by the movement of space-matter S_m.

BASIC REPULSIVE EQUATION

The basic equation of repulsion between antigravity and gravity matter galaxies is

$S_m * F_{ac} = G\, m_{1A}\, m_{2C} /r_{ac}^2$,

where m_{1A} represents an antigravity galaxy and m_{2C} represents a gravity galaxy. The distance r_{ac} between antigravity galaxy and the normal gravity galaxy is inversely proportional to the repulsion of the masses of galaxies m_{1A} and m_{2C}.

This equation can also be written as,

$S_m * F_{ac} = \text{-}6.6734E\text{-}11 \text{ N} * \text{m/kg}*m_{1A} \text{ m/kg}*m_{2C} /r_{ac}^2$,

where 6.6734E-11 N is the force provided by the motion of space-matter S_m.

Newton's basic acceleration and gravity equation govern the relationship between like types of galaxies when attracting each other where ordinary gravity galaxy attracts another gravity galaxy and where antigravity galaxy attracts another antigravity galaxy. The negative aspect of these basic equations shows us the repulsion formula of two different types of galaxies where antigravity galaxy repels a gravity galaxy and vice versa. This interaction between galaxies provides the homogenous distribution we see all over the universe[283].

Given the current configuration of the universe, in the form of sheets and filaments, observations have not yet confirmed if the entire gravity matter portion of the universe is rotating in unison in a specific direction or will begin to rotate any time soon. For now, it is reasonable to assume that its components are in a chaotic and balanced movement. In any case, the nature of the Cosmological Balance allows for such motion, whether uniformed or random, while still maintaining its isotropic and homogenous state.

---

[283] The existence of all three types of matter in the universe enables it to sustain a homogenous and isotropic state, allowing some galaxies to merge while keeping others apart.

Conclusion and Way Ahead

"The magic is only in what books say, how they stitched the patches of the universe together into one garment for us."

— Ray Bradbury, Fahrenheit 451

In part 1, we discussed matter and its properties, the micro and macro elements of the universe, from the building blocks to the largest galaxies and the awesome black holes within them, the super clusters of galaxies, and the sheets and filaments of the visible universe, and their associated physics, and natural laws defining each. We also reviewed the inexplicable elements that baffled astrophysicists and cosmologists alike, "dark matter" and "dark energy." In part 2, we recapped the natural state of balance found in the universe. No matter where we look, we find that everything in the universe moves or flows toward balance and symmetry. Then, we presented the Cosmological Balance Theory, summarized the theory and its assumptions, and then discussed mathematics outlining the basic theory. From there, we continued with the definition of Cosmological Balance, its application and some supporting observation tests, and its physics of creation, and a function of a field theory. Finally, in closing part 2, we presented and explained the two sides to the universe, the visible and the invisible. Invisible only in the sense that our current instruments and hence our eyes are not able to detect them directly. We talked about the visible side first and explained how black holes evaporated in another way besides Hawking Radiation, the excretion process. We then discussed the invisible side of the universe, the evaporation and excretion of white holes, and depicted the interaction of both sides of the universe. In part 3, chapter 7, we re-evaluated the Big Bang Tenants. To do this, we discussed the universe' expansion, the source of microwave background noise in deep space beyond the galaxy, the abundance of light elements, the lack of other remnant background radiation, the Big Crunch, the age of the universe, and finally the formation of the Milky Way galaxy and its age. In chapter 8, we mathematically determined exactly how much change of the sun's acceleration is necessary to maintain orbital speed at 828,000 km/h (230 km/s). And closed part 3 with Cosmological Balance principles.

The Cosmological Balance Theory has made bold claims of the existence of antigravity antimatter and space-matter through slightly different interpretation of data already collected by astrophysicists and cosmologists. Data used to support our claim are namely the accelerated orbital period of stars around galaxies, the size of the massive black hole at the center of the galaxy, the sustainment of black holes despite its ferocious appetite, the lack of exploding black holes, along with several other observations. Certain elements of the Cosmological Balance Theory, like the existence of space-matter and antigravity antimatter, which are undetectable by our instruments, can only be proven through detailed observations and careful interpretations of the data collected in the past and prescribing areas for future fact gathering. Unexplained data being used to justify "Dark Matter" and "Dark Energy" are worthwhile candidates to use to search for and to estimate the size of antigravity galaxies, based on repulsive gravity antigravity interactions. Movements of galaxies against known or expected gravitational interactions are other nominal examples.

The existence space-matter atoms are extremely difficult to prove for these atoms are just as illusive to us as antigravity antimatter, which are not in synch with our flow of time. Even Einstein himself through his equation $G_{\mu\nu} = 8\pi G/c^4\, T_{\mu\nu}$ admitted that space itself tells normal matter what to do, which implies that it has some amount of mass and energy in order to transfer force to normal matter, we call vacuum force. Einstein eventually threw in a constant into his equation to stabilize the expansion of space, and then later retracted it as his greatest blunder. Space-matter's existence can be extrapolated by careful examination of the "halo" around galaxies and the presence of dark matter and energy. This halo is but one example of the interactions between space-matter and normal matter or antigravity antimatter. Another example of the interaction between space-matter and normal matter is the extremely high temperatures of the corona around the sun, a natural barrier where heat builds up faster than it can escape. Space-matter, in this case, creates a barrier over another transparent layer of solar atmosphere above the visible and super-heated corona. This transparent layer space-matter causes the significant bending of starlight as it passes through this exotic dense clear and colorless atoms and molecules.

The solutions of the past are not always pertinent to the present or the future. In my analysis of time dilation, I discovered that Einstein omitted one important fact of reality in his imaginary experiments, distant light equates to past information or an after-image possessing time delay. Special and general relativity, based on misread folded-after-images or mental-mirages-of-reality, were successfully portrayed as the 'physics of light' to academia and to the world. Just as a pulled string lags behind while pushed strings obstinate, these after-image glimpses into the past are not perfect reflections of the now or of tomorrow. There is no time dilation only time delay caused by after-image effect. Sir Isaac Newton's gravity laws, therefore triumphant above Albert Einstein's theories and as such space and time separately remain universal, and relative in terms of Galilean Galileo, not Einstein.

The universe is trying to reveal to each one of us its secrets and to connect us to the universe itself, many of us are just not listening. As I stated in the beginning, mathematicians, physicists, and scientists strive to emulate reality in the form of equations, and in doing so, are able to redirect the four natural forces, excluding gravity, to accomplish specific work seen in machinery and instruments as scientific and technological advancements. These breakthroughs are made possible because we mostly understand, and therefore believe we know, how nature works. However, when it is all said and done, the four natural forces and its reality does not have to obey the math and science, because we really do not understand the why part. All we know is that the visible universe we observe around us exists because of gravity, which is by the grace of God.

We learned that the tenants supporting the Cosmological Balance, Theory of Balance are the following: First, massive black holes have not exploded despite the anti-matter created from the "soup" within it. Second, massive black holes release energy and mass in the form of micro white holes through wormhole tunnels. Third, the wormhole tunnels antigravity forces the stars in the galaxy's arms to orbit faster than with gravity alone while enhanced by the gravity-antigravity hum of the surrounding space-matter. Fourth, these ejected micro white holes travel outward in all directions in great numbers like sunlight moving away from the sun, and therefore their great numbers carry with it enough mass to gently nudge stars, like waves hitting a large ship on the ocean of space. Fifth, the micro

CONCLUSION AND WAY AHEAD

white holes' sudden drop below the speed of light as the wormhole tunnel collapses releases a microwave burst at a frequency around 160.2 GHz, which comes to us as background noise from outside the galaxy, and essential sub-atomic particles to create antigravity antihydrogen, antihelium, and other light elements. Sixth, the pattern of the white holes ejected from the massive black hole at the center of the galaxy determines the shape of the galaxy. Seventh, the representation of gravity using Einstein's fabric of space-time curvature is faulty, as Newtonian equations are sufficient to solve Mercury's orbital anomaly. Eighth, antigravity galaxies are invisible to us because their light never reaches us, or our instruments, they curve away, just as like magnetic poles avoid connecting, and flow backwards in motion. Ninth, the size of the universe is attributed to the repulsive interaction between galaxies with gravity and antimatter galaxies with antigravity. The lensing effect of antigravity galaxies and their repulsive force is what cosmologists and astrophysicists perceive as "dark matter" and "dark energy." Tenth, the first through ninth tenants above apply to the invisible antigravity galaxies where massive white holes generate micro black holes, which are expelled into wormholes that emerge and provide the hydrogen, helium, and other light element building blocks for all visible galaxies.

Other tenants supporting the Cosmological Balance Theory, include: eleventh, gravity waves are best represented with "bubble" layers, something like the orb levels of electrons, but at an infinite number of levels, with the strongest pull at the surface of the object in space. As seen from the North Star above, the orbiting planet "rolling" in a counterclockwise direction on the "bubble" layer gives the planet its prograde rotational direction, conservation of angular momentum, and is a balance between gravity and centrifugal force. Twelfth, there is a lack of remnant background radiation at cosmic or light frequency range supposedly left by the Big Bang. Thirteenth, evidence in Hubble Deep Field Image shows us that the universe is much older than 13.8 billion years old. The image depicts galaxies as developed as this galaxy, 12 to 13 billion years old, thereby, making the galaxy in the image at least 25 billion years old, which contradicts the supposed age of the universe. Fourteenth, this age evidence of galaxies imaged in the Hubble Deep Field is in agreement with simulations of how galaxies, like the Milky Way is formed, piece by piece, billions upon billions of years to build, like the rings of a tree growing and aging over time. The oldest part of the galaxy is the center, and the youngest part can be, therefore found at the edges, accumulated as the galaxy moves through space. Fifteenth, the slowing down of the pioneer spacecraft as it leaves the solar system can be attributed to the effects of antigravity wave generated by micro white hole orbs passing nearby. Sixteenth, the true age of the universe is certainly beyond 14 billion or even 23 billion years old, therefore unknown; for all we know it could be googolplex years old. Seventeenth, space-matter we call "*hygratium*" fills the void of outer space. It makes space what it is and keeps gravity and antigravity in balance. Finally, the super massive black hole at the center of the galaxy is not connected via one massive wormhole to a super massive white hole at some distant galaxy or point in space, as imagined, and calculated by Einstein. With all the evidence stated above, the Cosmological Balance Theory stands firm, justified, and factual. In the words of Arthur Conan Doyle, "Once you eliminate the impossible, whatever remains, no matter how improbable, must be the truth." Praise God! Glory be to the Father, and to the Son, and to the Holy Spirit, as it was in the beginning, is now, and ever shall be, world without end. Amen!

C.1 Adjunct Thoughts

"Two things are infinite: the universe and human stupidity; and I'm not sure about the universe."

— Albert Einstein

We are part of the universe we live in, and the universe is trying to teach us its secrets. Let us open our minds to the universe of stupendous possibilities to raise our intellectual level, and not perpetuate the human stupidity as Einstein said. Our current spaceship propulsion technology does not give us the ability to travel between planets effectively let alone the great distances between stars. We should look for ideas at the pioneer spacecraft anomaly and reexamine when it first encountered the slowing force. If this force is in fact the result of antigravity wormholes in close proximate to the solar system, then maybe there could be some way we can harness that force to propel us rapidly in space. Maybe someday our scientists may develop technology similar to some kind of antigravity sail to utilize and ride the "galactic wind" to travel at speeds close to the speed of light squared or potentially faster when considering the direction and galaxy's speed added to it. If sailboats can sail at an angle against or with the wind, then maybe antigravity sailing spaceships can harness the antigravity galactic wind to propel it forward through open space, thereby making intra-galactic voyages possible.

All in all, humankind has come a long way in the modern world, but we have yet to relearn all that our ancient ancestors knew about space and the universe. We as a human race cannot afford to continue going down the path imagined and established by Einstein's theories, and those that followed his logic, which were supported by no more than his eloquent mathematics, which were riddled with perpetuating errors. We need to spend more time and money solving problems and challenges than destroying each other. We should all work toward improving our livelihood, curing, and reversing diseases, and lengthening our lives. Life is precious. The book Chariots of the Gods (von Daniken, 1969) proposes questions, ideas, and concepts that some of us should review again to ensure we did not miss something important there to benefit humanity. I suspect that our ancestors may have had the knowledge to harness antigravity waves on earth. Usage of such waves could enable an individual the ability to move objects weighing tons but suspended with antigravity with just a push from their hands. Maybe there is some truth to the stories of Merlin the Wizard moving the stones to the Stonehenge and stories of Edward Leedskalnin doing a similar feat in the 1940s building the Coral Castle. Egyptians may have used the same techniques in building the Great Pyramids in Giza. How about Easter Island and all the other unexplained locations worldwide?

This book presented an answer to some of the greatest inexplicable elements of the universe. With it come more questions still waiting to be solved. It beckons the best of us to dream up solutions and share those possible ideas with the world. This book is but one theory that effectively tackles the unknown and solves most inexplicable if not the important questions concerning the universe. When validated, one can see that its contents combine the scientific breakthroughs of our greatest minds in one Theory of Balance and solve numerous dilemmas challenging our astrophysicists and cosmologists. The Cosmological Balance Theory presented in this book is the linkage of Newton's Gravity as presented in Principia, the original Galilean's Relativity Theory (with the refusal to use Einstein's general relativity work), and Quantum Mechanics with a proposed sub-atomic defined absolute for space and time. The resulting theory, the Cosmological Balance Theory, potentially qualifies as one of the candidates for consideration as the *Theory of Quantum Gravity, which is* sought after by the scientific community as the final *Theory of Everything* and the base trunk of the unification tree.

In any case, antigravity particles are a necessary release valve for the black hole, they are the source of the "galactic wind." Either the black hole somehow releases matter, or it explodes in a colossal death, resulting from all the anti-matter it creates within it. That has not happened in any of the billions of galaxies in the observable universe, so we can assume super massive black holes at the center of galaxies do not explode; they release their excess energy and pressure, and a minuscule amount of mass. Antigravity creation is that release. How do we know anti-matter is created within black holes? Well, we know that collisions within the Hadron Collider have shown us the momentary creation of anti-matter, and its quick annihilation. Therefore, it is logical to assume that the same anti-matter is created within the black hole but in much greater quantities. Numerous experiments of smashing particles using the Hadron Collider have not yielded the creation of antigravity particles, but that does not mean it cannot be created with the additional intense energy from surrounding anti-matter and matter annihilations within the depths of black holes. This same logic applies to super massive white holes in antigravity galaxies, in their creation of gravity sub-atomic particles and micro black holes travelling in wormholes from the center of their galaxy to the outer reaches beyond their edge. We see from this unique release of material that a black hole is an antimatter white hole, and a mirrored antigravity black hole is also a gravity white hole, each pair overlaid in time.

So, there you have it, the final unified Cosmological Balance Equation: space-matter and antigravity antimatter tells gravity matter what to do, and vice versa:

$$S_m * [4\pi^2/G] * [c^4 e^2/2GM_A \pi^2] * A_m = [4\pi^2/G] * [c^4 e^2/2GM_C \pi^2] * C_m,$$

$$S_m * [4\pi^2/G] * [c^4 e^2/2GM_C \pi^2] * C_m = [4\pi^2/G] * [c^4 e^2/2GM_A \pi^2] * A_m,$$

where S_m is space-matter and its motions, A_m is antigravity antimatter and its motions, and C_m is normal gravity matter and its motions. And $C_m = a_c^3/(T_c^2 * M_C)$, where M_C is the accrued mass of everything within the orbital radius a_c, whether it is luminous or not.

And the planetary and galactic orbital periods are derived from above:

$$T_c = [\sqrt{(4\pi^2/(GM_C)*a^3)}] * [c^2 * \sqrt{(1/(2GM_C))} * e/\pi],$$

where $[c^2 * e/\pi] * \sqrt{(1/(2GM_C))}$ is the acceleration resulting from the interaction between gravity and antigravity antimatter. In instances where the system is shielded from such interactions or where the wormhole tunnel loses energy and collapses, the second part eventually goes to "1", we get:

$$T_c = [\sqrt{(4\pi^2/(GM_C)*a^3)}]$$

which is the planetary orbital period known as Kepler's third law, and ultimately equates to Newton's gravity equation. The next quest is to design and build a working protype antigravity device and patent it.

C.2 Consciousness and Intelligence

In analyzing paired point particles and their attributes within the quark, we discover a hint of intelligence through the precise selective motions they undergo in generating the graviton effect. The patterns of these motions are clearly not random or spontaneous, but deliberate. Creator traits found within subatomic point particles contribute to qualities of the atom, which in turn flow with increase enhancements as it moves upward into the various levels from inanimate to animate kingdoms from microscopic life all the way to humans. Paired point particles have inherent self-awareness, the will to survive by learning to avoid collisions, the ability to stabilize and stay connected, while avoiding decay or being ripped apart. Their cooperation with other nearby particles allows them to synchronize efforts within their environment and enables them to create associations or bonds with other similar and supportive structures to build and interconnect into a larger "subatomic group or community." These simple aspects found within subatomic particles contribute to the "consciousness" of the atom (Bailey, 2013). Additionally, the primordial "consciousness" of paired point particle, creating the gravity effect; atoms progressively contribute and develop higher cognition traits, life, as they move upward within each level of the animal kingdom, through consumption and absorption by the higher entity, until they reach the highest known level within spiritually enlightened humans.

In essence, humans are created in the image and likeness of the creator, the Almighty God, and the creator is within everything and everywhere at the same time yet residing outside of space and time. The creator formed and energized point particles in quarks with enough intelligence and unceasing power to form an atom with gravity with extreme precision, which in turn assembled with innumerable other atoms into the universe. The motions of the atom and its subatomic particles are not random haphazard configurations but deliberate patterns and structures. The universe is within us and we are in it, just as much as the creator of the universe is within us and we are in and with the creator, if we choose to be. The universe is not the Creator, as pantheism would lead you to believe. That is to say God, the Creator, is above and beyond the universe and exists outside space and time, the very essence of the universe He created. We are not just humans living earthly lives to die into nothingness, we are spiritual beings living human lives waiting and hoping to return to heavenly lives. If you believe in everlasting life after death, then you should be convinced of the reality presented in the Cosmological Balanced Theory. If you do not believe this, then your view of birth, life, and termination by death, justly matches exactly what modern academic sciences and astrophysics teaches today, the grand birth of the universe in the big bang and its glum demise at the end of its lifespan, snuffed out of existence, *"solve et coagula"*. Trust the creator and not modern science's false interpretation of wisdom. Love triumphs all. For it is the Creator whom in His infinite wisdom and divinity instilled conscientiousness and intelligence within every particle in the universe, all for the Glory of God, exists per His design. This book is consecrated to Christ through Mary *ad majorem Dei gloriam*. Glory be to the Father (the Creator of the universe), and to the Son (the Redeemer), and to the Holy Spirit (the Holy Ghost, the Paraclete), as it was in the beginning, is now and ever shall be, world without end. Amen!

C.3 The Way Ahead

Though I authored this elaborate and extraordinary story with the help of few colleagues who requested omission, it is a knowledge I received in a dream; I am but a conduit, a messenger who listened to the constant whispers within, which provided me with the inspiration, wisdom, and courage to share this simple unified theory. As I stated earlier, mathematicians, physicists, and scientists strive to emulate reality in the form of equations, and in doing so, are able to redirect the four natural forces, excluding gravity, to accomplish specific work seen in machinery and instruments as scientific and technological advancements. These breakthroughs are made possible because we mostly understand, and therefore believe we know, how nature works. However, when it is all said and done, the four natural forces and its reality does not have to obey the math and science, because we really do not understand the why part. Specifically, we cannot fully comprehend why the natural forces exist or what energizes them; we just see them as existing and declare them as laws, physical laws. Here, science fails in answering why the natural laws work, leaving the door wide open to interpretations and credit to the power of God. These four natural forces are clearly four steadfast miraculous feats, among countless others, wrought for the benefit of all, good or bad. The energy, which powers the three natural forces work together in exactly the right proportions to produce gravity, and hence causes the universe to exist, as we know it. The whole of creation and the resulting cosmic activity is the work of three fundamental forces symbolized by three entities, which constitutes the Trinity. This Trinity concept is at the root of many religions worldwide. Some cultures, however, see nine parts, realms, or dimensions. The number nine within the universe is the intertwined influences between ordinary matter, antigravity antimatter, and space-matter; three dimensions times three types of matter equal nine.

Carl Boberg clearly articulated his religious belief as he wrote, "when I in awesome wonder, consider all the worlds Thy Hands have made; I see the stars, I hear the rolling thunder, Thy power throughout the universe displayed." Science and religion must coexist; one without the other is meaningless and impractical. We owe our existence to the Creator, who assigned characteristics and intelligence to point particles, and with fine-tuned self-perpetual motion within quarks generated gravity, and as such endlessly empowered the four natural forces science uses in technological advances. The visible universe we observe around us exists because of the creation of gravity within the quarks in each nucleus of every ordinary atom in the cosmos. Religions aspiring for peace, love, cooperation, and spiritual development throughout the world have every right to thank the creator for our existence, as the creator within us wields his work and his will through us. Just as every subatomic particle, atom, DNA strand, and part of the body works together to make the human body function properly. Any deviation of any part of the body can lead to sickness and even death, while corrective action and repair to that part leads to restoral of health of the whole body. Every person in the community is like the parts of the body. Spiritual enlightenment of one person is as beneficial to those around them and radiates or propagates good outward to make the community whole, productive, and healthy, while the presences of one person with evil thoughts and actions are just as contagious and destructive to the community, bringing chaos, disarray, injury and death. We all know this as the struggle between good and evil. The choice is yours. I hope you nurture and radiate the good within you.

Appendices

Appendix A: Units, Equations, and Constants

Appendix B: Commentaries and Future Challenges

Appendix A: Units, Equations, and Constants

1. Units:

QUANTITY, NAME, SYMBOL, AND OTHER NOTES.

Length, meter, m

Mass, kilogram, kg

Time, second, sec or s

Angle, radian, rad

Arc Angle, Arc second, arcsec

Frequency, Hertz, Hz, 1/s

Force, newton, N, kg m/sec^2 or kg m/s^2

Energy-work, joule, J, N m or kg m^2/s^2

Area, square meter, __, m^2

Volume, cubic meter, __, m^3

Speed-velocity, meter per sec, __, m/sec or m/s

Acceleration, meter per sec^2, __, m/sec^2 or m/s^2

Thermodynamic temperature, Kelvin, K

Density, kilogram per cubic meter, __, kg/m^3

2. Multiples, Prefix, and their symbol abbreviations:

10^9, giga, G

10^6, mega, M

10^3, kilo, k

10^2, hector, h

10^1, deca, da

10^{-1}, deci, d

10^{-2}, centi, c

10^{-3}, milli, m

10^{-6}, micro, µ

10^{-9}, nano, n

3. Equations:

Trigonometric equations for right angle triangles:

sin θ = opposite / hypotenuse

cos θ = adjacent / hypotenuse

tan θ = opposite / adjacent

Area of Circle:

Area = πr^2

Circle:

360 degrees = 2π radians

1 degree = 60 arc minutes = 360 arc seconds

1 radian = 206264.806247 arcsecond

1 degree = 0.0174532925 radian

Escape Velocity:

V = 2 times Gravity Constant times Mass = 2 x G x M

Newton's Gravity Equation:

F = Gravity Constant (mass$_1$ x mass$_2$) / (distance r)^2 = G (m$_1$ x m$_2$) / r^2

Euler's *e* constant is defined as the infinite series:

$$e = \sum_{n=0}^{\infty} \frac{1}{n!} = \frac{1}{0!} + \frac{1}{1!} + \frac{1}{2!} + \frac{1}{3!} + \frac{1}{4!} + \ldots$$

where *e* equals the sum of 1/n! (n factorial) from n=0 through n=infinity; *e*=2.718281828459.

4. Cosmological constants:

1. Gravitational constant, G, 6.6734E-11 N m^2/kg^2 or 6.6734 x 10^{-11} N m^2/kg^2
2. Speed of light in a vacuum, *c*, 2.998 x 10^8 m/s
3. Planck's constant, *h*, 6.63 x 10^{-34} J s
4. Mass of the Sun, M☉, 1.9891 x 10^{30} kg
5. Radius of the Sun, R☉, 695,800 km
6. Mass of the Earth, M⊕, 5.97 x 10^{24} kg
7. Radius of the Earth, R⊕, 6.37 x 10^6 m
8. Light-Year, ly, 9.461 x 10^{15} m
9. Parsec, pc, 3.086 x 10^{16} m or 3.26 ly
10. Exponential constant or Euler's #, *e*, 2.71828

11. Pi, π, 3.14159
12. Mega-electron volts, MeV, $1.60217657 \times 10^{-13}$ joules
13. Boltzmann's constant, k_B, $1.3806488 \times 10^{-23}$ kg m²/s² K

5. Estimated Cosmological constants:

1. Estimated Mass of the Milky Way Galaxy, __, 0.8 to 1.5×10^{12} M☉
2. Mass of black hole at center Milky Way Galaxy, __, 4.1 to 4.31×10^{6} M☉ or 8.2 to 9.34736×10^{36}

6. Greek Alphabet for Mathematics

Αα	Alpha	Νν	Nu
Ββ	Beta	Ξξ	Xi
Γγ	Gamma	Οο	Omicron
Δδ	Delta	Ππ	Pi
Εε	Epsilon	Ρϱ	Rho
Ζζ	Zeta	Σσς	Sigma
Ηη	Eta	Ττ	Tau
Θθ	Theta	Υυ	Upsilon
Ιι	Iota	Φφ	Phi
Κκ	Kappa	Χχ	Chi
Λλ	Lambda	Ψψ	Psi
Μμ	Mu	Ωω	Omega

Appendix B: Commentaries and Future Challenges

Writing this book has been an extreme challenge for me as a rusty mathematician spending countless hours researching and relearning techniques to help me present the concepts, ideas, hypothesis, and finally the theory in the best possible light of my abilities and understanding of the natural world. I hope you found this edition to be entertaining and informative. The Cosmological Balance Theory is a plausible unified theory, with an all-encompassing solution to the mysteries of the universe. As you know, I started out writing this book primarily for the general reader who possesses minimal or no formal schooling in the sciences. However, as I wrote along, I seemed to have wandered into highly technical and somewhat intense mathematical explanations; I hope I did not overwhelm or confused you. The basic intent of the story was to provide you insights of the universe. Although some ideas presented seemed outright impossible or deemed as pseudoscience by academia, they are practical, mathematically sound, and supported by known visible observations. The same observations that baffled our astrophysicists and cosmologists today provided the baseline for the Cosmological Balance Theory and the solution they desire, the ultimate and yet simple unification theory and base trunk of the unification tree contained here within.

Please note again that academia and mainstream science is adamantly opposed to any theory from an outside source, no matter how simple, that is not their own. As I said in the foreword, I have afforded several prominent representatives from academia and mainstream science numerous opportunities to read drafts of my theory to provide me comments and corrections, and the only reply I got was that they were too busy, not interested, or non-responsive to my requests for an audience with them. Popular notice of this book and or public talks may get their attention. I hope you can help spread the word and tell your friends and family to read this unique story of the universe, the Cosmological Balance Theory. I have already begun to work on my next endeavor to turn this theory into an actual functional design, a working model, and to patent it. It may take several years of numerous trials and errors, but I believe it is possible. Such accomplishment will open the doors to many other future possibilities. The ideas and projects along these lines are endless. Wish me luck.

Finally, I want to sincerely thank you again for investing resources and taking the time to read this book in its entirety. I apologize to the readers that I have offended in my discussions pointing out the errors made by prominent and popular scientists, physicists, and cosmologists alike. We all make mistakes. Please feel free to provide productive and critical feedback to improve the contents of this book. Two options are available to you: upload notes publicly onto the amazon.com/kindle book or in future releases to barnesandnoble.com comment page for author's private review. Please do not expect a reply. Usage of your comments and or additions will be acknowledged within the next edition of the book.

Acknowledgements

"And books, they offer one hope -- that a whole universe might open up from between the covers, and falling into that universe, one is saved."

— Anne Rice, *Blackwood Farm*

First, thanks a million to NASA for convincing the government to launch the Hubble Telescope into space and for gaining the necessary support to fix its faulty lens and the focusing issue it caused. The spectacular photos taken by Hubble have generated the interest of current and future scientists and cosmologists alike. These pictures have shown us our place in the visible universe and given us a perspective of how vast the universe is, and a glimpse of a realm of possibilities of what lies beyond. I am in awe because of it. Thanks to all the credible physicists and cosmologist who shared their knowledge to the public in terms the layperson could understand, particularly Steven Weinberg for his book Cosmology. Thank you to Wikipedia for enabling intelligent and credible users to contribute all the detailed and accurate data on the internet, without it, I could not verify information.

Second, many thanks to my friends and colleagues that took time to read the rough drafts of this book. They provided me with much needed input, corrections, and criticism for me to finalize the book. I would like to thank specifically Aaron Kawczk, Caolionn O'Connell, Craig Rivet, Malcolm Garland, David Farquharson, Charles Pangle, Maggie Kurtts, Jun Zhang, Laura and Brandon Meitz, Jeffrey Lepak, James Herring, Kevin Colyer, Kevin Jenne, Teresa Flahive, Ronald Bouchard, Stewart Crane, and Michael Moland and many others for their time and efforts to refine the contents of this book. Special thanks to the certified physicists who consistently and insistently tried to convince me to follow and yield to all of Einstein's work and many others; they challenged me to rise to the occasion and fully develop my theory. To protect their credentials, I left out their names.

Third, I also would like to thank the entertaining hosts of many of the shows about the universe and world we live in. They have presented a world of possible space exploration and intrigued the minds and hearts of millions of people worldwide. Their unique method of presentation has captured my interest and held it. Special thanks to Carl Sagan and Neil deGrasse Tyson. I would also like to thank the many authors who took the time to share their education, knowledge, and understanding of the universe.

Fourth, I would like to thank local students and enthusiasts for taking the time to hear my thoughts and give me ideas to improve this book. Their comments and edits have added enormous value to the theory presented within this book.

Finally, I would like to thank you the reader for your comments and corrections to improve this book. Your review is greatly appreciated and welcomed.

About the Author

Mark D. Calvo
Colonel (USA Retired)

Cambridge University, England
Photo © 2015 Mrs. Calvo

Mark graduated from the Guam public school system, and Kemper Military College, Missouri, earned an Associate of Arts degree in Liberal Arts. He then attended California State University at Sacramento earned a Bachelor of Arts Degree in Applied Mathematics and served over 30 years in the Army in various Command and Staff positions. While on active-duty service, Mark earned a master's degree in National Strategic Studies. After retirement, he worked as a Government contractor in Korea and in the U.S. Mark is the author of "Digitizing the Force XXI Battlefield" article in Military Review Magazine, pg. 68, May-June 1996 issue; the Parable Mysteries book, May 3, 2015; and numerous articles and memorandums for the Army. After settling into retirement, Mark returned to dream to expand his interests in Astronomy, Physics, and Cosmology.

Mark's life experiences, riddled with thought and wonder about the world we live in and the universe around us, gave him the background and insights to develop this theory. When he was a young child, Carl Sagan in the "Cosmos" series and his book by the same name inspired him with the desire to learn more about science, physics, and the universe (Sagan, Cosmos TV Series, 1980). Shows like Star Trek (Roddenberry, 1966-1969) and Star Wars (Lucas G. , 1977) opened a world of imagination about space travel. From time to time, he watched television shows about physics, science, astronomy, and the universe, shows like "How the Universe works (Abbas, 2010)," "The Universe (Thompson, 2007)," "Cosmos: A Spacetime Odyssey (Tyson, Cosmos: A Spacetime Odyssey, 2014)," "Extreme Universe (Blumenstein, 2010)," and the movie "The Theory of Everything (Marsh, 2014)" to name a few. Books like "Contact" by Carl Sagan 1985 (Sagan, Contact, 1985) and the "Dune" Trilogy by Frank Herbert 1979 (Herbert, 1979) also opened a world of possibilities. Arthur C. Clarke words echo in Mark's mind from time to time, he said *"Two possibilities exist: either we are alone in the universe or we are not. Both are equally terrifying."* The possibility I had in mind is of the survival of the species. So, I questioned the concepts presented in these shows and developed my own thoughts and hypothesis. The dream of solving the inexplicable part of the universe never stopped.

In early January 2015, he had a dream, one with a potential solution to one of the most baffling elements cosmologists have been trying to find of the universe, "Dark Matter" and

"Dark Energy" and why the universe is perceived as expanding. The dream alluded to the possibility that there are two parts to the universe, the visible and the invisible, separated in time. Mark's extensive research has led him to write and published this book. He spent countless hours relearning the necessary material and techniques, and developed the hypothesis and theory presented in these books solely for the benefit of scientific advancement. The clues that support his new theory are commonly known and recorded by science; it is just a matter of reading them a slightly different way.

To write this book, Mark conducted his own intense research on astrophysics and cosmology as a self-educating, and enlightening endeavor. A friend recommended Steven Weinberg's book Cosmology, among several other books, listed in the bibliography, as a reputable source in the field. Weinberg's book is widely used in universities to teach students in physics and cosmology. It is not an ideal book for beginners in the field. It is, however, an outstanding, and effectively presented gathering of current astrophysics and cosmology studies into one book. Mark enjoyed reading Weinberg's book, where he introduced various current equations, studies, and branches of ideology in the field. It is apparent, by his presentation, that this science is still developing, and from time to time, new and better theories override old theories, but observations remain constant, and stand the test of time. At times, some of the equations in Weinberg's book were confusing and misleading, but after additional research from other sources, Mark was able to make some sense of this incomprehensible material and made every effort to try to understand why our scientists are leading us down their path, which in his view is not necessarily the right road. Always remember that math and physics does not control how the universe exists, it only tries to capture and replicate it in numbers and equations. The same applies with math trying to emulate how nature operates and functions, even in the micro world of quantum physics. Question everything, and most of all do your own research; do not trust the word of our greatest scientists as absolute, for they too are as human as you and he and are prone to mistakes. The universe does not obey them; the universe is indifferent.

The sheer size of the universe amazes us all. Many of us begin our journey in search to understand our purpose and place in the universe we live in. For Mark, it began with a childlike fascination of the stars he saw at night at his hometown and continues to this very day. Although, he is new at cosmology and astrophysics, Mark has learned and will continue to learn from well-respected authorities on the subject. The more he reads and studies the detailed elegant mathematics supporting working theorems, the more he wonders about the validity of their baseline justifications. Rather than wait another year or two to conduct more research and learn the lingo, Mark decided not to sit back and watch our greatest minds continue taking us down a path based on a weak foundation not worth standing or building on, and the rabbit hole through which it has led us, seeking things that are simply not there, non-science. This book embodies the journey he has been on for the past thirty plus years to find the truth about the universe, and up until now, has become the greatest step for him. This book is his version of the universe's story. Who knows its acceptance might cause him to pursue a formal degree in physics or maybe not? Praises and thanks be to God!

"We are the cosmos made conscious and life is the means by which the universe understands itself."

— *Brian Cox*

Glossary

Absolute Space. It is Newton's perception of space, where it is unchanging and independent of its contents. In Cosmological Balance Theory, modifies Newton absolute space in that it is the push source of the force of gravity and as such can conform around objects in space to contain it.

Absolute Time. Newton's perception of time maintains a consistent unchanging duration pace. It is the same throughout the universe. It neither speeds up, nor slows down, in duration regardless of gravitational field or speed of observer.

Absolute Zero. The lowest possible temperature estimated to be about -273 degrees Celsius or 0 degrees Kelvin.

Acceleration. Increase change of motion or speed.

Accumulated Mass. Consecutively adding the mass contained within each expanding concentric circle.

Amplitude. Measured with the maximum height or peak of the wave, or the maximum depth or trough of the wave.

Angular Momentum. Law of Angular Momentum is defined as motion providing a natural conservation of energy.

Antigravity. One of four natural forces in invisible part of the universe. Although it is the weakest of all four in that part of the universe, it is the most dominate at the macro scale of the invisible universe. Antigravity definition in the Cosmological Balance Theory, the tendency of antigravity objects or antigravity bodies in the invisible part of the universe to come together at the atomic level, through either chemical or molecular bonds, has an infinite range, and its mechanical source of force is space-matter.

Anti-lepton. It is the category of sub-atomic particles, which includes positrons and anti-neutrinos as well as their medium and higher-level equivalents.

Antimatter. All particles have associated counterparts called anti-particles with the same mass but opposite charges. Anti-matter consists of these associated anti-particles in the form of atoms and molecules. Matter and anti-matter annihilate one another on contact.

Antiparticle. A particle of antimatter.

Anti-time. System where time runs in the opposite direction, backwards.

Astrophysics. The branch of astronomy dealing with the physics or nature of the universe, not the positions or motions in space.

Atmospheric pressure. The amount of pressure at the surface of earth resulting from the mass of atmosphere above. Normal pressure is about 14 lbs. per square inch.

Atom. The smallest complete element in the micro universe.

Barycenter. The non-rotating coordinate or point between two objects where they balance each other.

Big Bang. Theory where the universe began with an explosive start at time zero (t=0).

Big Crunch. Theory where all the matter in the universe comes together to an infinitely dense and hot state, into one singularity. One possible final event as the universe collapses.

Black hole. It is a region in space-time where the gravity force is strong enough to bend light back into it. No EM radiation can escape from within the black hole event horizon.

Boltzmann constant. Temperature constant usually written as $k = 1.4 \times 10^{-23}$ J/K. It is used to estimate the temperatures within a star.

Centrifugal Force: an apparent force that acts outward on a body moving around a center, arising from the body's inertia.

Centripetal Force: a force that acts on a body moving in a circular path and is directed toward the center around which the body is moving.

Chemistry. The study of atoms, molecules, and their bonding. It is the branch of physical science, which studies the composition, structure, properties and change of matter.

Compound. A Molecule that contains more than one element not chemically bonded.

Closed universe. State where the universe has enough energy density to pull itself together into a Big Crunch. Gravity wins.

Concentric Circles. Series of circles drawn from one central point with gradually increasing radii.

Cosmic Current. New term to define the movement of space-matter throughout the universe. Its motion causes galaxies to gradually deviate to a different direction.

Cosmological background radiation. The diffuse microwave radiation left over from the supposed Big Bang Theory.

Cosmological Balance Theory. A unified theory in which declares three types of matter exists to sustain the universe in its current homogenous and isotropic state, constantly renewing light elements for galaxies to grow to sizes capable of replenishing space-matter.

Cosmological Balance Equation. Equation applicable to all objects in the universe; it unifies Newton and Kepler laws and equations, the Schwarzschild radius and mass, quantum mechanics, and uses the principles of geometric and polar coordinate systems.

Cosmological Constant. Einstein's hypothetical energy and pressure, which uniformly fills space; its origin, composition, and source is unknown.

Cosmological decade. A unit of cosmological time expressed by time $t=10$ to the nth power (10^n) where n represents the cosmological decade.

Cosmology. This is the study of the origin and evolution of the entire universe.

Curvature. The deviation of an object or of space from a flat form, codified by the rules of Euclidean geometry.

Dark energy. The unknown energy permeated in space that causes the expansion of the universe to accelerate.

Dark era. The time-period where black holes have experienced the full Hawking evaporation timeline.

Dark matter. Matter in the universe that gives off no light (or little light). It contains most of the universe's matter and can be detected through indirect means via gravitational effects.

Density waves. Theory where sections of the galaxy have greater mass density. These sections overtake orbiting gases and compress them to birth new stars.

Doppler Effect. An increase or decrease in the frequency of sound, light, or other electromagnetic waves as the source and observer move toward or away from each other. The effect is heard as a sudden or gradual change in pitch noticeable in a passing siren or horn honking, as well as the red shift seen by astronomers.

Dwarf Planet. A spherical celestial body revolving or orbiting around the sun, like a planet but not large enough to gravitationally clear its orbital region of most or all other celestial bodies.

Electromagnetic force. One of the four natural forces. This force is composed of both the force between charged particles and that of magnetic fields.

Electromagnetic or EM radiation. This radiation is self-propagating and includes the entire spectrum between gamma rays, visible light, and all radio waves.

Electroweak Theory. The theory, which unifies the electromagnetic and the weak nuclear forces into one force, the electroweak.

Ellipse. Planets and objects orbit the mass of a body in an oval or circular path with the central body at one of the two foci.

Escape speed or escape velocity. The speed required to overcome the gravitational pull of an object and leave its surface.

Electrons. Negatively charged particles orbiting the nucleus. It also enables electric current to flow and photons to travel at the speed of light.

Entanglement or Quantum Entanglement. A phenomenon in which spatially distant particles have correlated properties, it is an instantaneous connection with infinite range.

Euclidean geometry. The mathematical system named after the Greek mathematician Euclid who described geometry concepts in his textbook.

Euler's constant. The constant $e=2.718281828459$ is the summation of one divided by n factorial for n equal zero to infinity.

GLOSSARY

Event horizon. The boundary where matter and light cannot escape from the grasp of the black hole. It separates the black hole from the rest of the universe.

Excretion Process. The process whereby a black or white hole ejects through wormhole tunnels antigravity or gravity particles respectively from their surface outward. The grand unification force (gravity-antigravity repulsion) powers black hole and white hole excretions.

First-Generation Stars. The first stars to develop directly from dense gaseous cloud in space composed primarily of hydrogen and some helium.

Fission. Nuclear reaction which large nuclei are split into smaller nuclei, usually triggering a chain reaction.

Flat Space. Spatial shape of open and intra-galactic space in which there is no curvature, where its particles and atoms are spread evenly throughout with uniform pressure, energy, and force.

Fractal. Geometry equation developed by French Mathematician Benoit Mandlebrot to computer-generate or draw realistic looking pictures of "self-similar" objects found in nature. His equations make mountains appear like the ones seen in actual landscapes.

Frequency. The total count of full wave cycles each second.

Fusion. Nuclear reaction which two or more nuclei are combined to form a larger nucleus, providing energy and heat.

Galactic Plane. Flat disk shape contains most of the galaxy's mass, its stars and other material.

Galactic Wind. Repulsion force exerted by antigravity antimatter emanating from the galaxy center and passing through the galactic system. This force maintains the orbital velocity of the entire galaxy by pushing normal matter, such as stars and their solar systems, other debris, and gases, along the way.

Galaxy. It is a gravitationally bound single system of stars and their orbiting planets, remnant of stars, and other matter. These systems come in numerous variations of three basic forms: elliptical, spiral, and barred galaxies.

Gaseous Cloud. Consolidation of particles and atoms in space resembling a cloud when lit up. It can consist of a variety of material to include microscopic debris from the remnants of exploded stars.

General relativity. A theory of gravity developed by Albert Einstein taking into account the curvature of space, time, and its effect on mass.

Gluons. Particles carrying information for the strong nuclear force.

Goldie locks region. Range of distance away from a star where the planet can have water existing in liquid form.

Googol. Number represented by 1×10^{100}

Googolplex. Number represented by 1 followed by googol number of zeros or 10^{googol}.

Grand Unification. Theory where the strong, and electroweak forces are unified into one. Cosmological Balance super grand unification also adds the fourth force, gravity, or graviton, into this unification, resulting in the gravity-antigravity repulsion force.

Graviton. The massless particle that provides gravity.

Gravity or Gravitation. One of the four natural forces. Although it is the weakest of all four, it is the most dominate at the macro scale of the universe. Definition of Gravity in the Cosmological Balance Theory, the tendency of objects or bodies in the visible universe to clump together at the atomic level, through either molecular or chemical bonds, its attractive range is infinite, instantaneous, and its mechanical source of force is space-matter.

Gravitational Constant. Identified by Sir Isaac Newton in Principia as 6.673×10^{-11} N $(m/kg)^2$.

Hadrons. Class of particles consisting of quarks and antiquarks, first discovered in the Hadron Collider.

Hawking radiation. The energy emitted by a black hole due to quantum effects, as one particle gets ejected away from the black hole and another particle gets pulled into the black hole.

Hubble Telescope. A space-based telescope that was launched into low Earth orbit in 1990, repaired lens focusing, and remains in operation through today.

Hygratium. A unique space-matter neutral-quark equivalent but with high gravity-antigravity repulsion ability "by grace" forms the vacuum of space. Given the pairs of letters "*u* and *v*" and "*m* and *n*" are interchangeable, the letters in the word "*hygratium*" yields both "*gravity*" and "*antigravity*" matter.

Hypothesis. Supposition or proposed explanation made based on limited evidence as a starting point for further investigation and observations.

Interstellar medium. The gas and dust the permeate space between stars in the galaxy.

Jeans Mass, Length, and Instability. Named after the British Physicist Sir James Jeans who considered the process of gravitational collapse within a gaseous cloud in space.

Jiggle. Quick omni-directional continuous vibrating action of particles of matter. Also defined as to move about lightly and quickly from side to side or up and down.

Kelvin. Temperature scale where absolute zero is zero Kelvin, which is equal to -273 Celsius degrees.

Kepler. Johannes Kepler developed and published the laws of planetary motion between 1609 and 1619. He among other select scientists paved the way for Sir Isaac Newton to solve gravity and publish Principia.

Kinetic energy. Energy contained in an object in motion.

Large Hadron Collider. Also, the LHC is the world's largest and most powerful particle accelerator. It first started up on 10 September 2008.

Lensing effect. The bending of light by the gravity force of a massive object so that the source appears to be in two opposite sides of the massive object.

Leptons. Category of sub-atomic particles which includes electrons and neutrinos as well as their medium and higher-level counterparts.

Machian theory or Mach's principle. The name given by Einstein to an imprecise hypothesis often credited to the physicist and philosopher Ernst Mach. A very general statement of Mach's principle is "Local physical laws are determined by the large-scale structure of the universe."

Magnetic barrier. Sphere composed of bubble of magnetic fields spun and twisted which protects and shields the solar system from some of the deadly cosmic radiation.

Magnetron. Device internal to the microwave oven the generate the microwave radiation to cook food within the cooking chamber of the oven.

Molecule. A group of two or more chemically linked atoms.

Multi-verse. State where more than one universe exists in different dimensions. Each instance is unaware of the other universes.

Neutron Star. Remnant of a star that went super nova, typically of small radius (30 km) and extremely high density, composed predominantly of closely packed neutrons.

Neutrons. The neutral charged particles within a nucleus.

Newton's Universal Gravity. Theory of gravity where the force of attraction between two objects or bodies in space is proportional to the product of their masses and inversely proportional to the square of their distances between their centers (point mass).

Nucleus. The core group of the atom, consisting of neutrons and protons held together by the strong nuclear force.

Observer. Person or instrument collecting and sending information to a person for comprehension of relevant properties of a physical system being investigated.

Oort cloud. Sphere of icy bodies surrounding the solar system, named after Jan Oort, who discovered it.

Open Universe. State where the universe does not have enough energy density to pull itself together. Therefore, it is destined to expand endlessly. Gravity fails.

Orbital Period. The time it takes an object or body to orbit once around the center of the system is resides.

Orbital Speed or Velocity. Speed necessary for an object to remain in orbit around another object in space.

Parsec. It is the abbreviation of the parallax of one arc second. A unit of distance used in astronomy, equal to about 3.26 light years (3.086×10^{13} kilometers). One parsec is the mean distance radius of the earth's orbit subtends an angle of one second of arc. In other words, a parsec is the distance from the Sun to an astronomical object that has a parallax angle of one arc-second.

Pascal Triangle. List of binomial coefficients arranged in a triangular configuration named after French mathematician Blaise Pascal.

Periodic table. Chart of known elements found or created on earth.

Photon. The messenger particle that corresponds to electromagnetic radiation or light. It travels at the speed of light through open space.

Photosphere. Gaseous atmosphere of the sun.

Plank's constant. The fundamental constant of nature that sets the scale for quantum mechanical processes.

Planet. is a body in space that orbits the Sun, is massive enough for its own gravity to make it round, and has "cleared its neighborhood" of smaller objects around its path of orbit.

Positron. The antiparticle of electron that has a positive charge.

Postulate. To suggest or assume the existence, fact, or truth of something as a basis for reasoning, discussion, or belief.

Potential energy. Energy stored within an object at rest.

Principle of Consistency. Demands that any replacement or new theory be non-contradictory and adhere to proven and established laws of physics. New theory must be as good as the old ones while fixing their errors and discrepancies.

Principle of Equivalence. Core principle of Einstein's general relativity theory declaring that equivalent accelerated motion is indistinguishable from an actual gravitational field.

Principle of Relativity. This is the core principle of special relativity where all constant velocity observers are subject to the same set of physical laws, as if they are at rest or in as stationary position.

Product. The result achieved when multiplying two numbers together.

Protons. The positively charged particles within a nucleus.

Quantum Entanglement. Particles and objects considered intertwined with other particles or objects respectfully as if they are one particle or object. Quantum mechanics tells us that we should treat the universe as one entangled object.

Quantum Mechanical Attraction. Particles attracting each other at the quantum entanglement level.

Quantum Gravity. Unifies quantum physics with Newtonian gravity and Galilean relativity. Cosmological Balance gravity equation defines each mass' inherent 1 m/kg attraction of sub-atomic particles exists at the level of quantum mechanics entanglement of objects over infinite distances and their tendency to function or merge as one.

Quarks. The fundamental particle that makes up protons, neutrons, and other composite particles.

Radiation. The energy carried by electromagnetic waves, heat waves, or particles.

Remnants of Stars. Stellar debris left over from the explosion of a star at the end of its lifespan.

Rotational Period. Time it takes an object in space to rotate on its axis one time or one revolution.

Singularity. A point in space-time where density is infinite, and where temperature and pressure are also infinite.

Second-Generation Stars. The classification of stars created from remnants of first-generation stars. Their building block material consists primarily of hydrogen, some helium, and some heavier elements, like carbon, oxygen, nitrogen, as well as much heavier elements.

Slingshot effect. Catapult or rapid increase of speed resulting from a fast-moving massive object's gravity pulling and accelerating a smaller object as it overtakes it during their orbits around a common central orb.

Solar System. Star orbited by one or more planets, asteroids, and comets, and other debris.

Solar Mass. Mass of the sun, 1.9891×10^{30} kilograms

Space-matter. It is a substantial super cooled gaseous matter with perfect almost frozen gas-liquid characteristics, which occupies the ocean of deep space between galaxies. It neither has gravity nor antigravity properties, and it makes space what it is, appearing void and empty. These substantial high-energy atoms form an infinite grid and sustain the void of space throughout the entire universe, and possess pressure, energy, and force.

Space-time. Any mathematical model, which combines space and time into one continuum, used to solve four dimensions problems: also written as spacetime.

Special Relativity. The first theory developed by Albert Einstein, claims that it is the most accurate model of motion at any constant speed. In it, space and time are not absolute; they are dependent on the relative motion between distinct observers.

Stellar black hole. A black hole with a mass comparable to a star but three to thirty solar masses.

Stellar remnants. Refers to all objects left over at the death of a star.

String Theory. Theory of the universe postulating that the ingredients of nature are in one-dimensional vibrating filaments of energy.

Strong nuclear force. One of the four natural forces. It holds protons and neutrons together within the nucleus.

Sub-atomic. Particles smaller than the proton or neutron.

Super-massive black hole. Large black holes with millions to billions of solar masses can be found at the centers of most galaxies.

Supernova. Violent death by explosion of a super-massive star.

Symmetry, Perfect Symmetry, or Super-symmetry. This is a characteristic where an orientation change of a physical system that leaves its appearance unchanged and has no effect on the laws describing it. A perfect sphere is the same sphere regardless of which way it is rotated.

Taijitu or Yin Yang. Concept in Taoism defining the Yang "sunny" and Yin "shady" sides of everything we know. For every side of something, there is an opposite side.

Thermodynamics. A branch of physics concerned with heat and temperature, and their relation to energy and work. It defines macroscopic variables, such as internal energy, entropy, and pressure, which partially describe a body of matter or radiation.

Third-Generation Stars are those stars formed from the remnants of second and first-generation stars.

Tidal forces. Where this is the difference between the gravitational force on one side of the object and that of the opposite side of the object.

Time dilation. An effect in which time runs slower. It occurs for objects moving close to the speed of light, especially near the black hole event horizon.

Unified Theory. A theory that applies to all forces and all matter in a single theoretical structure or concept.

Vacuum Energy. Energy in deep space caused by space-matter pressure distributed throughout the volume of space. Energy = Pressure * Volume.

Vacuum Pressure. Pressure created by the vacuum force of substantial space-matter atoms spreading out and maintaining distance between each other within a given volume of space. This force is primarily electrostatic in nature.

Velocity. The specific speed and direction together of an object's motion.

Weak nuclear force. One of the four natural forces. It mediates radioactive decay.

White holes. It is a region in space-time where the antigravity force is strong enough to bend antigravity light back into it. No antigravity EM radiation can escape from within the white hole event horizon. In essence, white holes are "black holes with antigravity properties." This term originally defined mathematically by Einstein as the opposite of black holes spewing out material for the universe.

World line. A line representing a particle's travel progress through space-time.

Wormhole. This structure makes a connection between two black holes and acts like a bridge between two points of space-time, usually of great distances.

Yin. Defines the Shady side in Taoism of everything we know.

Yang. Defines the Sunny side in Taoism of everything we know.

Bibliography

Abbas, Y. (Director). (2010). *How the Universe Works* [Motion Picture].

Adams, F. (2000). *The Five Ages of the Universe.* New York: Simon & Schuster.

Alexander. (2010). Recent developments in gravity-wave effects in climate models and the global distribution of gravity-wave momentum flux from observations and models. *Quaterly Journal of the Royal Meteorological Society*, 1103-1124.

Alexander, J. (2017). Sensitivity of Gravity Wave Fluxes to Interannual Variations in Tropical Convection and Zonal Wind. *Journal of the Atmospheric Sciences*, 2701-2716.

Bailey, A. A. (2013). *The Consciousness of the Atom.* Kindle Edition: Kindle Direct Publishing.

Baumann, D. (2014). *Cosmology: Part III Mathematic Tripos.* Cambridge England: damtp.cam.ac.uk.

Baumgarten, G. (2017, 09 20). *Gravity waves influence weather and climate.* Retrieved from Science News: https://www.sciencedaily.com/releases/2017/09/170920100043.htm

Blumenstein, R. (Director). (2010). *Extreme Universe* [Motion Picture].

Bondi, H. a. (1948). The Steady-State Theory of the Expanding Universe. *Royal Astronomical Society, Vol 108*, 252-270.

Boyle, R. (2013, 06 25). *Why Does the Sun's Corona Get So Hot?* Retrieved 09 02, 2015, from Scientific American: http://www.scientificamerican.com/article/iris-satellite-launch/

Buhler, O. (2003). Equatorward Propagation of Inertia–Gravity Waves due to Steady and Intermittent Wave Sources . *Journal of Atmospheric Science*, 1410-1419.

Carlip, S. (2011). *Does Gravity Travel at the Speed of Light?* Retrieved 11 2, 2015, from Physics Frequently Asked Questions: http://math.ucr.edu/home/baez/physics/Relativity/GR/grav_speed.html

CDN77. (2015). *Hubble Space Telescope.* Retrieved 11 14, 2015, from ESA/Hubble: www.spacetelescope.org

Chyla, W. T. (2012). *Simultaneous Gravitational and Refractive Lensing.* Warszawa, Poland: Applied Science Enterprise.

Cleonis. (2017, 06 18). *Gyroscope physics.* Retrieved from The Science of Physics Rotation: http://www.cleonis.nl/physics/phys256/gyroscope_physics.php

Collier, P. (2014). *A Most Incomprehensible Thing.* Kindle Edition: Incomprehensible Books.

Colville, C., Nurmohamed, A., & Taylor, D. (Directors). (2012). *Orbit: Earth's Extraordinary Journey* [Motion Picture].

Cramer, J. G. (2002, 1 27). *Faster-than-Light Laser Pulses?* Retrieved from Analog Science Fiction & Fact Magazine: Alternate View Columns by John Cramer: https://www.npl.washington.edu/av/altvw105.html

Cromie, W. J. (1999). Physicist Slow Speed of Light. *Gazette, the Harvard University*, http://news.harvard.edu/gazette/1999/02.18/light.html. Retrieved from http://news.harvard.edu/gazette/1999/02.18/light.html

Davies, P. (2002). That Mysterious Flow. *Scientific America*, 40-47.

Earth Science. (2018, 4 13). *Semidiurnal Tide.* Retrieved from ScienceDirect: https://www.sciencedirect.com/topics/earth-and-planetary-sciences/semidiurnal-tide

Ebrahimi, E. (2014). Higher Dimensional Spherically Symmetric Expanding Wormholes in Einstein's Gravity. *Springer Science+Business Media New York.*

Einstein, A. (1920). *Relativity, The Special and General Theory.* Kindle Direct Publishing Edition 2013.

Einstein/Golm/Potsdam. (2015). *Einstein online provided by Max Planck Institute for Gravitational Physics.* Retrieved 11 02, 2015, from E=mc^2: http://www.einstein-online.info/elementary/specialRT/emc

Exchange, P. S. (2012, 03 19). *Physics.* Retrieved 02 15, 2016, from physics.stackexchange.com: http://physics.stackexchange.com/questions/7250/gravitational-lensing-or-cloud-refraction

Francis, E. M. (2016, 01 01). *A Proof of Kepler's laws.* Retrieved 01 24, 2016, from Kepler's laws: http://www.alcyone.com/max/physics/kepler/index.html

Freudenrich, C. P. (2015, April 03). *How Atom Smashers Work.* Retrieved June 25, 2015, from Science.howstuffworks.com: http://science.howstuffworks.com/atom-smasher9.htm

Fritts, D. a. (2016). The Deep Propagating Gravity Wave Experiment (DEEPWAVE): An Airborne and Ground-Based Exploration of Gravity Wave Propagation and Effects from Their Sources throughout the Lower and Middle Atmosphere . *Bulletin of American Meteorological Society*, 425-453.

Gao, S. (2013). *Quantum Mechanics: A Comprehensible Introduction for Students*. Kindle Edition: Amazon Kindle Direct Publishing.

Gao, S. (2014). *Dark Energy: From Einstein's Biggest Blunder to the Holographic Universe*. Kindle Edition: Amazon Kindle Direct Publishing.

Gao, S. (2014). *Understanding Gravity: Newton, Einstein, Verlinde?* Kindle Edition: Amazon Kindle Direct Publishing.

Gao, S. (2014). *Understanding Relativity: An Advanced Guide for the Perplexed*. Kindle Edition: Amazon Kindle Direct Publishing.

Greene, B. (2003). *The Elegant Universe: Superstrings, Hidden Dimensions, and the Quest for the Ultimate Theory*. New York: W.W. Norton & Company.

Greene, B. (2004). *The Fabric of the Cosmos: Space, Time, and the Texture of Reality*. New York: Vintage Books, Division of Random House, Inc.

Guidry. (2011, 08 23). *Astrophysics Lecture*. Retrieved from Lecture Ch15: http://eagle.phys.utk.edu/guidry/astro616/lectures/lecture_ch15.pdf

Hawking, S. (1996). *A Brief History of Time, The Illustrated*. New York: Bantam Book.

Hawking, S. (2002). *The Theory of Everything: The Origin and Fate of the Universe*. Beverly Hills, CA: New Millenium Press.

Heisenberg, W. (2015). *Uncertainty Principle*. Retrieved June 13, 2015, from abyss.uoregon.edu: http://abyss.uoregon.edu/~js/21st_century_science/lectures/lec14.html

Helmenstine, T. (2015, January 21). *Printable Color Periodic Table Chart - 2015*. Retrieved June 2, 2015, from sciencenotes.org: http://sciencenotes.org/printable-color-periodic-table-chart-2015/

Herbert, F. (1979). *Dune Trilogy*. London: Guild Publishing.

Heritage, A. (2015, June). *Atom Definition*. Retrieved June 30, 2015, from dictionary.reference.com: http://dictionary.reference.com/browse/atom

Julie, C. a. (2005). *Personal Tao*. Retrieved 05 12, 2015, from Yin Yang: http://personaltao.com/teachings/questions/what-is-yin-yang/

Kelly, A. G. (1998). *Hafele & Keating Tests; Did They Prove Anything?* Celbridge, Ireland: HDS Energy Ltd.,.

Lincoln, D. (2015, 11 06). *Einstein's True Biggest Blunder (Op-Ed)*. Retrieved from Space.com News website: https://www.space.com/31055-removing-cosmological-constant-was-the-blunder.html

Lucas, G. (Director). (1977). *Star Wars original movie* [Motion Picture].

Lucas, T. (Director). (2015). *The Age of Hubble* [Motion Picture].

Marmet, P. (2012, 07 02). *Newtonphysics*. Retrieved from Einstein GR vs Classical Mechanics: https://www.newtonphysics.on.ca/einstein/chapter10.html

Marsh, J. (Director). (2014). *The Theory of Everything* [Motion Picture].

NASA. (2004, 07). *Hubble Site*. Retrieved 06 12, 2015, from News Center: http://hubblesite.org/newscenter/archive/releases/2004/07/image/c/format/web_print/

NASA. (2005, 3 21). *Kepler's Three Laws of Planetary Motion*. Retrieved 12 08, 2015, from http://www-spof.gsfc.nasa.gov/: http://www-spof.gsfc.nasa.gov/stargaze/Kep3laws.htm

NASA. (2012, 10 10). *Layers of the Sun*. Retrieved 09 02, 2015, from NASA Mission Pages: http://www.nasa.gov/mission_pages/iris/multimedia/layerzoo.html

NASA. (2016, 02 11). *Detects Gravitational Waves, Just as Einstein Predicted*. Retrieved 02 11, 2016, from INFOWARS: http://www.infowars.com/observatory-detects-gravitational-waves-just-as-einstein-predicted/

NASA. (2017, 08 23). *NASA Image Education Center*. Retrieved from What is the speed of the Earth's rotation?: https://image.gsfc.nasa.gov/poetry/

NASA. (2018, 01 02). *Solar System Evolution*. Retrieved from Present State and Basic Laws: https://history.nasa.gov/SP-345/ch2.htm

Navy, U. (2016, 12 24). *Summer solstice: twenty years of solstice dates and times*. Retrieved from The Guardian News Datablog: https://www.theguardian.com/news/datablog/2011/jun/21/summer-solstice-data

Newton, S. I. (1687). *Principia Mathematica*. Cambridge: Unknown.

NOAA. (2016, 11 29). *GFDL*. Retrieved from Geophysical Fluid Dynamics Laboratory: https://www.gfdl.noaa.gov/atmospheric-processes/

NOAA. (2017, 10 18). *Weather Underground Storm Data*. Retrieved from Hurricanes and Tropical Cyclones: https://www.wunderground.com/hurricane/at2005.asp

NOAA_Ross. (2017, 07 06). *Ocean Service Education*. Retrieved from Tides and Water Levels: https://oceanservice.noaa.gov/education/kits/tides/tides03_gravity.html

OregonUniv. (2015, 03 18). *Uncertainty Principle*. Retrieved 04 06, 2015, from Abyss University of Oregon: http://abyss.uoregon.edu/~js/21st_century_science/lectures/lec14.html

Overbye, D., & Corum, J. a. (2016, 02 11). *New York Times Science*. Retrieved 02 11, 2016, from Science Out There: http://www.nytimes.com/video/science/100000004200661/what-are-gravitational-waves-ligo-black-holes.html?mabReward=A7&module=WelcomeBackModal&contentCollection=Science&region=FixedCenter&action=click&src=recg&pgtype=article

Patel, N. V. (2016, 05 24). *Scientist Have Just Found a New Form of Light*. Retrieved 05 26, 2016, from Huffington Post Science: http://www.huffingtonpost.com/inverse/scientists-have-just-foun_b_10116834.html

Peck, M. (2013, 05 31). *The Theory of Everything: Part 1 - Ruling out an Expanding Universe*. Retrieved from YouTube: https://www.youtube.com/watch?v=4ItFWXAfDHY&feature=youtu.be&t=2m27s

Peck, M. S. (2013, 05 20). *The Theory of Everything*. Retrieved from The Theory of Everything: Foundations, Application and Corrections to General Relativity: http://vixra.org/pdf/1305.0138v1.pdf

Physics, I. o. (2015, 02 12). *Earth and the Solar System, useful facts*. Retrieved 05 10, 2016, from IOP Physics for all ages: https://www.iop.org/activity/outreach/resources/pips/topics/earth/facts/page_43079.html

Roddenberry, G. (Director). (1966-1969). *Star Trek TV Series Original* [Motion Picture].

Rouse, M. (2015, January). *Quantum Theory Definition*. Retrieved June 15, 2015, from whatis.techtarget.com: http://whatis.techtarget.com/definition/quantum-theory

Ryden, B. P. (2003, 02 19). *Rotation of Our Galaxy*. Retrieved 01 13, 2015, from Lecture 30: Astronomy 162: http://www.astronomy.ohio-state.edu/~ryden/ast162_7/notes30.html

s8int. (1992, January). *Ancient Atomic Knowledge*. Retrieved June 24, 2015, from Ancient High Technology: http://www.s8int.com/atomic1.html

Sagan, C. (Director). (1980). *Cosmos TV Series* [Motion Picture].

Sagan, C. (1985). *Contact*. New York: Simon & Schuster Ltd.

SXS. (2016, 02 14). *Collaborative Research Site*. Retrieved 02 15, 2016, from Simulating eXtreme Spacetimes: http://www.black-holes.org/

Thomas, A. (2012). *Hidden In Plain Sight: The simple link between relativity and quantum mechanics*. Kindle_Edition: Aggrieved Chipmunk Publications.

Thomas, A. (2013). *Hidden in Plain Sight 2: The equation of the universe*. Kindle Edition: Aggrieved Chipmunk Publications.

Thomas, A. (2014). *Hidden in Plain Sight 3: The secrets of time*. Kindle Edition: Aggrieved Chipmunk Publications.

Thompson, E. (Director). (2007). *The Universe* [Motion Picture].

timeanddate.com. (2018, 12 20). *Moon Phases 2017 – Lunar Calendar*. Retrieved from Moon Phases for (City): https://www.timeanddate.com/moon/phases/?year=2017

Tyson, N. d. (Director). (2014). *Cosmos: A Spacetime Odyssey* [Motion Picture].

Tyson, N. d. (Director). (2014). *The Inexplicable Universe: Unsolved Mysteries* [Motion Picture].

Unknown. (1996-2015). *The Physics Classroom: Circular Motion and Satellite Motion*. Retrieved 12 08, 2015, from Kepler's Three Laws: http://www.physicsclassroom.com/class/circles/Lesson-4/Kepler-s-Three-Laws

unknown. (2010, June). *Solar System Quick*. Retrieved June 30, 2015, from Astronomy Guide: http://www.solarsystemquick.com/universe/sirius-star.htm

Verlinke, E. (2018). *A New View on Gravity and the Cosmos. Stadium Generale Delft* (p. https://www.youtube.com/watch?v=8ovRZuv5Lo8). Delft: YouTube.

Vidmar, D. (2011). *Dirac antimatter paper.pdf.* Retrieved 10 29, 2015, from The Dirac Equation and the Prediction of Antimatter: www6.ufrgs.br

von Daniken, E. (1969). *Chariot of the Gods.* New York: Berkley Books.

Wang, X., Walker, M., & Pal, J. (2008). Model-Independent Estimates of Dark Matter Distributions. *Journal of the American Statistical Association, Vol 103, No. 483, Sep 2008*, 1070-1084.

Weinberg, S. (2008). *Cosmology.* Oxford: Oxford University Press.

West, J. A. (Director). (1990). *Secrets of Egypt & Hidden Pyramid Symbols Revealed by John West* [Motion Picture].

Whitney, C. (2016, 03 15). *Jet Propulsion Laboratory, California Institute of Technology.* Retrieved 03 17, 2016, from JPL, CIT: http://www.jpl.nasa.gov/news/news.php?feature=4753

wikidot. (6500 BC). *Ancient City Found in India, Irradiated from Atomic Blast.* Retrieved 06 20, 2015, from Veda Archeological History: http://veda.wikidot.com/ancient-city-found-in-india-irradiated-from-atomic-blast

Wikipedia. (2015, June 27). *Alchemy.* Retrieved June 29, 2015, from Definition of Alchemy: https://en.wikipedia.org/wiki/Alchemy

Wikipedia. (2015, June 24). *Atom definition.* Retrieved June 30, 2015, from Atom: https://en.wikipedia.org/wiki/Atom

wikipedia. (2015, June 29). *Barycentric Coordinates (Astronomy).* Retrieved June 30, 2015, from en.wikipedia.org: https://en.wikipedia.org/wiki/Barycentric_coordinates_(astronomy)

Wikipedia. (2015, June 28). *Chemistry.* Retrieved June 30, 2015, from Definition of Chemistry: https://en.wikipedia.org/wiki/Chemistry

wikipedia. (2015, June 2). *Democritus.* Retrieved June 30, 2015, from wikipedia.org: https://en.wikipedia.org/wiki/Democritus

wikipedia. (2015, June 23). *Dogon Knowledge of Stars.* Retrieved June 30, 2015, from Dogon People: https://en.wikipedia.org/wiki/Dogon_people

wikipedia. (2015, June 24). *Graviton definition.* Retrieved June 30, 2015, from wikipedia.org: https://en.wikipedia.org/wiki/Graviton

Wikipedia. (2016, 06 10). *Perihelion and Aphelion.* Retrieved 06 18, 2016, from Wikipedia the Free Encyclopedia: https://en.wikipedia.org/wiki/Perihelion_and_aphelion

Wikipedia. (2018, 03 04). *Angular momentum.* Retrieved from Wikipedia, the free encyclopedia: https://en.wikipedia.org/wiki/Angular_momentum

Wikipedia, e. (2001). *Various web pages that anyone can edit.* Retrieved 05 2-28, 2015, from en.wikipedia.org: https://en.wikipedia.org

Wilczek, F. (2000). QCD Made Simple. *Physics Today*, 23-28.

Wolff, M. (2008). *Schrodinger's Universe: Einstein, Waves, & the Origin of the Natural Laws.* Parker, CO: Outskirts Press.

Ye, X.-H. a. (2008). *Gravitational Lensing Analyzed by Graded Refractive Index of Vacuum.* Hangzhou, China: Department of Physics, Zhejiang University.

Yee, J. (2014). *The Particles of the Universe.* Kindle Edition: Amazon Kindle Direct Publishing.

Zagorski, P. (2012, 05). *Modeling disturbances influencing an Earth-orbiting satellite.* Retrieved from Pomiary Automatyka Robotyka: www.par.pl/2012/PAR_05_2012_Zagorski_98_103.pdf

Zhang, J. a. (2017, 05 16). *UPI Home/Science News.* Retrieved from Hurricane's atmospheric gravity waves help predict the storm's path: https://www.upi.com/Hurricanes-atmospheric-gravity-waves-help-predict-the-storms-path/2951494959547/

www.ingramcontent.com/pod-product-compliance
Lightning Source LLC
Chambersburg PA
CBHW081423220526
45466CB00008B/2253